AF323311

Bioluminescence & Chemiluminescence

Progress and Perspectives

Pacifico Yokohama, Yokohama, Japan
2 – 6 August 2004

ISBC 2004

Bioluminescence&
Chemiluminescence
Progress and Perspectives

edited by

Akio Tsuji
Showa University, Japan

Masakatsu Matsumoto
Kanagawa University, Japan

Masako Maeda
Showa University, Japan

Larry J Kricka
University of Pennsylvania School of Medicine, USA

Philip E Stanley
Cambridge Research & Technology Transfer Ltd, UK

World Scientific

NEW JERSEY · LONDON · SINGAPORE · BEIJING · SHANGHAI · HONG KONG · TAIPEI · CHENNAI

Published by

World Scientific Publishing Co. Pte. Ltd.

5 Toh Tuck Link, Singapore 596224

USA office: 27 Warren Street, Suite 401-402, Hackensack, NJ 07601

UK office: 57 Shelton Street, Covent Garden, London WC2H 9HE

British Library Cataloguing-in-Publication Data
A catalogue record for this book is available from the British Library.

Proceedings of the 13th International Symposium on
BIOLUMINESCENCE AND CHEMILUMINESCENCE
Progress and Perspectives

ISBN 981-256-118-8

Printed in Singapore.

WELCOME!

On behalf of the Organizing Committee of 13th International Symposium on Bioluminescence & Chemiluminescence 2004 in Yokohama (13th ISBC), I would like to welcome you to the Symposium. The first symposium was held in 1978 in Brussels, Belgium and the symposium has subsequently been held every two years in Europe and America. This symposium is the first to be held in Japan, as well as in the Asia region.

At the first day, the symposium begins with the Memorial Lecture of Late Professor Toshio Goto of Nagoya University presented by Professor H. Niwa, The University of Electro-Communications, and following the Plenary Lectures of Professor O. Shimomura, Marine Biology Laboratory and Professor W. Adam, the University of Puerto Rico. The plenary lectures will presented by Professor T. Nagano, the University of Tokyo, Professor of F. Tsuji, the University of California, and Professor Y. Umezawa, the University of Tokyo at the second to third day, respectively.

These plenary lectures will be followed by symposium sessions on Mechanism and Theoretical Study of Chemiluminescent Reactions, Color of Chemi- and Bioluminescence, Luminescent Bio-imaging and Biosensors developed with Nanomolecules and Nanoparticles, Recent Development in Environmental Fields, Biological Application with Fluorescent Biomolecules and Cells, Bioluminescence; Evolutional, Biological and Ecological Aspects, Synthesis of Compounds related to Chemiluminescence, Basic and Applied Studies of Beatle Bioluminescence, Development in Biological Fields, Frontiers of Industrial Application of Firefly Luciferase, Basic and Applied Studies of Marine and Bacterial Bioluminescence, Functional Chemi-Luminophore and Singlet Oxygen and Superoxide Bioluminescence Chemiluminescence Festival Plaza will be planed for free talk and discussion. Moreover, the education session "Firefly Sheds Light on New Technology" will be opened to the early teens as scientific education.

Poster presentations which cover and wide array of topics and expand on information presented during the oral sessions are the Exhibit Hall. In addition, the Exhibit Hall is full of the latest technology to make our studies. We hope you will take time to visit each of our exhibitors during the symposium.

Bus tour to Hakkeijima Sea Paradise (Aquarium) or Walking tour around historical town Kamakura or Scientific visit to Yokosuka City Museum is prepared to all participants.

All attending owe much to the diligent Program Committee and to the Session Chairs who powerful assisted in attracting keynote speakers and in giving logical organization to the submitted papers. Our generous sponsors, listed elsewhere, improved the quality of the amenities of the meeting.

On behalf of the entire organizing committee - welcome. We are glad you are here. We hope you will enjoy the next few days and that you have a valuable and rewarding learning experience as well.

Akio Tsuji, PhD

The President of 13th International Symposium on Bioluminescence and Chemiluminescence

PREFACE

These are the Proceedings of the 13th Symposium on Bioluminescence and Chemiluminescence held at Pacifico Yokohama, Yokohama, Japan from 2 – 6 August 2004.   This series of symposia started in Brussels in 1978, and a list of the other Proceedings volumes appears at the end of this Preface. As in previous symposia, delegates came from far and wide and in all 18 countries were represented. Communications between delegates and the Organising Secretariat was almost entirely via the Internet and email. In the interest of efficiency and the environment very little printed paperwork was used. Abstracts of presentations were made available on the web site www10.showa-u.ac.jp/~ISBC/index.html.

The Organising Secretariat was fortunate to have the continued association with the International Society for Bioluminescence & Chemiluminescence (http://www.unibo.it/isbc).   The International Society recognised the need to encourage young scientists in the disciplines and so it provided financial support for 11 young scientists under the age of 35 years

We also thank John Wiley & Sons for publishing the regular abstracts in the journal *Luminescence* Vol. 19(3) 2004.

Editorial Note

This volume was compiled without peer review from camera-ready manuscripts of lectures and posters presented at the Symposium. The Editors have, in the interest of rapid publication, made only minor stylistic changes. They take no responsibility for scientific or priority matters.

The Editors: A Tsuji, M Matsumoto, M Maeda, LJ Kricka, PE Stanley

THE MARLENE DELUCA PRIZES

These prizes were again generously given by Dr Fritz Berthold with his new company Berthold Technologies (http://www.bertholdtech.com). Dr Berthold has provided these prizes at each symposium since the 1988 Symposium in Florence. The prizes are open to all Symposium participants who are aged below 35 years by the first day of the Symposium. They are in memory of Marlene DeLuca who made such a major contribution to the science of bioluminescence (*see* Stanley PE. Dedication to Marlene DeLuca: *J Biolumin Chemilum* 4;1989:7-11 (includes list of her papers)). The President of the International Society, Professor John Lee (University of Georgia, Athens, GA, USA) chose a small Committee from the Society to judge the presentations.

The Prize winners were:

Akira Kanakubo (Laboratory of Organic Chemistry, School of Bioagricultural Sciences, Nagoya Univeristy, Nagoya, Japan) *Chemical studies on bioluminescence of Acorn worm,* Ptychodera flava: *Isolation and characterization of luminous substances*

Asami Kaihara (Department of Chemistry, School of Science, University of Tokyo, Tokyo, Japan) *Flashing a protein-protein interaction in living cells by a split* Renilla *luciferase complementation*

Christine Vanderlinden (Laboratoire de Biologie Marine, Université Catholique de Louvain, Louvain-la-Neuve, Belgium) *Pharmacological and electrophysiological studies of light emission in 3 ophiuroid species: preliminary results*

ORGANIZATION

PRESIDENT
A Tsuji

VICE PRESIDENTS
M. Aizawa and M. Matsumoto

GENERAL SECRETARY
M Maeda

INTERNATIONAL ADVISORY BOARD
ISBC Council Members
J. Lee (President), A. Roda (Past President), A. Szalay (President-Elect), P. Pasini
(Secretary), P.E. Stanley (Past Secretary), B. Branchini (Treasurer & Membership
Secretary), LJ Kricka (Publication Officer)
Councillors
A. Berthold , S. Haddock, P. Hill, V. Kratasyuk, J-Francois Rees, E. Widder,
B. Zomer
Scientific Advisory Board
S. Daunert, P. De Sole, T. Quickenden, P. Schaap, D. Shah, N Ugarova, V. Viviani,
Xiaolin Yang **and also** Jin-Min Lin, Shiao-Chun (David) Tu

NATIONAL ADVISORY BOARD
J. Goto, T. Masujima, T. Nagano, M. Ohashi, Y. Umezawa, H. Utsumi

SCIENTIFIC PROGRAM AND LOCAL ORGANIZING COMMITTEE
A. Tsuji (President), M. Aizawa (Vice President), M. Matsumoto (Vice President),
M. Maeda (General Secretary), N. Amino, Y. Ashihara, H. Arakawa, K. Fujimori,
M. Fukuoka, S. Harada, N. Hattori, K. Hayakawa, M. Hiramatsu, S. Hosaka,
H. Hosoda, I. Imada, K. Imai, M. Inoue, S. Inouye, M. Isobe, Y. Kasahara,
M. Kimura, N. Kuroda, K. Nakashima, H. Niwa, O. Nozaki, K. Ohmiya, N. Suzuki,
M. Totani, T. Toyooka, K. Tsujimoto, N. Wada, M. Yamaguchi

ACKNOWLEDGEMENTS

We wish to express our sincere appreciation to the following for their generous support of this symposium.

INSTITUTIONS & FOUNDATIONS
Commemorative Organization for Japan World Expositive '70
Japan National Tourist Organization
The Foundation of Pharmaceutical Manufacturers' Association of Japan
The Nagai Foundation
The Tokyo Biochemical Research Foundation
Mochida Memorial Foundation for Medical and Pharmaceutical Research
The Naito Foundation
The Research Foundation for Pharmaceutical Sciences
Tokyo Ohka Foundation for the Promotion of Science and Technology
Sankyo Life Science Foundation
Suntory Institute for Bioorganic Research
Uehara Memorial Life Science Foundation

COMPANIES

Abbott Japan Co. Ltd.	Lumica Corporation
Aventis Pharma Ltd.	Mitsubishi Pharma Corporation
Berthold Japan Co.	Nissin Scientific Corporation
Chemco Scientific Co. Ltd.	Novartis Pharma K.K.
Chisso Corporation	Otsuka Pharmaceutical Co. Ltd.
Fujirebio Inc.	Roche Diagnostics K.K.
Hamamatsu Photonics K.K.	Sankio Chemical Co., Ltd.
Hitachi Ltd.	Thermo Electron K.K.
Horiba Biotechnology Co. Ltd.	Tokken Inc.
IBL Co., Ltd.	Tokyo Rikakikai Co. Ltd.
JASCO Corporation	Tosoh Corporation
Kikkoman Corporation	Tohoku Electric Industrial Co. Ltd.
Kowa Company Ltd.	Wako Pure Chemical Industry Ltd.

DIRECT SUPPORT/AWARDS TO INDIVIDUALS
Marlene DeLuca Prizes
Berthold Technologies (Germany), http://www.berthold-ds.com
International Society for Bioluminescence & Chemiluminescence
http://www.unibo.it/isbc
Support for 11 young scientists

NEXT SYMPOSIUM

The 14th International Symposium on Bioluminescence and Chemiluminescence will be held in 2006 at the panoramic university city of Wuerzburg, Germany with scientific traditions of W. Rontgen, R. Virchow and T. Boveri.

The symposium Chairman will be Professor Aladar A.Szalay, Virchow Center for Experimental Biomedicine. University of Wuerzburg, 97074 Wuerzburg, Germany.

Contact address:aladar.szalay@virchow.uni-wuerzburg.de.

Details of the 14th BL&CL Symposium will be posted on the Society website, http://www.isbc.unibo.it/.

Presentations as lectures and posters are invited. There will be an exhibition and workshops.

PROCEEDINGS OF PREVIOUS SYMPOSIA

1ˢᵗ   1978   Brussels, Belgium

International Symposium on Analytical Applications of Bioluminescence and Chemiluminescence. Proceedings 1978. Editors: Schram E, Stanley PE. Westlake Village, CA: State Printing & Publishing, Inc., 1979, pp. 696. (Privately published).

2ⁿᵈ   1980   San Diego, CA, USA

Bioluminescence and Chemiluminescence: Basic Chemistry and Analytical Applications. Editors: DeLuca MA, McElroy WD. New York: Academic Press 1981. pp.782. ISBN: 0-12-208820-4.

3ʳᵈ   1984   Birmingham, UK

Analytical Applications of Bioluminescence and Chemiluminescence. Editors: Kricka LJ, Stanley PE, Thorpe GHG, Whitehead TP. London: Academic Press 1984. pp. 602. ISBN: 0-12-426290-2.

4ᵗʰ   1986   Freiburg, Germany

Bioluminescence and Chemiluminescence: New Perspectives. Editors: Schölmerich J, Andreesen R, Kapp A, Ernst M, Woods WG. Chichester: Wiley 1987. pp. 600. ISBN: 0-471-91470-3.

5ᵗʰ   1988   Florence, Italy

Bioluminescence and Chemiluminescence: Studies and Applications in Biology and Medicine. Editors: Pazzagli M, Cadenas E, Kricka LJ, Roda A, Stanley PE. Chichester: Wiley 1989. pp. 646. (published as volume 4, issue 1 of the *Journal of Bioluminescence and Chemiluminescence*, 1989). ISBN: 0-471-92264-1.

6ᵗʰ   1990   Cambridge, UK

Bioluminescence and Chemiluminescence: Current Status. Editors: Stanley PE, Kricka LJ. Chichester: Wiley 1991. pp. 570. ISBN: 0-471-92993-X.

7ᵗʰ   1993   Banff, Canada

Bioluminescence and Chemiluminescence: Status Report. Editors: Szalay AA, Kricka LJ, Stanley PE. Chichester: Wiley. 1993, pp. 548. ISBN: 0-471-94164-6.

8th   1994   Cambridge, UK

Bioluminescence and Chemiluminescence: Fundamentals and Applied Aspects. Editors: Campbell AK, Kricka LJ, Stanley PE. Chichester: Wiley 1994. pp. 672. ISBN: 0-471-95548-5.

9th   1996   Woods Hole, MA, USA

Bioluminescence and Chemiluminescence: Molecular Reporting with Photons. Editors: Hastings JW, Kricka LJ, Stanley PE. Chichester: Wiley 1997. pp. 568. ISBN: 0-471-97502-8.

10th   1998   Bologna, Italy

Bioluminescence and Chemiluminescence: Perspectives for the 21st Century. Editors: Roda A, Pazzagli M, Kricka LJ, Stanley PE. Chichester: Wiley 1999. pp. 628. ISBN: 0-471-98733-6.

11th   2000   Monterey, CA, USA

Proceedings of the 11th International Symposium on Bioluminescence & Chemiluminescence. Editors: Case JF, Herring PJ, Robison BH, Haddock SHD, Kricka LJ, Stanley PE. Singapore: World Scientific 2001. pp. 517. ISBN 981-02-4679-X.

12th   2002   Cambridge, UK

Bioluminescence & Chemiluminescence: Progress & Current Applications. Editors: Stanley PE, Kricka LJ. Singapore: World Scientific 2002. pp. 520. ISBN 981-238-156-2.

CONTENTS

PART 2. BEETLE BIOLUMINESCENCE

PART 3. MARINE BACTERIA BIOLUMINESCENCE

PART 4. CYPRIDINA (VARGULA) BIOLUMINESCENCE

PART 5. CHEMILUMINESCENCE

PART 6. 1,2-DIOXETANES

PART 9. ANTIOXIDANTS, REACTIVE OXYGEN SPECIES & PHAGOCYTOSIS

PART 10. APPLICATIONS IN MICROBIOLOGY, ECOLOGY, AND ENVIRONMENTAL & FOOD TESTING

PART 1

BIOLUMINESCENCE

BIOLUMINESCENCE AND MATING BEHAVIOUR IN PONY FISH, *Leiognathus nuchalis*

N AZUMA[1], C FURUBAYASHI[1], T SHICHIRI[1], M WADA [2],
N MIZUNO[3], Y SUZUKI[3]

[1]*Faculty of Agriculture and Life Science, Hirosaki University,
Hirosaki 036-8561, Japan*
[2]*Ocean Research Institute, The University of Tokyo, Tokyo 164-8639, Japan*
[3]*Fisheries laboratory, The University of Tokyo, Shizuoka 431-0211, Japan*

INTRODUCTION

It is well known that the light emission of the firefly is used as a means of the species specific communication. It is also thought to have a similar function in some marine organism, because there are several patterns of the flashing in one species, for example *Photobelepharon palpebratus.* [1]

Leiognathus nuchalis is widely distributed in the coastal zone around Japan and has light organ containing luminous bacteria, *Photobacterium leiognathi.*[2] The circumesophageal light organ appears as a ring shape which surrounds the gullet, with the symbiotic luminous bacteria packed in the inside. The outside of luminous organ is covered with the epidermal tissue in which the melanophore develops. The light discharged by luminous bacteria is emitted in the inside of a swim bladder through the transparent part of the luminous organ. The interior surface of a swim bladder becomes a reflector. The whole area is almost covered for the silver albedo by the guanine pigment, and it reflects the light from the luminous organ. Based on these features of the internal structure of the pony fish, *Leiognathus nuchalis* seems to be a luminous fish. However, the light emission has not been observed from this species which is the most common species in spite of confirming light emission in some Leiognathidae fishes recently.[3]

In this study, we have studied light emission from fish and their behaviour and have focused on *Leiognathus nuchalis.*

MATERIALS AND METHODS

The fish were obtained by field collection in Enshu-nada Sea and Suruga-Bay in Japan. They were kept in the 2000 L circuit type tank before the experiment. A 500 L circular transparent tank set in dark condition was used in the experiment. The light emission was measured under constant dark conditions using ultrasensitive light quantum meter which was installed by the water tank. The measurement of one experiment was done continuously for almost 24 h, the water in the tank was exchanged after each experiment. The light emission measurements were carried out 4 d during breeding season and 3d after the breeding season, and the control experiment (sea water without fish) was also carried out in the breeding season. In each experiment, the five fish were introduced into the water tank (sex ratio not biased). The signals from the photometer were recorded by a data-recorder and counted in each 30 min interval. Simultaneously, a video recording using the infrared light video camera was also carried out.

RESULTS

Light signals were detected in all experiments during the breeding season. The emission of the light gradually increased from the sunset. Frequency of light signals increased at night, especially midnight (20:00- 3 :00) during the spawning period.

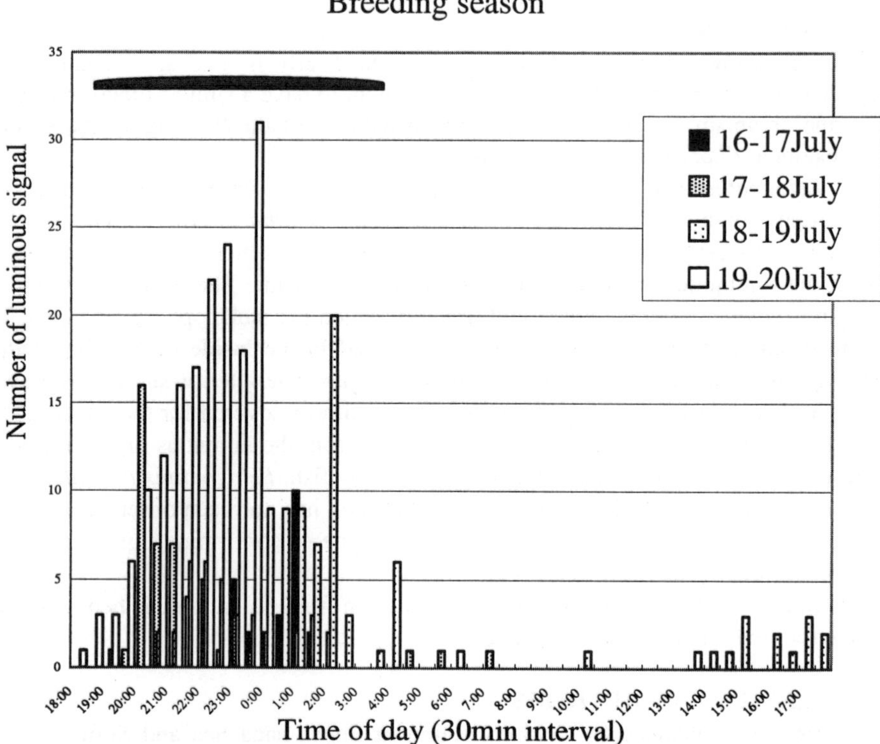

Figure 1.   Diel rhythm of Bioluminescence from the pony fish in breeding season
Horizontal closed bar indicates the night period

Little light emission was observed in daytime but was most common around midnight (Figure 1). In the experiment after the breeding season, light emission was only observed 2 times (Figure 2). The light was not detected in the control experiment. Though in the breeding season, the infrared light videotaping was carried out, the light emission from the fish was not observed by CCD cameras.

After breeding season

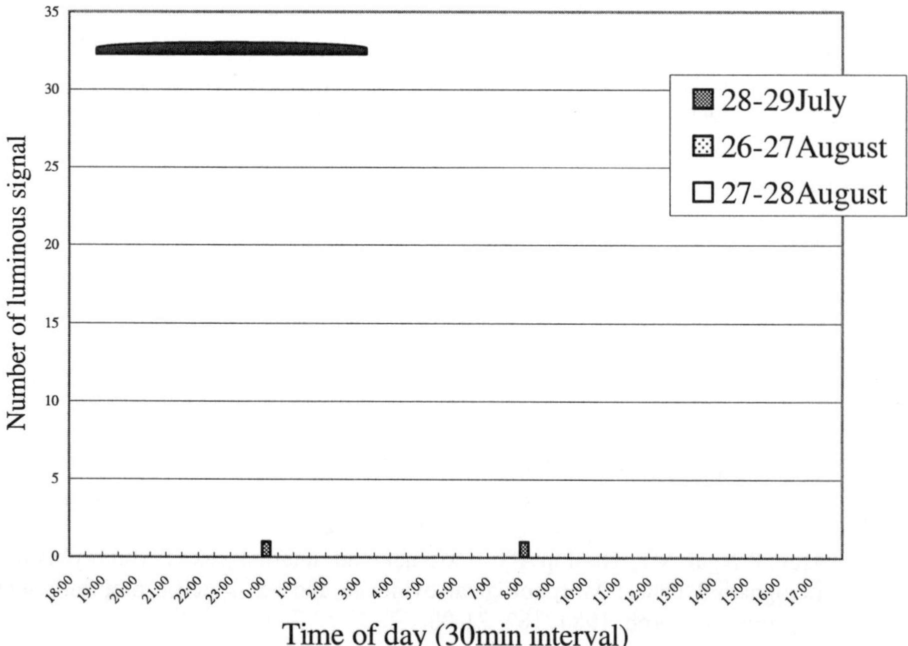

Figure 2. Diel rhythm of bioluminescence from the pony fish after breeding season
Horizontal closed bar indicates the night period

DISCUSSION

It has been considered that there may be a relationship between reproduction and light emission of *Leiognathus nuchalis*, because there is sexual dimorphism of the light organ.[4] Some Leiognathidae fishes flash, and it is known that there are sexual dimorphism in the luminous organ.[5] In our previous field study, we suggested the existence of communication based on the light signal in *Leiognathus elongatus*.[3]

However, it is first record to clarify the light emission of diel pattern and seasonal change in *Leiognathus nuchalis*. The light emitting rhythm must be regulated by circadian rhythm of fish. The light emission was remarkably active in the breeding season. In addition, it has been proven that spawning occurs around midnight in other experiments (unpublished data). These results suggest that the light signal must be used as significant information for mating and spawning behaviour. However the light of this species is very weak, because we could not detect the light by eye or using CCD camera.

ACKNOWLEDGEMENTS
We thank the members of the laboratory of Wildlife management and Animal ecology,   Hirosaki University and Fisheries laboratory, The University of Tokyo for assistance in collecting animals and for helpful discussions.

REFERENCES

1.   Morin JG, Harrington A, Nealson K, Krieger N, Baldwin TO, Hastings JW. Light for all reasons: Versatility in behavioral repertoire of flashlight fish. Science 1970; 190:74-5.
2.   Haneda Y, Tsuji FI. The luminescent systems of pony fishes. J. Morph 1976; 150: 539-52.
3.   Sasaki A , Ikejima K , Aoki S, Azuma N, Kashimura N, Wada M. Field evidence for bioluminescent signaling in the pony fish *Leiognathus elongatus* . Environ Biol Fishes 2003; 66: 307-11.
4.   Ikejima K, Ishiguro B, Wada M, Kita-Tsukamoto K, Nishida M. Molecular phylogeny and possible scenario of pony fish (Perciformes: Leiognathidae) evolution. Mol. Phylogenet Evol 2004; 31: 904-9.
5.   McFall-Ngai MJ, Dunlop PV. External and internal sexual dimorphism in Leiognathid fishes: Morphological evidence and sex-specific bioluminescent signaling. J Morph. 1984; 182: 71-83.

STUDIES ON THE BIOLUMINESCENT MECHANISM OF SYMPLECTIN PHOTOPROTEIN

M ISOBE[1], T MATSUDA[1], M KUSE[2],
H MORI[1], T FUJII[1], N KONDO[1], Y KAGEYAMA[1]

[1]Graduate School of Bioagricultural Sciences and [2]Chemical Instrument Center
Nagoya University, Nagoya 464-8602, Japan
Email: isobem@agr.nagoya-u.ac.jp, kuse@cic.nagoya-u.ac.jp

INTRODUCTION

Tobiika (*Symplectoteuthis oualaniensis* L.) is a flying squid that emits blue light (470 nm) in the presence of mono-cation (Na^+, K^+) and molecular oxygen.[1] We have investigated the molecular mechanism of symplectin bioluminescence, and confirmed that the light comes from its photoprotein 'symplectin'. Symplectin is a 60 kDa protein and needs dehydrocoelenterazine (DCT) as an organic luminous substance.[2] DCT exists as a thiol bound form with the symplectin active site cysteine to construct the chromophore for bioluminescence as shown in Fig. 1.[3]

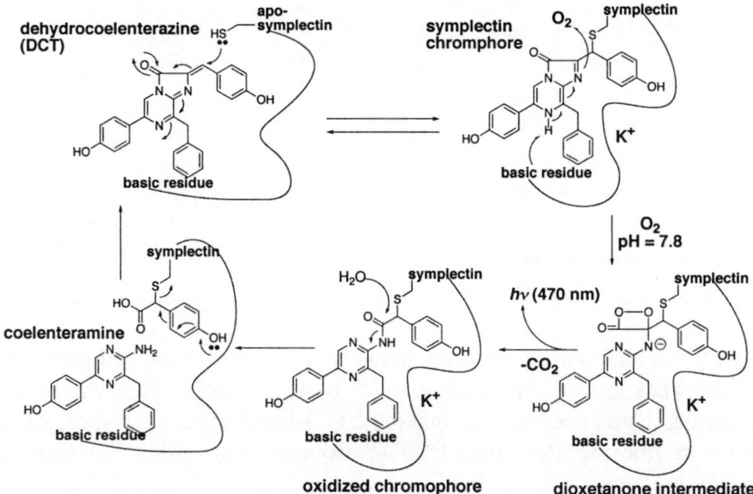

Figure 1. Postulated mechanism of symplectin bioluminescence

Under basic condition (pH 7.8), this chromophore reacts with oxygen to produce a dioxetanone intermediate.[4, 5] The dioxetanone collapses to *alpha*-thio-amide with producing blue light (470 nm) and CO_2. After the oxidized chromophore decomposes to a coelenteramine, the free active site cysteine becomes available again.

Now we focus on the structural analysis of symplectin active site and the molecular mechanism of symplectin bioluminescence.[6, 7]

RESULTS AND DISCUSSION
Symplectin analysis
From the sequence analysis of symplectin with LC-Q-TOF-MS, MS/MS and cDNA, symplectin has 501 amino acids sequence. Partial degradation of symplectin with trypsin afforded a 40 kDa protein (symplectin A'), which is the C-terminal part of symplectin and still has bioluminescent activity. We suppose that the active site of symplectin exists in the 40 kDa symplectin A'.[8]
Proof of chromophoric structure
[13]C-labeled DCT analog (^{13}C-DCTa) was synthesized to prove the chromophoric structure of symplectin. Dithiothreitol and glutathione (apo-symplectin models) reacted with ^{13}C-DCTa to afford the luminescent active chromophores, which were analyzed with NMR and MS.[9] We demonstrated that DCT binds with the sulfhydryl residues at the 2'-^{13}C-labeled carbon as shown in Fig. 2, and we succeeded in reproducing a model symplectin bioluminescence with ^{13}C-DCTa.[10]

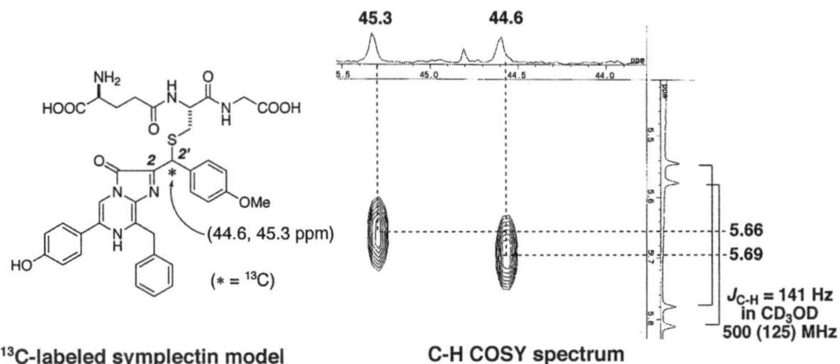

Figure 2. ^{13}C-labeled symplectin model and its NMR spectrum

We also found the equilibrium between DCT and thiol adducts that is the origin of the weakness of chromophoric C-S bond. To detect the symplectin active center cysteine, it was necessary to make the C-S bond tight. Therefore, we decided to introduce a fluorine atom into DCT 2'-aromatic ring instead of hydroxyl and methoxy group.
Fluorinated dehydrocoelenterazine
Fluorinated dehydrocoelenterazine (F-DCT) was synthesized as a probe to investigate the active center cysteine of symplectin. We found that F-DCT strongly bound to the sulfhydryl residue of cysteine to afford a stable chromophore, which was proved with NMR and MS. There was also no equilibrium between F-DCT and its thol adducts as expected. Three F-DCTs (*ortho, meta, para*) were synthesized and

checked its bioluminescent activity.[11] As shown in Fig. 3, 2-*ortho*-F-DCT was the most active substance for symplectin bioluminescence.

Reconstituted symplectin (Recon-symplectin) was prepared from 2-*ortho*-F-DCT and apo-symplectin. The Recon-symplectin was proteolytically digested with trypsine to obtain a chromophoric peptide, which contains both the active center cysteine and F-DCT. Nano-LC-MS analysis afforded plausible data for the chromophoric peptide of the symplectin active center. But, we could not perfectly demonstrate the active center cysteine of symplectin with MS/MS analysis.

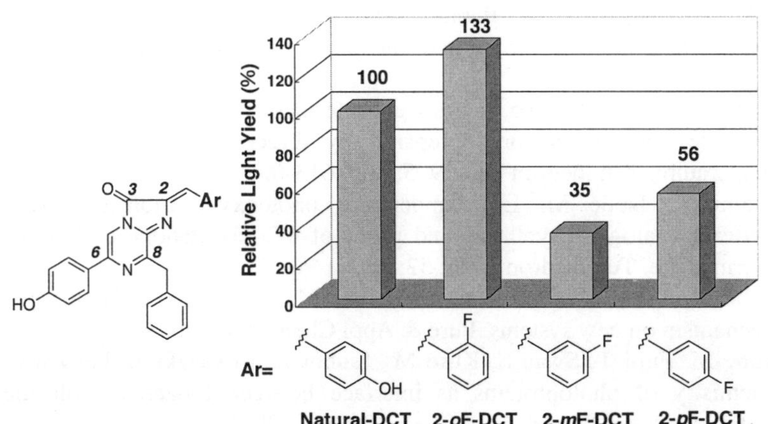

Figure 3. Comparison of the bioluminescent activities of fluorinated DCTs

Photoaffinity Labeling
It is attractive to irreversibly bind DCT against apo-symplectin through covalent bond for analyzing symplectin active site, therefore, photoaffinity labeling is a suitable method for such a purpose. A photoprobe (Azide-F-DCT) was synthesized to label the active site of symplectin with photo-irradiation. As a model chromophore of symplectin, azide-fluoro-coelenterazine was photo-irradiated in a solution. Hexafluoroisopropanol was the best solvent for converting the azide to a nitrene intermediate from the MS analysis data of the photo-irradiated products.

Now we investigate the symplectin active site by using a photoaffinity labeling and the nano-LC-Q-TOF-MS analysis[12] and also study the structure and activity relationship (SAR) between DCT structures and symplectin bioluminescent activities. For the SAR study, we also developed a novel synthetic method for DCT analogs.[13]

ACKNOWLEDGEMENTS
We acknowledge to financial support from JSPS-RFTF 96L00504, and Gant-in Aid for Scientific Research from the Ministry of Education, Culture, Sports, Science and

Technology, Japan. MK also special thanks to SUNBOR Grant and Naito Foundation.

REFERENCES

1.   Tsuji FI, Leisman G. K^+/Na^+-triggered bioluminescence in the oceanic squid *Symplectoteuthis oualaniensis*. Proc Natl Acad Sci USA 1981; 78: 6719-23.
2.   Takahashi H, Isobe M. *Symplectoteuthis* bioluminescence (1) --- structure and binding form of chromophore in photoprotein of a luminous squid. BioMed Chem Lett 1993; 3: 2647-52.
3.   Takahashi H, Isobe M. Photoprotein of luminous squid, *Symplectoteuthis oualaniensis* and reconstruction of the luminous system. Chem Lett 1994; 843-6.
4.   Usami K, Isobe M. Two luminescent intermediates of coelenterazine analog, peroxide and dioxetanone, prepared by direct photo-oxygenation at low temperature. Tetrahedron Lett 1995; 36: 8613-6.
5.   Usami K, Isobe M. Low-temperature photooxygenation of coelenterate luciferin analog --- synthesis and proof of 1,2-dioxetanone as luminescence intermediate. Tetrahedron 1996; 52: 12061-90.
6.   Isobe M, Takahashi H, Usami K, Hattori M, Nishigohri Y. Bioluminescence mechanism on new systems. Pure & Appl Chem 1994; 66: 765-72.
7.   Isobe M, Fujii T, Swan S, Kuse M, Tsuboi K, Miyazaki A, Feng MC, Li J. Chemistry of photoproteins as interface between bioactive molecules and protein function. Pure & Appl Chem 1998; 70: 2085-92.
8.   Fujii T, Ahn JY, Kuse M, Mori H, Matsuda T, Isobe M. A novel 60 kDa-photoprotein from oceanic squid (*Symplectoteuthis oualaniensis*) with sequence similarity to mammalian carbon-nitrogen hydrolase domains. Biochem Biophys Res Commun 2002; 293: 874-9.
9.   Isobe M, Kuse M, Yasuda Y, Takahashi H. Synthesis of ^{13}C-Dehydrocoelenterazine and model studies on *Symplectoteuthis* squid bioluminescence. BioMed Chem Lett 1998; 8: 2919-24.
10. Kuse M, Isobe M. Synthesis of ^{13}C-dehydrocoelenterazine and NMR studies on the bioluminescence of a *Symplectoteuthis* model. Tetrahedron 2000; 56: 2629-39.
11. Isobe M, Fujii T, Kuse M, Miyamoto K, Koga K. ^{19}F-Dehydrocoelenterazine as probe to investigate the active site of Symplectin. Tetrahedron 2002; 58: 2117-26.
12. Kurahashi T, Miyazaki A, Suwan S, Isobe M. Extensive investigations on oxidized amino acid residues in H_2O_2-treated Cu, Zn-SOD protein with LC-ESI-Q-TOF-MS, MS/MS for the determination of the copper-binding site. J Am Chem Soc 2001; 123: 9268-78.
13. Kuse M, Kondo N, Ohyabu Y, Isobe M. Novel synthetic route of aryl-aminopyrazine. Tetrahedron 2004; 60: 835-40.

CHEMICAL STUDIES ON BIOLUMINESCENCE OF ACORN WORM, *PTYCHODERA FLAVA*: ISOLATION AND CHARACTERIZATION OF LUMINOUS SUBSTANCES

A KANAKUBO, K KOGA, M ISOBE

Laboratory of Organic Chemistry, School of Bioagricultural Sciences, Nagoya University, Chikusa, Nagoya 464-8601, Japan
E-mail: isobem@agr.nagoya-u.ac.jp

INTRODUCTION

Luminous acorn worm, *Balanoglossus biminiensis* (*B. biminiensis*) was found to emit light by a luciferin – luciferase reaction by Cormier and Dure.[1-4] They found that the luciferase was a kind of peroxidase. Their report was the first example to require the H_2O_2 for a bioluminescence reaction. No further details have been reported on the structure of the luciferin and luciferase.

Ptychodera flava (*P. flava*), smaller acorn worm, was found in Kattore bay, Kohama Island, Okinawa by Higa.[5-6] Only the fact that *P. flava* emitted light by H_2O_2 was known. For the understanding of the molecular mechanism of bioluminescence, we determined the structures of luminous compounds and a possible light emitter.

MATERIALS AND METHODS

Instrumentation

The HPLC analyses were carried out using a JASCO PU-980 Intelligent HPLC pump systems equipped with a JASCO UV-970 Intelligent UV/VIS detector, a JASCO FP-920 Intelligent Fluorescence Detector, a JASCO CL-925 Intelligent CL detector and a JASCO 807-IT integrator. Samples were analysed by ODS-5 (4.6 id x 250 mm), Nomura Chemical Co. Ltd. Aichi, Japan and purified by Cosmosil 5C$_{18}$-AR (10 id x 250 mm), Nacalai tesque Inc. Kyoto, Japan. Luminescence and fluorescence spectra were measured with a JASCO FP-770 spectrofluorometer. UV/Vis spectra were measured with a JASCO Ubest-50 UV/VIS Spectrophotometer. Proton NMR spectra were recorded on a Bruker AMX-600 at 600 MHz. Chemical shifts (δ) are given in parts per million relative to DMSO-d_6 (δ 2.49) as internal standard. Carbon NMR spectra were recorded on a Bruker AMX-600 at 150.9 MHz. Chemical shifts are given in parts per million relative to DMSO-d_6 (δ 39.7) as internal standard. A Q-TOF Mass Spectrometer instrument equipped with a Z-spray ESI source (Micromass, Manchester, UK) was used.

RESULTS AND DISCUSSION

P. flava emits a green light from whole body by adding diluted H_2O_2 solution and intermittently continued emitting light around a minute. The bioluminescence spectrum was recorded by using a live specimen and showed emission centered at

528 nm (Fig. 1). It was supposed that the mechanism of bioluminescence was different between *B. biminiensis* and *P. flava* because it was found that *P. flava* did not show the luciferin – luciferase reaction in a previous study by this laboratory.

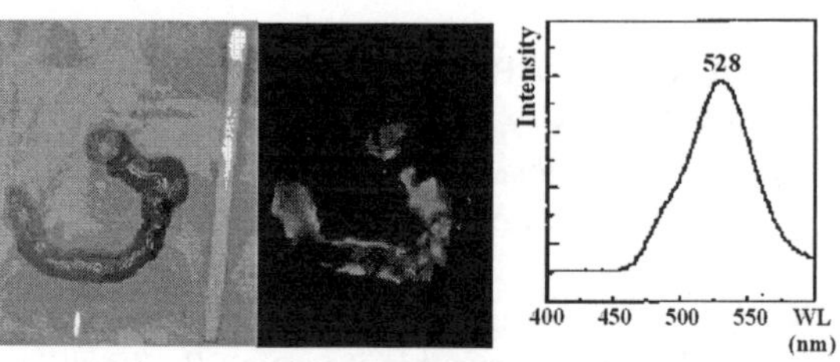

Figure 1. *Ptychodera flava* and bioluminescence spectrum

Luminous compounds were successfully extracted by ethyl acetate from lyophilized sample. HPLC separated three luminescent peaks, **Fr.1**, **2** and **3** as shown in Fig. 2. Interestingly, no peak was observed at fluorescence detector.

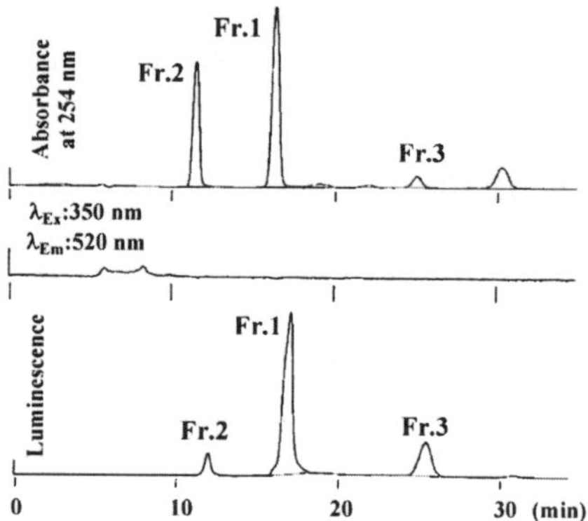

Figure 2. HPLC chromatogram of ethyl acetate extract

Major luminescent peak (**Fr.1**) was purified twice on an ODS column to obtain as crystals. ^{1}H NMR gave one peak at 9.92 ppm and ^{13}C NMR gave two peaks at 115.6 and 146.6 ppm. NMR analyses did not give enough information. X-ray crystallographic analysis of **Fr. 1** indicated the tetra halogenated hydroquinone structure. ESI-Q-TOF-MS measurement gave two sets of five peaks in negative mode. These peaks and isotopic abundance were derived from four bromines. Finally we determined the structure of **Fr. 1** as 2,3,5,6-tetrabromohydroquinone (TBHQ) (Fig. 3). Minor luminescent peaks, **Fr.2** and **3**, gave similar spectra as **Fr. 1**. From NMR and MS analyses, the elemental components of **Fr.2** and **3** were determined to be $C_6H_3O_2Br_3$ and $C_{12}H_4O_4B_6$, respectively (Fig. 4). **Fr.2** was 2,3,5-tribromohydroquinone. Structure of **Fr.3** has not been determined yet. There were still remained the 2 possible structures of **Fr.3** as shown in Fig.4.

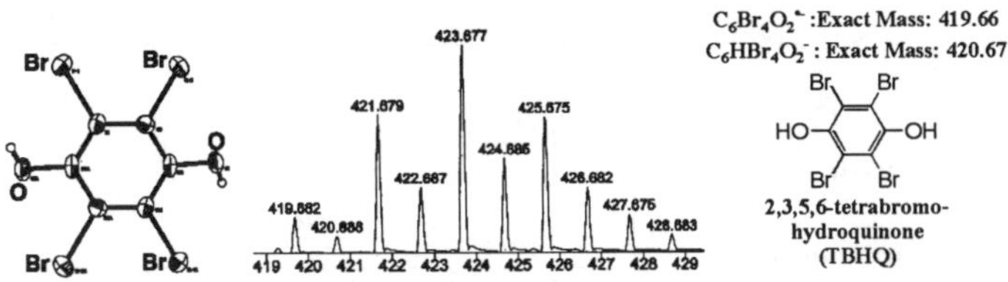

Figure 3. X-ray structure, mass spectrum and structure of isolated **Fr.1**

Hydroquinone is easily oxidized and gives quinone. Proton peaks due to hydroxyl groups of **Fr.1**, **2** and **3** on ^{1}H NMR indicated that each compound existed as a hydroquinone structure, not quinone.

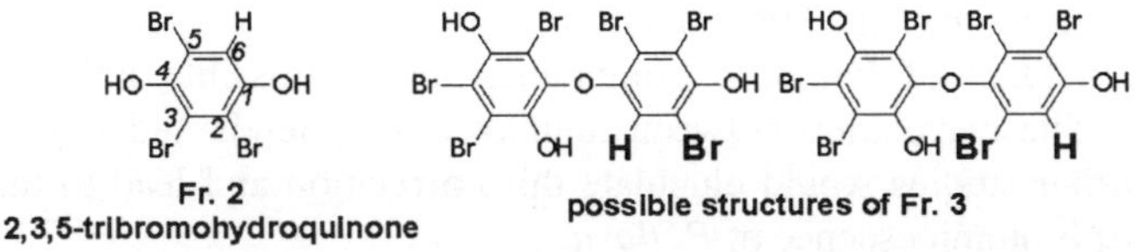

Figure 4. Structure of **Fr.2** and possible structures of **Fr.3**

These compounds had simple but highly brominated structures. Authentic TBHQ was subjected to the luminescence assay and showed an almost equivalent intensity as the isolated **Fr.1**. This result confirmed that TBHQ emitted the light.

No chemiluminescent spectrum was obtained even under the conditions of maximum light intensity. Besides, the absence of fluorescence of these compounds indicated the existence of a light emitter. A methanol extract showed strong green fluorescence and this fluorescent compound was purified by ODS column chromatography. UV absorption, excitation and emission spectra of the light emitter were identical with the riboflavin. ESI-Q-TOF-MS/MS spectra of light emitter and riboflavin were shown in Fig.5. A molecular ion was observed at *m/z* 377. From these results, the light emitter was identified as riboflavin. There was a minor fluorescent fraction in the methanol extract and UV and fluorescent spectra were similar to that of riboflavin. Besides, mass spectrum showed the peak at *m/z* 375 and fragment at *m/z* 243 which was derived from isoalloxazine moiety. Based on MS/MS analysis, the minor fluorescent compound is a dehydro-riboflavin.

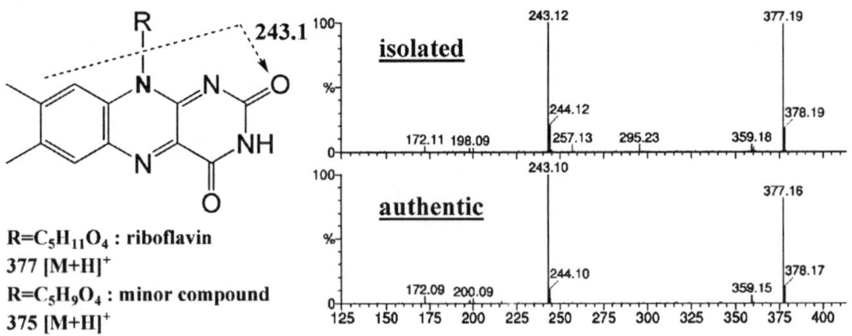

Figure 5. Structure and mass spectra of isolated and authentic riboflavin

In order to identify the structure of TBHQ after the luminescence, it was mixed with H_2O_2 at pH 8.5, and resultant solution was first treated with Na_2SO_3 and secondly extracted with ethyl acetate. The products were very unstable and decomposed during the final HPLC analysis.

The structures of luminous compounds and a possible light emitter were determined. The correlation between luminous compounds and riboflavin is still unclear. Further studies would elucidate this correlation and lead to the molecular mechanism of bioluminescence of *P. flava*.

ACKNOWLEDGEMENTS

We thank Mr. Kenji Yoza at Japan Bruker Co. Inc. for X-ray analysis and JSPS fellowship for financial support.

REFERENCES

1. Dure LS, Cormier MJ. Requirements for luminescence in extracts of a balanoglossid species. J Biol Chem 1961; 236: PC48-50.
2. Cormier MJ, Dure LS. Studies on the bioluminescence of *Balanoglossus biminiensis* extracts. J Biol Chem 1963; 238: 785-9.
3. Dure LS, Cormier MJ. Studies on the bioluminescence of *Balanoglossus biminiensis* extracts. J Biol Chem 1963; 238: 790-3.
4. Dure LS, Cormier MJ. Studies on the bioluminescence of *Balanoglossus biminiensis extracts*. J Biol Chem 1964; 239: 2351-9.
5. Higa T, Fujiyama T, Scheuer PJ. Halogenated phenol and indole constitute of acorn worm. Comp Biochem Physiol 1980; 65B: 525-30.
6. Higa T, Sakemi S. Environmental studies on natural halogen compounds. J Chem Ecol 1983; 9: 495-502.

pH REGULATION OF LUCIFERASE ACTIVITY IN DINOFLAGELLATES INVOLVES A NOVEL ENZYMATIC MECHANISM

L LIU[1], W SCHULTZ[2], JW HASTINGS[1]

[1]*Dept of Mol. & Cellular Biology, Harvard Univ., Cambridge, MA 02138, USA*
[2]*Hauptman-Woodward Institute, 73 High Street, Buffalo, NY 14203-1196, USA*

INTRODUCTION

Regulation of luciferase (LCF) activity in dinoflagellates is unique, without precedent in other enzymes. In brief, it involves control of substrate binding. At pH 8 its conformation prevents substrate from binding. At pH 6 the LCF conformation changes, allowing the tetrapyrrole luciferin (LH_2) to access the binding site. The reaction then occurs giving a bright, brief flash of less than 100 msec. In at least one species, (*Lingulodinium polyedrum*, formerly *Gonyaulax polyedra*), regulation is enhanced by the action of a second protein, luciferin binding protein (LBP), which binds luciferin at pH 8 but releases it at pH 6.[1] Such a large and rapid pH jump is possible by virtue of the fact that the light emitting system is contained in small (~0.5 μm), cortically located novel organelles named scintillons. They occur as outpocketings of the cytoplasm, projecting into the cell vacuole, connected like a balloon (Fig. 1), and containing only LCF, LH_2 and LBP. They can be identified by

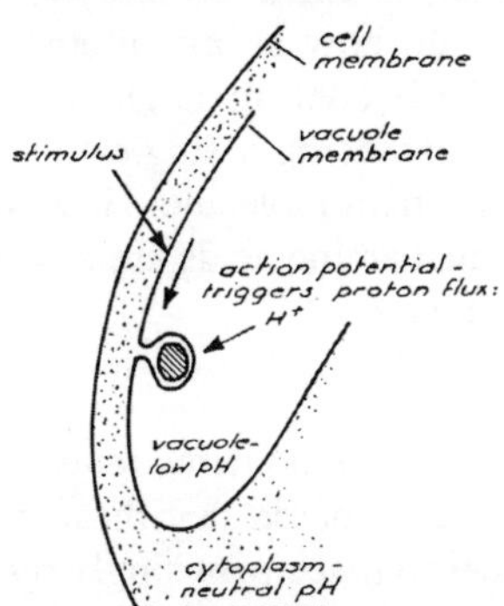

Figure 1. Schematic scintillon

immunolabeling with antibodies raised against LCF or LBP, and visualized by their bioluminescent flashing following stimulation, which is co-localized with the fluorescence of luciferin. Based on the effects of pH on the activities of purified LCF and LBP, and also on isolated scintillons, *in vivo* flashing was postulated to result from a transient pH change in the scintillons, triggered by a mechanically initiated action potential in the vacuolar membrane which opens ion channels that allow protons from the acidic vacuole to enter. As they are effectively isolated from the parent cytoplasm, very few protons are needed to change the pH from 8 to 6.

THREE ACTIVE SITES IN A SINGLE PROTEIN

The full-length *lcf* cDNA (4,037 bp) has an open reading frame of 3,723 bp and encodes the 136,994 Da protein (Fig. 2), being comprised of three contiguous intramolecularly homologous domains, D1, D2 and D3, with no intervening nucleotides.[2] Upon cloning and expressing each domain individually, it was found that each is catalytically active as a luciferase. The domains are ~ 75% identical overall, and ~95% identical in the more central active site regions.

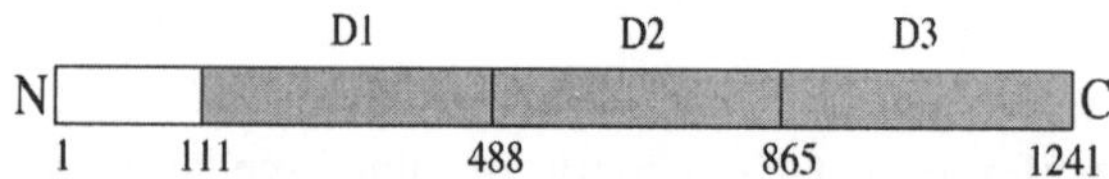

Figure 2. The structure of *L. polyedrum* luciferase

We hypothesize that the three contiguous luciferase domains, with their catalytic sites associated separately with an LBP and LH_2, are structured as a supra molecular unit, and that scintillons contain many such units. This structure is itself unique. The presence of three repeated conserved sequences in one enzyme molecule is not unprecedented, but to our knowledge, this is the only enzyme in which each of the domains has been shown to be separately active, and to be serviced by a protein holding the substrate. A possible reason for such a structure is that the presence of three active sites on a single molecule allows activity to be greater without an increase in the osmotic pressure of the scintillon.[2]

It was early observed that ~35kDa proteolytic fragments of the soluble luciferase have activity, which can now be interpreted as coming from individual domains. But fragments lacked pH control; might this be a property of the full-length molecule? No. Peptides embracing a single full domain exhibit pH activity curves similar to that of the full-length molecule. However, single domain peptides, in which about 50 to 70 N-terminal amino acids are absent, exhibit higher activity at pH 8, like the proteolytic fragments.

CONSERVED HISTIDINES

Inspection of the N-termini of the three domains revealed four histidines conserved in all three domains, and it was considered that these might be responsible for the low activity at pH 8. Indeed, their replacement singly by alanine, using site-directed mutagenesis, resulted in luciferases with much higher activity at pH 8, and even greater with multiple replacements (Table 1).[3] From this result we proposed that pH-dependent charge changes of these non-catalytic histidine residues regulate luciferase activity through conformational shifts in protein structure.

The four conserved histidines are found in all seven species examined

We cloned and sequenced the *lcf* genes from six additional species of luminous dinoflagellates and found that the histidine residues identified in *L. polyedrum* are conserved in all three domains of all seven species (Fig. 3), indicating that the mechanism of activity regulation by pH is similar in all.

Table 1. Activities of mutants

Mutants	Activity at pH 8*
WT	4
H35A	43
H45A	30
H60A	8
H66A	9
H1,2	82
H1,2,3	108
H1,2,3,4	98

* % of activity at pH 6.3

```
           30      ▼    40      ▼    50           60  ▼        ▼
Aa_D1   FKDGLHKPKW DSEGLHKPHT IGGKTYETGF HYLLEAHELG
At_D1   FKDGLHKPKW DSEGLHKPHT IGGKTYETGF HYLLEAHELG
Lp_D1   FKGGLHRPKF DSEGLHKPHT SGGKTYETGF HYLLEAHELG
Pf_D1   FKGGLHKPNF HSEGLHMPHT SGDKTYDTGF HYLLEMHELG
Pl_D1   FREGLHQPKF HTDGLHMPHT SGEKTYETGF HYLLEVHDLG
Pn_D1   FKGGLHKPNF HSEGLHMPHT SGGKEYETGF HYLLEMHELG
Pr_D1   FKGGLHKPDW DKEGLHKPHT IGGKTYDTGF HYLLEAHDLG
Aa_D2   FKDGLHQPKF HEEGLHKPME AGGKVYTTGF HYLLEAHELG
At_D2   FKDGLHQPKF HEEGLHKPME AGGKVYTTGF HYLLEAHELG
Lp_D2   FKNGLHAPNF HDDGLHKPME AGGKVYSTGF HYLLEAHDLG
Pl_D2   FEDGLHKPKF HDDGLHKPME AGGKVYETGF HYLLEAHELG
Pn_D2   FEGGLHKPKF HDDGVHGPMT AGGKEYETGS HYLLEAHELG
Pr_D2   FEGGLHKPKF HEEGLHKPME AGGKVYTTGF HYLLEAHELG
Aa_D3   FKNGLHQPTF HAEGLHKEME VNGKTYGSGF HYLLECHELG
At_D3   FKNGLHQPTF HPEGPHKEME VNGKKYDSAF HYLLECHELG
Lp_D3   FKNGMHKPEF HEDGLHKPME VGGKKFESGF HYLLECHELG
Pf_D3   FKDGLHQPTF HEEGLHKPVE AGGRVYETGF HYLLECHELG
Pl_D3   FKDGMHQPTF HDEGLHKPME AGGKTFESGF HYLLECHELG
Pn_D3   FQDGLHQPTF HEEGLHKPVE AGGKVYETGF HYLLECHELG
Pr_D3   FKNGLHQPTF HPEGAHKEME VNGKTYESGF HYLLECHELG
```

Figure 3. Alignments in the region of the conserved histidines of the three domains of the seven luciferases. Pl, *Pyrocystis lunula*; Pn, *P. noctiluca*; Pf, *P. fusiformis*; At, *Alexandrium tamarense*; Aa. *A. affine*; Pr; *Protoceratium reticulatum*.

Structure of domain 3 of L. polyedrum giving the location of histidines

We determined the 3-D structure of LCF domain 3 of *L. polyedrum*.[4] The histidines are in a region where protonation could regulate substrate binding (Fig. 4).

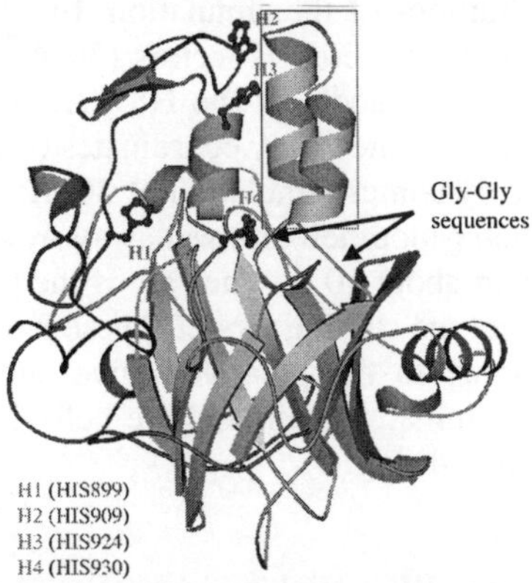

Figure 4. Crystal structure, domain 3

The structure at pH 8 reveals a putative active site pocket, but in that structure there is neither enough room to bind the substrate nor an opening to the interior of the barrel. The presence in the pocket of several polar residues that could participate in catalysis, including histidine, tyrosine, and glutamate, is suggestive of an active site. Site directed mutagenesis of five such residues has been found to result in loss of luciferase activity. Proline rich loops surround the β-barrel and may serve to stabilize the active site by tying the protein together like a standing rib roast.

MOLECULAR MECHANISM OF pH REGULATION

Our hypothesis regarding the mechanism for the pH-controlled regulation of LCF activity is that the histidine residues make contacts that stabilize the N-terminal domain and the helix-loop-helix. Disruption of such contacts by protonation in the wild-type, or in his to ala mutants at pH 8, causes the N-terminal domain and the helix-loop-helix to move and open the catalytic active site. This appears to be altogether novel in enzyme chemistry as a mechanism for regulation of activity.

The D3 domain structure reveals that these histidine residues are at an interface with the helix-loop-helix motif that covers the putative active site (Fig. 4). H899 acts as a core residue in the N-terminal subdomain forming a hydrogen bond with Y925. There are very few interactions that tether the N-terminal subdomain to the rest of the protein and the hydrogen bond of H899 to the main chain carbonyl of V1087. This is an important residue, as it is positioned on a stable turn of the β-barrel and may serve to anchor the N-terminal subdomain to the β-barrel. H909 is in van der Waals contact with A1052 and forms a hydrogen bond with the main chain carbonyl of L1050, both of which lie in the loop of the helix-loop-helix. H924 forms a hydrogen bond with S921 and is also in van der Waals contact with I1045 of the helix-loop-helix. Lastly, H930 forms a hydrogen bond with Q1037 of the helix-loop-helix and is resting in a hydrophobic pocket created by A1088, A1038 and M1070.

Preliminary molecular dynamics calculations (Sybyl, Tripos engine) indicate that the N-terminal domain and the helix-loop-helix are more mobile than the rest of the protein under the conditions of the simulation. In addition, the residues in this area of the protein have higher overall B-factors (39 Å^2) when compared with the rest of the protein (23 Å^2). A model of the H899A, H909A, H924A and H930A variant of D3 was created using the X-ray coordinates of the native D3 structure and used in a 1 ns molecular dynamics simulation at 300K. During the simulation, the N-terminal domain and helix-loop-helix moved away from each other about 5 Å and both away from the protein about 10 Å. The rest of the protein remained stable and deviated less than 1 Å from the starting model. These movements served to open up a solvent-accessible channel to the putative active site. Three separate Gly-Gly sequences within the N-terminal domain and the helix-loop-helix served as hinges about which the chains rotated (Fig. 4).

REFERENCES

1. Wilson T, Hastings JW. Bioluminescence. Ann Rev Cell Dev Biol 1998;14:197-230.
2. Li L, Hong R, Hastings JW. Three functional luciferase domains in a single polypeptide chain. Proc Natl Acad Sci USA 1997;94:8954-8.
3. Li L, Liu L, Hong R, Robertson RL, Hastings JW. N-terminal intramolecularly conserved histidines of three domains in *Gonylaulax* luciferase are responsible for loss of activity in the alkaline region. Biochemistry 2001;40:1844-9.
4. Schultz W, Hastings JW, Liu L. The structure of domain 3 of *Lingulodinium polyedrum* luciferase. Proc Natl Acad Sci USA 2004 (to be submitted).

BIOLUMINESCENCE IN OPHIUROIDS (ECHINODERMATA): A MINIREVIEW

J MALLEFET

Laboratoire de Biologie Marine, Center for biodiversity study, UCL,
Place Croix du Sud, 3, B-1348 Louvain-la-Neuve, Belgium.
Email : mallefet@bani.ucl.ac.be

LUMINESCENCE IN ECHINODERMS

In the marine life, 700 genera representing 13 phyla contain luminous specimens,[1] Echinoderms represent a major phylum of benthic organisms and although their ability to produce light has been reported for decades, little is known about bioluminescence in this phylum. In echinoderms, four of the five classes contains luminous representatives (Fig. 1A), a total number of 91 luminous species are now reported.[2--7]

A

Classes	Species number	
	Total	Luminous
Echinoidea	950	0
Ophiuroidea	2000	38
Holothuroidea	900	30
Asteroidea	1500	20
Crinoidea	550	3

B

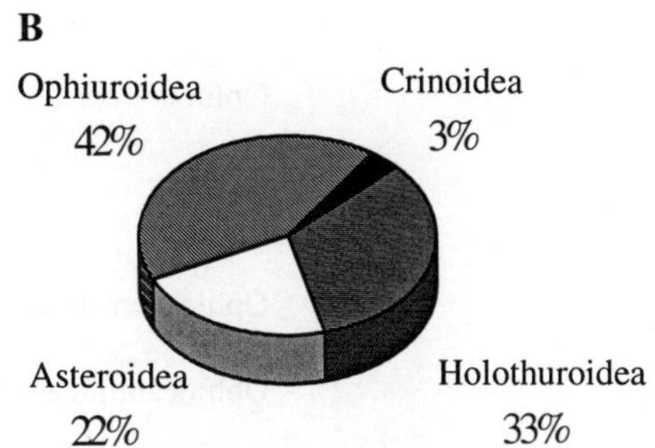

Figure 1. (**A**) Total number of species and known luminous species in echinoderms; (**B**) Relative abundance of species within four classes of echinoderms known to contain luminous representatives.

Despite the reasonable number of luminous species, a. literature survey of the last thirty years provided us with limited information about echinoderm luminescence. Inspection of luminous echinoderms listing reveals that within the echinoderm phylum bioluminescence is not uniformly distributed in each classes : with 38 and 30 species, ophiuroids and holothurians represent 78 % of the echinoderm luminous species while asteroids and crinoids correspond respectively to 23 and 3% of the phenomenon. The total absence of luminescent species in echinoids remains unsolved (Fig. 1B). It must be pointed out that when available, echinoderm luminescence studies have been mostly carry out on the ophiuroids largely because some species are easily observable *in situ*, collected and kept in captivity allowing experimental works.

LUMINESCENCE IN OPHIUROIDS

The first description of a luminous ophiuroid was done by Viviani in 1805; since then lists of luminous echinoderms have been compiled by different authors. [2-7] In 1995 Herring mentioned 33 ophiuroids species in a list of luminous echinoderms[4], since then we performed field surveys in shallow waters of California, South Australia and New Zealand in order to study *Ophiopsila californica, Ophionereis schayeri and O. fasciata* luminous capabilities. During these surveys, 35 ophiuroid species were also tested for luminescence.[6,7] Four new luminous ophiuroid species (species not known to be luminous) were described and one new ophiuroid species was discover, this species referred as *Amphipholis* sp. being also luminous. These results prompted us to add five species to the list of luminescent ophiuroids; an updated list is provided in Table 1 adapted from.[2-7]

Class/Order	Family	*Genera species*
Ophiuroidea/Ophiurida	Ophiomyxidae	*Ophioscolex glacialis*
	Ophiuridae	*Ophiomusium lymani, Homalophiura tesselata, Ophiura mundata, O. concreta, Ophioplocus bispinosus*
	Ophiocomidae	*Ophiopsila annulosa, O. aranea, O. californica, O. riisei*
	Ophionereididae	*Ophionereis schayeri, O. fasciata*
	Ophiodermatidae	*Ophiarachnella ramesayi*
	Ophiocanthidae	*Ophiocantha bidentata, O. abyssicola, O. aculeata, O. bairdi, O. densa, O. aristata, O. simulans, O. smitti, O. crassidens, O. cuspidata, O. simulans, Ophioplinthaca chelys, Ophiomitra spinea, Ophiomitrella sp.*
	Ophiactidae	*Ophiopholis cf. longispina,*
	Amphiuridae	*Acrocnida brachiata, Amphipholis squamata, Amphipholis sp., Amphiura filiformis, A. josephinae, A. grandisquama, A. kandai, A. constricta, A. arcystata, A. magellanica*
	Ophiothricidae	*Ophiothrix fragilis*

Table 1. Luminous ophiuroids, new luminous species added to the list are underlined.

The increased occurrence of the bioluminescence in ophiuroids suggests that luminescence is more widespread than initially thought, raising questions about the functional importance of luminescence in this echinoderms class.

Study of control mechanisms of luminescence in ophiuroids represents a major research program carried out in our laboratory since more than 14 years. Physiological research allowed us to describe control mechanisms of ophiuroids light emissions (nervous control, receptors, neuromodulators and second messengers). Pharmacological results have shown acetylcholine (Ach) to induce luminescence through muscarinic cholinergic receptors in *Amphipholis squamata*[8], through muscarinic and nicotinic cholinergic receptors in *Amphiura filiformis*[9]. Evidences for a cholinergic mechanism were also found in *Amphipholis* sp. and *Amphiura arcystata*.[6] In *Ophiopsila aranea,* on the other hand, none of the neurotranmitters tested so far are involved in the luminous control.[10] In the case of *Ophionereis* species, a Gabaergic as well as a cholinergic controls are proposed.[7] In *Amphipholis squamata*; several neuromodulators have being described.[11-13] Recently, synergetic effects of tryptamine (Tryp) and octopamine (Octo) were described for *Amphiura filiformis*[14] while a transmitter role has been proposed for tryptamine in *O.californica*.[14]

In light of these results, bioluminescence nervous control mechanisms in ophiuroids appear to be extremely diversified and study of intrinsic control mechanisms is now under investigations.[15] New research efforts are considered to (i) investigate the luminous status of more ophiuroid species; (ii) provide some keys in order to understand luminescence evolution in ophiuroids; (iii) extend our knowledge to other echinoderm classes. Access to deep sea species in good physiological conditions will represent one of our next goals.

ACKNOWLEDGMENTS

This research program was possible thanks to financial support from FNRS, FSR-UCL, Communauté française, Fonds Léopold III, Fondation Agathon de Potter, Petra och Karl Erik Hedborg Foundation, EEC LSF & ARI grants. My deepest recognition to directors of Station marine de Langrune sur Mer, France; Arago Laboratory, Banyuls-sur-Mer, France; Kristineberg Marine Station, Fiskebäckskil, Sweden; Marine Sciences Institute Santa Barbara, California, USA; Portobello marine station, Dunedin, New Zealand; Melbourne museum, Australia; Marine Sciences, Sydney University, Australia. Special thanks to diving officers and collectors from various stations who helped me during scuba collections. J. Mallefet is Research Associate of the FNRS (Belgium).

                                                          Mallefet J

REFERENCES
1.   Hastings JW, Morin JG. Bioluminescence. In: Prosser C.L. ed. Neural and integrative animal physiology, fourth edition. New York: John Wiley & Sons 1991: 131-71.
2.   Harvey EN.. Echinodermata. In: Bioluminescence, New York: Academic Press 1952: 472-9.
3.   Herring PJ. Systematic distribution of bioluminescence in living organisms. J. Biolumin Chemilumin 1987; 1: 147-63.
4.   Herring PJ. Bioluminescent echinoderms: Unity of function in diversity of expression? In Emson RH, Smith AB, Campbell AC, eds. Echinoderm Research 1995 Rotterdam: Balkema 1995: 9-17.
5.   Mallefet, J. Physiology of bioluminescence in echinoderms. In: Candia Carnevali, MD, Bonasoro F. eds. Echinoderm Research 1998. Rotterdam: Balkema 1999 : 93-102.
6.   Mallefet J, Hendler G, Herren CM., McDougall C, Case J. A new bioluminescent ophiuroid species from the coast of California. In press. In Heinzeller T. Nebelsick J. eds. Echinoderms 2003. Rotterdam: Balkema.
7.   Mallefet J, Barker M, Byrne M, O'Hara T. First study of bioluminescence in Ophionereis. In press In: Heinzeller T, Nebelsick J. eds. Echinoderms 2003. Rotterdam: Balkema.
8.   De Bremaeker N, Mallefet J, Baguet F. Luminescent control in the brittlestar *Amphipholis squamata*: effect of cholinergic drugs. Comp Biochem Physiol 1996; 115C: 75-82.
9.   Dewael Y, Mallefet J. Luminescence in ophiuroids (Echinodermata) does not share a common nervous control in all species. J Exp Biol 2002; 205 : 799-806.
10.   Mallefet J, Dubuisson M, Preliminary results of luminescence control in isolated arms of *Ophiopsila aranea* (Echinodermata). Belg J Zool 1995 125: 167-73.
11.   De Bremaeker N, Baguet F, Thorndyke MC, Mallefet J. Modulatory effects of some amino acids and neuropeptides on luminescence in the brittlestar *Amphipholis squamata*. J Exp Biol 1999; 202: 1785-91.
12.   De Bremaeker N, Mallefet J, Baguet F. Effects of catecholamines and purines on the luminescence of *Amphipholis squamata* (Echinodermata). J Exp Biol 2000; 203: 2015-23.
13.   Dupont S, Mallefet J, Vanderlinden C. Effect of b-adrenergic antagonists on bioluminescence control in 3 species of brittlestars (Echinoderms). In press Comp Biochem Physiol C 2004.
14.   Vanderlinden C, Mallefet J. Synergic effects of tryptamine and octopamine on ophiuroid luminescence (Echinodermata). In press J Exp Biol 2004.
15.   Vanderlinden C, Dewael Y, Mallefet J. Screening of second messengers involved in photocyte bioluminescence control of three ophiuroid species (Ophiuroidea, Echinodermata). J Exp Biol 2003; 206: 3007-14.

LUMINESCENT BEHAVIOUR IN THE NEW ZEALAND GLOWWORM, *ARACHNOCAMPA LUMINOSA* (INSECTA; DIPTERA; MYCETOPHILIDAE)

N OHBA[1], VB MEYER-ROCHOW[2]

[1]*Yokosuka City Museum, Yokosuka 238-0016, Japan*
E-mail: QGB00523@nifty.ne.jp
[2]*International University Bremen (IUB), Faculty of Engineering & Science, D-28725 Bremen, P.O.Box 750561, Germany*
E-mail: b.meyer-rochow@iu-bremen.de

INTRODUCTION

The larva of the mycetophilid *Arachnocampa luminosa*, commonly known as the New Zealand glowworm, emits a seemingly continuous bluish light. *A. luminosa* larvae secrete numerous vertical silk threads covered with droplets of sticky mucus, which hang from the ceiling of the limestone cave. The larvae occupy a nest made of thin silk and with their light the glowworms attract small insects and other invertebrates that subsequently get stuck on the mucus-covered silk threads.[1-5] Since to date there are no studies that concern themselves with the luminescence activity rhythm of the larvae (let alone the adults), we set out to record in the field any luminescence biorhythmicity the larvae might display.

MATERIAL AND METHODS

Recordings

Luminescence activities of the larval glowworms, *Arachnocampa luminosa,* were recorded and analyzed in the Waitomo Caves on the North Island of New Zealand on March 6th, 1998, with the aid of a night-scope, installed in a VTR camera (Sony Digital Video camera recorder DCR-TR V9). Periods of glowworm luminescence in the cave were compared between day (13:44:21 – 15:45:43) and nighttime periods (17:24:12 – 19:03:32). The analyses of the luminescence activity periods were based on the assumption that the pictures represented fixed-point continuities and the glowworms were not changing their nest sites. The recorded pictures permitted us to conclude whether or not individuals had been emitting light (30 individuals marked at daytime, 20 at night) throughout the period of investigation.

Habitat

Subterranean water-flows exist in the Waitomo limestone cave; the latter is humid and there is no artificial light (at least not in the part of the cave, in which our observations took place). Larvae of *A. luminosa* occur together in large aggregations, but keep fixed minimal distances between neighbouring individuals to reduce interference with the mucus-covered and up to 30 cm long fishing lines that hang down from the ceiling of the cave, i.e., the nests of the glowworms. Because of the underground water in caves occupied by glowworms, there is always a high level of humidity and ceiling as well as cave walls are usually damp. The temperature of the Waitomo Caves does not fluctuate greatly and is usually around 12 °C.

RESULTS
Population density, growth stages, and glowworm nests
Nests of glowworms were spaced several cm apart from each other and glowworm densities were very high. Colonists of such glowworm patches included a mixture of very young, first larval instars through to last larval instars. Pupae and adults were present, but in much smaller numbers. Glowworm nests, constructed under the ceiling of the cave, had on average 10-14 mucus-coated up to 30 cm long silk strands. The total number of the mucus strands was fewer in the younger larvae and the strands of the latter were also shorter.

Larval movements in the nest
When prey got stuck to or became entangled in the mucus strands, the larva, resting upside down in its nest, quickly headed towards the direction where the prey was caught. The body of the glowworm was bent 180° when turnabouts on a string were necessary. Such turnabouts occurred freely and frequently. When stationary in their nests, the larvae emitted light from virtually fixed positions. Because of this regularity, it was possible to individually distinguish glowworms on the recorded pictures.

Luminescence activity
As a group, the glowworms under observation in the cave always emitted light regardless of day and night. Light emission, therefore, continued without apparent blinking, but group luminescence at night differed between cave and outside populations. The luminescence of the few adult glowworm flies present was not distinguishable from larval light, although we know it to be continuous as well.

Cessation of luminescence
On March 6th, 1998, luminescence activity was recorded between 13:44:21 and 15:45:43 (i.e., 2 hours, 1 minute, and 22 seconds = 7282 seconds). There were 24 marked individuals that had continued emitting light for 7282 seconds. The luminescence exhibited fluctuations in six individuals. Three individuals completely stopped emitting light once. Individuals that stopped emitting light twice during the recording period were not observed. Total individual luminescence stoppage times ranged from 339 seconds to 1500 seconds. Total luminescence stoppage time ratio (total luminescence stoppage time/total luminescence time = 2741/7282 x 30) amounted to 0.013.

On March 6, 1998: Luminescence activity was recorded between 17:24:12 and 19:03:32 (1hour, 39 minutes, and 20 seconds – 5960 seconds). Seven individuals continued emitting light for 5960 seconds. That was 35% of all 20 marked individuals. The luminescence became weak once in two individuals, the percentage of maximal activity being 10%. Six individuals, corresponding to 30%, completely stopped emitting light only once; ten individuals stopped twice. The luminescence stoppage times were 15 – 1487 seconds or less (being about 25 minutes). Total luminescence stoppage ratio (i.e., total luminescence stoppage time/total luminescence time) was 0.03-0.24 and the luminescence stoppage ratio of the entire group was 0.06.

DISCUSSION
Prey items in the native habitat
About 21 kinds of separate prey items such as insects, other arthropods like arachnids and millipedes, and even annelids and small gastropods are know from previous investigations around the habitat of the Waitomo Caves to serve as food for the glowworms.[3] However, our observations, having been very limited in time and scope, were unable to confirm this variety of food items in the Waitomo Caves. We have to assume that the vast majority of the food items that the Waitomo Caves glowworms have been feeding on, was made up of small insects from the subterranean stream.

Larval population densities and growth stages
Larval densities can be very high in suitable places, because of inter-individual nest distances amounting to only a few cm. The high density is maintained by the empty spaces becoming the next generation's living place, for instance when a larva has grown, matured, and turned its space over to a new generation of larval glowworms. We believe the chance of catching prey is not only simply increased in group living, because of the brighter light a group emits, but predict that in places where one finds larvae of various developmental stages and in which older larvae possess stronger luminescence than the younger larvae, the latter also benefit from the former's stronger bioluminescence. This could be one reason why young and immature larvae, actually in danger of being accepted as prey by the older larvae, tend to occur together.

Larval nest and fishing lines
Vertical threads tend to be less well developed outside the cave, because of the presence of leaves, branches, and wind. Larvae have to react quickly to the presence of prey and move freely and rapidly on the strings of silk, capable of 180° turnabouts. A larva must possess sensitive mechanoreceptors allowing it not only to sense that prey has been caught, but also to identify which of the vertical strands it has to descend (or haul in). It is known that chemoreception is also involved in accepting a prey item.[3]

Luminescence activity
Luminescence activity is more or less continuous in the cave, but emission of light outside the cave takes place only at nighttime. Individuals in caves keep emitting light for two hours or longer. This appears to be the longest uninterrupted luminescence period of any luminescent organism (other than fungi and bacteria). It was demonstrated that most of the individuals that stopped emitting light for a while, or continued to emit light at a reduced and weaker rate, were doing that at night

rather than during the day. Perhaps luminescence behaviour changes at night and activities other than prey catching co-occur at night. However, more likely glowworms have optimized light emission to coincide with the emergence of small aquatic prey, which inside the cave may take place less frequently at night than during the day. Alternatively, it is possible that cessation of luminescence occurs when prey has been caught or is being digested and no new prey items are needed.

Luminescence stoppages

Total stoppages of luminescence in the cave are so short that this fact seems to underscore the importance, not to say indispensability, of the light for the glowworm to attract prey. Surprisingly, mycetophild larvae with virtually identical prey-catching life styles, but living in caves of the neotropical realm, are incapable of producing light, but survive very well (probably because of the much higher number of small flying insects in the tropical caves and no need for photic lures).[6] As for the total luminescence stoppage time ratios of the groups: they amounted to 0.013 during the day and 0.06 at night. It is thought that in the New Zealand the glowworms' light emission periodicities have adaptive value, because emitting light is indispensable to attract prey in the caves. What exactly the adaptive value for the more frequent luminescence breaks at night in the cave is, remains somewhat obscure (see above), but it is hoped that further research will be able to shed some light on this question.

ACKNOWLEDGEMENTS

We wish to express our gratitude to Drs. H. Niwa and Y. Omiya, who cooperated in the field investigation. This study was supported through a grant from the Japan Ministry of Education, Science, and Culture, i.e., International Science Research No. 0941100.

REFERENCES

1. Gatenby JB. Note on the New Zealand Glowworm, *Bolitophila (Arachnocampa) luminosa*. Trans Roy Soc NZ 1959; 87: 291-314.
2. Stringer IAN. The larval behaviour of the New Zealand glowworm *Arachnocampa luminosa*. Tane 1967; 13: 107-17.
3. Meyer-Rochow VB. The New Zealand Glowworm. Waitomo Caves Museum Society Inc., Waitomo Caves, New Zealand, 1990: 1-60.
4. Broadley RA, Stringer IAN. Prey attraction by larvae of the New Zealand glowworm, *Arachnocama luminosa*. Invert Biol 2001; 120: 170-7.
5. Ohba N. External morphology and feeding habits of the New Zealand glowworm *Arachnocampa luminosa* (Diptera: Mycetophilidae). Sci Rept Yokosuka City Mus 2002; 49: 13-22.
6. Stringer IAN, Meyer-Rochow VB. Attraction of flying insects to light of different wavelengths in a Jamaican cave. Mém Biospéol 1994; 21: 133-9.

AEQUORIN AND GFP: AN HISTORICAL ACCOUNT

O SHIMOMURA

324 Sippewissett Road, Falmouth, MA 02540, USA

Aequorin and the green fluorescent protein (GFP) were discovered from the jellyfish *Aequorea* in 1961.[1] Since then, the research on these proteins has gradually, but steadily, progressed, eventually reaching to the present state of comprehensive knowledge. Both proteins are now being used as important tools in research, aequorin as a calcium indicator and GFP as a marker protein. Such progress, however, might not have happened if the information on the *Cypridina* luminescence system were unavailable.

The study on *Cypridina* luciferin was started in April 1955, when I was a teaching assistant at the Pharmacy School of Nagasaki. My Professor, Dr. Shungo Yasunaga, wished to broaden my knowledge and sent me to the laboratory of Professor Yoshimasa Hirata, at Nagoya University, to work as a visiting researcher. Professor Hirata, an expert in the chemistry of natural products, gave me dried *Cypridina* stored in a large vacuum desiccator and told me to purify and crystallize the luciferin. Crystallization was the only practical means to prove the purity at the time, prior to the structural study of a compound. The luciferin of *Cypridina* had been studied at Newton Harvey's lab, Princeton University, for almost 30 years, but no significant information was obtained on the chemical nature of the luciferin.

Cypridina luciferin is extremely unstable in air. After some test extractions of the luciferin from dried *Cypridina*, I made a plan to extract 500 g of dried *Cypridina* in the absence of air to obtain 2-3 mg of purified material for crystallization. It was 10 times the amount used at Princeton, and I thought the plan must succeed if the compound was crystallizable. Professor Hirata agreed to the plan and his glass blower made an over-sized soxhlet apparatus for me.

The luciferin was extracted in an atmosphere of purified hydrogen and then purified under nitrogen. This process involved day-and-night work for seven days. I tried to crystallize purified luciferin with all combinations of solvents and salts I could think of, but all my efforts ended up with amorphous precipitates. After six or seven batches of purified luciferin were vainly spent in unsuccessful efforts, the first crystals of *Cypridina* luciferin were finally obtained, rather unexpectedly, on one cold morning in February 1956. On the previous night, I had some leftover purified luciferin after my fruitless crystallization attempts. I could not think of any further idea of crystallization, so I decided to spend the material for amino acid analysis. Thus, I added an equal amount of concentrated hydrochloric acid to the solution of luciferin. The color of the solution instantly changed from yellow to dark red. Next morning, I saw that the solution was discolored to light orange, with a small pinch of dark precipitate at the bottom of the test tube. Examination under a microscope revealed, to my surprise, that the precipitate was indeed crystals - fine red needles.

The successful crystallization developed into the elucidation of the structures of *Cypridina* luciferin and its luminescence reaction products in 1966.[2] That information made it possible to determine the structure of the aequorin luminophore several years later.

In 1959, I received an invitation from Dr. Frank H. Johnson, Princeton University. In August 1960, I was among the Fulbright Travel Grant grantees onboard Hikawa-maru leaving Yokohama for Seattle. After 13 days to cross the Pacific and 3 nights by rail to cross the continent, I arrived at Princeton with great excitement for my new life. Shortly after my arrival, Dr. Johnson asked me if I would be interested in studying the bioluminescence of the jellyfish *Aequorea aequorea*. I was quite impressed by Dr. Johnson's description of the brilliant luminescence of live jellyfish and the great abundance of specimens at the Friday Harbor Laboratory in the State of Washington.

In the early summer of 1961, we traveled from Princeton to Friday Harbor in Dr. Johnson's station wagon that was fully loaded with equipment and chemicals. The jellyfish were indeed abundant there. We carefully scooped up the jellyfish using a shallow dip-net. The jellyfish *Aequorea* are shaped like hemispherical umbrellas, measuring 3-4 inches in diameter. The light organs -- 100-150 granules -- are located along the edge of the umbrella. Thus, the margin of the umbrella containing light organs could be easily cut off with a pair of scissors, yielding a thin strip called a "ring". When the rings obtained from 20-30 jellyfish were squeezed through a rayon gauze, a liquid called "squeezate" was obtained. The squeezate was only dimly luminescent, but when diluted with water, its luminescence increased significantly, as the granular light organs were cytolyzed.

We tried to extract luminescent substance from the squeezate by various methods, but all failed, and we ran out of ideas after only a few days of work. Convinced that the cause of our failure was the luciferin-luciferase hypothesis that dominated our thinking, I suggested to Dr. Johnson that we should forget the idea of extracting luciferin and luciferase and, instead, try to isolate the luminescent substance, whatever it might be. However, I was unable to convince him. I spent the next several days soul-searching, trying to imagine the reaction that occurs in luminescing jellyfish and searching for a way to extract the luminescent principle. One afternoon, a thought suddenly struck me -- a thought so simple that I should have had it much sooner: "Even if a luciferin-luciferase system is not involved in the jellyfish luminescence, an enzyme or protein is probably involved in the light-emitting reaction. If so, the activity of this enzyme or protein can probably be altered by a pH change. There might be a certain level of acidity at which an enzyme or protein is reversibly inactivated."

I immediately went to the lab and made a squeezate, and tested a small portion of it with acetate buffer solutions of various acidities. The squeezate was luminous at pH 6.0 and pH 5.0, but not at pH 4.0. I filtered the rest of the squeezate, and mixed the solid part containing granular light organs with pH 4.0 buffer. After 2-3 minutes,

I filtered the mixture. The filtrate, now free of cells and debris, was nearly dark, but it regained its luminescence upon neutralization with a small amount of sodium bicarbonate. Indeed, the experiment showed that the luminescence substance of the jellyfish was extracted into the solution at pH 4. But my real surprise came next. When I added a small amount of sea water to the solution, its luminescence became explosively strong. Because the composition of seawater is known, I quickly discovered that the activator is Ca^{2+}. The discovery of Ca^{2+} as the activator in turn suggested that EDTA should serve as a better inhibitor of luminescence than acidification. Based on this information, we devised a method of extracting the light-emitting principle. We collected and extracted about 10,000 jellyfish in that summer.

After returning to Princeton, we began purification of the light-emitting principle in the extract by repeated chromatography on various kinds of columns. It was completed in early 1962, with a total yield of 5 mg, of which only 1 mg was highly purified. The substance was found to be a protein with a molecular weight of about 20,000. It emitted light when a trace of Ca^{2+} was added, even in the absence of oxygen[1]. We named the protein "aequorin." During the purification of aequorin, another protein with a brilliant green fluorescence (λ_{max} 508 nm) was separated and also purified[1] (named GFP later). We were greatly interested in studying the mechanism of the luminescence of aequorin because the reaction did not fit in the luciferin-luciferase hypothesis. However, our every attempt to extract the luminophore resulted in an intramolecular reaction that destroyed the luminophore. We therefore postponed further study on aequorin.

In 1966, we discovered in the tube worm *Chaetopterus* another unusual bioluminescent protein that did not fit in the luciferin-luciferase hypothesis. We proposed a general term "photoprotein" to designate these bioluminescent proteins.[3]

In 1967, Ridgway and Ashley reported their observation with aequorin of transient Ca^{2+} signals in single muscle fibers of barnacle.[4] Considering the importance of aequorin that became obvious, we decided to resume our study of aequorin. Because there was no way to extract the native luminophore, we aimed at a fragment of luminophore that was formed when aequorin is denatured with urea in the presence of 2-mercaptethanol. The compound was named AF-350 based on its absorption maximum at 350 nm.

To obtain 1 mg of AF-350, 100-200 mg of pure aequorin would be needed, which would originate from about 50,000 jellyfish (2.5 tons). To process 50,000 jellyfish in one summer, we would have to collect, cut and extract at least 3,000 jellyfish each day, allowing for days of bad weather and poor fishing. This workload could not be achieved by collecting jellyfish at the lab dock and cutting rings with scissors at a rate of one ring per minute. Thus, we exploited new fishing

Table 1. Major events of development in the research of aequorin and GFP.

Year	Description
1961-2	Aequorin and GFP discovered[1]
1966	A new general term "photoprotein" proposed[3]
1967	The first intracellular use of aequorin[4]
1971	Förster-type energy transfer from aequorin to GFP suggested[14]
1972	Structure determination of AF-350 (coelenteramine), a partial chromophore of aequorin[5]
1973	Structure determination of coelenteramide, the light-emitter[6]
1974	Hypothetical structure of coelenterazine[7]
1974	Crystallization of GFP, and the first demonstration of the *in vitro* Förster-type energy transfer from aequorin to GFP[16]
1975	Coelenterazine isolated from the squid *Watasenia* and synthesized[8]
1975	Regeneration of aequorin[9]
1978	Coelenterazine-2-hydroperoxide moiety suggested in aequorin[10]
1979	Structure of GFP chromophore elucidated[17]
1985-7	Cloning and expression of aequorin[12]
1986	Coelenterazine-2-hydroperoxide moiety confirmed in aequorin[11]
1988	Preparation of various semisynthetic aequorins[13]
1992	Cloning of the cDNA of GFP[21]
1993	Chromophore structure of GFP confirmed[18]
1994	Expression of GFP in various organisms[22]
1996	X-ray structure of GFP solved[19]
1999	Discovery of red-fluorescent GFP-like protein in coral[23]
2000	X-ray structure of aequorin solved[14]

grounds, and Dr. Johnson constructed two sets of jellyfish cutting machines with which two hired workers could cut 1,200 rings per hour. Our daily routine started at 6 AM with my family and one or two assistants collecting jellyfish. After breakfast, two assistants cut rings until noon. Afternoon was for the extraction of aequorin from the rings. After supper, we again collected jellyfish until 9 PM for the work of the next day.

After five years of hard work, the chemical structure of AF-350 was finally determined in 1972.[5] The result astonished me. The structure contained the skeleton of a 3,5-disubstituted 2-aminopyrazine, identical with that in *Cypridina* oxyluciferin and etioluciferin, although the substituents at 3 and 5 positions were different (Fig. 1). The resemblance suggested a close relationship between the luminescence systems of *Aequorea* and *Cypridina*. In order to obtain the compound corresponding to oxyluciferin, an aequorin solution was luminesced with Ca^{2+}, and

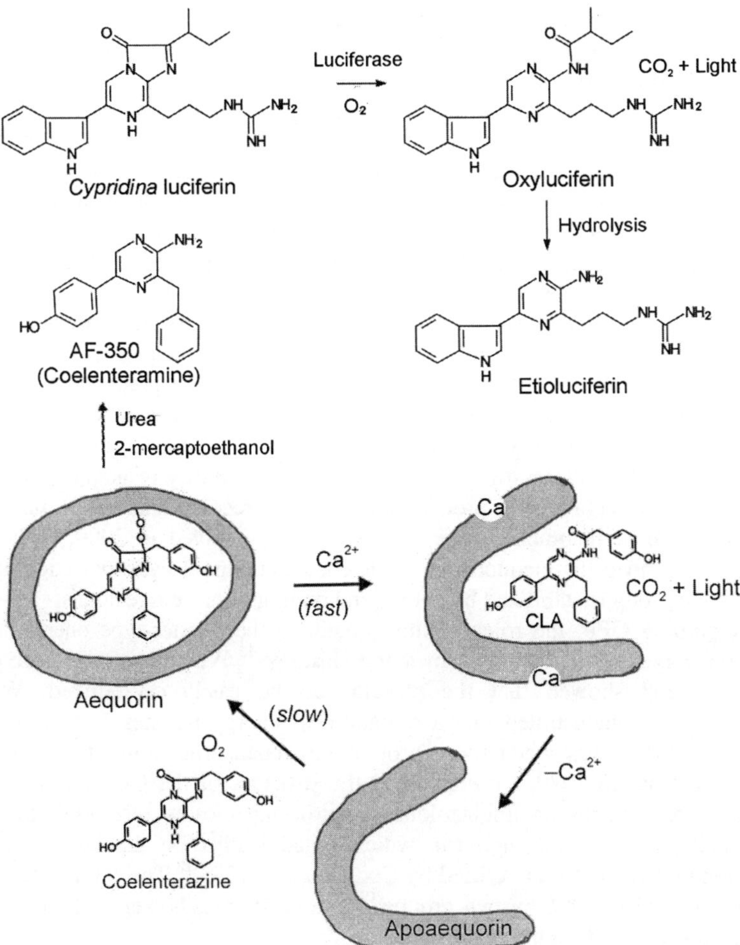

Figure 1. Luminescence reactions of *Cypridina* luciferin and aequorin, in which *Cypridina* luciferin is oxidized to oxyluciferin and etioluciferin, whereas aequorin is decomposed into coelenteramide (CLA), CO_2 and Ca^{2+}-bound apoaequorin. Note the structural resemblance between AF-350 and etioluciferin.

a blue fluorescent product in the spent solution was extracted with ether. The structure of the blue fluorescent oxyluciferin was determined to be *p*-hydroxy-phenylacetylated AF-350 (coelenteramide).[6] The result pointed to an hypothetical compound, now called coelenterazine, as the luciferin. It was a pivotal step in the research of aequorin. In the absence of the information on *Cypridina* luminescence, our progress would have stopped. Coelenterazine was actually isolated from the squid *Watasenia* and chemically synthesized in 1975 by Inoue *et al.* under the name of *Watasenia* preluciferin.[8]

In 1975, we demonstrated that treating spent aequorin with coelenterazine in the presence of oxygen results in the regeneration of original aequorin,[9] consequently proving that aequorin contains a coelenterazine moiety. It was an important discovery that provided the basis for producing recombinant aequorin from apoaequorin. In 1978, we proposed that aequorin contains coelenterazine-2-hydroperoxide,[10] based on the chemical properties and reactions of aequorin. The peroxide structure was confirmed by ^{13}C-NMR spectrometry in 1986.[11]

The cloning and expression of apoaequorin cDNA were accomplished in 1985-1987 by two independent groups.[12] The modified forms of aequorin, "semisynthetic aequorins", with widely different calcium sensitivities, were prepared by replacing the coelenterazine moiety of aequorin with various analogues of coelenterazine.[13] The attempts at X-ray crystallography with natural aequorin and recombinant aequorin were all unsuccessful, due to the insufficient purity of aequorin used. We developed a new technique to produce a high purity recombinant aequorin, then the X-ray structure of recombinant aequorin was finally solved in 2000.[14] The X-ray structure confirmed that aequorin contains coelenterazine-2-hydroperoxide and that the peroxide group is stabilized by hydrogen bonding to protein residues.

Regarding GFP, the roles of this protein in the Förster-type energy transfer were first suggested in 1971 by Morin and Hastings.[15] We characterized the protein in 1974,[16] and showed that the protein can be easily crystallized. We also experimentally demonstrated that a radiationless energy transfer from aequorin to *Aequorea* GFP occurs when these proteins are coadsorbed onto the particles of DEAE cellulose. In 1979, we elucidated the structure of the GFP chromophore.[17] The structure contains an imidazolone skeleton and closely resembles the model compounds of *Cypridina* luciferin, which I had studied in the late 1950s. The chromophore structure was verified by Cody *et al.* in 1993.[18] The crystal structure of GFP was solved in 1996 by two groups.[19] The GFP crystals were found to show a striking anisotropy of fluorescence emission.[20]

The cDNA of *Aequorea* GFP was cloned by Prasher *et al.* (1992).[21] It was soon followed by the expression of GFP in living organisms by two different groups in 1994.[22] It is really remarkable that the cyclic structure of imidazolone chromophore has been spontaneously formed during the expression. The successful expression of GFP has established the basis of utilizing GFP as a marker for gene expression, resulting in the present popularity of GFP. In 1999, a red-fluorescent (λ_{max} 583 nm) GFP-like protein was discovered in the anthozoan coral *Discosoma* sp.,[23] thus further increasing the usefulness of the fluorescent proteins.

Finally I would like to pay my tributes to Dr. Frank Johnson who initiated the study of the *Aequorea* bioluminescence, and to Dr. Yoshimasa Hirata who introduced me to the chemistry of the *Cypridina* luminescence.

REFERENCES

1. Shimomura O, Johnson FH, Saiga Y. Extraction, purification and properties of aequorin, a bioluminescent protein from the luminous hydromedusan, *Aequorea*. J. Cell. Comp. Physiol. 1962; 59:223-39.

2. Kishi Y, Goto T, Hirata Y, Shimomura O, Johnson FH. *Cypridina* bioluminescence I: structure of *Cypridina* luciferin. Tetrahedron Lett. 1966; 3427-36.

3. Shimomura O, Johnson FH. Partial purification and properties of the *Chaetopterus* luminescence system. In: Johnson FH., Haneda Y, Eds. Bioluminescence in Progress. Princeton University Press, Princeton, N.J. 1966:495-521.

4. Ridgway EB, Ashley CC. Calcium transients in single muscle fibers. Biochem. Biophys. Res. Commun. 1967; 29:229-34.

5. Shimomura O, Johnson FH. Structure of the light-emitting moiety of aequorin. Biochemistry 1972; 11:1602-8.

6. Shimomura O, Johnson FH. Chemical nature of light emitter in bioluminescence of aequorin. Tetrahedron Lett. 1973; 2963-6.

7. Shimomura O, Johnson FH, Morise H. Mechanism of the luminescent intramolecular reaction of aequorin. Biochemistry 1974; 13:3278-86.

8. Inoue S, Sugiura S, Kakoi H, Hashizume K. Squid bioluminescence II. Isolation from *Watasenia scintillans* and synthesis of 2-(p-hydroxybenzyl)-6-(p-hydroxyphenyl)-3,7-dihydroimidazo[1,2-a]pyrazin-3-one. Chem. Lett. 1975; 141-4.

9. Shimomura O, Johnson FH. Regeneration of the photoprotein aequorin. Nature 1975; 256:236-8.

10. Shimomura O, Johnson FH. Peroxidized coelenterazine, the active group in the photoprotein aequorin. Proc. Natl. Acad. Sci. USA 1978; 75:2611-2615.

11. Musicki B, Kishi Y, Shimomura O. Structure of the functional part of photoprotein aequorin. Chem. Commun. 1986; 1986:1566-8.

12. Inouye S, Noguchi M, Sakaki Y, Takagi Y, Miyata T, Iwanaga S, Miyata T, Tsuji FI. Cloning and sequence analysis of cDNA for the luminescent protein aequorin. Proc. Natl. Acad. Sci. USA 1985; 82:3154-8; Inouye S, Sakaki Y, Goto T, Tsuji FI. Expression of apoaequorin complementary DNA in *Escherichia coli*. Biochemistry 1986; 25:8425-9; Prasher D, McCann RO, Cormier MJ. Cloning and expression of the cDNA coding for aequorin, a bioluminescent calcium-binding protein. Biochem. Biophys. Res. Commun. 1985; 126:1259-68.

13. Shimomura O, Musicki B, Kishi Y. Semi-synthetic aequorin: an improved tool for the measurement of calcium ion concentration. Biochem. J. 1988; 251, 405-410; Shimomura O, Musicki B, Kishi Y. Semi-synthetic aequorins with improved sensitivity to Ca^{2+} ions. Biochem. J. 1989; 261:913-20; Shimomura O, Musicki B, Kishi Y, Inouye S. Light-emitting properties of recombinant semi-synthetic

aequorins and recombinant fluorescein-conjugated aequorin for measuring cellular calcium. Cell Calcium 1993; 14:373-8.

14. Shimomura O, Inouye S. The *in situ* regeneration and extraction of recombinant aequorin from *Escherichia coli* cells and the purification of extracted aequorin. Protein Expres. Purif. 1999; 16:99-5; Head JF, Inouye S, Teranishi K, Shimomura O. The crystal structure of the photoprotein aequorin at 2.3A resolution. Nature 2000; 405:372-6.

15. Morin JG, Hastings JW. Energy transfer in a bioluminescent system. J. Cell. Physiol. 1971; 77:313-8.

16. Morise H, Shimomura O, Johnson FH, Winant J. Intermolecular Energy Transfer in the bioluminescent system of *Aequorea*. Biochemistry 1974; 13:2656-62.

17. Shimomura O. Structure of the chromophore of *Aequorea* green fluorescent protein. FEBS Lett. 1979: 104:220-2.

18. Cody CW, Prasher DC, Westler WM, Prendergast FG, Ward WW. Chemical structure of the hexapeptide chromophore of the *Aequorea* green-fluorescent protein. Biochemistry 1993; 32:1212-8.

19. Ormö M, Cubitt AB, Kallio K, Gross LA, Tsien RY, Remington SJ. Crystal structure of the *Aequorea victoria* green fluorescent protein. Science 1996; 273:1392-5; Yang F, Moss LG, Phillips Jr GN. The molecular structure of green fluorescent protein. Nature Biotechnology 1996; 14:1246-51.

20. Inoue S, Shimomura O, Goda M, Shribak M, Tran PT. Fluorescence polarization of green fluorescence protein. Proc. Natl. Acad. Sci. USA 2002; 99:4272-7.

21. Prasher DC, Eckenrode VK, Ward WW, Prendergast FG, Cormier MJ. Primary structure of the *Aequorea victoria* green fluorescent protein. Gene 1992; 111:229-33.

22. Chalfie M, Tu Y, Euskirchen G, Ward WW, Prasher DC Green-fluorescent protein as a marker for gene expression. Science 1994; 263:802-5; Inouye, S, Tsuji FI. *Aequorea* green fluorescence protein. Expression of the gene and fluorescence characteristics of the recombinant protein. FEBS Lett. 1994; 341:277-80.

23. Matz MV, Fradkov AF, Labas YA, Savitsky AP, Zaraisky AG, Markelov ML, Lukyanov SA. Fluorescent proteins from nonbioluminescent anthozoa species. Nature Biotechnology 1999; 17:969-73.

BIOLUMINESCENCE REACTION IN THE FIREFLY SQUID, *WATASENIA SCINTILLANS*

FREDERICK I TSUJI

Marine Biology Research Division, Scripps Institution of Oceanography,
University of California, San Diego, La Jolla, California 92093-0202, USA

INTRODUCTION

The bioluminescence of the deep-sea squid, *Watasenia scintillans* (average mantle length, 6 cm), a species that is indigenous to northern Japan, was first described by Watasé.[1] Each spring, females carrying fertilized eggs come inshore by the hundreds of millions in and around Toyama Bay to lay their eggs. The squid has ~800 tiny (<1 mm diam) dermal light organs on its ventral side, 5 prominent organs beneath each eyeball and 3 tiny (<1 mm) pigmented organs at the tip of the fourth pair of arms. The light emitted by the ventral and eye organs is weak, but the arm organs are able to emit brilliant flashes of light. The periodic nature of the flashes has led to the squid being called "hotaru-ika" or "firefly squid." Following the discovery of its luminescence, many attempts were made to determine the source of the light.[2-7] The light was initially thought to come from symbiotic luminous bacteria because of the presence of rod-shaped bodies in the light organs and luminous bacteria could be cultured from them, but following micro-chemical and electron microscopic studies revealed that the squid is self-luminous.[6,8] This view was later supported by the studies of Goto and his coworkers, who isolated from the arm organs two compounds which they called *Watasenia* oxyluciferin and luciferin.[9-12] The chemical structures of these compounds and that of *Watasenia* preluciferin (coelenterazine) are shown below.

$$CH_2 \text{—} OSO_3H \quad +O_2 \rightarrow \quad NH\text{—}CCH_2 \text{—} OSO_3H \quad +CO_2 + h\nu_{blue}$$

Watasenia luciferin

Watasenia oxyluciferin

Watasenia preluciferin : SO₃H=H

In subsequent studies, using dark homogenates of the arm organs, it has been shown that the luminescence reaction of *Watasenia* is ATP-dependent.[13] The injection of ATP solution into a homogenate of the arm organs results in a sharp increase in light intensity, after which the light decays according to pseudo-first order kinetics. The reaction has an optimum pH of 8.8. Light is not emitted if ATP is injected into the supernatant of a homogenate, nor is light emitted by injecting ATP into a suspension of the pellet. Light is emitted only when the pellet is resuspended in the supernatant and injected with ATP, demonstrating that the *Watasenia* reaction is a luciferin-luciferase reaction involving a soluble luciferin present in the supernatant and a luciferase bound to membrane in the pellet. Using experimental protocols based on these results, it has been shown that the bioluminescence reaction has an emission peak of 470 nm, an absolute requirement for molecular oxygen and a luciferin with a chemical structure identical to that of coelenterazine disulfate, in agreement with the postulated structures of Goto and co-workers. Further, the *Watasenia* reaction involves a base/luciferase-catalyzed enolization of the C-3 keto oxygen of coelenterazine disulfate, adenylation of the keto group by ATP, removal of the AMP, addition of molecular oxygen to the C-2 carbon and formation of a dioxetanone intermediate, which spontaneously decomposes to yield light.[14]

METHODS

The methods used were previously described.[13, 14] Squid specimens were caught at night using nets set along the shore of Toyama Bay and the squids were immediately transferred to a holding tank with running, oxygenated sea water controlled at 2 °C. Under these conditions the squids lived for 3-5 d. The light organs of 15 squids were removed with a pair of ophthalmic scissors and homogenized in an all-glass homogenizer in an ice-bath in 5.0 mL of 0.001 M $MgCl_2$. The following two solutions were prepared: (1) 400 μL of homogenate + 44 μL of 1.0 M Tris-HCl, pH 8.26 and (2) 200 μL of 0.005 M ATP (in 0.001 M $MgCl_2$) + 22 μL of 1.0 M Tris-HCl, pH 8.26. Compound to be tested was dissolved in 200 μL of 0.001 M $MgCl_2$ and control consisted of 0.001 M $MgCl_2$ without the test compound. To initiate the luminescence reaction, solution (2) kept in a syringe was injected into solution (1) contained in a small reaction vial placed in the sample compartment of a Mitchell-Hastings photomultiplier photometer calibrated with a light standard.[15]

RESULTS AND DISCUSSION

Panel A shows the result of injecting solution (2) containing ATP into solution (1) containing homogenate. It is seen that the injection of ATP causes a rapid rise in the light intensity, followed by a gradual decay. After the light intensity had decayed for ~3.6 min, the shutter of the photomultiplier was closed and the first syringe was replaced with a syringe containing 1.0 μg of synthetic coelenterazine disulfate dissolved in 200 μL of 0.001 M $MgCl_2$. The shutter was again opened and coelenterazine disulfate was injected directly into the luminescing mixture. A sharp increase in light intensity occurred, followed by a rapid decay. The initial light intensity after coelenterazine disulfate injection was greater by an order of magnitude than the intensity produced by the injection of ATP. Considering that the arm organs of 15 squids were homogenized in 5.0 mL of 0.001 M $MgCl_2$, which calculates to 1.2

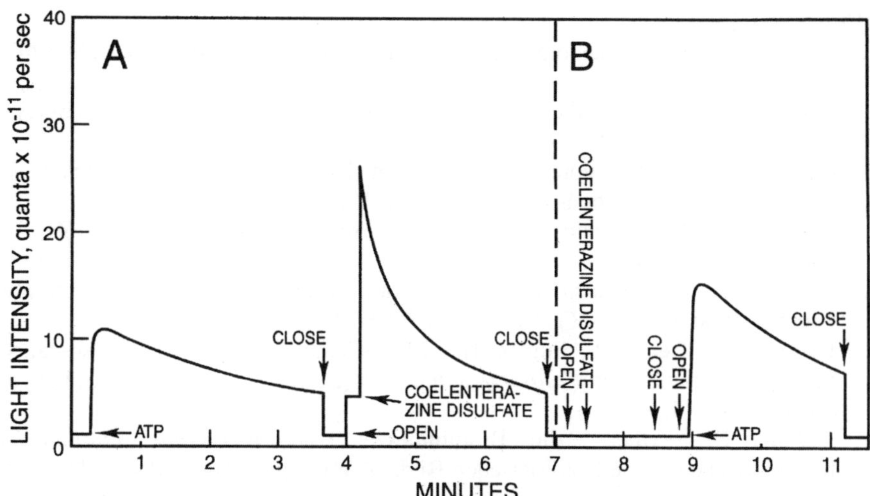

specimens per 400 μL of homogenate, and considering that the measured light intensity of 400 μL of homogenate was 10^{11} quanta/sec, which is the intensity readily visible to the dark-adapted eye, the light of the homogenate is extremely bright when dilution is taken into account. This marked stimulation by 1.0 μg of synthetic coelenterazine disulfate strongly indicates that the luciferin in the supernatant is coelenterazine disulfate. The rapid decay in light intensity is probably due to a depletion of molecular oxygen. Membrane-bound luciferase was found to be highly unstable, losing activity in a matter of hours, whereas luciferin was stable when kept at -20 °C. Panel B shows the result of injecting coelenterazine disulfate into a duplicate solution of (1). No light was detected, indicating an absence of ATP. When the mixture was then injected with ATP, the initial light intensity was only moderately higher than that of the mixture injected with ATP in Panel A, suggesting that the reaction mixture already contained a high concentration of luciferin. From these and other data, it is concluded that the *Watasenia* reaction requires luciferin (coelenterazine disulfate), membrane-bound luciferase, ATP and molecular oxygen for light emission.

ACKNOWLEDGMENTS

The author is very grateful to the Toyama Prefectural Fisheries Research Institute for providing laboratory facilities and to the following individuals: Prof. S. Inoue, Meijo University, Nagoya, for a gift of synthetic coelenterazine disulfate and Mr. N. Nakura, Mr. A. Imamura, Dr. S. Hayashi, Mr. H. Wakabayashi, Mr. T. Miyazaki, and Dr. K. Kawasaki for assisting in the research in various ways. The research was supported in part by a grant from WAVE Namerikawa.

REFERENCES

1. Watasé S. The luminous organ of the firefly squid. Dobutsugaku Zasshi 1905; 17: 119-23.
2. Hayashi S. Studies on the luminous organs of *Watasenia scintillans* (Berry). Folia Anat Jpn 1927; 5417-27.
3. Shima G. Preliminary note on the nature of the luminous bodies of *Watasenia scintillans* (Berry). Proc Imp Acad (Tokyo) 1927; 3: 461-4.
4. Kishitani T. On the luminous organs of *Watasenia scintillans*. Annot Zool Jpn 1928; 11: 353-67.
5. Takagi S. Mitochondria in the luminous organs of *Watasenia scintillans* (Berry). Proc Imp Acad (Tokyo) 1933; 9: 651-4.
6. Okada Y K, Takagi S, Sugino H. Microchemical studies on the so-called photogenic granules of *Watasenia scintillans* (Berry). Proc Imp Acad (Tokyo) 1934; 10: 431-4.
7. Hasama B. Über die Biolumineszenz bei *Watasenia scintillansim* bioelektrischen sowie histologischen Bild. Z Wiss Zool Abt A 1941; 155: 109-28.
8. Okada Y K. Observations on rod-like contents in the photogenic tissue of *Watasenia scintillans* through the electron microscope. In: Johnson F H, Haneda Y. eds. Bioluminescence in Progress. Princeton University Press, 1966: 611-25.
9. Goto T, Iio H, Inoue S, Kakoi H. Squid bioluminescence. I. Structure of *Watasenia* oxyluciferin, a possible light-emitter in the bioluminescence of *Watasenia scintillans*. Tetrahedron Lett 1974: 2321-4.
10. Inoue S, Sugiura S, Kakoi H, Hasizume K, Goto T, Iio H. Squid bioluminescence. II. Isolation from *Watasenia scintillans* and synthesis of 2-(p-hydroxybenzyl)-6-(p-hydroxyphenyl)-3,7-dihydroimidazo [1,2-a] pyrazin-3-one. Chem Lett 1975: 141-4.
11. Inoue S, Kakoi H, Goto T. Squid bioluminescence. III. Isolation and structure of *Watasenia* luciferin. Tetrahedron Lett 1976: 2971-4.
12. Inoue S, Taguchi H, Murata M, Kakoi H, Goto T. Squid bioluminescence. IV. Isolation and structural elucidation of of *Watasenia* dehydropreluciferin. Chem Lett 1977: 259-62.
13. Tsuji F I. ATP-dependent bioluminescence in the firefly squid, *Watasenia scintillans*. Proc Natl Acad Sci USA 1985; 82: 4629-32.
14. Tsuji F I. Bioluminescence reaction catalyzed by membrane-bound luciferase in the "firefly squid," *Watasenia scintillans*. Biochim Biophys Acta 2002; 1564: 189-97.
15. Mitchell G W, Hastings J W. A stable, inexpensive, solid-state photomultiplier photometer. Anal Biochem 1971; 39: 243-50.

PHARMACOLOGICAL AND ELECTROPHYSIOLOGICAL STUDIES OF LIGHT EMISSION IN 3 OPHIUROID SPECIES: PRELIMINARY RESULTS

C VANDERLINDEN[1], M VANHEMELEN[1], B NILIUS[2],
P GAILLY[3], J MALLEFET[1]

*[1]Laboratoire de Biologie Marine, UCL, Place Croix du Sud, 3,
B-1348 Louvain-la-Neuve, Belgium*
[2]Afd. Fysiologie, O. & N.,Kul, Herestraat, 49, B-3000 Leuven, Belgium
*[3]Unité de physiologie générale des muscles, UCL, Avenue Hippocrate, 55,
B-1200 Brussels, Belgium*
Email : vanderlinden@bani.ucl.ac.be

INTRODUCTION

Bioluminescence is a widespread phenomenon in the marine environment. In ophiuroids, it was observed that the control mechanisms of light emission differ between species. Pharmacological studies have shown acetylcholine (ACh) to induce luminescence through muscarinic and nicotinic cholinergic receptors in *Amphiura filiformis*[1] and through muscarinic cholinergic receptors in *Amphipholis squamata*[2]; several neuromodulators being described.[2-5] In *Ophiopsila aranea* and *Ophiopsila californica,* on the other hand, none of the neurotranmitters tested so far are involved in the luminous control.[1] Although there are heterogeneities in the signal transduction pathways leading to photogenesis, the requirement of extra-cellular calcium to induce light emission is conserved in all the species studied so far.[6] The aim of this work is to study the properties of ionic channels involved in the photogenesis using pharmacology and electrophysiological techniques (microspectrofluorometry, patch-clamp) on the luminous cells (photocytes).

METHODS

In this work we studied 3 different ophiuroid species: *Amphiura filiformis* (Müller 1776), *Ophiopsila aranea* (Forbes 1843) and *Ophiopsila californica* (Clark 1921). Experiments were carried out exclusively on isolated luminous cells (photocytes). Cells were kept in artificial seawater (ASW) after enzymatic digestion and differential centrifugation.[7] Light was measured with a FB12 Berthold luminometer (Pforzheim, Germany) linked to a personal computer. For each pharmacological experimental protocol, control stimulations were performed by using 200 mmol l^{-1} KCl, which is know to trigger the maximal light emission. The tested preparations were first immersed in ASW containing the drug for 10 minutes before KCl stimulation. For the electrophysiological experiments, photocytes were kept in a modified L-15 Leibovitz medium.[8] The patch-clamp technique allowed us to perform whole cell recordings and a fura-2 probe was used in the microspectrofluorometry experiments. The following drugs were used in this study: 5-nitro-2-(3-phenylpropylamino)-bezoate (NPPB, Sigma-RIB), Aconitine (Sigma), Procaïnamide

HCl (Sigma), Tetrodotoxin (TTX, Sigma-Aldrich), Veratridine (Sigma), Apamin (Sigma), Diazoxide (Sigma), Glibenclamide (RIB), Minoxidil (Sigma), 2'hydroxy-5'-trifluoromethyl-2(3H)benzimidazolone (NS1619, Sigma), Tetraethylammonium (TEA, Sigma), Ouabain (Sigma).

RESULTS
Pharmacology
Several blockers and activators were used to study the involvement of Na^+, Cl^-, K^+ channels as well as the Na^+/K^+ pump in luminescence control. In *O. aranea*, treatment with ouabain ($10^{-5}M$), a drug blocking the Na^+/K^+ pump increased the total amount of light produced. Moreover, the drug blocking Cl^- channels (NPPB, $10^{-4}M$) increased both the maximal light intensity and the total amount of emitted light. Finally, in this species, results obtained with drugs affecting Na^+ channels show that when *O. aranea* photocytes are treated with aconitin (activator of TTX sensitive Na^+ channels), they produce a weak luminescence (about 3.2% of KCl control). Moreover, KCl induced luminescence (Lmax) is decreased by a pretreatment with aconitin. Another interesting result was obtained in all 3 species for drugs affecting K^+ channels.   Indeed, apamin ($10^{-6}M$), a blocker of $K^+(Ca^{2+})$ channels completely inhibits light emission triggered by 200mM KCl (Fig. 1). The activator of those channels, $10^{-4}M$ NS1619, induces a weak light emission and decreases KCl induced luminescence. None of the other drugs such as TEA had any effect on photogenesis.

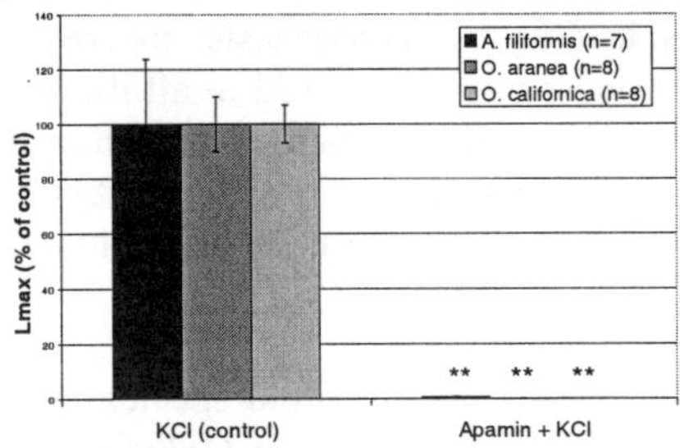

Figure 1. Effect of $10^{-6}M$ apamin on the maximal light emission (Lmax) triggered by 200mM KCl in all 3 species. Mean ± std error of mean, n= number of repetitions

Electrophysiology
Microspectrofluorometry measurements on clusters of about 30 photocytes indicated intracellular calcium variations during KCl stimulations in all 3 studied species. Indeed, an increase of intracellular calcium is observed during stimulation and can be decreased again when photocytes are rinsed with ASW (Fig. 2A). The patch-clamp technique allows current recordings at the photocyte membrane level. Preliminary results obtained in whole cell recordings highlight the presence of voltage-dependent currents. Fig. 2B shows both outward (a) and inward (b) currents.

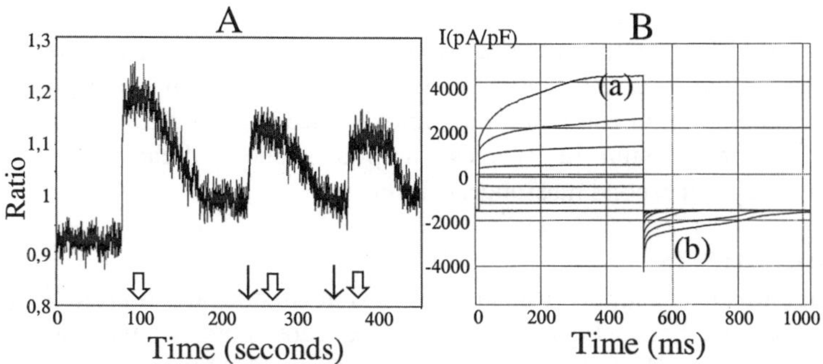

Figure 2. A. Recording of calcium variations in a photocyte cluster of *O. aranea*. (↓) 200mM KCl stimulation, (⇓) rinsing with ASW. Values are expressed as a ratio between Ca^{2+} bound fura-2 and free fura-2. **B.** Whole cell patch-clamp recording on *O. californica* photocyte. Stimulation protocol: holding potential of −100mV, steps of 25mV from −100mV to +100mV; (a) outward current, (b) inward current

DISCUSSION

Pharmacological results show that Na^+, Cl^-, K^+ channels and the Na^+/K^+ pump seem to be involved in luminescence control of *O. aranea*. On the opposite, in the 2 other species, only clear results were observed for K^+ channels.

In *O. aranea,* blocking of the Na^+/K^+ pump increases light emission, which could be due to an ionic disturbance increasing therefore depolarisation. Cl^- channels, on the other hand, seem to be involved in photocyte membrane repolarisation since their blocking maintain a higher and long lasting luminescence. Moreover, in this species, activation of Na^+ channels induces a weak luminescence probably due to a Na^+ influx with a subsequent depolarisation. Ionic channels involvement seems thus quite important in *O. aranea* luminous control while this does not appear for the 2 other species. In all 3 species, drugs affecting K^+ channels have highlighted their importance in bioluminescence control. More precisely, Ca^{2+} dependent K^+ channels might act as "brakes" for depolarisation. As a consequence, blocking of these channels would induce an uncontrolled depolarisation, which could modify normal photogenesis. On the contrary, the activator of this channel would maintain it open and therefore inhibit a proper depolarisation process. The involvement of this $K^+(Ca^{2+})$ channel is in the line of previous studies that have shown the importance of Ca^{2+} in light production of all 3 species.[6] Moreover, it has been shown that cAMP plays some role in the light emission mechanism[9]; cAMP, which increases Ca^{2+} influx through Ca^{2+} channels. A hypothesis for photogenesis can be postulated: depolarisation of the photocyte would trigger Ca^{2+} influx through activation of the cAMP pathway for instance; this Ca^{2+} would then bind and activate $K^+(Ca^{2+})$ channels. These channels would thus be involved in a negative feedback control of light emission. Electrophysiological experiments are in the line of these results since they show calcium variations inside the photocytes during KCl depolarisation. The patch-clamp technique, which shows both inwards and outwards currents, should, in the future, allow us to identify the nature of these currents. Finally, our results highlight once more that luminescence ionic control mechanisms are different between ophiuroid species, a phenomenon already shown at neurotransmitters[1] and on second messengers[9] levels.

ACKNOWLEDGMENTS

We acknowledge financial support from an EEC ARI at Kristineberg Marine Station, Fiskebäckskil, Sweden; Fonds Léopold III at Arago Laboratory, Banyuls-sur-Mer, France. Special thanks to Prof. J. Case, S. Anderson and D. Divins for invaluable help during scuba collections at UCSB Marine Sciences Institute (USA). Part of this work was also supported by an F.R.F.C. grant (2.4516.01). J.M. is a Research Associate F.N.R.S., Belgium. Research supported by a FRIA grant for CV.

REFERENCES

1. Dewael Y, Mallefet J. Luminescence in ophiuroids (Echinodermata) does not share a common nervous control in all species. J Exp Biol 2002; 205 : 799-806.
2. De Bremaeker N, Mallefet J, Baguet F. Luminescent control in the brittlestar *Amphipholis squamata*: effect of cholinergic drugs. Comp Biochem Physiol 1996; 115C: 75-82.
3. De Bremaeker N, Baguet F, Thorndyke MC, Mallefet J. Modulatory effects of some amino acids and neuropeptides on luminescence in the brittlestar *Amphipholis squamata*. J Exp Biol 1999; 202: 1785-91.
4. De Bremaeker N, Mallefet J, Baguet F. Effects of catecholamines and purines on the luminescence of *Amphipholis squamata* (Echinodermata). J Exp Biol 2000; 203: 2015-23.
5. Dupont S, Mallefet J, Vanderlinden C. Effect of b-adrenergic antagonists on bioluminescence control in 3 species of brittlestars (Echinoderms). Accepted for publication in Comp Biochem Physiol C 2004.
6. Dewael Y, Mallefet J. Calcium involvement in the luminescence control of three ophiuroid species (Echinodermata). Comp Biochem Physiol C 2002; 131: 153-60.
7. De Bremaeker N, Dewael, Y, Baguet F, Mallefet J. Involvement of cyclic nucleotides and IP3 in the regulation of luminescence in the brittlestar *Amphipholis squamata* (Echinodermata). Luminescence 2000; 15: 159-63.
8. Moss C, Beesley PW, Thorndyke MC, Bollner T. Preliminary observations on ascidian and echinoderm neurons and neuronal explants in vitro. Tissue and Cell 1995; 30: 517-24.
9. Vanderlinden C. Dewael Y, Mallefet J. Screening of second messengers involved in photocyte bioluminescence control of three ophiuroid species (Ophiuroidea, Echinodermata). J Exp Biol 2003; 206: 3007-14.

PART 2

BEETLE BIOLUMINESCENCE

IMPORTANCE OF FIREFLY LUCIFERASE C-TERMINAL DOMAIN IN BINDING OF LUCIFERYL-ADENYLATE

K AYABE[1], T ZAKO[2], H UEDA[1]

[1]*Dept of Chemistry and Biotechnology, School of Engineering, University of Tokyo, Bunkyo-ku 113-8656, Japan*
[2]*Dept of Biotechnology and Life Sciences,*
Tokyo University of Agriculture and Technology, Koganei 184-8588, Japan
Email: ayabe@bio.t.u-tokyo.ac.jp

INTRODUCTION

The crystal structural studies[1,2] revealed that firefly *Photinus pyralis* luciferase is composed of a large N-terminal domain (1-435aa, N-domain) and a smaller C-terminal domain (441-550aa, C-domain) linked by a flexible linker region, and that the presumptive active site is surrounded by the residues locating on the N-domain except Lys529. So far Lys529 is the only active site residue identified in the C-domain, which is conserved among acyl-adenylate forming enzyme superfamily that catalyses the formation of acyl-adenylate intermediate from carboxylate substrate and ATP. Mutational study revealed that Lys529 is only essential for the adenylation step, and not for the following oxidative reaction.[3]

Recently we reported that *Photinus pyralis* luciferase lacking the whole C-domain still retained its luminescent activity.[4] The luminescence from the N-domain luciferase had a peak wavelength around 620 nm independent of pH, and was not enhanced by CoA. Most interestingly, the rise time, which was defined as the time until the emission reaches its maximum, was extremely long for the N-domain compared to wild-type (WT) or K529A mutant lacking the Lys529 in the C-domain (Table 1). Here we present another finding that the specific activity of the N-domain depended on its enzyme concentration. We tried to clarify the reason for these unique kinetic properties from several points of view, and found that the obtained N-domain luminescence faithfully reflected the concentration of free reaction intermediate, probably luciferyl adenylate (LH_2-AMP) in the reaction mixture. In addition, we also found that the N-domain can be utilized as a sensitive sensor to specifically detect presumptive LH_2-AMP in the presence of excess luciferin and ATP in solution.

Table 1. Kinetic properties of the WT luciferase and the mutants

Enzyme	Relative activity [-]	Rise time [sec]	Relative maximum LH_2-AMP conc. [-]
WT	100	0.4	100
K529A	0.06	0.5	2.8
N-domain	0.0018	250	15

METHODS

General methods

The specific activities of luciferase and its mutants were determined with a luminometer AB-2100 (Atto, Tokyo, Japan) and 96-well white microplate (Nalge-Nunc, Tokyo, Japan). The reaction mixture contained 100 mM Tricine, 10 mM $MgSO_4$, 300 μM D-luciferin (LH_2), 10 mM Na-ATP, and 1 mg/ml BSA, pH 8.0 unless otherwise indicated. Enzyme concentrations used were 0.1 nM for WT, 10 nM for K529A, and 1 μM for the N-domain. The specific activity was determined by a peak-height-based assay based on the maximum intensity for the N-domain, and the peak intensity within the first 10 s for WT and K529A, respectively.

Detection of LH_2-AMP

The first reaction with WT or mutant luciferase was performed in the buffer containing 100 mM MOPS, 10 mM $MgSO_4$, 10 mM Na-ATP, 300 μM LH_2 and 1 mg/ml BSA, pH 7.0 at room temperature. At several time points of the incubation (called '1st reaction' to be distinguished from the latter reaction), 40 μL aliquot of the reaction mixtures was taken and mixed with 4 μL of 1N HCl, and the enzyme wherein was removed by spin-filtration (13 krpm, 5 min, 4°C) using Microcon YM30 (Millipore, Tokyo, Japan), while this step was omitted later. The 5 μL aliquot of the filtrates were mixed with 100 μL of the 2nd reaction mixture containing 100 mM Tricine, 1 μM N-domain and 1 mg/ml BSA, pH 8.0 on a 96-well white microplate well at room temperature, and the light intensity was obtained as an integration of 4 seconds (from 1 to 5 s after mix).

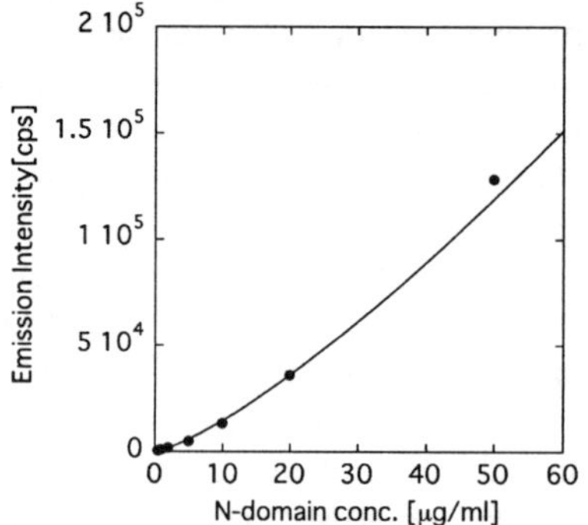

Figure 1. The enzyme concentration dependency of the N-domain emission

RESULTS

Concentration dependency of the N-domain activity

When we tried to determine the specific activity of the N-domain, we found that the peak luminescence intensities from the N-domain did not increase linearly along with the enzyme concentration. Since the addition of BSA did not show any positive effect, it was not likely due to the inactivation of the enzyme at lower concentration. Also, because there observed no difference between the specific activities of dimeric

GST-luciferase and WT enzyme (data not shown), dimerization/ oligomerization was not likely to explain this non-linear relationship.

Sensitive detection of LH$_2$-AMP using the N-domain

Since the chemically synthesized LH$_2$-AMP as a substrate did not show any rise time in the N-domain reaction (data not shown), it was possible that the luminescence from the N-domain is influenced by the amount of a reaction intermediate LH$_2$-AMP in the reaction mixture. To analyze the reaction more precisely, we tried to compare the concentrations of LH$_2$-AMP produced during the reaction of WT, K529A and the N-domain. Previously,[5] Dukhovich *et al.* estimated the concentration of LH$_2$-AMP in the reaction mixture by stopping the reaction with HCl, diluting by 100-fold, and measuring the emission from the aliquot mixed with fresh WT luciferase[5]. We speculated that if we use the N-domain instead of the WT in the detection step, we will not have to dilute the reaction mixture since the N-domain does not react instantly with ATP and LH$_2$ but with LH$_2$-AMP. By taking advantage of this unique character of the N-domain, improvement in sensitivity over the previous method was expected. In fact, the emission profiles observed in the detection step using the filtrates of the 1st N-domain reaction and the additional N-domain luciferase had a fast rise time of around 1sec, suggesting that almost all the initial light intensity corresponded to the amount of a reaction intermediate, most likely LH$_2$-AMP in solution (data not shown).

Using this method, we could observe a considerable time dependency of LH$_2$-AMP concentration in the N-domain reaction (Fig.2). The relative concentrations of LH$_2$-AMP thus measured and the emission intensities of the first reaction showed very similar curve, that increased for 20 minutes and became constant, indicating that the emission intensity from the N-domain well reflects the free LH$_2$-AMP concentration in the reaction mixture.

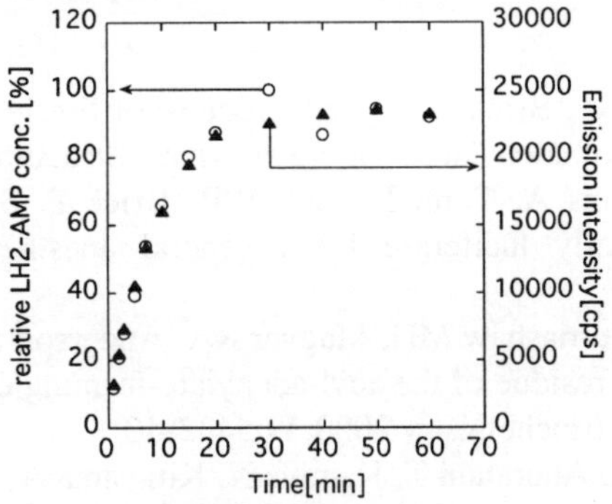

Figure 2. Time dependency of the emission and relative LH$_2$-AMP concentration in the N-domain reaction

The concentrations of free LH$_2$-AMP produced during the reactions of WT and K529A were also measured. LH$_2$-AMP concentrations in WT and K529A reactions increased for around 5 and 20 minutes before reaching plateau, respectively, which were quite distinct from the decay curves of the emission observed for both enzymes (data not shown). Relative concentrations of maximum LH_2-AMP formed by WT, K529A and N-domain under the same condition are summarized in Table 1. Interestingly, compared with K529A, the N-domain produced 5-times free LH_2-AMP, while its specific activity remained 30-fold lower.

DISCUSSION

By taking advantage of the N-domain's differential sensitivity against the substrates and the reaction intermediate, we could significantly improve the sensitivity of LH_2-AMP detection by at least two orders in magnitude. The validity of the method was also supported by the fact that the time course of emission from the N-domain in the 1st reaction well reflected that of LH_2-AMP accumulated during the same reaction. Also, the amount of LH_2-AMP detected in the N-domain reaction was considerably higher than that of K529A, suggesting that large part of LH_2-AMP was released from the N-domain and accumulated in solution rather than directly consumed in the following reaction. Probably, the lack of the C-domain leads to fast dissociation of LH_2-AMP, meaning that the C-domain has a very important role independent of Lys529, on stabilization of the enzyme-intermediate complex.

This hypothesis also explains the non-linear enzyme concentration dependency of the reaction. In the event of concurrent adenylation and oxidation reactions in parallel, not in sequential, as a result of the intermediate dissociation, the observed luminescent activity with excess substrates will be rather the square of, than in proportional to, the enzyme concentration. In reality, it was proportional to the enzyme concentration powered by ~1.3, probably due to the neighboring effect and/or slow accumulation/ fast degradation of LH_2-AMP at neutral pH.

REFFERENCES

1.   Conti E, Franks NP, Brick P. Crystal structure of firefly luciferase throws light on a superfamily of adenylate-forming enzymes. Structure 1996;4:287-98.
2.   Franks NP, Jenkins A, Conti E, Lieb WR, Brick P. Structural basis for the inhibition of firefly luciferase by a general anesthetic. Biophys J 1998; 75:2205-11.
3.   Branchini BR, Murtiashaw MH, Magyar RA, Anderson SM. The role of lysine 529, a conserved residue of the acyl-adenylate-forming enzyme superfamily, in firefly luciferase. Biochemistry 2000;39:5433-40.
4.   Zako T, Ayabe K, Aburatani T, Kamiya N, Kitayama A, Ueda H, Nagamune T. Luminescence and substrate binding activities of firefly luciferase N-terminal domain. Biochim Biophys Acta 2003;1649:183-9.
5.   Dukhovich A, Sillero A, Sillero MA. Time course of luciferyl adenylate synthesis in the firefly luciferase reaction. FEBS Lett 1996;395:188-90.

COMPARISON OF KINETIC PROPERTIES OF FIREFLY LUCIFERASE FROM *PHOTINUS PYRALIS* AND *LUCIOLA MINGRELICA*

LY BROVKO[1], OA GANDELMAN[2], IB KERSHENGOLZ[3], NN UGAROVA[3]

[1]*Canadian Research Institute for Food Safety, University of Guelph, Guelph, Ontario N1G 2W1, Canada, E-mail: lbrovko@uoguelph.ca*
[2]*Lumora Ltd., Institute of Biotechnology, University of Cambridge, Cambridge, UK*
[3]*Dept of Chemical Enzymology, Moscow State University, Moscow 119899, Russia*

INTRODUCTION

Firefly luciferases catalyse reaction of firefly luciferin (LH_2) oxidation by molecular oxygen in the presence of adenosine-5'-triphosphate magnesium salt ($ATP\text{-}Mg^{2+}$). The generally accepted mechanism of the reaction includes several consecutive stages: 1) formation of ternary complex of the enzyme with two substrates – luciferin and ATP; 2) adenylation of the carboxylate group of LH_2 resulting in the formation of luciferyl adenylate, and release of AMP and pyrophosphate (PP_i); 3) oxidation of luciferyl adenylate by molecular oxygen via the intermediate dioxitanone formation; 4) decarboxylation of dioxitanone producing oxyluciferin ($OxyLH_2$) in an electronically excited state, which then decays to the ground state with emission of visible light in the green-orange range of the spectrum. This mechanism can be represented with the following scheme:

$$E + S_1 \Leftrightarrow E\,S_1$$
$$E\,S_1\,S_2 \Leftrightarrow EP_1 \xrightarrow{O_2} EP_2 \Leftrightarrow E + P + h\nu$$
$$E + S_2 \Leftrightarrow E\,S_2 \qquad AMP + CO_2 + PP_i$$

where E is luciferase, S_1 and S_2 are LH_2 and $ATP\text{-}Mg^{2+}$, respectively, P_1 is luciferyl adenylate, P and P_2 are $OxyLH_2$ in ground and excited state respectively.

It was noted before that the percentage of conversion of the substrates to the products is rather low suggesting inactivation of the enzyme during the reaction. In our previous work it was observed for *Luciola mingrelica* firefly luciferase that the total amount of product formed during the entire course of reaction was proportional to the initial amount of active enzyme present and had hyperbolic dependence on luciferin concentration.[1] This type of behaviour is characteristic of the so-called suicidal substrates when inactivation of the enzyme during reaction is induced by a substrate.[2] The kinetic scheme for the luciferase reaction was supplemented with the stages of enzyme and intermediate enzyme-substrate/product complexes inactivation and kinetic constants for all stages were evaluated using the simplified analytical solution for the derived system of differential equations. The resulting kinetic parameters were introduced to the postulated equations and compared with experimental data. There was good correlation of experimental and calculated curves indicating the validity of this approach.

In this age of powerful computers, it is no longer even necessary to find analytical solution to differential equations. There are many software packages available that carry out numerical integration of differential equations followed by non-linear regression to fit the model and assess its quality by comparing with experimental data. In this study we have used a numerical integration approach to compare kinetic properties of *Photinus pyralis* and *Luciola mingrelica* firefly luciferases.

MATERIALS AND METHODS

Photinus pyralis firefly luciferase (PP) was obtained from Sigma, *Luciola mingrelica* firefly luciferase (LM) was isolated and purified according to.[3] Time-course of bioluminescent reaction rate (v) was monitored as the intensity of light (I) in time according to equation 1 using a luminometer model 1251 (LKB Sweden).

$$v = d[P]/dt = d[hv]/dt = I \tag{1}$$

After initiation of the reaction by the injection of luciferin or ATP, bioluminescent intensity was registered every 1s for the first 5 min and every 12 s afterwards for total time of 2 h. Endpoint of reaction was defined as a time when intensity of light decreased 100 times compared with the maximum. The area under the resulting curve was determined and was used as a measure of the product yield $[P_\infty]=P$.

Numerical integration was performed using in-house software package provided by Drs. A. Abramenkov and D.Rassokhin (MSU, Russia).

RESULTS

Dependence of product yield on enzyme and substrate concentration

Unlike for common reaction rate monitoring, when concentration of the product formed in time is registered and reaction rate is calculated as an instantaneous slope of [P] vs time (t) curve, for bioluminescent reactions the reaction rate is obtained directly from experimental data as an instantaneous value of light intensity. Time course of light intensity (I) represents the differential of regular progress curve for the reaction. The progress curve can be obtained by integration of the I vs t curve, and the yield of the product is represented in this case by an area under this curve. The yield of the product was determined for both PP and LM luciferases for a wide range of enzyme substrate concentrations (the second substrate was kept at saturation level) (Fig,1A,B,C). There was direct proportionality of reaction yield to enzyme concentration for both enzymes (Fig. 1A). Slopes of the lines were 0.94 and 1.04 for PP and LM, respectively. Dependence of reaction yield on substrate concentrations was hyperbolic (Fig. 1B,C). This pattern is characteristic of the enzymes that are inactivated on the stages of reaction following the formation of the enzyme-substrate complex. Addition of PP_i (one of the intermediate products) to the reaction mixture increased the final product yield 2-5 times (Fig. 1A) and decreased the slopes of P vs E lines to 0.83 and 0.70 for LM and PP, respectively. This can be explained by the increased turnover numbers for luciferases in the presence of PP_i due to enhanced dissociation of the enzyme-product complex. However the addition of pyrophosphate did not change the whole pattern indicating the significance of the inactivation steps in the kinetic mechanism of luciferases.

Based on the obtained data the original kinetic scheme was supplemented with the steps involving inactivation of the enzyme-substrate/product complexes as well as formation of the ternary complex enzyme-product-pyrophosphate (scheme 2):

$$(EP\text{-}PP_i)_{in} \xleftarrow{6} EP\text{-}PP_i \xrightarrow{8} E + P + PP_i$$

$$ES_1 + S_2 \underset{-1}{\overset{1}{\rightleftharpoons}} ES_1 S_2 \overset{2}{\longrightarrow} EP_1 \overset{3}{\longrightarrow} EP \overset{4}{\underset{-4}{\rightleftharpoons}} E + P$$

with vertical steps 5 (from ES_1 to $ES_{1\,in}$), 5 and 7 (from $ES_1 S_2$ to $ES_1 S_{2\,in}$, giving AMP, CO_2, light), 7 (PP$_i$), and 6 (from EP to EP_{in}). (2)

Taking into account that under experimental conditions the bioluminescent reaction was initiated by injection of the second substrate to the mixture of enzyme with the first substrate, the starting point of the reaction scheme was enzyme substrate complex ES_1. This kinetic problem was solved using a numerical integration approach, the estimations of kinetic constants for separate stages were partly obtained from our previous work[2] and partly from the fitting of experimental and theoretical kinetic curves (Table). After the fitting procedure there was a good correlation between calculated and experimental kinetic curves in a wide range of substrate concentration. The difference in the time-course of light intensity did not exceed 6-8%. The resulting values of kinetic constants are presented in the Table. It was observed that kinetic curves obtained by the injection of ATP were practically identical for PP and LM luciferase. However, initiation of the reaction by luciferin resulted in slightly different kinetic curves described by a higher reaction rate constant for dissociation of the luciferin-luciferase complex (k_{-1}) for PP compared with LM (85 and 10 c^{-1}, respectively). The obtained set of kinetic constants for luciferase reaction was quite stable in terms of quality of fitting. The changes in any constant of more than 30% resulted in a significant decrease in similarity with experimental kinetic curves.

The main conclusion from this study is that both PP and LM firefly luciferases have similar kinetic mechanisms characterized by significant inactivation of the enzyme induced by its interaction with the substrates. The difference in kinetic properties for both enzymes is mainly in the reaction rates for formation and dissociation of the luciferin-luciferase comlex. The addition of pyrophosphate to the reaction mixture increases the reaction yield due to enhanced regeneration of active enzyme from the enzyme-product complex.

                                                          Brovko LY et al.

Table 1. Kinetic constants for the LM and PP firefly luciferases (numbering corresponds to the numbering of reaction steps in scheme 2)

Rate constant	LM		PP	
	Initiation by LH_2	Initiation by ATP	Initiation by LH_2	Initiation by ATP
k_1, $M^{-1} c^{-1}$	10^6	10^5	1.5×10^6	10^5
k_{-1}, c^{-1}	10	10	85	10
k_2, c^{-1}	30	30	30	30
k_3, c^{-1}	10	10	10	10
k_4, c^{-1}	1	1	1	1
k_{-4}, $M^{-1} c^{-1}$	10^7	10^7	10^7	10^7
k_5, c^{-1}	2.6×10^{-5}	2.6×10^{-5}	2.6×10^{-5}	2.6×10^{-5}
k_6, c^{-1}	1.1×10^{-3}	1.1×10^{-3}	1.1×10^{-3}	1.1×10^{-3}
k_7, c^{-1}		2.3×10^7		2.3×10^7
k_8, c^{-1}		20		20

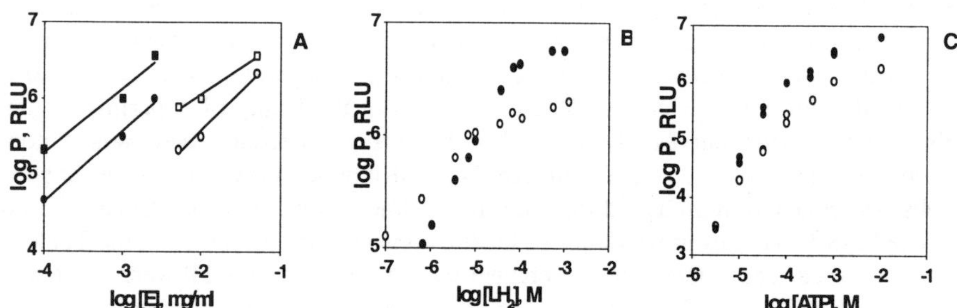

Figure 1. Dependence of the yield of the product on enzyme concentration (A), concentration of luciferin (B) and ATP (C) for PP (●) and LM luciferase (○) and effect of addition of PP_I (1 μM) on the reaction yield (A: □ and ■ for PP and LM, respectively). Saturation concentrations of [ATP] and luciferin 1 mM.

REFERENCES

1. Brovko LY, Gandelman OA, Polenova TE, Ugarova NN. Kinetics of bioluminescence in the firefly luciferin-luciferase system, Biochemistry (Moscow) 1994; 59; 195-201.
2. Vrzheshch PV, Varfolomeev SD. Steady-state kinetics of multisubstrate enzymatic reactions – inactivation of the enzyme in the course of reaction. Biochemistry (Moscow) 1985;50: 125-32.
3. Dementieva EI, Kutizova GD, Ugarova NN. Biochemical properties and stability of homogeneous luciferase of fireflies Luciola mingrelica. Vestn Mosk U Khim 1989;30: 601-6.

FIREFLY LUCIFERASE AND *DROSOPHILA CG6178* GENE PRODUCT ARE FATTY ACYL-COA SYNTHETASES

Y OBA[1], M OJIKA[1], S INOUYE[2]

[1]Graduate School of Bioagricultural Science, Nagoya University, Nagoya 464-0831, Japan
[2]Chisso Co., 5-1 Okawa, Kanazawa-ku, Yokohama 236-8605, Japan
Email: oba@agr.nagoya-u.ac.jp

INTRODUCTION

In firefly luciferase reaction, the luminescence activity is enhanced by addition of Coenzyme A (CoA)[1] and this phenomenon is explained by release of product inhibition.[2] Also, firefly luciferase shows the sequence similarity to mammalian fatty acyl-CoA synthetase (AcCoAS)[3] and plant 4-coumarate:CoA ligase (4CL).[4] They are classified as an adenylation enzyme for synthesizing acyl-CoA derivatives from carboxylic acid compounds in the presence of CoA, ATP and Mg^{2+} (Scheme 1). Furthermore, it was reported that the luminescence activity of firefly luciferase is inhibited competitively by various long-chain fatty acids.[5] We have determined that firefly luciferase is a bi-functional enzyme, catalyzing both the luminescence reaction and fatty acyl-CoA synthetic reaction.[6]

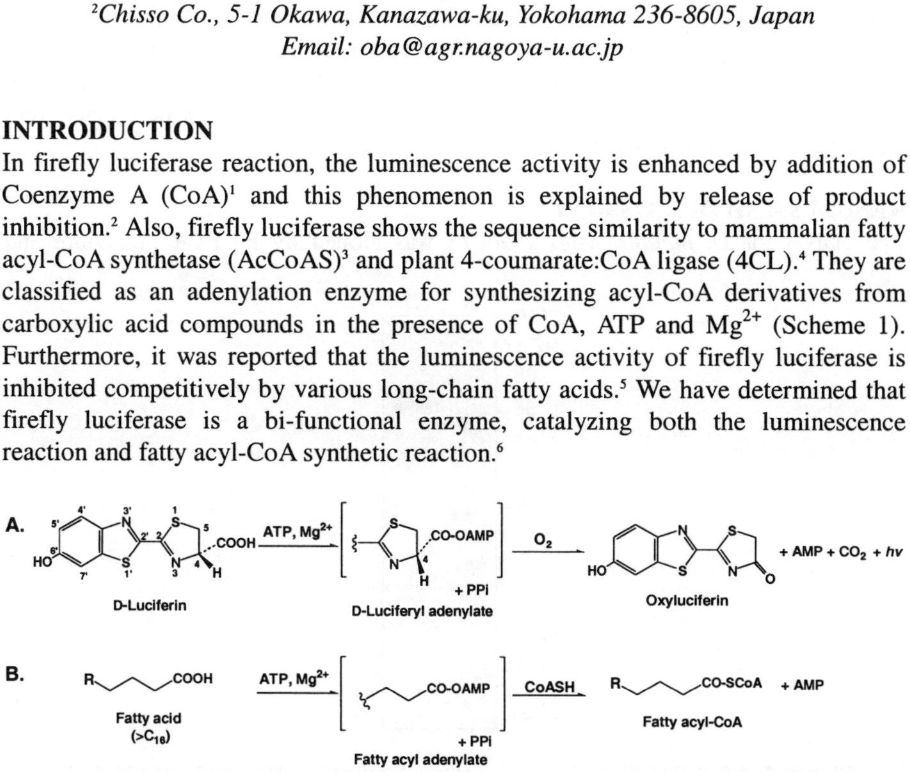

Scheme 1. Bioluminescence reaction (A) and fatty acyl-CoA synthetic reaction (B)

A search of the *Drosophila melanogaster* genome database revealed that several genes similar to firefly luciferase are present. The most similar gene, *CG6178*, is predicted to be an orthologue to firefly luciferase by phylogenetic analysis.[7] In this study, we described the properties of CG6178 and firefly luciferase; requirement factors, pH dependency, temperature preference and substrate specificity.

MATERIALS AND METHODS

His-tagged CG6178 was expressed in *E. coli*, and then was purified using

Ni-chelating chromatography as previously described.[7] For pH dependency and temperature preference analyses, fatty acyl-CoA synthetic activity was determined using [1-^{14}C]oleic acid as a substrate in the presence of 250 μM ATP, 250 μM CoA, 5mM MgCl$_2$ with 1.2 μM recombinant *Photinus pyralis* luciferase (Promega) or 362 nM purified CG6178 in 200 mM Tris-HCl (pH 7.8). After reaction for 20 min, the [^{14}C]oleoyl-CoA was separated by TLC and then the radioactivity was measured using imaging analyzer BAS2500 (Fuji film). Assay for acyl adenylation activity was performed in the reaction mixtures of [α-^{32}P]ATP, 250 μM CoA, 5mM MgCl$_2$ and 50 nM *P. pyralis* luciferase or purified CG6178. The [^{32}P]AMP formed was separated by TLC, the radioactivity was measured.[7] The value for each substrate (n=3) was obtained by subtracting the background. The other experiments were performed as previously described.[7]

RESULTS AND DISCUSSION

The cDNA for *D. melanogaster CG6718* was cloned by RT-PCR procedure and

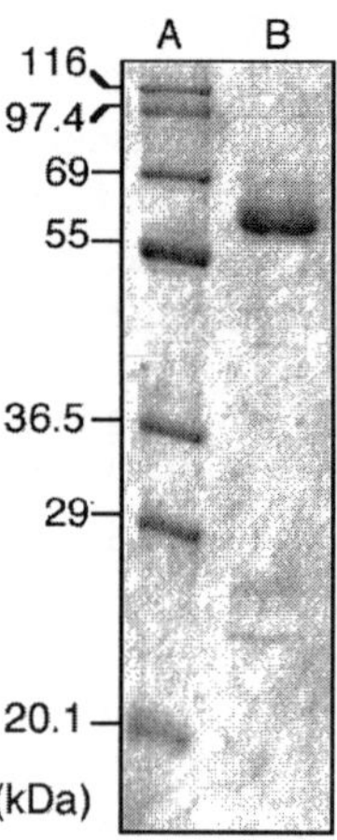

Figure 1. SDS-PAGE analysis. A: molecular marker, B: purified CG6178

expressed in *E. coli*. The gene product was purified, showing 62 kDa on SDS-PAGE (Fig. 1). Acyl-CoA synthetic activity was detected in CG6178 by TLC using [1-^{14}C]oleic acid. The fatty acyl-CoA was separated by HPLC and identified by MALDI-TOF-MS analyses.[6,7] The enzymatic properties for fatty acyl-CoA synthesis between firefly luciferase and CG6178 are characterized as follows; (i) ATP, Mg^{2+} and CoA are essential for both reactions. (ii) Other nucleotides, GTP, CTP, TTP, UTP and ITP, did not stimulate the formation of oleoyl-CoA in both reactions (data not shown). (iii) The optimum pH is 7.5-8.5 in both reactions. (iv) The optimum temperatures for both catalytic activities are at 20-25°C (Fig. 2): Optimal pH and

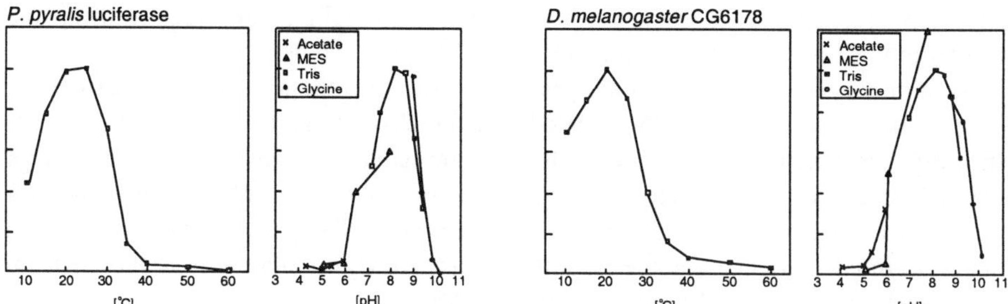

Figure 2. Optimum temperature and pH

temperature). These results are very similar to that of mammalian AcCoAS[8] and the luminescence reaction of luciferase. (v) The effect of fatty acid concentration on acyl adenylation activity was examined by AMP formation. More than 20 μM of fatty acids showed the inhibitory effect on acyl-CoA synthesis of *P. pyralis* luciferase, as same as in CG6178.[7] However, the high concentration of firefly luciferin did not affect on the formation of luciferyl-CoA by firefly luciferase. (data not shown). (vi) Substrate specificity was investigated by AMP-formation (Fig. 3). The results showed that short chain acids (acetic acid, propionic acid) and middle chain fatty acid (hexanoic acid) were not used for substrates. Phenylpropionic acids (4-coumaric acid, caffeic acid, ferulic acid), the typical substrates for plant 4CL, were also not for substrate. On the other hand, long-chain fatty acids (palmitic acid,

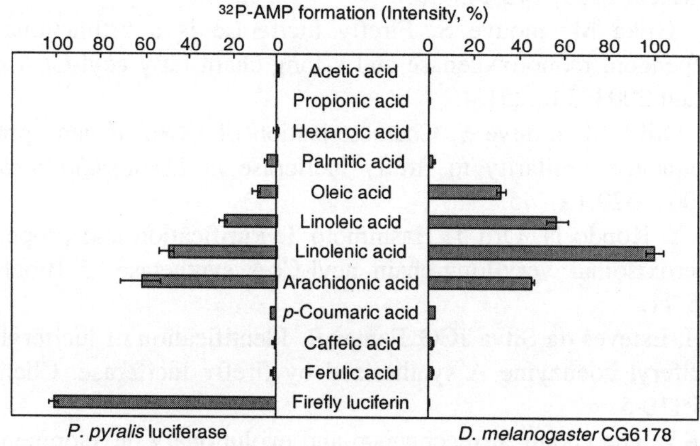

Figure 3. Substrate specificity

oleic acid, linoleic acid, linolenic acid, arachidonic acid) were significantly effective. Firefly luciferin is a best substrate for adenylation activity in *P. pyralis* luciferase. In contrast, CG6178 has no luminescence activity. Also the formation of luciferyl-CoA and dehydroluciferyl-CoA[9] were not detected by HPLC analyses during the incubation of firefly luciferin with CG6178, ATP, CoA and Mg^{2+}.[7]

In summary, the comparison studies of acyl-CoA synthetic properties between *P. pyralis* luciferase and CG6178 revealed that the characteristics of two enzymes are strikingly similar in each other, and also to mammalian long-chain AcCoAS, except for luminescence property. Here, we conclude that CG6178 in *D. melanogaster* is a long-chain AcCoAS gene and is not a luciferase gene. The present studies may give the substantial evidence for the hypothesis that firefly luciferase is diverged from CoA synthetase,[10] especially long-chain AcCoAS.

REFERENCES

1.	Airth RL, Rhodes WC, McElroy WD. The function of coenzyme A in luminescence. Biochim Biophys Acta 1958; 27:519-32.
2.	Rhodes WC, McElroy WD. The synthesis and function of luciferyl-adenylate and oxyluciferyl-adenylate. J Biol Chem 1958; 233:1528-37.
3.	Suzuki H, Kawarabayasi Y, Kondo J, Abe T, Nishikawa K, Kimura S, Hashimoto T, Yamamoto T. Structure and regulation of rat long-chain acyl-CoA synthetase. J Biol Chem 1990; 265:8681-5.
4.	Schröder J. Protein sequence homology between plant 4-coumarate:CoA ligase and firefly luciferase. Nucleic Acids Res 1989; 17:460.
5.	Matsuki H, Suzuki A, Kamiya H, Ueda I. Specific and non-specific binding of long-chain fatty acid to firefly luciferase: cutoff at octanoate. Biochim Biophys Acta 1999; 142:143-50.
6.	Oba Y, Ojika M, Inouye S. Firefly luciferase is a bifunctional enzyme: ATP-dependent monooxygenase and a long chain fatty acyl-CoA synthetase. FEBS Lett 2003; 540:251-4.
7.	Oba Y, Ojika M, Inouye S. Characterization of *CG6178* gene product with high sequence similarity to firefly luciferase in *Drosophila melanogaster*. Gene 2004; 329:137-45.
8.	Uchida Y, Kondo N, Orii T, Hashimoto T. Purification and properties of rat liver peroxisomal very-long-chain acyl-CoA synthetase. J Biochem 1996; 119:565-71.
9.	Fraga H, Esteves da Silva JCG, Fontes R. Identification of luciferyl adenylate and luciferyl coenzyme A synthesized by firefly luciferase. ChemBioChem 2004; 5:110-5.
10.	Wood KV. The chemical mechanism and evolutionary development of beetle bioluminescence. Photochem Photobiol 1995; 62:662-73.

SOLVENT EFFECT ON THE NMR AND ABSORPTION SPECTRA OF FIREFLY LUCIFERIN IN TETRAHYDROFURAN

K ODAI[1], S NISHIYAMA[2], R SHIBATA[3], Y YOSHIDA[2,3], N WADA[4]

[1]Dept. of Informatics and Media Technology, Shohoku College, Atsugi, Kanagawa 243-8501, Japan
[2]Bio-Nano Electronics Research Center and [3]Fac. of Engineering, Toyo University, Kawagoe, Saitama 350-8585, Japan.
[4]Fac. of Life Sciences, Toyo University, Itakura-machi, Gunma 374-0193, Japan.
Email: bhwada@itakura.toyo.ac.jp

INTRODUCTION

The bio- and chemiluminescence of firefly luciferin (Ln; see Fig. 1a)) derivatives had been investigated by many scientists.[1,2] With these studies, Ln moiety was found to be oxygenated to form a dioxetanone, from which an excited state-oxyluciferin (Oxyln*) moiety and CO_2 are produced. Then, yellow-green or red Oxyln*-luminescence[1,2] is observed. The protonation and deprotonation of Oxyln* is said to play an important role for luminescence of Oxyln* in aqueous solution and in dimethyl sulfoxide (DMSO; water and DMSO are highly hydrogen-bonding solvents). In this paper, the role of solvent and its hydrogen-bonding to carboxylic acid and phenolic part of Ln was studied in tetrahydrofuran (THF; a weakly hydrogen-bonding solvent) in terms of ^{1}H NMR, absorption, fluorescence spectra, and *ab initio* calculations. In previous Molecular orbital (MO) calculations[3,4] for Ln, the intermediate oxetane structures were highlighted, and to our knowledge, no calculation concerning hydrogen-bonding between Ln and solvent had been done previously.

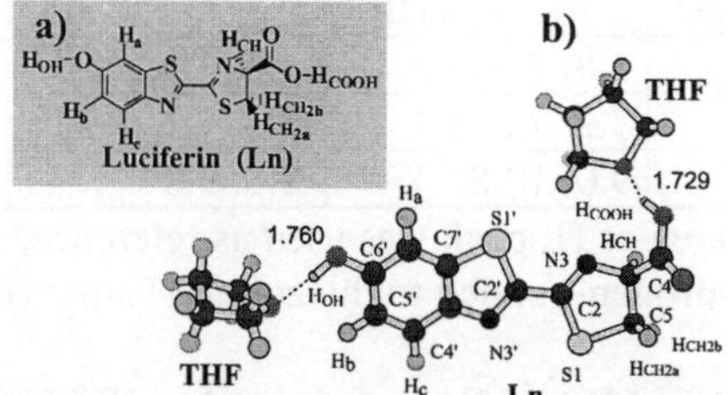

Figure 1. **a)** Molecular structure of Ln (in gray square) and **b)** the optimized geometry of Ln interacting with THF molecules (Ln-THF complex)

EXPERIMENTAL METHODS

A small portion of Ln (Aldrich; Lot:101K9255) was dissolved into deuterated THF-d_8 (99.5 atom% D, Acros Organics) (7×10^{-3} M) under air. All solvents and chemicals were used as supplied. ^{1}H NMR spectrum was recorded with JEOL EX-400 at 25 ~ 45 °C. Relative signal area of each ^{1}H peak was normalized by the area of H_a. Fluorescence and absorption spectra of Ln in THF (99.8%, WAKO Pure Chemical Industries, Ltd.) were also recorded with Hitachi F-4010 fluorescence spectrophotometer and Hitachi UV-3200 spectrophotometer, respectively.

COMPUTATIONAL METHODS

Ln in vacuum and Ln-THF complex model containing hydrogen bonds, $-OH\cdots O_{THF}$ and $-COOH\cdots O_{THF}$, were optimized and compared using the *ab initio* method (GAUSSIAN98)[5] at the B3YLP/6-311+G(2d,p) level. ^1H chemical shifts were also computed at the same level using the GIAO (gauge-invariant atomic orbital) method. The absorption spectra of the Ln-THF were calculated by INDO/S method (mos-f program) with the 20 occupied MOs and the 20 unoccupied MOs in singly-excited configuration interaction. Binding energy was calculated as the energy of the complex minus the sum of the energies of the isolated monomers.

RESULTS AND DISCUSSION

The observed and calculated ^1H NMR δ of Ln were listed in Table 1. The observed ^1H NMR spectrum of Ln in THF was shown in Fig. 2a). The intensity of H_{COOH} was very low (~0.05) even at 25 °C. This could be due to the water residue, which would be unstabilized in THF solution as the solvent's low hydrogen-accepting characteristics or low polarity. Each ^1H peak of Ln was assigned by literature.[6,7] In general, though H_{OH} and H_{COOH} are easy to be shifted with the microenvironment of the molecule, $\delta_{calc.}$ agreed with $\delta_{obs.}$. The reproducibility of δH_{OH} and δH_{COOH} was better with Ln-THF model optimized structure, which appeared in Fig. 1b).

Table 1. Observed and calculated ^1H NMR chemical shifts of Ln

Assigned Hydrogen	$\delta_{obs.}$ in THF	$\delta_{calc.}$ in vacuum	$\delta_{calc.}$ of Ln-THF
H_{CH2a}, H_{CH2b}, H_{CH}	3.8, 3.9, 5.5	3.3, 3.5, 5.2	2.9, 4.0, 5.1
H_b, $H_a{}^{\ddagger}$, H_c	7.2, 7.5, 8.1	6.6, 7.2, 7.9	6.9, 7.2, 7.8
H_{OH}, H_{COOH}	9.0, 10.8	3.8, $9.0^{\dagger}$	8.9, 11.2

$\ddagger$The integrated area of H_a peak was used as reference, and was set to 1
$\dagger$Intramolecular-hydrogen-bonded to thiazoline-N atom (free H_{COOH} δ ~5.9)

Absorption (4×10^{-5} M) and fluorescence (4×10^{-6} M) spectra of Ln in THF were shown in Fig. 2b). Slightly shorter than the observed $\lambda_{max.}$ at 328 nm, the calculated $\lambda_{max.}$ was at 322.5 nm. The binding energy of Ln-THF complex model is 0.59 eV lower than that of isolated Ln in vacuum, due to the influence of the hydrogen-bonding formation. The observed emission peak at 401 nm would be reasonable for neutral Ln, as Jung *et al.*[8] reported that the $\lambda_{max.}$ of Ln at RT appeared at ~400 nm in *p*-dioxane, and the $\lambda_{max.}$ of Ln and its anion (Ln⁻) in ethanol at 77 K also appeared at 404 nm (444 nm at RT) for Ln, and at 450 nm (532 nm at RT) for

Ln⁻, respectively. Preliminary measurements had done on emission spectra. Methanol (hydrogen-donating solvent), acetone (hydrogen-accepting solvent) or $NaPF_6$ (salt for increasing polarity, but not to increase hydrogen-bond strength) were added to the Ln/THF solution up to less than 1 M additives. Though no significant change was seen for methanol and acetone (less than 1 nm, data not shown), the addition of salt showed a significant change of emission properties of Ln. The emission peak appeared at 10~17 nm (in cases of (2) and (3) in Fig. 2b)) longer than the original ((1) in Fig. 2b)). Of course it is just a preliminary result, the polarity of the solution is supposed to be more important than the hydrogen-bonding properties, restricted to the emission wavelength. In addition, the emission intensity was several times increased only in the case of $NaPF_6$. As the fluorescent state of Ln has CT characteristics[8], the wavelength would be red shifted with the addition of $NaPF_6$.

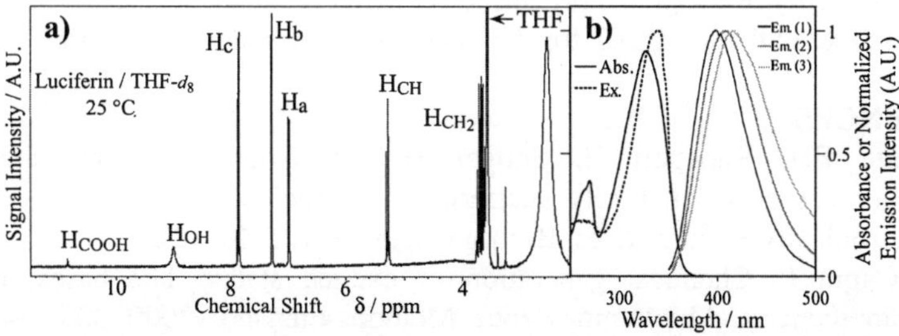

Figure 2. **a)** ^{1}H NMR of Ln in THF-d_8, and **b)** absorption and fluorescence spectra of Ln in THF. NMR peaks are symbolized according to Fig. 1a)

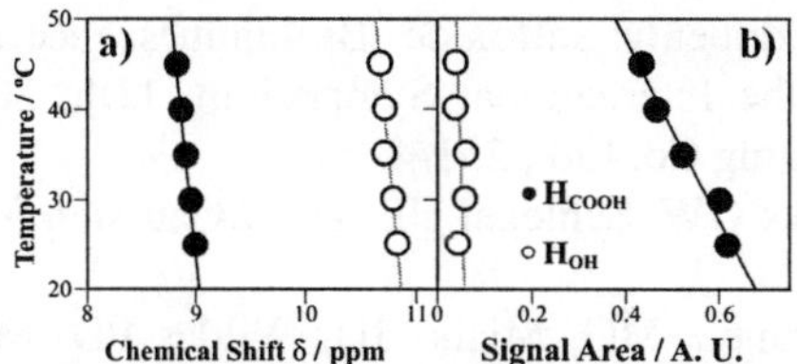

Figure 3. Changes in **a)** δ and **b)** signal area of H_{COOH} and H_{OH} with temperature

The temperature dependence of a) δ of H_{COOH} and H_{OH}, and b) signal area of Ln in THF-d_8 was shown in Fig. 3. The H_{COOH} and H_{OH} peaks were shifted to the higher magnetic field region with increasing temperature, and no new peak appeared at higher temperatures. The intensity of H_{COOH} looked like constant, contrary to H_{OH}, which decreased with increasing temperature. Though the δs were in line with the Ln-THF complex model, the difference of the temperature dependence between H_{COOH} and H_{OH} is in question. Other types of hydrogen-bondings, for example, 1:n (n=1~3) complex[9] with THF or other kind of clusters could be possible around Ln.

CONCLUSION

The interaction between Ln and THF was studied by using ^1H NMR, absorption, fluorescence spectra and the *ab initio* calculations. The calculated chemical shifts for Ln-THF agreed with the observed ^1H peaks. As the reason of the difference between signal area change of H_{COOH} and that of H_{OH} with increasing temperature is in question, it is necessary to study hydrogen-bondings of Ln in detail.

ACKNOWLEDGMENTS

This work was partly supported by Grants-in-Aid (Nos. 11470246 and 13740261) from the Japan Society for the Promotion of Science to K.O., by a Grant-in-Aid for Scientific Research (C) for N.W. from the Ministry of Education, Culture, Sports, Science and Technology (MEXT) of Japan and by a grant for the 21st Century's Center of Excellence Programs organized by MEXT, Japan, science 2003. Y.Y. and S.N. would like to thank MEXT for the opportunity of financial support.

REFERENCES

1. White EH, Rapaport E, Seliger HH, Hopkins TA. The chemi- and bioluminescence of firefly luciferin: An efficient chemical production of electronically excited states. Bioorg Chem 1971; 1: 92-122.
2. McCapra F. Chemical generation of excited states: The bases of chemiluminescence and bioluminescence. Methods Enzymol. 2000; 305: 3-47.
3. Itoh S, Nameda N. Molecular orbital calculations for dioxetane as a part of the intermediate of firefly luciferin. Kagoshima Daigaku Kenkyu Houkoku 1996; 38: 257-60.
4. Wada N, Sameshima K. Ab initio calculation for D-(-)-luciferin and its intermediates in dimethyl sulfoxide. Bioluminescence & Chemiluminescence, Proceedings of the International Symposium, 11th, 2001; Singapore: World Scientific Publishing Co. Ltd., 251-4.
5. Frisch MJ, Trucks GW, Schlegel HB, *et. al.* Gaussian 98. Pittsburgh: Gaussian Inc., 1998.
6. White EH, Steinmetz MG, Miano JD, Wildes PD, Morland R. Chemi- and bioluminescence of firefly luciferin. J Am Chem Soc 1980; 102: 3199-208.
7. White EH, Wörther H, Field GF, McElroy WD. Analogs of firefly luciferin. J Org Chem. 1965; 30: 2344-8.
8. Jung J, Chin C-A, Song P-S. Electronic excited states of D-(-)-luciferin and related chromophores. J Am Chem Soc. 1976; 98: 3949-54.
9. Khutsishvili VG, Serebryanskaya AI, Bogachev YuS, Kurenkova VM, Shapet'ko NN, Shatenshtein AI. Study of proton-transfer processes by the NMR method applied to various nuclei. VII. Trifluoroacetic acid-tetrahydrofuran system. J Gen Chem USSR 1983; 83: 628-33.

AN EVOLUTIONARY HISTORY OF THE JAPANESE AQUATIC FIREFLIES INFERRED FROM MITOCHONDRIAL DNA SEQUENCES

H SUZUKI, Y SATO, N OHBA

[1]*Bioscience Division, Olympus Co., Tokyo 192-8512, Japan*
[2]*National Institute of Vegetables and Tea Science, Shizuoka 428-8501, Japan*
[3]*Yokosuka City Museum, Kanagawa 238-0016, Japan*
Email: hirobumi2_suzuki@ot.olympus.co.jp

INTRODUCTION

Two aquatic firefly species, *Luciola cruciata* and *L. lateralis,* are very popular and common in Japan. They adapt to water in larval stage. *Luciola cruciata* is an endemic species distributed throughout the three major islands (Honshu, Shikoku and Kyushu) of Japan. *Luciola lateralis* is widely distributed throughout the Korean Peninsula, northeast China, Sakhalin, and the four major islands (Hokkaido, Honshu, Shikoku and Kyushu) of Japan. Ecology and flash communication system of these species have been studied intensively,[1,2] and two flash types (slow- and fast-flash types) were recognized in each species. The two types differ in behavior, but not morphologically. In *L. cruciata*, the inter-flash interval of the mate-seeking males in the slow-flash type is about 4 sec, while that in the fast-flash one is about 2 sec. The slow- and fast flash fireflies have indigenous distributions in the east and west area, respectively, and the boundary corresponds to the great rupture zone, Fossa Magna.[3] On the other hand in *L. lateralis*, the inter-flash interval in Hokkaido is about 1 sec (slow-flash type), while that in Honshu is about 0.5 sec (fast-flash type).[4]

The flashing-time differentiation must function as an important factor in reproductive isolation and speciation, because the flash pattern is strictly related to mating approaches.[5] To elucidate the origin and inter-relationships of these flash types in each species, mitochondrial (mt) cytochrome oxidase II (CO II) gene was surveyed by restriction fragment length polymorphism (RFLP) and sequence analyses. The evolutionary history of these species is discussed together with the related species, *L. owadai.*

METHODS

Luciola cruciata specimens were collected from 62 sites in Japan covering almost all the insect's distribution areas. *Luciola lateralis* from 46 sites in Japan and two sites in Korea. *Luciola owadai* from Kume-jima Island, the Okinawa Islands, Japan.

Mitochondrial CO II region was amplified by means of PCR, and the product was digested with restriction endonucleases (*Ase* I, *Dra* I, *Hae* III, *Hinf* I, *Hpa* II, *Mva* I, *Pst* I and *Rsa* I). Haplotypes were determined based on digestion-product electrophoretic patterns. To evaluate the phylogenetic relationships among haplotypes, one individual from each population locations was subjected to DNA sequencing. Phylogenetic analyses were conducted in PAUP* and PHYLIP.

RESULTS AND DISCUSSION

Nineteen haplotypes were detected by RFLP analysis in *L. cruciata*. Based on the nucleotide sequence comparison of the haplotypes, six haplotype-groups (I to VI) were recognised, and their distributions were indigenous to local areas (Fig. 1 and 2). Namely, Group I occurs in north Honshu area, Group II in Kanto to north Chubu area, Group III in Chubu area, Group IV in west Japan area, Group V in north Kyushu area, and Group VI in south Kyushu area. But the boundary between Group III and IV is overlapping around southwestern part of the Chubu area. Group I and II, III and IV, and V and VI make east Japan, west Japan, and Kyushu lineages, respectively. The east Japan lineage is more closely related to the west Japan lineage than Kyushu lineage. Nucleotide divergences between haplotypes range from 0.1 to 4.8 % with a mean value of 2.8 %.

The distributions of the flash types are concordant with the haplotype-groups that the slow-flash type fireflies belong to Group I and II, while the fast-flash type ones to Group III to VI. If the flash type characteristics are assigned to the phylogenetic tree parsimoniously (Fig. 2), the slow-flash type is considered a derived form of the fast-flash form.[6]

In *L. lateralis*, eleven haplotypes were detected. Phylogenetic tree of the haplotypes was separated into two clades (Japanese and Korean lineages). In Japanese lineage, no monophyletic clades could not be identified, and nucleotide divergences between the haplotypes were quite small (0.3 to 1.4 %), although those between Japan and Korea were much more differentiated (8.1 %).

By RFLP analysis, Hokkaido populations (slow-flash type) show no discernible differences from Honshu populations (fast-flash type), and the phylogenetic tree shows that the haplotypes between Hokkaido and Honshu populations are not separated as different clades (Fig. 2). The two ecological types cannot be segregated into groups from a phylogenetic standpoint, but must have evolved independently from their phylogenies. Hokkaido is located at a high latitude in a subfrigid region, and annual air temperature in Hokkaido (8.5 °C   in Sapporo) is much lower than that in Honshu (15.9 °C in Tokyo). The flashing-time difference in *L. lateralis* might have evolved through physiological adaptation to the colder climate of Hokkaido.

Both *L. cruciata* and *L. lateralis* show similar ecologies and life histories, and have two flash types in each species. Although the geographical differentiation pattern of CO II haplotypes in *L. cruciata* is congruent with its flash types, we could not recognize such a situation for *L. lateralis*. The possibility exists that the CO II gene markers are not informative enough to uncover differences between flash types of this species. Fig. 2 shows UPGMA tree under distance p (%) with some selected haplotypes of the three species. *Luciola cruciata* is separated into six haplotype-groups and is more closely related to *L. owadai* than *L. lateralis*. To estimate divergence times, the evolutionary rate constancy was tested by likelihood-ratio test, and the gene tree can be considered to fit a molecular clock. If the evolutionary rate of a chrysomelid beetle (0.76 % per million years)[7] is assigned to the UPGMA tree,

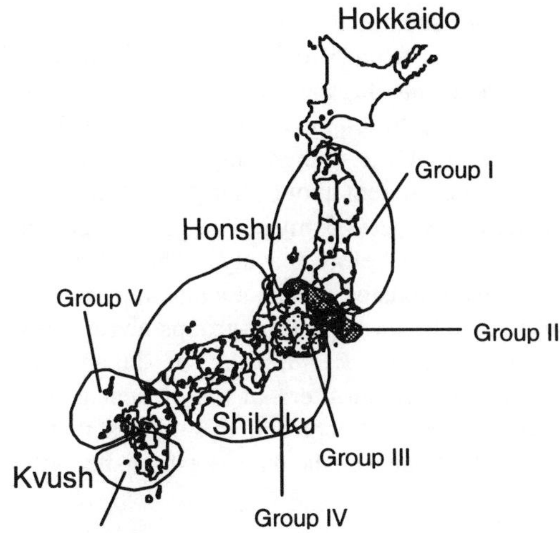

Figure 1. Distribution pattern of CO II haplotype-groups in *Luciola cruciata*.

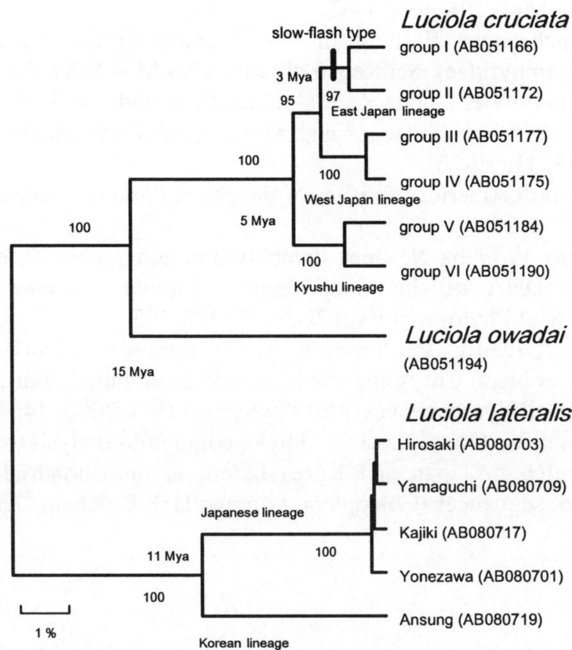

Figure 2. UPGMA tree of the Japanese aquatic fireflies inferred from CO II gene sequences.

L. cruciata and *L. owadai* separated about 15 million years (mya). *Luciola cruciata* diverged into three lineages from 5 to 3 mya and six haplotype-groups at 1 mya. *Luciola lateralis* separated into Japanese and Korean lineages at 11 mya, then later haplotypes diverged from 1.2 to 0.4 mya in the Japanese lineage.

It has been known that the ancient Japan land mass split from the eastern periphery of the Eurasian continent about 16 mya, and the protoform of the Japanese islands was established at about 4.5 mya. The present geological structures of the Japanese islands were formed before 0.7 mya. Therefore, the ancestral stock of *L. cruciata* must have already existed on the ancient Japanese landmass before 5 mya and diverged vicariantly into six haplotype-groups about 1 mya. On the other hand, gene diversity of the current *L. lateralis* in Japan was established almost simultaneously during the formation era of the present geological structures of the Japanese islands about 1 mya. If correct, this means that the distribution of the current *L. lateralis* on the Japanese islands was established after that of *L. cruciata*.[8]

REFERENCES

1.   Kanda S. The firefly. Association of Luminous Organisms of Japan, Maruzen, Tokyo, 1935
2.   Ohba N. Studies on the communication system of Japanese fireflies. Sci Rept Yokosuka City Mus 1983; 30: 1-62.
3.   Ohba N. Synchronous flashing in the Japanese firefly, *Luciola cruciata* (Coleoptera: Lampyridae). Sci Rept Yokosuka City Mus 1984; 32: 23-32.
4.   Ohba N, Tsumuraya T, Honda K, et al. Ecological study of the firefly, *Luciola lateralis*, of the Kushiro Shitsugen and Akkeshi, Hokkaido. Sci Rept Yokosuka City Mus 1993; 41: 15-26.
5.   Barber H S. North American fireflies of the genus *Photuris*. Smithsonian Musc Coll 1951; 117: 1-58.
6.   Suzuki H, Sato Y, Ohba N. Gene diversity and geographic differentiation in mitochondrial DNA of the Genji firefly, *Luciola cruciata* (Coleoptera: Lampyridae). Mol Phylogenet Evol 2002; 22: 193-205.
7.   Gómez-Zurita J, Juan C, Petitpierre E. The evolutionary history of the genus *Timarcha* (Coleoptera, Chrysomelidae) inferred from mitochondrial CO II gene and partial 16S rDNA sequences. Mol Phylogenet Evol 2000; 14: 304-317.
8.   Suzuki H, Sato Y, Ohba N, et al. Phylogeographic analysis of the firefly, *Luciola lateralis*, in Japan and Korea based on mitochondrial cytochrome oxidase II gene sequences (Coleoptera: Lampyridae). Biochem Genet 2004; 42: 287-300.

BIOLUMINESCENCE SPECTRA OF NATIVE AND MUTANT FIREFLY LUCIFERASES AS A FUNCTION OF pH

NN UGAROVA, LG MALOSHENOK, IV UPOROV

Dept of Chemistry, Moscow State University, 119992, Moscow, Russia
E-mail:unn@enz.chem.msu.ru

INTRODUCTION

The relationship between the luciferase structure and the color of light emission is a particularly intriguing problem of firefly bioluminescence. According to the accepted photo-physical concepts about the influence of medium on bioluminescence spectra, the effects observed may be divided into general (non-specific) and specific ones.[1] General effects result from changes in polarizability of emitter microenvironment and manifest themselves as shifts of bioluminescence maxima without changes in the spectrum shapes. The spectral shapes and bioluminescence maxima are changed due to specific effects of microenvironment on the structure of the emitter. The effects of these two types are difficult to distinguish in real systems but this approach simplifies the discrimination of the prevailing mechanism. The observed variations of bioluminescence spectra of native and mutant firefly luciferases show that specific influence of mutations on bioluminescence spectra predominates. In literature one may find different opinions about the structure of the emitter in the firefly luciferase system. Some authors claim that the keto-form of oxyluciferin is the only emitter and that changes in the bioluminescence spectra result from changes in its conformation. According to other authors, the enolate form of oxyluciferin is the emitter. In many cases, bioluminescence spectra are non-symmetric; this is typical of systems where light is emitted from several electronically excited particles rather than from a single one. This is especially apparent for bioluminescence spectra obtained at different pH. The goal of this work was to study the pH-dependence of the bioluminescence spectra of the wild-type recombinant *Luciola mingrelica* firefly luciferase and its mutant form with the His433Tyr point mutation and a bioluminescence maximum at 606 nm. The analysis of the experimental data allowed us to propose a new approach to the interpretation of bioluminescence spectra for firefly luciferases.

METHODS

Plasmid mpLR containing *L. mingrelica* firefly luciferase gene with a single nucleotide change equivalent to His433Tyr protein mutation was obtained from the pLR plasmid[2] using the PCR method. *E. coli* cells, strain LE 392, were transformed with the mpLR plasmid and used for the preparation of the mutant luciferase. The proteins were isolated and purified to homogeneity as described in.[2]

To obtain bioluminescence spectra, 2 mL of 0.05 mol/L of Tris-acetate buffer solution containing 2 mmol/L EDTA, 30 mmol/L $MgSO_4$, 1 mmol/L ATP, and 0.25 mmol/L luciferin (pH 5.6–10.2) were put into a fluorimetric cell, 100 µL of $5 \cdot 10^{-6}$

mol/L luciferase solution were added, and bioluminescence spectra were recorded on a LS 50B spectrofluorimeter (Perkin-Elmer, UK) at 450–650 nm. The spectra were corrected for the photomultiplier sensitivity using the instrument software.

RESULTS

His433Tyr mutation had no significant effect on the catalytic activity of the luciferase and on K_m both for luciferin and ATP. Normalized bioluminescence spectra of the wild-type and mutant luciferases are shown in Fig. 1.

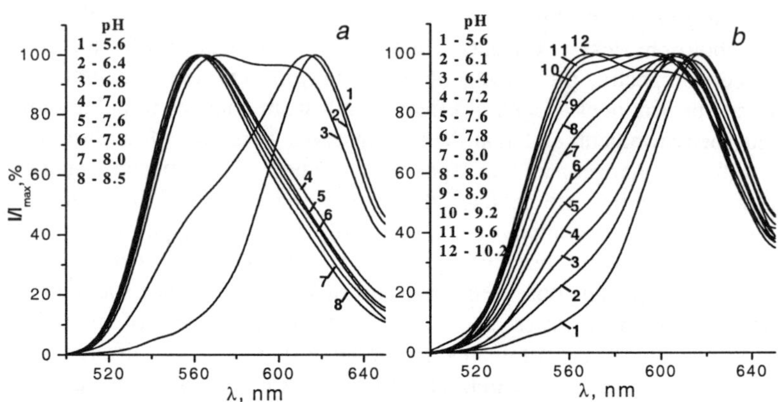

Figure 1. Bioluminescence spectra of the native (*a*) and mutant (*b*) *L. mingrelica* firefly luciferases at different pH

For the wild-type luciferase, yellow-green bioluminescence (λ_{max} = 570 nm) was observed at pH ≥ 7.0 and red bioluminescence (λ_{max} = 618 nm) at pH = 5.6. Both forms are present at intermediate pH values (Fig. 1*a*). For the mutant luciferase, λ_{max} of bioluminescence at pH < 10.0 is in the red region (higher than 600 nm). However, a band in the yellow-green region appears at pH ≥ 7 and its intensity increases with an increase in pH. The yellow-green bioluminescence prevails only at pH ~10.2 (Fig. 1*b*). Analysis of the pH-dependence of bioluminescence spectra of the wild-type and mutant luciferases indicates that the shift in λ_{max} of bioluminescence observed at the pH-optimum of the catalytic activity (pH = 7.8) is explained by a specific mechanism of changes in bioluminescence spectra.

The electronically excited oxyluciferin molecule is known to exist in two forms: ketone (λ_{max} = 618 nm) and enol (λ_{max} = 550–570 nm). It was tacitly assumed that the emission spectra of the enol and the enolate-ion are identical. However, the Gauss multi-peak fit with only two forms of the emitter occurred with a very low correlation coefficient. In this connection we proposed that the spectra of the enolate-ion and the enol have different maxima and the observed bioluminescence spectrum is a sum of three forms of oxyluciferin rather than two. Assuming that λ_{max} = 556 nm belongs to the enolate-ion and λ_{max} = 618 nm is the maximum for the ketone, the Gauss multi-peak fit of the bioluminescence spectra allowed us to identify the third form of oxyluciferin, i.e., the enol, with λ_{max} = 587 nm. In this case the correlation coefficient was 0.999, which confirms the validity of our interpretation of the bioluminescence spectra.

Integration of the bioluminescence spectra for the three forms of oxyluciferin under the assumption that the bioluminescence quantum yields of all the three emitter forms are equal gave the relative content of each form at different pH values. As can be seen from Fig. 2, at pH $\geq$ 7.0 the enolate-ion form prevails in the wild-type luciferase, whereas the ketone and enol forms prevail in the mutant luciferase. The relative content of the enol reaches its maximum at pH 7.0 for the wild-type luciferase and at pH 8.6 for the mutant one. Thus, it can be concluded that the observed changes in bioluminescence spectra of the mutant luciferase are due to shifts in the ketone $\leftrightarrow$ enol $\leftrightarrow$ enolate equilibria.

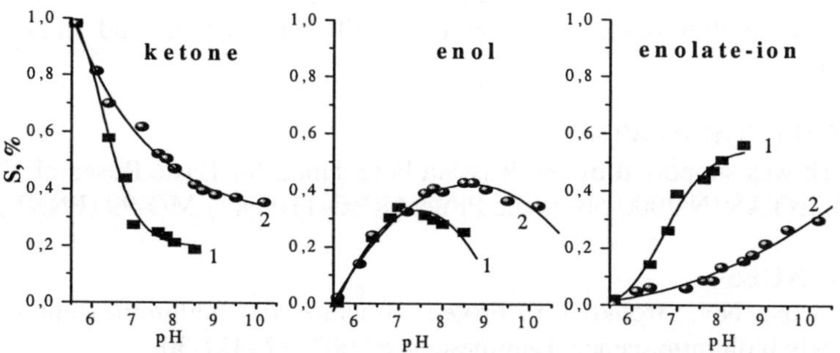

Figure 2. The pH dependence of the relative content (%) of ketone, enol, and enolate-ion for the wild-type (1) and mutant (2) firefly luciferases

The difference in the relative content of different emitter forms for the wild-type and mutant luciferases can be explained by a change in the molecular dynamics in the luciferase active site as a result of the mutation. According to the model of the firefly luciferase active center,[3] the absolutely conservative amino acid residues of *Luciola mingrelica* firefly luciferase His247, Thr345, and Lys531, which are, probably, involved in the keto-enolic tautomerization and enolization processes, are located close to the oxyluciferin thiazole ring. A change in the localization of these groups in the active site of the mutant luciferase results in a shift in the ketone $\leftrightarrow$ enol $\leftrightarrow$ enolate equilibrium towards ketone and enol.

The His433 residue, which is highly conservative for firefly luciferases, is located at a distance of 12 Å from the enzyme active site. The residues Tyr427 – Phe435 form molecular hinge located in proximity to the flexible polypeptide loop binding the N and C-domains. Analysis of thermal fluctuation amplitudes of the atoms engaged in this molecular hinge for the structures of *P. pyralis* firefly luciferase[4] and N-terminal adenylation subunit of gramicidin S synthetase[5] revealed the high flexibility of this hinge. The imidazole ring of His433 forms a hydrogen bond with the carboxyl group of the Glu431 residue from this loop. This bond increases somewhat the rigidity of the loop and stabilizes the structure of the enzyme-product complex. This hydrogen bond disappears upon replacement of His by Tyr, which decreases the rigidity of the hinge and increases the amplitude of thermal fluctuations of the N and C-domains with respect to each other. In turn, this makes the emitter environment more flexible and impedes the keto-enolic tautomerization. It results in a shift of the maximum of the bioluminescence spectrum towards the red region. The influence of the C domain on the bioluminescence spectra is confirmed by experimental data: the N domain in the absence of the C domain generates only red bioluminescence.[6]

In conclusion it should be stressed that the approach proposed, which is based on the photo-physical concepts on the correlation between bioluminescence and structure of the emitter and which takes into account the contribution of three (rather than two) forms of the emitter to bioluminescence spectra, can be rather fruitful for analysis of bioluminescence spectra of different native and mutant firefly luciferases.

ACKNOWLEDGMENTS
This work was supported by the Russian Foundation for Basic Research (N 02-04-48-961), INTAS (N 2000-562), and Project RBO-11009(1)-MO-99 (PNNL).

REFERENCES
1.   Ugarova NN, Brovko LY. Protein structure and bioluminescent spectra for firefly bioluminescence. Luminescence 2002; 17: 321-30.
2.   Lundovskich IA, Leontieva OV, Dementieva EI, Ugarova NN. Recombinant *Luciola mingrelica* firefly luciferase. Folding *in vivo*, purification and properties. In: Roda A, Pazzagli M, Kricka LJ, Stanley PE. eds. Bioluminescence and Chemiluminescence. Perspectives for 21st Century. Chichester: Wiley, 1999: 420-4.
3.   Sandalova TP, Ugarova NN. Model of the active site of firefly luciferase. Biochemistry (Moscow) 1999; 64: 962-7.
4.   Conti E, Franks NP, Brick P. Crystal structure of firefly luciferase throws light on a superfamily of adenylate-forming enzymes. Structure 1996; 4: 287-98.
5.   Conti E, Stachelhaus T, Marahiel MA, Brick P. Structural basis for the activation of phenylalanine in the non-ribosomal biosynthesis of gramicidin S. EMBO J 1997; 16: 4174-83.
6.   Zako T, Ayabe K, Aburatani T, Kamiya N, Kitayama A, Ueda H, Nagamune T. Luminescent and substrate binding activities of firefly luciferases N-terminal domain Biochim Biophys Acta, Proteins & Proteomic 2003; 1649, 183-9.

INTERACTION OF OXYLUCIFERIN ANALOGS, DIMETHYL OXYLUCIFERIN AND MONOMETHYL OXYLUCIFERIN, WITH FIREFLY LUCIFERASE

TN VLASOVA, OV LEONTIEVA, NN UGAROVA

Dept of Chemistry, Moscow State University, 119992, Moscow, Russia

E-mail:unn@enz.chem.msu.ru

INTRODUCTION

Firefly luciferase catalyzes oxidation of luciferin with oxygen in the presence of MgATP. Luciferin (substrate) and oxyluciferin (reaction product) are molecules with pronounced fluorescent properties, therefore, fluorescent methods are widely used to study interactions of luciferase with the substrate, the product, and their analogs.[1] Oxyluciferin is extremely unstable in aqueous solutions, however, one may expect that oxyluciferin analogs, dimethyl oxyluciferin (DMOL) and monomethyl oxyluciferin (MMOL), are more stable. Previously we have studied spectral and fluorescence properties of DMOL in aqueous solutions[2] and it was shown that DMOL at alkaline pH undergoes decomposition to form a product with λ_{abs} = 350 nm and λ_{em} = 500 nm. The goal of this work was to study absorption and fluorescence spectra of MMOL and stability of MMOL and DMOL in aqueous solutions, and in the complexes with the wild-type and mutant (His433Tyr) *L. mingrelica* firefly luciferases.

METHODS

Recombinant and mutant (His433Tyr) *L. mingrelica* firefly luciferases were isolated form *E.coli* cells and purified to homogeneity as described in.[3] MMOL and DMOL were synthesized and kindly provided to us by Dr. D. Weiss.[4] Absorption and fluorescence spectra of MMOL and DMOL in 0.05 Tris-acetate, containing 2 mM EDTA, 10 mM $MgSO_4$ were obtained at different pH on a Shimadzu UV 1202 spectrophotometer, and a LS 50B Perkin Elmer spectrofluorimeter, respectively.

RESULTS

Absorption and fluorescence spectra of MMOL

Spectral characteristics of MMOL (and, for comparison, of DMOL) in buffer solutions within 6.0–9.0 pH interval are given in Table 1. The changes in the absorption spectra of MMOL upon changes of pH are explained with the existence of the three forms of MMOL under these conditions (Fig.1). MMOL is a dianion at pH > 8.0. At pH < 8.0, two groups of MMOL, hydroxythiazolic and phenolic, can be protonated. According to the literature data[5], the hydroxythiazolic group is more acidic as compared with the phenolic one, therefore, the phenolic group is protonated first. Thus, the peaks with λ_{abs}= 390 and 375 nm correspond to the monoanion with protonated phenolic group and to the neutral MMOL molecule, respectively. pK of

                                                      Vlasova TN et al.

the dianion↔monoanion transition is 7.8. At pH<6.0, rapid changes in the absorption spectra of MMOL indicate low stability of MMOL at acid pH.

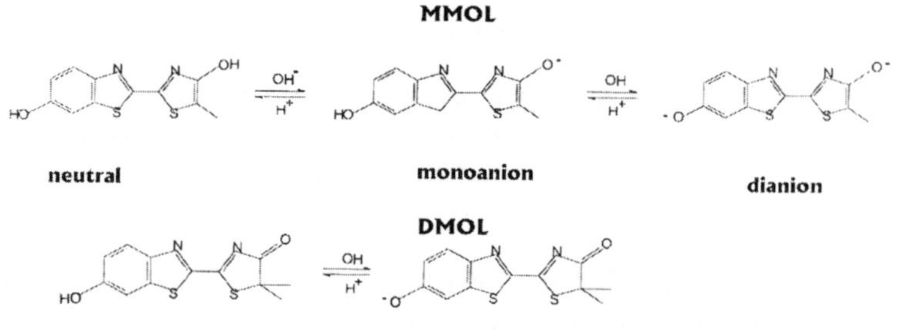

Figure 1. DMOL and MMOL structures at different pH

Table 1. Spectral properties of MMOL, DMOL, and product of their decomposition (P) in buffer solution at different pH

MMOL			DMOL			P		
pH	$\lambda_{max, abs}$, nm	$\lambda_{max, em}$, nm	pH	$\lambda_{max, abs}$, nm	$\lambda_{max, em}$, nm	pH	$\lambda_{max, abs}$, nm	$\lambda_{max, em}$, nm
8.0-9.0	440	550	6-9	383 and 485	639	> 8.8	310	520
7.3-7.7	440 and 390	550				< 8.8	350	500
6.0-7.0	375	550						

Fluorescence spectra of MMOL have a peak with λ_{em}=550 nm at λ_{ex}=440 nm, which does not change within 6.0–9.0 pH interval. At λ_{ex} = 375 nm, the observed $\lambda_{max, em}$ is shifted towards short-wave region with the pH increase due to MMOL decomposition, because MMOL and the product (P) have rather close spectral properties (Table 1) and the fluorescence spectrum of MMOL is a sum of that of MMOL and the product (P).

Stability of DMOL and MMOL in buffer solutions

As was indicated above, MMOL and DMOL gradually decompose in aqueous solutions. We have determined rate constants for the decomposition of MMOL and DMOL at different pH (Table 2). MMOL appeared to be more stable at alkaline pH and DMOL – at acid pH. At pH 7.8 (the pH optimum of the luciferase catalytic activity), DMOL are several times more stable than MMOL. The half-life period ($\tau_{1/2}$) is 65 min for MMOL and about 6 h for DMOL that allows one to use MMOL and DMOL as fluorescence markers to study the luciferase active center.

The product of MMOL and DMOL decomposition was isolated by us using thin-layer chromatography. Its spectral properties (Table 1) are close to those of the product of decomposition of oxyluciferin with broken C-S-bond.[6] The fluorescence spectrum of this product lies in the shorter wavelength region and this indicates the decrease in the system of conjugated bonds that becomes possible only at the break of the thiazole ring.

Interaction of MMOL and DMOL with the wild-type and mutant luciferases

Binding constants (K_s) of MMOL and DMOL with the wild-type and mutant luciferases at different pH were determined by quenching of the fluorescence of the single Trp residue upon binding of the oxyluciferin analogs with the protein (Fig. 2).

Table 2. Stability of MMOL and DMOL in buffer solutions at different pH

pH	k_{in}, min^{-1}	pH	k_{in}, min^{-1}
MMOL		DMOL	
6.0	0.02	7.0	0.0007
7.8	0.017	7.8	0.001
9.0	0.002	9.0	0.002

Lower values of K_s were observed for the protonated forms of the effectors indicating an important role of the effector's charge in the formation of the complex. DMOL binds with the wild-type luciferase more effectively than MMOL at all studied pH. MMOL and DMOL are bound more effectively with the mutant luciferase as compared with the wild-type one. The His433Tyr mutation causes, probably, loosening of the luciferase active site and this results in the decrease of steric hindrances for the binding of MMOL and DMOL with the active site.

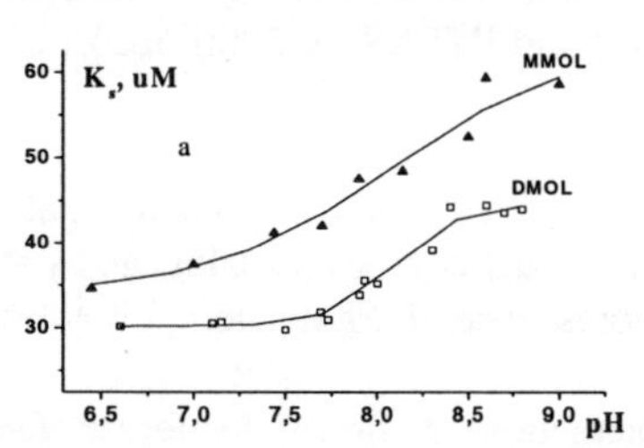

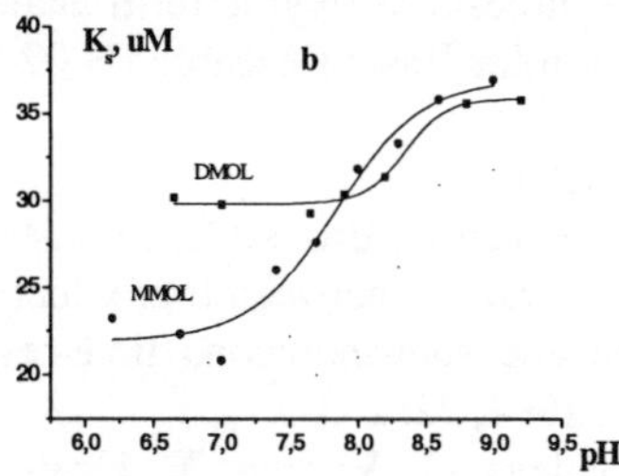

Figure 2. Binding constants of MMOL and DMOL with the wild-type (*a*) and mutant (*b*) luciferases as a function of pH

The fluorescence maxima of DMOL and MMOL shift to the short-wave region upon their binding the enzyme: 20 nm for DMOL and 100 nm for MMOL, and the fluorescence intensity significantly increases (Fig. 3). The observed changes in the fluorescence emission spectra of DMOL can be explained by the increase in hydrophobicity of microenvironment of the emitter on its binding with the protein. Excitation spectrum of MMOL-luciferase complex (λ_{em}=450 nm) corresponds to the absorbance spectra of MMOL-monoanion. In buffer solution we did not seen the fluorescence spectra of this form due to the fact that in water solution electronically excited MMOL-monoanion exists in form of phenolate-ion only (it is in dianion form). The luciferase microenvironment stabilizes MMOL-monoanion by protecting its phenolic group from dissociation. It shows that in vicinity of MMOL bound to the enzyme there is an aminoacid residue that forms H-bond with phenolic group and prevent it from dissociation. It can be explained by increased hydrophobicity of the luciferase active site, too.

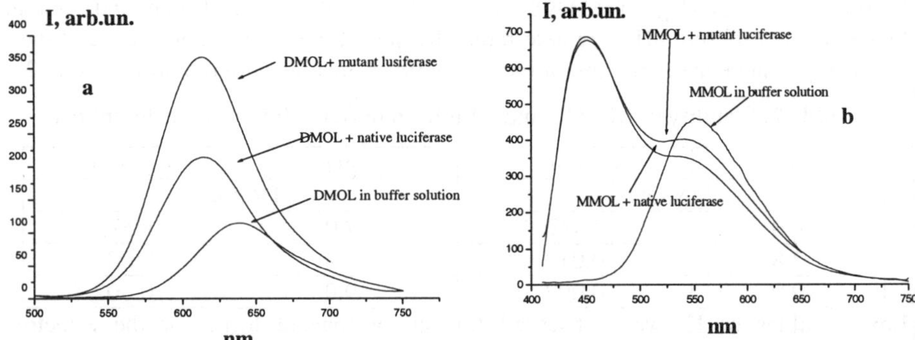

Figure 3. Fluorescence spectra of DMOL (*a*) and MMOL (*b*) in buffer solution and in the complexes with the wild-type and mutant luciferases at pH 7.8.

ACKNOWLEDGMENTS

The authors are grateful to Dr Dieter Weiss (Inst. fur Org. Chemie, Jena, Germany) for the synthesis of oxyluciferin analogs. This work was supported by the Russian Foundation for Basic Research (N 02-04-48-961) and INTAS (N 2000-562).

REFERENCES

1. Gandelman O, Brovko L, Chikishev A, Shkurinov A, Ugarova N. Investigation of the interaction between firefly luciferase and oxyluciferin or its analogues by steady state and subnanosecond time-resolved fluorescence. J Photochem Photobiol B; Biol 1994; 22: 203-9.
2. Leontieva O, Vlasova T, Ugarova N. Interaction of firefly luciferase *Luciola mingrelica* with dimethyloxyluciferin. In: Stanley P, Kricka L. eds. Bioluminescence & Chemiluminescence. Progress and Current Applications. Singapore: World Scientific. 2002: 41-4.
3. Ugarova N, Maloshenok L, Uporov I. Bioluminescence spectra of native and mutant firefly luciferases as a function of pH. This volume.
4. Weiss D, Beckert R, Lamm K, Baader W, Bechara E, Stevani C. Playing with luciferin – new results on the luminescence of a well-known molecule. In: Case J, Herring P, Robinson B, Haddock S, Kricka L, Stanley P. eds. Bioluminescence & Chemiluminescence 2000. Singapore: World Scientific. 2001: 197-200.
5. White E, Roswell D. Analogs and derivatives of firefly oxyluciferin, the light emitter in firefly bioluminescence. J Photochem Photobiol 1991; 53: 131-6.
6. Suzuki N, Sato M, Okada K, Goto T. Studies on firefly bioluminescence–II. Identification of oxyluciferin as a product in the bioluminescence of firefly lanterns and in the chemiluminescence of firefly luciferin. Tetrahedron 1972; 28: 4065-74.

PART 3

MARINE BACTERIA BIOLUMINESCENCE

EFFECT OF OXYGEN AND HYDROGEN ION ON THE MODULATION OF THE BIOLUMINESCENCE FROM LUMINOUS BACTERIA

H KARATANI, S YOSHIZAWA, S HIRAYAMA

*Dept of Polymer Science and Engineering, Kyoto Institute of Technology,
Sakyo-ku, Kyoto 606-8585, Japan
Email: karatani@kit.ac.jp*

INTRODUCTION

Marine luminous bacteria produce light in the luciferase catalyzed reaction with reduced flavin mononucleotide ($FMNH_2$), molecular oxygen (O_2) and a long-chain aliphatic aldehyde.[1] Some species alter the color of bioluminescence (BL), in which an endogenous fluorescent protein participates as a secondary emitter. *Vibrio fischeri* strain Y1 in the logarithmic phase emits the yellow BL peaking around 540 nm that is shifted by about 50 nm from the normal blue-green light arising from the luciferase reaction ($\lambda_{max} \sim 490$ nm).

Recently, we reported that the yellow BL production is facilitated by a supply of O_2 to the O_2-limited culture and that the enhanced yellow BL adversely falls to the original level after halting the forced aeration.[2] Such a BL modulation can be regarded as a reversible response to the variation of O_2 concentration. It is postulated i) that the reversible BL color change is associated with the activity of the accessory yellow fluorescent protein (YFP) and ii) that the YFP activity is governed by the redox state of the respiratory protein complexes within the cell membrane.

In this study, we examined the effect of the O_2 concentration on BL with varying hydrogen ion (H^+) concentrations and cell density. Moreover, the effect of O_2 on the BL of *Photobacterium phosphoreum*, carrying the lumazine protein responsible for the blue-shifted BL,[3] was also examined.

MATERIALS AND METHODS

Luminous cells were grown in a seawater complete (SWC) liquid medium at about 16 °C unless otherwise specified. A 250-mL of the glowing medium in the logarithmic phase was used to measure BL emissions under various conditions of O_2 concentration, pH and cell population.

BL spectrum, entire photon flux, O_2 concentration and pH were recorded by a home-made system reported previously with some modification.[2] In this system, a Keyence NR-2000 DATA acquisition system (Osaka, Japan) was employed for the real-time data monitoring. The O_2 concentration was measured by a Clark type O_2 electrode, of which calibration was carried out by using air-saturated water and 0.4 mol/L sodium sulfite.

At regular time intervals, a 2.5-mL of culture was taken out by a catheter for the measurements of the cell density. The cell density was determined by the conventional spread plate method.

RESULTS AND DISCUSSION

During the middle of the logarithmic phase, the yellow BL band of the *V. fischeri* Y1 becomes markedly distinct. As the cell-growth comes closer to the stationary phase, the yellow BL band is sharply weakened, whereas the blue-green BL band does not develop a tendency to drop.[4] Under such conditions where the yellow BL is no more a primary light component, it was found that a supply of O_2 to the culture triggers the enhancement of the yellow BL band remarkably (Fig. 1). The yellow BL band enhancement is distinctly observed when the cell population was considerably high in the stationary phase.

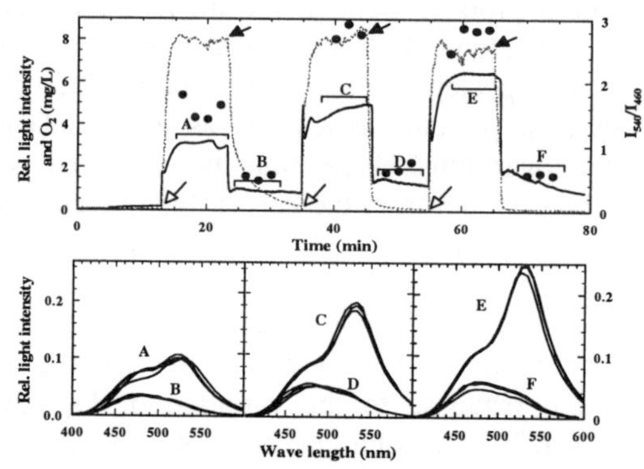

Figure 1. Time-courses for *V. fischeri* Y1 BL intensity and O_2 concentration (upper panel) and spectral distribution (lower panel). Open arrowheads, aeration on and closed arrowheads, off. Capital letters signify the spans, employed for the spectral measurements. Closed circle (upper panel), ratio of intensity at 540 nm to that at 460 nm. Cell density (/mL), 2.0×10^9. Culture   temperature, 16 °C.

Noticeably, the rise and fall of the yellow BL band reversibly occurs in accordance with the aeration cycle (Fig. 1 lower panel). The steady BL level ascended during the forced aeration, as shown in Fig. 1 upper panel, seems mainly to be attributed to the intensified yellow BL. The initial flash appeared immediately after the aeration is begun can be explained by that $FMNH_2$-luciferase complex accumulated in the cell is intensively subject to the oxidation with O_2 fed by the aeration.[5] Contrary to the yellow BL band susceptible to varying O_2 concentration, the blue-green BL emission does not seem to be influenced by the forced aeration.

As shown in Fig. 2, by lowering the culture pH with a successive addition of 0.1 mol/L HCl, the BL intensity likewise decreased step by step. Based on the spectral analysis, it was found that the BL intensity at 540 nm (I_{540}) decreases in accordance with a decrease of pH, whereas the effect of pH does not apply to the case of the intensity at 460 nm (I_{460}). In relation to this, the I_{540}/I_{460} ratio decreased gradually with a decrease in culture pH, *i.e.*, the increase in the H^+ concentration.

In the case of BL of *P. phosphoreum* strain bmFP, as a result of supplying O_2 to the cells under the O_2 limited conditions, the BL was strongly intensified (Fig. 3

upper panel). However, no change in the wavelength distribution was present (Fig. 3 lower panel). Indeed, the BL spectra recorded in the presence of the forced aeration was substantially superimposable on that recorded in the absence of aeration and normalized. The results indicate that a supply of O_2 solely acts on the entire BL intensity but not on its wavelength distribution.

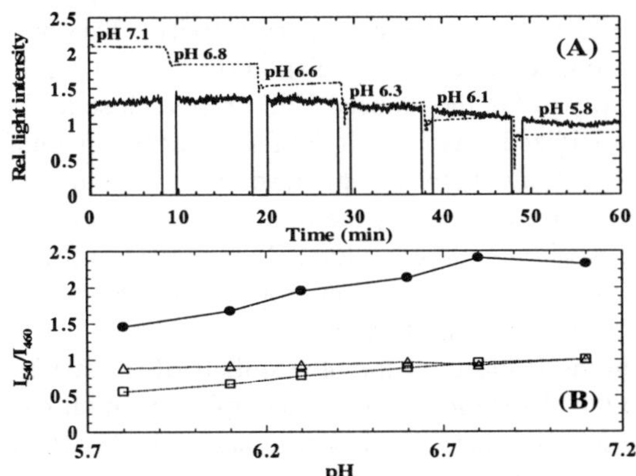

Figure 2. Time courses of *V. fischeri* Y1 BL intensity and culture pH in the presence of aeration (A) and plots of BL intensities at 540 nm (□) and 460 nm (△) and the intensity ratio (I_{540}/I_{460} ratio) (●)vs. pH. Culture temperature, 18 °C.

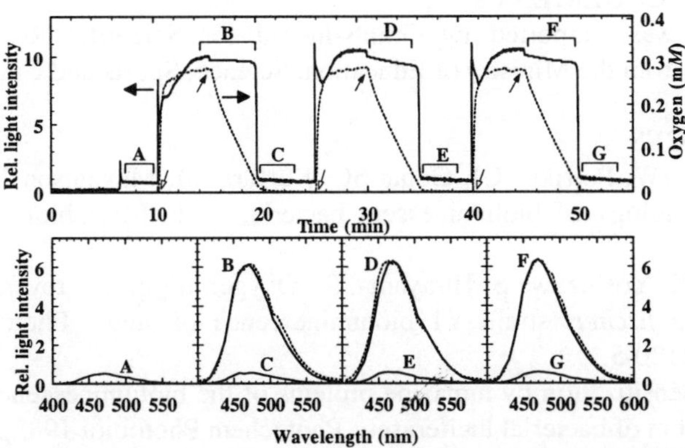

Figure 3.   Time-courses for *P. phosphoreum* bmFP BL intensity and O_2 concentration (upper panel) and spectral distribution (lower panel). Open arrowheads, aeration on and closed arrowheads, off. Capital letters signify the spans, employed for the spectral measurements. Dotted curves in lower panel; normalized BL spectra recorded without aeration.   Cell density (/mL) at $t = 0$ min, ca. $1.8_2 \times 10^9$ and at $t = 30$ min, ca. $1.8_6 \times 10^9$. Culture temperature, 15 °C.

In a previous report, we proposed a hypothesis that the yellow BL of *V. fischeri* Y1 is governed by the redox states of the respiratory complexes within the cell membrane: *V. fischeri* Y1 cells efficiently produce the yellow BL when the respiratory electron flow is active, whereas once the respiratory complexes are fallen into their reduced states under the O_2 limited conditions, YFP is likewise subject to reduction, resulting in the conversion to the non-fluorescent YFP that is less active in BL modulation. At higher cell population, the dissolved O_2 to be shared out among the cells would be strictly restricted. In such a state, the respiratory electron flow may be stagnant so as to lower the respiratory activity. This possibly explains that the yellow BL is absent at higher cell population (Fig. 1). On stimulating the aerobic respiration by a supply of O_2, the reduced YFP present close to the respiratory chain is expected to be reoxidized. As a result, YFP can get BL modulation activity back. The postulated relationship between the respiratory activity and the yellow BL intensity can possibly be explained by the finding that the yellow BL intensity alone is weakened in the weakly acidified culture (Fig. 2), where the proton pump in the cell membrane is expected not to function properly because of an increase in the H^+ concentration on the external cell membrane.

The change in the intensity of the blue-shifted BL of *P. phosphoreum* (Fig. 3) may be ascribed to the change in apportionment of the respiratory reducing power to the luciferases reaction. Furthermore, it seems that the highly fluorescent property of lumazine protein is hardly influenced by the redox state of the respiratory components unlike YFP. That is, the cellular redox state may not be influential in the electronic excitation transfer interaction responsible for the blue-shifted BL.

ACKNOWLEDGEMENTS

This study was supported by Grants-in-Aid for Scientific Research to H. K. (15510173) from the Ministry of Education, Science, Sports and Culture of Japan.

REFERENCES

1. Hastings JW, Potrikus CJ, Gupta SC, Kurfürst M, Makemson JC. Biochemistry and Physiology of bioluminescent bacteria. Adv Microbiol Physiol 1985;26: 235-91.
2. Hajime K, Yoshizawa S, Hirayama, S. Oxygen triggering reversible modulation of *Vibrio fischeri* strain Y1 bioluminescence *in vivo*. Photochem Photobiol 2004;79:120-5.
3. Lee J. Sensitization by lumazine proteins of the bioluminescence emission from the reaction of bacterial luciferases. Photochem Photobiol 1982;36:689-97.
4. Karatani H, Chiba T, Hirayama S. Relationship between spectral distribution of *Vibrio fischeri* strain Y1 bioluminescence and intracellular level of its fluorescent proteins. In: Stanley PE, Kricka LJ editors. Bioluminescence and Chemiluminescence: Progress and current applications. New Jersey: World Scientific Publishing Co. 2002, 81-4.
5. Nealson KH, Hastings JW. Low oxygen is optimal for luciferase synthesis in some bacteria. Arch Microbiol 1977; 112:9-16.

KINETIC INVESTIGATION OF BACTERIAL LUCIFERASE

VV MEZHEVIKIN, IE SUKOVATAYA, NA TYULKOVA
Institute of Biophysics, Russian Academy of Science, Siberian Branch,
Krasnoyarsk, 660036, Russia
E-mail: biotech@ibp.ru

INTRODUCTION

Bacterial luciferase catalyzes the oxidation of long chain aldehydes in the presence of reduced flavin mononucleotide ($FMNH_2$). During the reaction blue-green light is emitted: $RCHO + FMNH_2 + O_2 \rightarrow RCOOH + FMN + H_2O + h\nu$.[1,2] A feature of this reaction is its non-stationary character. We have the enzymatic system containing the unstable participant - $FMNH_2$, as $FMNH_2$ is quickly oxidized, forming FMN and peroxide of hydrogen. Usually the kinetic features of the bacterial bioluminescent reaction, the interaction of luciferase with inhibitors or activators are described by the classical Michaelis-Menten model of steady-state kinetic. However, there are particular kinetic features and theoretical understandings of luciferases that cannot be explained within the context of a steady-state kinetics. The aim of this work was to describe the non-steady-state kinetic of bioluminescent reaction catalysed by bacterial luciferase and to produce the method for the estimation of elementary reaction constants, which characterize the interactions of luciferase with its substrates and also inhibitors or activators.

THEORY

Proceeding from the available literary data,[1-3] the following kinetic scheme of luciferase functionality (Fig. 1) can be derived:

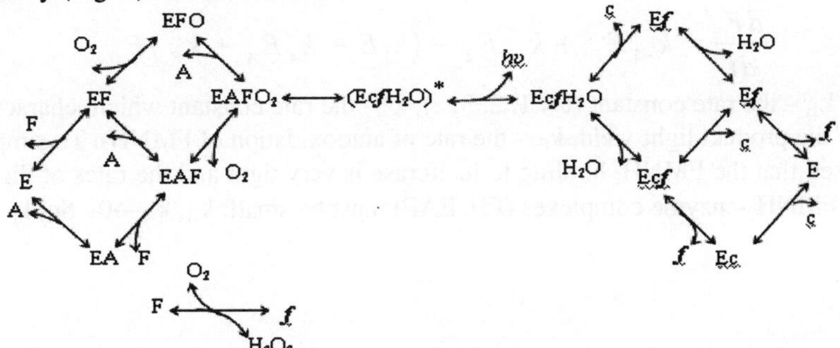

Figure 1. Hypothetical reaction scheme for the bioluminescent reaction catalyzed by bacterial luciferase

                                                                   Mezhevikin VV et al.

Here E, A, c, F and *f* is enzyme, aldehyde, fatty acid, $FMNH_2$ and FMN, respectively. EA, EF, EAF, Ec, E*f*, Ec*f*, EAFO and Ec*f*H_2O are the respective complexes. We assume here that the concentration of O_2 in the reaction media is usually high and the interaction with O_2 is rapid. In this case the rate of bioluminescent reaction will be depends by stage of formation of excited emitter. The following simpler kinetic scheme can be derived:

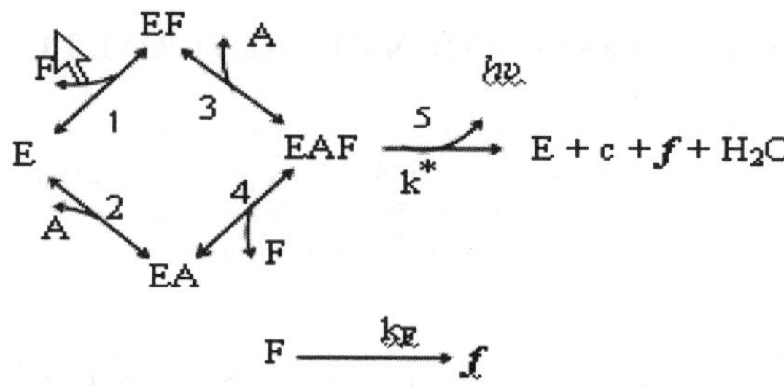

Figure 2. Simple reaction scheme for the bacterial bioluminescent reaction

The kinetic behaviors of the system can be described as the non-linear differential equations as follows:

$$\frac{dE}{dt} = k_{-1}E_F + k_{-2}E_A - k_1 E \cdot F - k_2 E \cdot A + k*E_2$$

$$\frac{dE_F}{dt} = k_1 E \cdot F + k_{-3}E_2 - k_{-1}E_F - k_3 E_F \cdot A$$

$$\frac{dE_A}{dt} = k_2 E \cdot A + k_{-4}E_2 - k_{-2}E_A - k_4 E_A \cdot F$$

$$\frac{dE_2}{dt} = k_3 E_F \cdot A + k_4 E_A \cdot F - (k_{-3} + k_{-4} + k*)E_2 \tag{1}$$

$$\frac{dA}{dt} = k_{-2}E_A + k_{-3}E_2 - k_2 E \cdot A - k_3 A \cdot E_F$$

$$\frac{dF}{dt} = k_{-1}E_F + k_{-4}E_2 - (k_1 E + k_4 E_A + k_F)F$$

where $k_{\pm i}$ – the rate constant (i = 1, 2, 3...), k* - the rate constant which characterize the rate of the product light yield, k_F - the rate of autooxidation of $FMNH_2$. To simplify, it is supposed that the $FMNH_2$ binding to luciferase is very tight and the rates of dissociation of the $FMNH_2$-enzyme complexes (EF, EAF) must be small: k_{-1}, $k_{-4} \rightarrow 0$. So, $k_F \gg k_1 E +$

$k_4 E_A$[4], and from last equation of system (1) we get $F = F_0 e^{-k_F t}$. In this case the following kinetic equation can be derived:

$$\frac{dE}{dt} = k_{-2} E_A - k_1 E \cdot F_0 e^{-k_F t} - k_2 E \cdot A + k * E_2$$

$$\frac{dE_F}{dt} = k_1 E \cdot F_0 e^{-k_F t} + k_{-3} E_2 - k_3 E_F \cdot A$$

$$\frac{dE_A}{dt} = k_2 E \cdot A - k_{-2} E_A - k_4 E_A \cdot F_0 e^{-k_F t} \qquad (2)$$

$$\frac{dE_2}{dt} = k_3 E_F \cdot A + k_4 E_A \cdot F_0 e^{-k_F t} - (k_{-3} + k*) E_2$$

$$\frac{dA}{dt} = k_{-2} E_A + k_{-3} E_2 - k_2 E \cdot A - k_3 A \cdot E_F$$

Assuming that the initial concentration of aldehyde is much higher then that of the initial concentration of luciferase ($A_0 \gg E_0$) and using the initial condition $A = A_0$ we get from system (2) the linear system with variable coefficients:

$$\frac{dE}{dt} = k_{-2} E_A - (k_2 \cdot A_0 - k_1 F_0 e^{-k_F t}) E + k * E_2$$

$$\frac{dE_F}{dt} = k_1 E \cdot F_0 e^{-k_F t} + k_{-3} E_2 - k_3 E_F \cdot A_0$$

$$\frac{dE_A}{dt} = k_2 E \cdot A_0 - (k_{-2} + k_4 F_0 e^{-k_F t}) E_A \qquad (3)$$

$$\frac{dE_2}{dt} = k_3 E_F \cdot A_0 + k_4 E_A \cdot F_0 e^{-k_F t} - (k_{-3} + k*) E_2$$

For this system the conservation relationship is $E + E_A + E_F + E_2 = E_0$.

Initiated by photorecovered $FMNH_2$, the bioluminescent reaction is a short flash of light with pronounced maximum. In this case the k_F value is very large[4] and the enzyme makes, at this, one cycle.[1,2] It means, that interaction of enzyme with $FMNH_2$ very quickly decreases. So the kinetic behavior of the system can be described as the system of ordinary linear differential equations with constant coefficients:

$$\frac{dE}{dt} = k_{-2}E_A - k_2 \cdot A_0 E + k^* E_2$$

$$\frac{dE_F}{dt} = k_{-3}E_2 - k_3 E_F \cdot A_0$$

$$\frac{dE_A}{dt} = k_2 E \cdot A_0 - k_{-2}E_A$$

$$\frac{dE_2}{dt} = k_3 E_F \cdot A_0 - (k_{-3} + k^*)E_2$$

(4)

Its solution is $E_2 = C_1 e^{\lambda_1 t} + C_2 e^{\lambda_2 t}$, where $\lambda_{1,2} = -\frac{k_3 A_0 + k_{-3} + k^*}{2} \mp \sqrt{\frac{(k_3 A_0 + k_{-3} + k^*)^2}{4} - k_3 k^* A_0}$

are the roots of the characteristic equation $\lambda^2 + (k_3 A_0 + k_{-3} + k^*)\lambda + k_3 A_0 k^* E_2 = 0$ corresponding to the simultaneous differential equations (4) which describe Fig. 1.

According to the experimental results: $\lambda_1 < \lambda_2$, it can be seen that the experimental plot of bioluminescence decay allows to estimate by variation of the A_0 concentration the follows constants k_3, k_{-3}, k^*. Moreover the present kinetic analysis may be used for identification of inhibitors, which have a competitive character with respect to the aldehyde or $FMNH_2$ from the shape of bioluminescent flash.

ACKNOWLEDGEMENTS

This work was supported by the grant from Ministry of Education of Russian Federation, grant PD 02-1.4-315; and Civilian Research and Development Foundation, grant Y1-B-02-17.

REFERENCES

1. Baldwin T, Nicoli M, Becvar J, Hastings J. Bacterial luciferase: binding of oxidized flavin mononucleotide. J Biol Chem 1975; 250: 2763-8.
2. Hastings J, Potrikus C, Gupta S, Kurfurst M, Makemson J. Biochemistry and physiology of bioluminescent bacteria. Advan Microb Physiol 1985; 26: 235-91.
3. Lin L, Szittner R, Meighen E. Binding of flavin and aldehyde to the active site of bacterial luciferase. In: Stanley P, Kricka L. eds. Bioluminescence & Chemiluminescence: Progress & Current Applications. Singapore: World Scientific, 2002: 89-92.
4. Gibson Q, Hastings J. The oxidation of reduced flavin mononucleotide by molecular oxygen. Biochem J 1962; 83; 368-76.

A MNDO-PM5 STUDY OF THE ENZYME-FREE NADH GENERATION FROM NAD⁺ IN THE PRESENCE OF ELECTRON-TRANSFER MEDIATOR AND AN EVALUATION OF THE PROCESS BASED ON BACTERIAL BIOLUMINESCENCE

T SUGIMOTO[1], N WADA[2], H KARATANI[3]

[1]*Dept of Applied Material and Life Science, Kanto-gakuin University,*
Kanazawa-ku, Yokohama 236-8501, Japan
[2]*Faculty of Life Science Toyo University, Itakura-machi, Gunma 374-0193, Japan*
[3]*Dept of Polymer Science and Engineering, Kyoto Institute of Technology,*
Sakyo-ku, Kyoto 606-8585, Japan
E-mail:sugimoto@kanto-gakuin.ac.jp

INTRODUCTION

The redox system consisting of oxidized and reduced forms of nicotinamide-adenine dinucleotide (phosphate), $NAD(P)^+$ and $NAD(P)H$, functions to feed electrons to the bacterial luciferase via the catalysis of NADH/FMN oxidoreductase.[1] The construction of an artificial conversion system between $NAD(P)^+$ and $NAD(P)H$ and its characterization may be useful in an explanation of bacterial bioluminescence from a viewpoint of bioenergetics. We have fabricated an enzyme-free system for regeneration of NADH from NAD^+ by use of polyxylylviologen (PXV^{2+}), in which the viologen moiety (referred to as V) acts as an electron-transfer mediator.[2] Cyclic voltammetric and spectropotentiometric studies using KCl as a supporting electrolyte showed (i) that the viologen monocation radical ($V+^{\bullet}$) and its dication ($V2+$) act as electron-transfer mediator and (ii) that monocation radical $V+^{\bullet}$ reacts with NAD+,leading to generation of the enzymatically active NADH. The following two reactions (1) and (2) have been proposed to be responsible for the generation of NADH from NAD^+:

$$NAD^+ + 2V^{+\bullet} + H^+ \rightarrow NADH + 2V^{2+} \qquad (1)$$

$$NAD^{\bullet} + V^{+\bullet} + H^+ \rightarrow NADH + V^{2+} \qquad (2)$$

where $NAD^{\bullet}$ is the one electron reduced form of NADH.

However, the underlying processes for proposed reactions (1) and (2) have not yet been elucidated. We have recently shown that the computer simulation of the redox reaction of N,N'-dimethyl-4,4'-bipyridinium (dimethyl viologen, MV) is applicable to characterize the regeneration of NADH from NAD^+ by catalysis with PXV^{2+}.   In the present study, based on a molecular orbital calculation, we analyzed the underlying processes for the reactions, (1) and (2).

METHODS

The geometrical and electronic structures of each molecule in (1) and (2) were optimized with respect to the total energy of those reaction systems by the MNDO-PM5 molecular orbital method.   The potential energy curve of reactions was obtained by calculating the total energy of a system consisting of two molecules, by varying all the geometrical parameters.   The phosphate group of dinucleotide was assumed to be a dianion (Pi^{2-}).

RESULTS AND DISCUSSION

From the calculation, it was expected that the reaction (1) consists of three elementary processes (3)-(5) and that the processes (4) and (5) are also present in (2);

$$NAD^+ + X^{+\cdot} \;\rightarrow\; NAD^{\cdot} + X^{2+} \qquad\qquad (3)$$
$$NAD^{\cdot} + YH^+ \rightarrow (NAD^{\cdot})H^+ + Y \qquad\qquad (4)$$
$$(NAD^{\cdot})H^+ + X^{+\cdot} \;\rightarrow\; NADH + X^{2+} \qquad\qquad (5)$$

where $X^{+\cdot}$ is an electron-transfer mediator, YH^+ a proton donor and $(NAD^{\cdot})H^+$ a protonated $NAD^{\cdot}$.   The processes (3) and (5) include an electron-transfer reaction and a proton-transfer reaction is included in (4).

The elementary processes (3) to (5) were further found to consist of the following reactions by the calculation.

Elementary process (3): In the following explanation, V is taken as X,

$[NAD^+ + V^{+\cdot} + Cl^-]$

$\qquad\quad\downarrow \qquad\qquad$ $V^{+\cdot}$ and cloride anion Cl^- are connected (Fig.1)

$[NAD^+ + V^{\cdot}\;{\cdot}\;Cl]$

$\qquad\quad\downarrow \qquad\qquad$ an electron is transferred from $V^{\cdot}\;{\cdot}\;Cl$ to NAD^+

$[NAD^{\cdot} + V^{+\cdot}\;Cl]$

$\qquad\quad\downarrow\;\;+2e^- \qquad$ two electrons are supplied from anode to $V^{+\cdot}\;Cl$

$[NAD^{\cdot} + V^{\cdot}\;{\cdot}\;+ Cl^-]$

$\qquad\downarrow$ $-2e^-$ two electrons are extracted by cathode from $V^{\cdot\,\cdot}$

$[NAD^{\cdot} + V^{2+} + Cl^-]$

Elementary process (4): hidronium ion $(H_3O)^+$ is chosen as YH^+,

$[NAD^{\cdot} + (H_3O)^+]$

$\qquad\downarrow$ a proton is transferred from $(H_3O)^+$ to $NAD^{\cdot}$

$[(NAD^{\cdot})H^+ + H_2O]$

Elementary process (5): V is taken as X,

$[(NAD^{\cdot})H^+ + V^{+\cdot} + Cl^-]$

$\qquad\downarrow$ $V^{+\cdot}$ and Cl^- are connected (Fig.1)

$[(NAD^{\cdot})H^+ + V^{\cdot\,\cdot}Cl]$

$\qquad\downarrow$ an electron is transferred from $V^{\cdot\,\cdot}Cl$ to $(NAD^{\cdot})H^+$

$[NADH + V^{+\cdot}Cl]$

$\qquad\downarrow$ $+2e^-$ two electrons are supplied from anode to $V^{+\cdot}Cl$

$[NADH + V^{\cdot\,\cdot} + Cl^-]$

$\qquad\downarrow$ $-2e^-$ two electrons are extracted by cathode from $V^{\cdot\,\cdot}$

$[NADH + V^{2+} + Cl^-]$

In light of the potential energy of reaction, the aforementioned elementary processes are barrierless irrespective of the difference between PXV and MV. The elementary processes, (3)–(5), are explained as follows. The electron to reduce NAD^+ originates from Cl^- and the proton to reduce $NAD^{\cdot}$ is supplied by H_3O^+. The 1-e^- reduced X^{2+} ($X^{+\cdot}$) at first forms a complex with Cl^- and subsequently transfers the excess electron of the complex $(NAD^{\cdot})H^+$, resulting in NADH formation and X^{2+} regeneration.

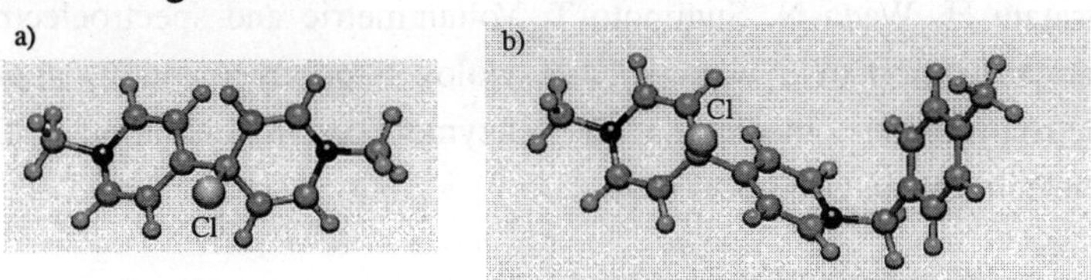

Figure 1.   Energy-optimized structures of (a) $MV^{\cdot\,\cdot}Cl$ and (b) $PXV^{\cdot\,\cdot}Cl$ determined by MNDO-PM5 method

Secondly, it should be noted that MV$^{\cdot}$ $^{\cdot}$Cl homo-dimer was formed more readily than PXV$^{\cdot}$ $^{\cdot}$Cl homo-dimer. This may be because the polyxylyl group of PXV prevents PXV from dimerization in the polymer microenvironment. The V$^{\cdot}$ $^{\cdot}$Cl homo-dimer can transfer its electron to NAD$^+$ only when the relative configuration of the reactants is suitable for the transfer, whereas V$^{\cdot}$ $^{\cdot}$Cl monomer can easily change its orientation to fit the suitable configuration. This may explain the observation that the PXV mediated reaction is more efficient than the case of MV. The calculation result is also in good agreement with the experimental proposal.[2]

It should be noted that both NAD$^+$ and NAD$^{\cdot}$ easily form homo-dimer (Fig. 2). As a result, it is not expected that electron-transfer occurs from V$^{\cdot}$ $^{\cdot}$Cl to those homo-dimers, leading to interfering with NADH generation.

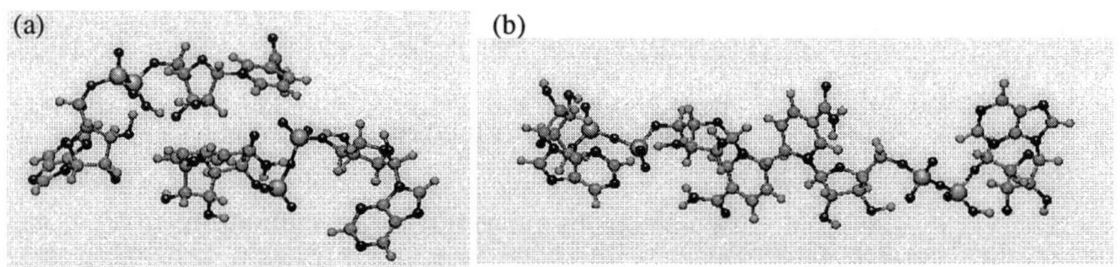

Figure 2.   Energy-optimized structures of (a) NAD$^+$ homo-dimer and (b) NAD$^{\cdot}$ homo-dimer, calculated by MNDO-PM5 method

REFERENCES

1.   Hastings JW, Potrikus CJ, Gupta SC, Kurfürst M, Makemson JC. Biochemistry and physiology of bioluminescent bacteria. Adv Microbiol Physiol 1985;26: 235-91.

2.   Karatani H, Wada N, Sugimoto T. Voltammetric and spectroelectrochemical characterization of a water-soluble viologen polymer and its application to electron-transfer mediator for enzyme-free regeneration of NADH. Bioelectrochem, 2003;60:57-64.

EFFECTS OF ORGANIC SOLVENTS ON BIOLUMINESCENCE EMISSION SPECTRA OF BACTERIAL LUCIFERASE FROM *PHOTOBACTERIUM LEIOGNATHI*

IE SUKOVATAYA, NA TYULKOVA

*Institute of Biophysics, Russian Academy of Science, Siberian Branch,
Krasnoyarsk, 660036, Russia
E-mail: biotech@ibp.ru*

INTRODUCTION

Bacterial luciferase is a flavin monooxygenase that catalyzes the oxidation of reduced flavin mononucleotide ($FMNH_2$) and a long chain aldehyde by molecular oxygen to give the corresponding acid, FMN, and blue-green light.[1] A number of flavin isomers and analogues were studied with regard to their activity with luciferase and the emission spectrum of the resulting bioluminescence.[2] All were found to be active with the enzyme to a greater or lesser extent, and to result in a shift in the colour of the emitted light. Mutationally altered bacterial luciferases have been found which exhibit a light emission whose colour is shifted.[3-5] For example, significant red shifts in light emission of 3-10 nm were measured for mutant luciferases.[6] In earlier reports,[7-9] we described the effects of organic solvents on luciferase activity. If the spectral properties of the emitter are influenced by not only structure of substrates and luciferase, but properties of reaction mixture (dielectric constant, logP, viscosity and et. al.), addition of organic solvents might affect the bioluminescence spectrum. In this work, we have studied the effects of increasing the concentrations of several water-miscible organic solvents on the spectra of luciferase from *Photobacterium leiognathi in vitro* with the chemical reduced $FMNH_2$.

METHODS

The luciferase from *P. leiognathi* (strain 208) used in the work has undergone high purification by ion-exchange chromatography.[10] To measure the control, a reaction was carried out in the mixture of the following composition: 10 μL of $(0.07–0.13)\times10^{-5}$ mol/L luciferase, 50 μL of 9.4×10^{-4} mol/L aqueous solution of tetradecanal - C_{14} (Merck, Germany), 840 μL of 2×10^{-2} mol/L phosphate buffer, 0.5 mL of 7.2×10^{-4} mol/L aqueous solution of $FMNH_2$ (Sigma, USA), 0.5 mL of 0.1 mol/L NADH, pH 7. The reaction was initiated by the injection of the chemical recovered $FMNH_2$. In this case the light is emitted as a glow with a slow decay. Measurements were carried out with a bioluminometer designed at the Institute of Biophysics (Russian Academy of Sciences, Siberian Branch) at a temperature of 25 °C. Reaction parameters were recorded with a 2210 ("LKB-Wallac", Finland) recorder. In experiments, the phosphate buffer was substituted for a water-organic mixture. Methanol (Sigma, USA), acetone, acetonitrile, ethanol, dimethyl sulfoxide (DMSO) (Serva, Germany), glycerol and formamide (Serva, Germany) were used as organic solvents. They were used at the highest purity grade. Concentrations of solvents were expressed in volume %. Emission spectra were recorded on a scanning spectrometer Aminco®Bowman Series 2 (Thermo Spectronic, USA) with a 2 nm bandwidth and a spectral range of 300 - 650 nm. Bioluminescence intensity values are normalized. The experimental results obtained have been statistically processed by Excel for Windows 98.

RESULTS

Bioluminescence catalyzed by *P. leiognathi* luciferase *in vitro* and at pH 7 is blue-green with a wavelength of maximal emission (λ_{max}) at about 506 nm. The value of this λ_{max} is different from it reported usually in literature. This is probably due to the concentration of $FMNH_2$ which was not saturated in the reaction media. The light intensity of luciferase dropped sharply with increasing concentration of solvent (more then 2% v/v), as reported for activity of other enzymes.[11-13] The λ_{max} values at the addition of glycerol, formamide, methanol, as well as acetone, in the concentration, which smaller C_{50} (at which half enzymatic activity is displayed), were almost identical, being 506 nm. At both low and higher (which smaller C_{50}) concentrations of glycerol and formamide, the addition of these solvents showed an inhibitory effect without spectral change. At all these concentrations of formamide and glycerol the shape of the emission spectra was not changed. In contrast, with acetone and methanol addition induced a change of spectral shape. As seen from the Fig. 1 (line 1 for acetone), in this case the spectra were a narrower between 506 and 630 nm. In the presence of same concentrations of acetone and methanol the spectra were identical between 300 and 506 nm. The reversible inhibition at concentrations (< C_{50}) of ethanol and DMSO shows the spectra shifted to the blue by about 4 nm. Addition of these solvents concentration did not affect on the shape of spectra.

Then, bioluminescent spectra of luciferase were investigated with addition of organic solvent concentrations at which half or less of the maximum activity was observed. In the presences of these concentrations of DMSO, acetone, glycerol and formamide the greatest difference was 10 nm. The addition of 2.83% v/v of ethanol in the reaction medium induced a spectral shift of the light emission from 506 to 512 nm and the addition of 15% v/v of methanol in the reaction medium induced a spectral shift of the light emission from 506 to 525 nm. In the presence of 2.38 % v/v glycerol the λ_{max} and right shoulder of spectrum is shifted parallel to the red by about 4 nm (Fig. 1, line 3). The addition of similar formamide concentration into the reaction mixture did not have an effect on the shape of spectra.

In the presence of 4.76 % v/v acetone the luciferase produced emission spectra with increasing maximal emission wavelengths from 506 to 516 nm (Fig. 2). The methanol-induced bioluminescence change was bimodal with two emission maxima at 502 and 525 nm. In the presence of these methanol concentrations spectrum was a narrow in their spectrum between 450 and 570 nm.

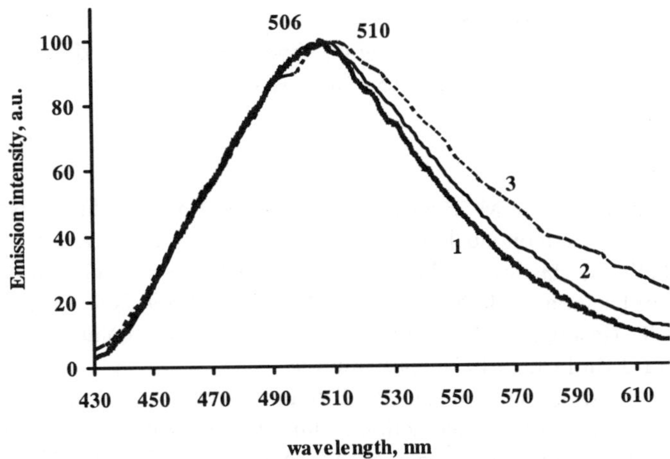

Figure 1. Bioluminescent emission spectra of bacterial luciferase from *P. leiognathi* in 1.48 % v/v acetone (1), 0.02 M pH 7.0 phosphate buffer (2) and 2.38 % v/v glycerol (3). Wavelengths of spectral maxima are indicated.

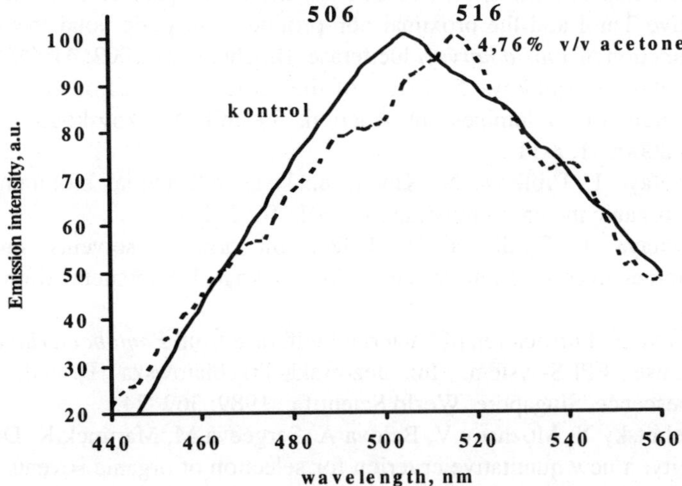

Figure 2. Bioluminescent emission spectra of bacterial luciferase from *P. leiognathi* in 4.76 % v/v acetone and 0.02 M pH 7.0 phosphate buffer. Wavelengths of spectral maxima are indicated.

ACKNOWLEDGEMENTS
This work was supported by the grant from Ministry of Education of Russian Federation, grant PD 02-1.4-315; and Civilian Research and Development Foundation, grant Y1-B-02-17.

REFERENCES
1. Baldwin T, Nicoli M, Becvar J, Hastings J. Bacterial luciferase: binding of oxidized flavin mononucleotide. J Biol Chem 1975; 250: 2763-8.
2. Mitchells G, Hastings J. The effect of flavin isomers and analogues upon the color of bacterial bioluminescence. J Biol Chem 1969; 244: 2572-6.
3. Cline T, Hastings J. Mutated luciferases with altered bioluminescence emission spectra. J Biol Chem 1974; 249: 4668-9.
4. Lee J, O'Kane D, Gibson B. Bioluminescence spectral and fluorescence dynamics study of the interaction of lumazine protein with the intermediates of bacterial luciferase bioluminescence. Biochemistry 1989; 28: 4263-71.
5. Eckstein J, Cho K, Colepicolo P, Ghishla S, Hastings J, Wilson T. A time-dependent bacterial bioluminescence emission spectrum in an *in vitro* single turnover system: Energy transfer alone cannot account for the yellow emission of *Vibrio fischeri* Y-1. Proc Natl Acad Sci USA 1990; 87: 1466-70.
6. Lin L, Sulea T, Szittner R, Kor C, Purisima E, Meighen E. Implications of the Reactive Thiol and the proximal non-proline cis-peptide bond in the structure and function of *Vibrio harveyi* luciferase. Biochemistry 2002; 41: 9938-45.
7. Sukovataya I, Tyulkova N. Effect of dielectric properties of media on kinetic parameters of bioluminescent reaction. Vestnik Moskovskogo Universiteta: Khim 2000; 41: 8-11.
8. Sukovataya I, Tyulkova N. Kinetic analysis of bacterial bioluminescence in water-organic media. Luminecsence 2001; 16: 271-3.
9. Sukovataya I, Tyulkova N. Effect of organic solvents on bacterial bioluminescence reaction. Vestnik Moskovskogo Universiteta: Khim 2003; 44: 9-12.
10. Tyulkova N. Purification of bacterial luciferase from *Photobacterium leiognathi* with use FPLS-system. In: Jezowska-Trzebiatowska B. ed. Biological luminescence. Singapore: World Scientific, 1989: 369 -74.
11. Khmelnitsky Y, Mozhaev V, Belova A, Sergeeva M, Martinek K. Denaturation capacity: a new quntitative criterion for selection of organic sovents as reaction media in biocatalysis. Eur J Biochem 1991; 198: 31-41.
12. Gupta M. Enzyme functions in organic solvents. Eur J Biochem 1992; 203: 25-32.
13. Rodakiewicz-Nowak J, Kasture S, Dudek B, Haber J. Effect of various water-miscible solvents on enzymatic activity of fungal laccases. J Mol Catal B: Enzym 2000; 11: 1–11.

FORMATION OF H_2O_2 IN BACTERIAL BIOLUMINESCENCE REACTION WITH FLAVINMONONUCLEOTIDE ACTIVATED WITH N-METHYLIMIDAZOLE ON THE PHOSPHATE GROUP WITHOUT ADDITION OF THE EXOGENOUS ALDEHYDE

NA TYULKOVA, OI KRASNOVA

Institute of Biophysics, Russian Academy of Science,
Siberian Branch, Krasnoyarsk, 660036, Russia
E-mail: biotech@ibp.ru

INTRODUCTION

Bacterial luciferase catalyzes the oxidation of reduced flavin mononucleotide ($FMNH_2$) and a long chain aldehyde by molecular oxygen to yield FMN, the corresponding acid, H_2O and blue-green light.[1] The high specificity of luciferase in relation to the reducing flavin mononucleotide is exhibited that any modifications of the isoalloxazine ring and the residual ribitil of the flavin influence on the catalytic property of enzyme.[2,3] It is known that the phosphate group of $FMNH_2$ is extremely important for binding the flavin with protein, which has the phosphate site, located very close to active center.[4,5] It is considered that the negative charge the phosphate is necessary for optimum orientation of the substrate on luciferase and for stabilization active conformation of the enzyme - substrate complex.

For study of the role of phosphate groups for many enzymes the phosphamide derivatives with phosphorilic activity are using. Such N-methylimidazole derivatives have increased the reactability in relation to a number of nucleophiles, especially to amine, in which role they cannot act as functional aminoacid residuals of the active center of an enzyme or the site of the flavin bindings.[6,7]

In this work the interaction of bacterial luciferase from *Photobacterium leiognathi* with the flavin mononucleotide activated on phosphate group by N-methylimidazole, without addition of the exogenous aldehyde was study.

METHODS

The luciferase from *P. leiognathi* (strain 208) used in the work has undergone high purification by ion-exchange chromatography.[8] Preparation of flavin mononucleotide FMN (Sigma, USA) used after purification by chromatography on a DE-cellulose column. Activation of the FMN was realized N- methylimidazole in presence 3-phenylphosphine and dipyridyldisulfide, observing molar ratio on a method offered for updating oligonucleotide.[6] FMN was transferred in the threeethylamine form by realization chromatography on DE-cellulose column pre-equilibrated three-ethylamine (TEA), pH 7.0. The dry FMN-TEA was dissolved in dimethylsulfoxide also was adding N- methylimidazole. The product of the reaction was reacted with 2 % LiClO4 in acetone. The state of the activation was checked by a response transamination by 1,6-dihexoethyldiamine. The analysis of amide by a method of chromatography on anion sorbent Polysil-A has shown, that phosphamide eluted as substance which have a charge per unit of smaller, than native FMN. Activation of flavin was 100%. This derivative was used as substrate in measuring light emission of luciferase without exogenous aldehyde addition. Measurement of the H_2O_2 was conducted; using by luminol method.[9] For a standard H_2O_2 the calibrating curve of dependence luminol chemiluminescence intensity from H_2O_2 concentration was obtained.

To measure the chemiluminescence intensity, a reaction was carried out in the mixture of the following composition: 200 μL of 1.6 x 10^{-6} mol/L luciferase, 50 μL of $8x10^{-4}$ mol/L native FMN (or 50 μL of $6x10^{-4}$ mol/L activated FMN), 50 μL of 10^{-1} mol/L dithiotreitol (DTT) and 700 of μL $2x10^{-2}$ mol/L phosphate buffer, pH 7.0. 100 μL from these probes in certain period of time was added in bioluminometer cuvette, containing 400 μL of 10^{-1} mol/L NaOH, 100 μL of $2x10^{-4}$ mol/L luminol, 100 μL of $2.5x10^{-6}$ mol/L ferritin and intensity of chemiluminecsence was registered. The reaction mixture without enzyme was used as control. For chemical reduction of FMN the DTT was used. Formation of peroxide was found as the difference between chemiluminescence intensity of probe and control. The constant of inhibition was second order and calculated from straight line slope, approximately dependence of the rate constant of inhibition of the first order from concentration of the activated flavin derivative.

RESULTS

Incubation of luciferase with the not reduced activated FMN derivative results in irreversible inactivation of the enzyme. The activated FMN derivative, obviously, modifies no more one functional groups of luciferase, because the coefficient characterizing a number of modified protein residues was determined as 1.05. The addition in an incubation mixture the second substrate of luciferase - tetradecanal before adding the activated FMN derivative, does not protect the luciferase from inactivation. The addition of the protector of SH-groups - DTT result in essential inhibition decreasing. The activated FMN derivative reacts most likely with SH-groups of luciferases, which, as is known, are good nucleophile.[10]

The activation of the flavin phosphate group for order decreases affinity of FMN to luciferase. Nevertheless activated FMN which was chemically recovered by DTT participates in bioluminescence reaction and competes with native substrate for the luciferase active center. In this case the long luminescence was observed, about 25 % from intensity of the luminescence with the native FMN.

Further, we demonstrated, that when native FMN is used as substrate in the bioluminescence reaction without addition of exogenous aldehyde, the formation peroxide is observed (Fig. 1, line 1), as shown by earlier authors.[9] When the activated derivate was used as substrate, the formation of H_2O_2 is not observed (Fig. 1, line 2), but in case activated FMN the bioluminescence was registered. The value of light intensity was 7-10 % as compared with native flavin.

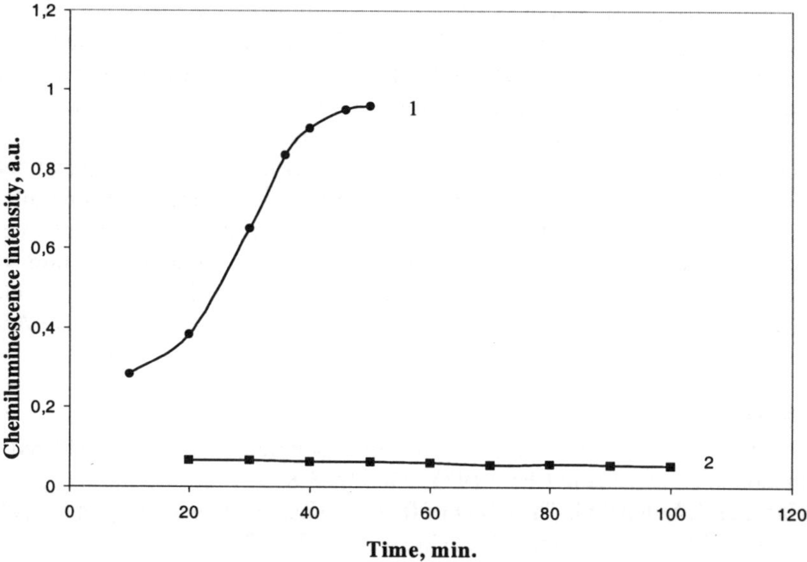

Figure 1. Formation H_2O_2 in luciferase reaction with native FMN (1) and flavin mononucleotide activated by N- methylimidazole on the phosphate (2) without exogenous aldehyde addition

Figure 2. The proposed scheme of flavin mononucleotide activated by N-methylimidazole on the phosphate group

The introduction N-methylimidazole probably has a twofold effect on the properties of the phosphate group (Fig. 2). Replacing one atom of oxygen can result in formation of a new bond. On the other hand redistribution of an electronic density increases electro negativity of free oxygen atom phosphate, which can result in an amplification of electrostatic interactions of the activated substrate with the luciferase. It is impossible to exclude also the formation of a covalent bond between the activated phosphate group and functional groups of luciferase. This possibility is quite highly, due to the available amino acid residuals in the active center being good nucleophile.[10] Thus, the absence of peroxide formation in luciferase reaction without exogenous aldehyde with activated flavin substrate may be the confirmation of derivative covalent binding close to or inside the luciferase active center. And this does not prevent from a light emission.

REFERENCES
1. Baldwin T, Nicoli M, Hastings J. Bacterial luciferase: binding of oxidized flavin mononucleotide. J Biol Chem 1975; 250: 2763-8.
2. Hastings J, Mitchel G, The effect of flavin isomers and analogs upon the color of bacterial bioluminescence. J Biol Chem 1969; 244: 2572-76.
3. Watanabe T, Matsui K, Kasai S, Nakamura T. Studies on luciferase from *Photobacterium phosphoreum*. XI. Interaction of 8-substituted $FMNH_2$ with luciferase. J Biochem 1978; 4: 1441-6.
4. Horizman T, Baldwin T. The effects of phosphate on the structure and stability of the luciferases from *Beneckea harveyi*, *Photobacterium fischeri*, and *Photobacterium phosphoreum*. Biochem Biophys Res Commun 1980; 94:1199-206.
5. Meighen E, Mac-Kenzie R, Flavin specificity of enzyme-substrate intermediates in the bacterial bioluminescent reaction. Structural requirements of the flavin side chain. Biochem 1973; 12: 1482-91.
6. Godovicova T, Zarytova V, Khalimskaya L. Reactive phosphamidates of mono- and dinucleotides. Bioorg Chem 1986; 12, 4: 475-8.
7. Buneva V, Godovicova T, Zarytova V. Modification of RNAse by di- and oligodeoxyribonucleotide 5-phospho-n-methylimidazolide derivatives. Bioorg Chem 1986; 12: 906.
8. Tyulkova N. Purification of bacterial luciferase from *Photobacterium leiognathi* with use FPLS-system. In: Jezowska-Trzebiatowska B. ed. Biological Luminescence. Singapore: World Scientific, 1989: 369 -74.
9. Hastings J, Balny C, The oxygenated bacterial luciferase-flavin intermediate. J Biol Chem 1975; 250, 18: 7288-93.
10. Torchinsky Yu. Ed. Serum in Proteins. Moscow: Mir, 1977: 5-18.

LUMINOUS BACTERIA: BIOTECHNOLOGICAL ASPECTS

GA VYDRYAKOVA, YuV CHUGAEVA, NA TYULKOVA, SE MEDVEDEVA,
AM KUZNETSOV, EK RODICHEVA
*Institute of Biophysics, Russian Academy of Science,
Siberian Branch, Krasnoyarsk, 660036, Russia
E-mail: biotech@ibp.ru*

INTRODUCTION

The Culture Collection of the Institute of Biophysics of the Siberian Branch of the Russian Academy of Sciences (acronym CCIBSO, wdcm 836) is a unique collection. It consists of about 700 strains of luminous bacteria with specific properties and which have been isolated from different regions of the World Ocean.[1,2] They are: mesophiles and psychrophiles, free-living and associated with various marine inhabitants. CC IBSO deposits also genetically modified *Escherichia coli* strains with marker *lux*-gene.

The present work makes a review of investigations of luminous bacteria stored in the Culture Collection IBSO.

LUMINOUS BACTERIA AS PRODUCERS OF ENZYMES

Luminous bacteria from CCIBSO are used in different scientific fields (Fig.1). Wide distribution of luminous bacteria in marine biocenoses is indicative of their possible contribution to transformation of natural polymers, such as chitin, cellulose, alginates, lipids, and other polymers contained in marine algae and animals. When microorganisms degrade particular biopolymers as a source of carbon, they synthesize requisite enzymes of the hydrolase class. Hydrolases are the greatest class of enzymes produced by luminous bacteria. Lipases were found in *V. fischeri, P. leiognathi, V. harveyi*, but not in *P. phosphoreum*. Strains *V. harveyi* produce amylases. Our results support conclusions made by Baumann P and Baumann L.[3]

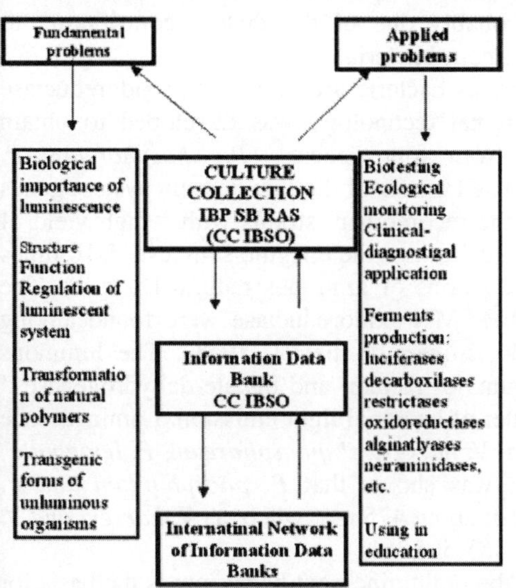

Cellulases are enzymes contributing to biodegradation of algal biomass. Cellulases production has been recorded in representatives of all the four investigated species of luminous bacteria from CC IBSO. Alginate lyase production is recorded in the cultures of luminous bacteria *V. harveyi* IBSO isolated both from algal surface and, in some cases, from water. Chitinases are enzymes participating in degradation of chitin, the polymer which is annually synthesized in great quantities by crustaceans, insects, and unicellular plant organisms. The tested cultures of *V. fischeri, V. splendidus*, and *P. phosphoreum* exhibited no chitinase activity. Only in 17% of the tested strains of *P. leiognathi*, and 25% - *V. harveyi* from CC IBSO chitinase activity was recorded. Luminous bacteria *P. phosphoreum, P. leiognathi* and *V. harveyi* accumulate polyoxybutyrate in intracellular granules.[1,2] That is one of group of polyesters produced by a large number of bacteria. These thermoplastic polymers have drawn great interest since their discovery due to their degradability and the potential to produce them from renewable carbon sources.[4]

The presence of site-specific endodeoxyribonucleases (restrictases) is essential for construction of recombinant DNA molecules, determination of the primary structure of DNA, and some other studies of genetic material. . The producers of the following restriction endonucleases: BanI, HaeIII, MboI, AfIII, Asu I, Pst I, are found among the strains CC IBSO.[5] Other strains can utilize aspartic and glutamic acids, valine, glycine, histidine, proline, and other amino acids.[6]

Most cultures from Collection IBSO produce lyases: L-ornithine, L-arginine, and L-lysine decarboxylases.[6] Neuraminidase (sialidase, or mucopolysaccharide - N-acetylneuraminilhydrolase) is the enzyme of the hydrolase group. As is usual neuraminidase activity is a property of pathogenic organisms. We found for the first time that luminous bacterial cultures of the species *V. harveyi* possess low neuraminidase activity.[7] It may be probably one of the factors contributing to contamination of marine animals by luminous bacteria.

The main target enzymes of luminous bacteria are luciferase, oxidoreductase and different dehydrogenases. The original technology was developed to obtain highly purified bacterial luciferases from four species (*P. phosphoreum, P. leiognathi, V. fischeri and V. harveyi*). The transfer of the luminescent system genes into *E. coli* made it possible to obtain the recombinant strains with a high yield of the luminescent system basic enzyme – luciferase (the enzyme synthesis 5-10 times in comparing with the most productive strains of luminous natural bacteria). The strains with a high activity of NAD(P)H:FMN-oxidoreductase were found among the strains belonging to the species *V. fischeri* of the CC IBSO. The luminous bacteria also contain considerable amounts of lactate- and malate-dehydrogenases.[6] There is relatively little data about emitter of bacterial light emission. Luminescence spectra of luminous bacteria (*V. fischeri, V. harveyi, P. phosphoreum, P. leiognathi*) have been studied in our research. It was shown that *P. phosphoreum and P. leiognathi* wavelength spectrum peaked at about 475 nm, where as *V. harveyi and V. fischeri* wavelength spectrum peaked at 485-505 nm.

The inhibition of light emitted by the bioluminescent bacterium is the basis for different toxicity bioassays. High sensitivity of the luminescent system even to micro-quantities of toxicants, rapid attainment of results, and exact quantification of changes in luminescence level make bioluminescent analysis one of the express methods usable for biological monitoring of the environment.[1, 8-11]

Strains that produce luciferase and oxidoreductase are widely used for development and implement of bioluminescent bioassays. In CC IBSO the basic integral bioassays have been developed to analyse the quality of different chemical solutions, natural waters and effluents.[8,11] Bioluminescent bioassays based on lyophilised marine luminous bacteria *P. phosphoreum* (Microbiosensor-B17-677F) and bacteria *Escherichia coli Z905* with *lux*-gene (Microbiosensor-ECK) are designed as well as the kit for bioluminescent analyses (based on the luminescent system isolated from luminous bacteria). The general characteristics of assays were studied to elucidate possibilities of using in bioluminescent analysis.[8,11] The set of assays can be used to monitor quality of water in the regions with different ecology situations. The analysis of model toxic agents, PMW effluent waters and surface waters was shown that it could be used to analyse the quality of natural waters and effluents of industrial enterprises. Luminescence intensity was measured 10 minutes after addition samples of the waste waters of pulp-and-mill works (PMW) to assay cell suspension. Our studies demonstrate that luminescence is inhibited by 50% at the 10-fold dilution. The plot produced can form the grounds to conclude that the water is not toxic when diluted 1000 times, the bioassay retains 80-120% of its luminescence activity. Both bioassays had the same characteristics as "Microtox" assay of "AZUR Environmental", which was widely used in field and laboratory studies to determine the water contamination of different water reservoirs or the acute toxicity of pure chemical substances, antimicrobial agents, etc.

DATABASE OF CULTURE COLLECTION IBSO

The culture collection IBSO forms a basis for reception of *lux*-genes and genetically modified or transgenic (TM) microorganisms. The CC IBSO is developing an information system "BiolumBase» on basic properties of luminous bacterial cultures, peculiarities of their metabolism, cultivation and storage conditions, and applied use.[12] It contains two sections: 1) natural luminous bacteria and 2) genetically modified or transgenic microorganisms (TM), bearing marker *lux*-genes as the reporter. The complete description of characteristics of the wild types of strains-recipients of *lux*-genes, and vectors and the promoters used for cloning of bioluminescent marks will be contained in "BiolumBase". Such characteristics can be obtained from experimental data and partially from the international databases in existence. The data about the properties will be assembled into the interacting databanks of collections available in the world to provide users free Internet access to the database and meet requirements of employing these bacterial groups in fundamental research, biotechnology, ecological monitoring. There is the work version of DB "BiolumBase" of natural and transgenic luminous microorganisms that are maintained in microbial collections of IBP SB RAS (http://lux.ibp.ru). It contains information about their main morphological, physiological-biochemical and genetic features and bioluminescent systems of different bacteria. The reference guide is to represent general literature on the subject and the authors' publications.

REFERENCES

1. Rodicheva E, Vydryakova G, Medvedeva S. Catalogue of Luminous Bacteria Cultures. Novosibirsk: Nauka, 1997.
2. Gitelson J, Rodicheva E, Medvedeva S et al.: The Luminous Bacteria. Moscow: Nauka, 1984 (in Russian).
3. Baumann P, Baumann L. The Marine Gram-Negative Eubacteria: Genera *Photobacterium, Beneckea, Alteromonas, Pseudomonas*, and *Alcaligenes*: The Prokaryotes. In: Starr M, Stolp H, Trüper H, Balows G, Schlegel H, eds. A Handbook on Habitats, Isolation, and Identification of Bacteria. Berlin, Heidelberg, New York: Springer-Verlag, 1981, vol. II.
4. Steinbuchel A, Valentin H. Diversity of bacterial polyhydroxyalkanoic acids: FEMS Microbiol Lett 1995; 128:219-23.
5. Repin V, Puchkova L, Rodicheva E, Vydryakova G. Luminous bacteria – the producers of specific restriction endonucleases. Microbiology 1995; 64:751-53.
6. Rodicheva E, Medvedeva S, Vydryakova G, Chugaeva Yu, Kuznetsov A. Gene pool maintenance and prospects for using a special-purpose luminous bacteria collection. Appl Biochem Microbiol 1998; 34:75-82.
7. Vydryakova G, Chugaeva Yu, Tyulkova N. Transformation of biopolymers by luminous bacteria. Sib Ecol J 2002; 3:137-44.
8. Kuznetsov A, Rodicheva E, Medvedeva S. Analysis of river water by bioluminescent biotests. Luminescence 1999; 14:263-65.
9. Ulitzur S, Lahav T, Ulitzur N. A novel and sensitive test for rapid determination of water toxicity. Environ Toxicol 2002; 17:291-96.
10. Wang C, Yediler A, Lienert D, Wang Z, Kettrup A. Toxicity evaluation of reactive dyestuffs, auxiliaries and selected effluents in textile finishing industry to luminescent bacteria *Vibrio fischeri.* Chemosphere 2002; 46:339-44.
11. Kuznetsov A, Rodicheva E, Medvedeva S. Using of genetically modified luminous strain Escherichia coli in biotesting. Probl Environ Natur Resourc 2000; 10:67-73.
12. Medvedeva S, Kotov D, Rodicheva E. Database on properties of natural luminous microorganisms. In: Lima N, Smith D. eds. Biological Resource Centres and the Use of Microbes. Braga: Micoteca da Universidade do Minho, 2003: 347-56.

EXPULSION OF SYMBIOTIC LUMINOUS BACTERIA FROM PONY FISH, *LEIOGNATHUS NUCHALIS*

M WADA[1], G BARBARA[1], N MIZUNO[2], N AZUMA[3], K KOGURE[1], Y SUZUKI[2]

[1]*Ocean Research Institute, University of Tokyo, Tokyo, 1648639, Japan*
[2]*Fisheries Laboratory, The University of Tokyo, Shizuoka 431-0211, Japan*
[3]*Agriculture & Life Sciences Faculty, Hirosaki University, Hirosaki 03-8224, Japan*
Email: mwada@ori.u-tokyo.ac.jp

INTRODUCTION

Bacteria and eukaryotes have many mutualistic associations, however a "symbiotic" partnership requires the bacteria to transfer to future host generations with fidelity. This requirement is generally met by either vertical or horizontal symbiont transfer. Vertical transfer is the direct transfer of symbionts to host offspring via the host's reproductive system, whereas horizontal transfer involves host's releasing cells into the environment, and subsequent host offspring infection by released bacteria.

In light organ symbioses, morphological and experimental evidence support horizontal transfer. The light organs (LO) of host organisms, known so far, possess pores through which symbionts are expelled to and/or recruited from surrounding seawater.[1] Experimental evidence, for horizontal transfer of symbionts, on the other hand, is mostly derived from Hawaiian sepiolid squid studies, *Euprymna scolopes* and symbiotic *Vibrio fischeri*.[2,3] Juvenile *E. scolopes* hatch as apo-symbiotic, then become infected by *V. fischeri*. The diurnal rhythm of adult *E. scolopes* controls the daily release of ca. 95 % of their symbiotic bacteria.[4] This daily release reportedly contributes directly to maintaining high bacterial abundance in the natural environment, increasing the chances of successful colonization of host offspring.[4]

In contrast to the bacterial-squid story, little is known about the horizontal transfer of luminous symbionts in bacteria-fish symbioses. Recently, however, we successfully raised *Leiognathus nuchalis*, pony fish juveniles, and infected them with symbiotic *Photobacterium leiognathi*.[5] Juvenile *L. nuchalis* raised separately from adults are apo-symbiotic and therefore not luminous. When the juveniles were kept with adults, a significant number became infected and luminescent. These results strongly support, the horizontal transfer of expelled *P. leiognathi*.

Despite observations that adult *L. nuchalis* turn aquaria seawater luminous, no quantitative assessment of symbiotic cell expulsion from leiognathid fish were made.[5] In this study, we aimed to reveal timing, duration and extent of symbiont expulsion from adult *L. nuchalis* by monitoring light intensity and bacterial numbers in seawater. We also examined the symbiont expulsion from host juveniles infected with *P. leiognathi*.

MATERIALS AND METHODS
Collection and maintenance of *Leiognathus nuchalis*
During 1998-2000, June-July, we collected 48 adult *L. nuchalis* from Hamanako Bay and kept them in 1,000 L aquaria for use in experiments. Seawater from Hamanako Bay (25°C, Sal 30-33 ‰) was introduced into aquaria at 2 L/min, after passing through charcoal and glass fiber filters. Laboratory bred apo-symbiotic juvenile *L. nuchalis* were raised as described.[5] Adults and juveniles were fed newly hatched *Artemia* nauplii and particulate formula feed (Kyowa Co. Japan). Stock aquaria were illuminated 08:30 to 18:00 by fluorescent light (>500 lux).
Experiments with *L. nuchalis* infected by symbiotic luminous *P. leiognathi*
Six adult fish (84-99 mm TL) were starved for 3 d to clear the gut, prior to each trial. Carry-over of bacteria from stock aquaria was reduced by rinsing fish with 10-20 L of SW, before transfer to a 6 L aerated aquarium 30 min prior to the 12D period. To reduce containment stresses, 3 L of freshly prepared SW was carefully exchanged at 47 h. The SW (20 mL) was periodically examined for light emission (Photometer, Model 3000, SAI Tech,Co.), two consecutive readings were averaged to estimate light intensity. In all cases, no light was detected from aquaria without fish. After each reading sample SW was returned to aquaria, to reduce water loss.

Juvenile *L. nuchalis* 82-116 d old (14-32 mm TL) were infected by symbiotic luminous *P. leiognathi*.[5] Only juveniles emitting >10^9 photons/min/fish 24 h prior to trials were used. In the 1st trial 11 juveniles were rinsed with 5 L of SW, and placed in a 2 L aquarium 3.5 h after the light period began. In the 2nd trial, another set of infected juveniles were thoroughly rinsed using 0.22 μm filter-sterilized SW, and each placed into 120 ml of filter-sterilised SW, 2 h after the light period began. Non-luminous juveniles (N>10) were kept as a control under the same conditions.

All trials ran for 3 d; the photo period was 06:00 to 18:00 (12L); controlled by fluorescent light (>500 lux), and dark period 18:00 to 06:00 (12D); aquaria and aeration tubes were pre-sterilized with 70 % ethyl alcohol.
Bacterial numbers from SW, fecal matter and juvenile fish LO
Colony luminescence and counts were made 24 h after spreading aquaria SW samples, fecal matter and light organs, onto SWC minus-glycerol agar (25 °C). [5]

RESULTS AND DISCUSSION
Seawater luminescence was used to examine *L. nuchalis* symbiont expulsion. The light intensity was below detection limit (< 300 photons/min/ml) during 12D, but increased to more than 10^6 photons/min/ml in 12L. This increase in SW luminescence normally occurred a few hours after the start of 12L (Fig 1). Plus, SW from fish kept in darkness for 3 d did not produce detectable light. This suggests adult *L. nuchalis* periodically expel *P. leiognathi* in response to environmental light, which is similar to findings for *E. scolopes.*

Overgrowth of non-luminous bacteria on SWC plates, meant we were unable to obtain direct evidence of adult *L. nuchalis* increasing *P. leiognathi* numbers in SW.

Nonetheless, the consistent and periodic SW light emission in repeated trials can be best explained by the expulsion of the symbiotic luminous bacteria .

Assuming expelled bacteria's luminous ability was the same as that in the light organ, we estimated the number of *P. leiognathi* in SW.[5] The total number of *P. leiognathi* cells in adult *L. nuchalis* light organs, averaged $1.4\pm0.3\times10^7$ cells (n=4), therefore 1% of the LO bacteria may have been expelled.

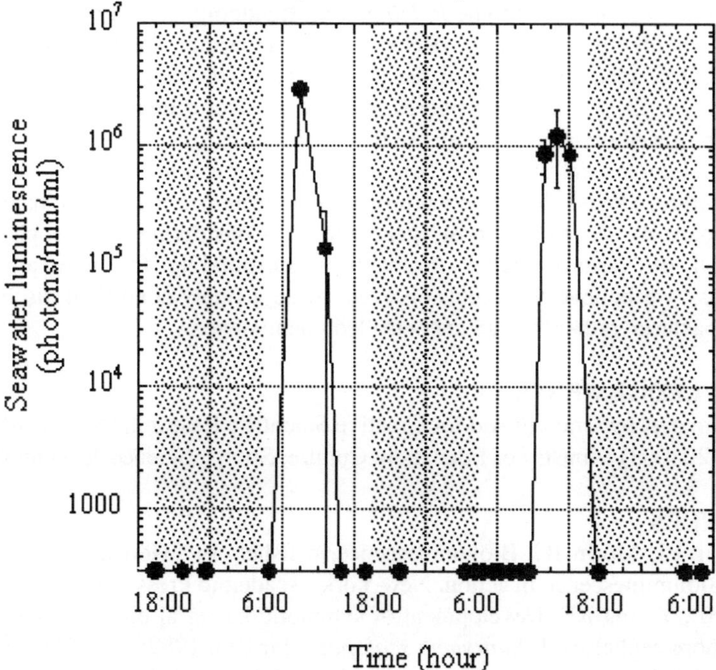

Figure 1. Diurnal pattern of SW luminescence from adult *L. nuchalis*. Shaded areas represent 12D. Bars represent ranges of two consecutive photometer readings.

We also found *L. nuchalis* defecated luminous feces during the 12L period, the release of luminous feces indicate that environmental light may trigger defecation of luminous bacteria. Fecal pellet light emissions were 8.3×10^7 to 2.2×10^{10} photons/min/pellet during 12L and feces collected at mid 12L (11:45) were up to 3,000x brighter than feces at the start of 12L. However, sub-sampled pellet light intensity fell below 10 % initial values, within 3 h and no light was detected from feces collected during 12D (data not shown). Thus, it is unlikely that fecal luminescence was due to growth of luminous bacteria on feces after defecation. Total viable *P. leiognathi* in feces ranged from 4×10^3 to 3.4×10^4 cells/pellet (n=5), up to 3 % of the total viable cells in fecal matter. Therefore just over 5 pellets could deliver 1% of LO *P. leiognathi* to surrounding SW.

Juvenile *L. nuchalis* infected with *P. leiognathi* were found to diurnally expel LO symbionts into SW, in a similar manner to adult *L. nuchalis*. With SW emitting higher luminescence mostly during 12L than in 12D, and SW from non-infected juveniles having no detectable light (data not shown). Light intensity and luminous cell numbers peaked simaltaneously, 5.2×10^5 to 4.1×10^9 photons/min/ml, and luminous bacteria 6×10^1 to 4.3×10^3 cells/mL. More than 70% of colonies that grew were luminous, and appeared to be *P. leiognathi*. Randomly picked colonies were all later positively identified using 16SrRNA. By comparing SW luminous bacteria counts to LO counts, we estimated the percentage of *P. leiognathi* expelled from LOs to be 17 - 43 %. Yet these figures, may be an overestimate since there could have been bacterial division after release from fish.

For bioluminescent symbiotic fish and squid, expulsion of luminous bacteria into surrounding SW is a fundamental process that ensures horizontal transfer of the bacterial partner to future host generations. Considering the sophisticated coordination of various shutter mechanisms surrounding the internal LO of leiognathids,[6] we believe the expulsion event to be a finely controlled discharge of symbiotic bacteria, rather than an uncontrolled phenomenon.

ACKNOWLEDGEMENTS

Supported in part by Grant in Aid for Creative Scientific Research (No. 12NP0201 and No. 14208063) Ministry of Education, Culture, Sports, Science &Technology.

REFERENCES

1. Herring PJ, Morin JG. Bioluminescence in fishes. In : Herring PJ, Morin JG. eds. Bioluminescence in action. New York: Academic Press, 1978; 272-93.
2. Wei SL, Young RE. Development of symbiotic bacterial bioluminescence in a nearshore cephalopod, *Euprymna scolopes*. Mar Biol 1989; 103: 541-46.
3. Ruby EG. Lessons from a cooperative, bacterial-animal association: the *Vibrio fischeri-Euprymna scolopes* light organ symbiosis. Ann Rev Microbiol 1996; 50:591-624.
4. Lee KH, Ruby EG. Effect of squid host on the abundance and distribution of symbiotic *Vibrio fischeri* in nature. Appl Environ Microbiol 1994 60:1565-71.
5. Wada M, Azuma N, Mizuno N, Kurokura H. Transfer of symbiotic luminous bacteria from parental *Leiognathus nuchalis* to their offspring. Mar Biol 1999; 135:683-87.
6. Dunlap PV, McFall-Ngai MJ. Initiation and control of the bioluminescent symbiosis between *Photobacterium leiognathi* and leiognathid fish. In: Lee JJ, Frederick JF. eds. Endocytobiology. III. New York: New York Academy of Sciences, 1987: 269-83.

RFLP ANALYSIS OF THE *LUX* A GENES OF *PHOTOBACTERIUM LEIOGNATHI* ISOLATES DERIVED FROM THE SYMBIOTIC LIGHT ORGAN OF LEIOGNATHID FISH, *LEIOGNATHUS RIVULATUS*

M WADA[1], A KAMIYA[1], K KITA-TSUKAMOTO[1],
K IKEJIMA[2], M NISHIDA[1], K KOGURE[1]

[1] *Ocean Research Institute, The University of Tokyo, Tokyo 164-8639, Japan*
[2] *Asian Institute of Technology, Pathumthani 12120, Thailand*
E-mail: mwada@ori.u-tokyo.ac.jp

INTRODUCTION

Photobacterium leiognathi is a symbiotic luminous bacterium found as a high density, mono-specific culture in the light organ of leiognathid fish.[1,2] *P. leiognathi* is maintained extracellularly within the light organ tubules, that communicate with the esophagus via small ducts, thus allowing the symbiotic bacteria to eventually pass into surrounding seawater.[3] The released cells of *P. leiognathi* in seawater can act as a bacterial inoculum for the apo-symbiotic juvenile fish, presumably entering the juvenile light organ via the ducts.[4] The light organ population of *P. leiognathi* is readily cultivated on artificial media from most leiognathid species. The resultant isolates, even from the same specimen of a single species of leiognathidae, can however be variable in terms of luminescence intensity,[5] and other phenotypic traits.[2, 6] Although the mechanisms by which such strain variations are generated in the symbiotic *P. leiognathi* are unknown, both phenotypic and genetic evidence suggest a multi-clonal nature of the light organ bacterial population.[7] As one of the possible factors affecting such strain variation of the symbiotic bacteria, sex-related dimorphism of the host's light organ should be noted. In many species of leiognathid, sexual dimorphism is apparent in size, shape and pigmentation of the internal light organs, as well as the external skin patch through which the emitted light passes; male organ are much larger or apparent in contrast to the female light organ. [8, 9] The light organ dimorphism can be so significant that > 10-times difference in the light organ volume is seen in a leiognathid species, [9] it may also have impacts on the light organ bacterial population. In order to examine the possibility, we initiated analyses of the luciferase gene (*lux*A) in isolates from the light organ of *L. rivulatus*, one of the leiognathid species that exhibit the most significant sexual dimorphism.

METHODS

Isolation of *P. leiognathi* from the light organ of *L. rivulatus*

Specimens of *L. rivulatus* were collected at a fishing dock in Odawara, Kanagawa prefecture for the 3-month period, from June through August 2002. The light organs were dissected from 8 to 10 specimens of *L. rivulatus,* at each sampling, briefly rinsed with 70 % ethanol, to remove possible non-symbiotic bacteria attached to the external surface of the organ, and each rinsed organ homogenized in 500µL sterile saline (0.8% NaCl). The homogenate was plated on agar of either SWC or half strength of Zobell 2216, and incubated at 25 °C for 3 days. A total of 4 to 20 luminescent colonies per light organ homogenate were randomly picked up from the plates. All the luminescent bacterial colonies were further purified by another plating and kept as glycerol stocks under −80 °C for later use.

DNA extraction and PCR amplification of the *lux*A gene of *P. leiognathi*

The bacterial isolates were cultivated at 25 °C with shaking in 5 mL broth (media described above). One to two mL of the cell suspension was centrifuged for 2 min at 10 °C, 10,000 rpm. DNA was extracted from the cell pellets with DNeasy tissue kit (Quiagen), according to the manufacturer's instruction. Polymerase chain reaction (PCR) amplification of the *lux*A gene was conducted with a primer pair, either 5'-TTCTCATACCAYCCACCAGG-3'(luxAF03) or
5'-CATGATTTGGGCGAAAACCT-3'(luxAF05), and
5'-TCAGAACCGTTTGCTTCAAAACC-3'(luxAR01).
The primer luxAF03 locates 261 bp downstream from the 3' end of luxAF05.

Sequencing and RFLP analysis of the luxA genes

The double-stranded PCR products amplified with a primer pair, luxAF05 and luxAR01, were used for direct cycle sequencing with dye-labelled terminators (Applied Biosystems). Labelled fragments were analysed on a Model 3100 DNA sequencer (Applied Biosystems). The PCR products amplified with luxAF03 and luxAR01 were digested with two restriction enzymes, *Hae*III and *Msp*I for restriction fragment length polymorphism (RFLP) analysis. The enzymes were selected by the aid of a computer program for virtual digestion of the DNA sequences (GENETYX-MAC, ver 11). The digested DNA fragments cut with each enzyme were electrophoresed on a 4.0% NuSieve agarose gel (Cambrex BioScience. Rockland, ME USA), and EtBr stained for band characterization via ultraviolet transillumination.

RESULTS AND DISCUSSION

Sex of all the *L. rivulatus* specimens was easily distinguished by the presence of the external skin patch. Male specimens always exhibited a rectangular-shaped skin patch, which was much darker in coloration than the otherwise silvery lateroventral skin. In contrast, females did not have a visible skin patch on their flank.

In June, the *lux*A gene sequences of the isolates derived from male hosts were found to be clearly distinct from the *lux*A genes of female hosts' isolates. Comparison of the two *lux*A gene sequences allowed us to design a pair of restriction enzymes, *Hae*III and *Msp*I, each of which produced distinct banding patterns between the male and female isolates (Fig. 1). The PCR-amplified luxA gene products from male isolates were not cut by HaeIII, but by MspI, resulting in two major bands, 565bp and 408 bp. While, the luxA gene from female isolates were cut by both *Hae*III and *Msp*I, resulting in bands of 623, 350 bp, and of 458, 417 and 99 bp, respectively.

However, female and male specimens collected in July and August did not show a distinct sex-related separation in the *lux*A-RFLP patterns of their symbiotic *P. leiognathi* isolates (Table 1). Except for 5 bacterial isolates from one male specimen in August, all male isolates showed RFLP patterns that were similar to those found in isolates from females in June. Female isolates showed consistently similar patterns, regardless of the sample month. These results indicate that symbiotic *P. leiognathi* in the light organs of *L. rivulatus* consists of at least two subpopulations with distinctive *lux*A gene sequences, and that the dominance of one subpopulation over another is likely to be related to either sex or seasonality, or both.

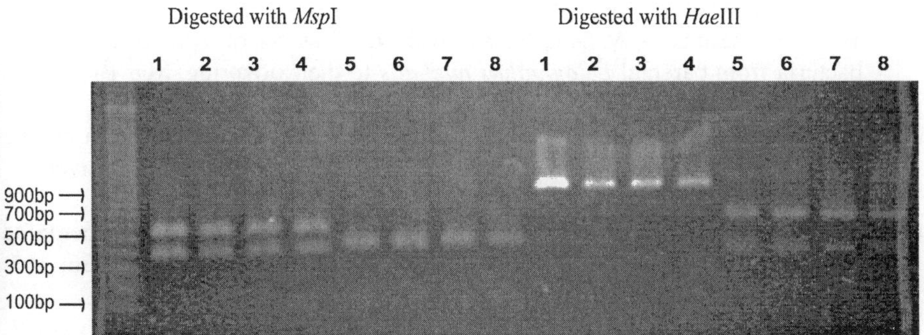

Figure 1. The *lux*A-RFLP patterns of symbiotic *P. leiognathi* from male (lane 1 to lane 4) and female (lane 5 to 8) *L. rivulatus.*

Table 1. Summary of the RFLP analysis of the *lux*A genes in the symbiotic *P. leiognathi* isolates from *L. rivulatus* collected in 2002.

Date	Sex of *L. rivulatus* (No. of Specimens)	No. of isolates tested	No. of bands produced after digestion	
			With *Hae*III	With *Msp*I
June 25	M (4)	4	0	2
	F (4)	4	2	3
July 23	M (4)	21	2	3
		2	2	2
	F (4)	23	2	3
August 26	M (5)	20	2	3
		5	0	2
	F (1)	5	2	3

ACKNOWLEDGEMENTS
We thank Mr. T Yamamoto for helping sample collection. We also thank Dr. G M Barbara for revising the manuscript. This study was supported in part by Ministry of Education, Culture, Sports, Science and Technology, Grant in Aid for Creative Scientific Research (No. 12NP0201 and No. 14208063)

REFERENCES

1   Hastings JW, Mitchell G. Endosymbiotic bioluminescent bacteria from the light organ of pony fish. Biol Bull 1971; 141: 261-8.

2   Reichelt, JL, Nealson KH, Hastings JW. The specificity of symbiosis: pony fish and luminescent bacteria. Arch Microbiol 1977; 112: 157-61.

3   Dunlap PV. Physiological and morphological state of the symbiotic bacteria from light organs of ponyfish. Biol Bull 1984; 167: 410-25.

4   Wada M, Azuma N, Mizuno N, Kurokura H. Transfer of symbiotic luminous bacteria from parental *Leiognathus nuchalis* to their offspring. Mar Biol 1999; 135: 683-7.

5   Silverman M, Martin M, Engebrecht J. Regulation of luminescence in marine bacteria In: Hopwood DA, Chater KF. eds. Genetics of bacterial diversity. London: Academic Press, 1989: 71-86.

6   Dunlap PV, Steinman, HM. Strain variation in bacteriocuprein superoxide dismutase from symbiotic *Photobacterium leiognathi*. J Bacteriol 1986; 165: 393-8.

7   Dunlap PV, Jiemjit A, Ast JC, Pearce MM, Marques RR, Lavilla-Pitogo CR. Genomic polymorphism in symbiotic populations of *Photobacterium leiognathi*. Environ Microbiol 2004; 6: 145-58.

8   Haneda Y, Tsuji F. The luminescent system of pony fishes. J Morphol 1976; 150: 539-52.

9   McFall-Ngai MJ, Dunlap PV. External and internal sexual dimorphism in Leiognathid fishes: morphological evidence for sex-specific bioluminescent signaling. J Morph 1984; 182: 71-83.

SUPPLEMENTS FOR *PHOTOBACTERIUM PHOSPHOREUM* RL-1 CULTURE MEDIUM TO ENHANCE THE LUMINESCENCE ACTIVITY

R YU[1], C IMADA[2], M WADA[3], T KOBAYASHI[2],
N HAMADA-SATO[2], E WATANABE[2]
*[1]Palace Chemical Co., Ltd., 1-11-16 Fukuura, Kanazawa-ku,
Yokohama 236-0004, Japan*
*[2]Graduate School of Marine Science and Technology, Tokyo University of Marine
Science and Technology, Minato-ku, Tokyo 108-8477, Japan*
[3]Ocean Research Institute, University of Tokyo, Nakano-ku, Tokyo 164-8639, Japan
Email: reiko-yu@edu.s.kaiyodai.ac.jp

INTRODUCTION

Luminous bacteria are bioluminescent microorganisms whose luciferase genes (*lux*), proteins and intact cells are widely used in applied research and commercial products. Acknowledging the commercial value of luminescent cells also in entertainment and education, we have conducted research on luminous bacteria from marine samples and have isolated *Photobacterium phosphoreum* (strain RL-1) from coastal marine sediment. In order to maximize the luminescence activity of RL-1, we examined a series of extracts prepared from dried marine foodstuff. Because chitinous compounds and some amino acids are known to be abundant in dried squid and shrimp, we also tested the effects of those compounds on the luminescence activity. Among the supplemental compounds tested, chitosan, cysteine, and aspartic acid were found to enhance the luminescence activity of RL-1. The present results indicate that some amino acids and chitinous compounds are effective supplements for further enhancing bacterial light production in an enriched medium (SWC[1]).

METHODS

Strains and culture conditions

Luminous bacteria were isolated from seawater, sediment and marine organisms around the coastal areas of Japan. Some of the isolates having strong luminescence were identified by their 16S-rDNA sequences. Luminous bacterial isolates were usually cultured in half-strength SWC[1] medium at 20 °C.

Preparation of supplements

The dried marine foodstuff used in the present study included seafood (squid, shrimp, bonito, scallop and sardine) and seaweed (wakame, tangle and hijiki), and the extracts were prepared as follows. One gram of each dried foodstuff was immersed in 100 mL of aged seawater, autoclaved at 121°C for 15 min, and filtered through filter paper to remove insoluble particles. The extracts were used for preparing the half-strength SWC medium instead of aged seawater. Amino acids and chitosan solutions (20 mg/mL) were added to the SWC medium to make a final concentration of 200 μg/mL.

Culture and measurement of luminescence activity

One-hundredth volume of a pre-culture was inoculated into medium containing supplemental material(s), and incubation was carried out overnight at 20 °C. Relative luminescence unit (RLU) and cell density were measured with a luminometer (Gene Lights 55, Microtec Nition Co., Ltd.) and a microplate reader (Model 550, BIO-RAD Co., Ltd.) at a wavelength of 630 nm, respectively.

RESULTS

Effect of marine food extracts on RLU of RL-1

The extracts from squid or shrimp increased the RLU of RL-1, whereas those from seaweed (wakame, tangle and hijiki) had little effect on the RLU (data not shown).

Effect of amino acids and chitosan on RLU of RL-1

Among the various amino acids tested, cysteine and aspartic acid were found to strongly enhance the RLU of RL-1 (Table 1), whereas arginine showed an adverse effect on the luminescence, as reported previously[2] (Table 2). The effects of the amino acids were independent of their isomeric forms (D-type and L-type). Chitosan also enhanced the RLU of RL-1.

Table 1. Effect of amino acids and chitosan on RLU of RL-1

Additive	RLU ($\times 10^8$)	%
SWC	2.46	100
+ L-Cysteine	3.48	141
+ L-Aspartic acid	4.99	203
+ Glycine	2.67	109
+ L-Methionine	2.69	109
+ L-Phenylalanine	2.87	117
+ L-Proline	2.67	109
+ L-Serine	2.54	103
+ L-Taurine	2.88	117
+ L-Sarcosine	2.67	109
+ D-Aspartic acid	4.46	181
+ D-Alanine	2.28	93
+ Chitosan	4.78	194

L-Cysteine, L-aspartic acid and chitosan at concentrations of 300 µg/mL, 200 µg/mL and 500 µg/mL respectively, worked synergistically, resulting in the highest RLU of RL-1 (Fig. 1).

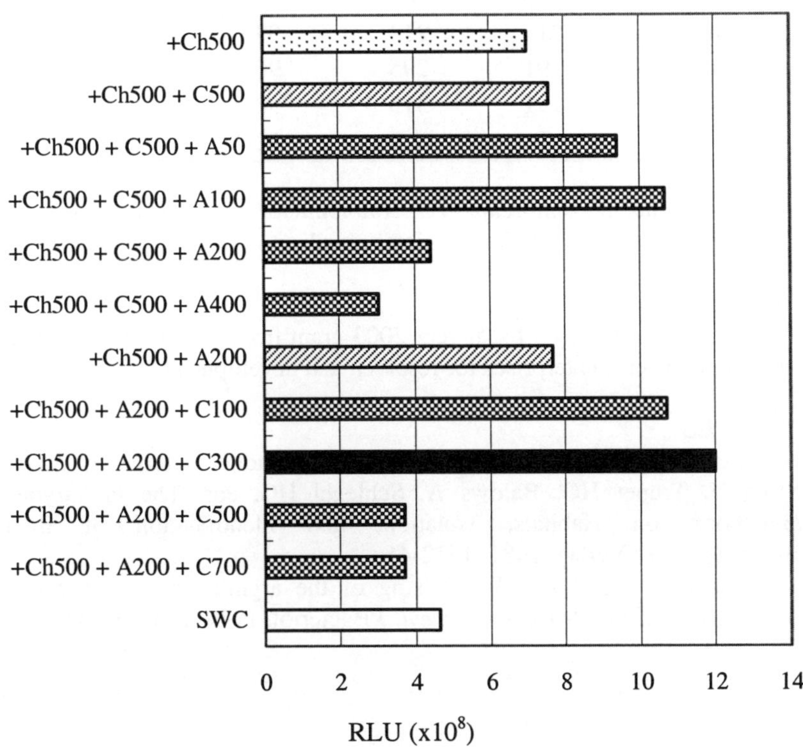

Figure 1. Synergistic effect of supplements on RLU of RL-1

+Ch500: 500µg/mL chitosan added; +Cn: n µg/mL L-cysteine added; +An: n µg/mL L-aspartic acid added.

Effect of amino acids and chitosan on RLU of other strains

Enhancement of luminescence activity by amino acids and chitosan was also observed in another *P. phosphoreum* isolate, RL-27 (Table 2) and type strains NCIMB844 and IFO1396 (data not shown). However, the effects of the supplements were not obvious in the isolates of other luminous bacterial species.

Table 2. Effect of amino acids and chitosan on RLU of RL-1 and other strains

	P.p. RL-1	*P.p.* RL-27	*P.l.* RL-5	*V.f.* RL-12	*V.h.* RL-16
SWC	100	100	100	100	100
+ Chitosan	215	213	99	86	89
+ L-Aspartic acid	191	295	96	155	87
+ L-Cysteine	191	315	96	133	80
+ L-Arginine	14	N.D.	72	130	90

P.p.: *Photobacterium phosphoreum*; *P.l.*: *Photobacterium leiognathi*; *V.f.*: *Vibrio fischeri*; *V.h.*: *Vibrio harveyi*; N.D.: Not determined.

ACKNOWLEDGEMENT

This study was supported by a fiscal year 2003 grant from the City of Yokohama to small and medium-scale enterprises for research and development.

REFERENCES

1. Hastings JW, Nealson KH. The symbiotic luminous bacteria. In: Starr MP, Stolp H, Trüper HG, Balows A, Schlegel HG. eds. The Prokaryotes: A Handbook on Habitats, Isolation, and Identification of Bacteria. Berlin:Springer-Verlag, 1981: 1322-45.
2. Makemson JC, Hastings JW. Poising of the arginine pool and control of bioluminescence in *Beneckea harveyi*. J Bacteriol, 1979, 140: 533-42.

PART 4

CYPRIDINA (VARGULA) BIOLUMINESCENCE

SUPEROXIDE OR SINGLET OXYGEN: THE CHEMILUMINESCENCE OF *CYPRIDINA* LUCIFERIN ANALOGUES IN PHOTODYNAMIC SOLUTIONS

M BANCÍŘOVÁ[1], I ŠNYRYCHOVÁ[2]

[1] *Dept of Physical Chemistry, Faculty of Science, Palacký University, Tř. Svobody 26, 771 46 Olomouc, Czech Republic*
[2] *Dept of Experimental Physics, Faculty of Science, Palacký University, Tř. Svobody 26, 771 46 Olomouc, Czech Republic*
Email: bancir@aix.upol.cz

INTRODUCTION

Reactive oxygen species (ROS) are presently thought to play important role in an increasing number of physiological and pathological processes in living organisms.

The photodynamic effect[1] involves the combination of light, photosensitizer and molecular oxygen. Upon irradiation by laser, the photosensitizer is excited to the first excited singlet state, which can react in two ways. A *Type I* mechanism involves hydrogen-atom abstraction or electron-transfer reactions to yield free radicals and radical ions. A *Type II* results from an energy transfer and generates singlet oxygen. The photodynamic generation of reactive oxygen species is the base of the cancer treatment known as photodynamic therapy (PDT).

Various chemiluminescent compounds (e.g. luminol) have been studied in order to find suitable and specific probes for detection of ROS. Recently *Cypridina* luciferin analogues CLA and MCLA are thought to emit light only when reacting with superoxide anion and singlet oxygen.[2, 3]

One of the sources of singlet oxygen is the chloroplast, where chlorophyll acts as a photosensitizer. It was shown, that minimal photon emission of singlet oxygen (1268 nm) can be observed with illuminated isolated photosystem II reaction centres without secondary electron acceptors. These acceptors under physiological conditions stabilize charge separation of a special pair of chlorophyll molecules in the reaction centre. When the electron transport chain is damaged or when its capacity is insufficient according to the high rate of excited chlorophyll formation, chloroplast is endangered by singlet oxygen. The absorbed energy that cannot be used for charge separation can be transferred by oxygen via excited triplet state of chlorophyll.[4] Naturally occurring carotenoids such as β-carotene are highly important for photosynthetic organisms because they act as protectors against photooxidative damage.

The aim of this work was to study the chemiluminescence of CLA and MCLA in various ROS generating systems and use it in the photodynamic system (e.g. phthalocyanines, chlorophyll *a*) after irradiation by laser.

MATERIAL AND METHODS

Reagents: $CuSO_4$, H_2O_2 luminol (5-amino-2,3-dihydro-1,4-phthalazinedione), CLA (2-methyl-6-phenyl-3,7-dihydroimidazo[1,2-a]pyrazin-3-one), MCLA (2-methyl-6-(p-methoxyphenyl)-3,7-dihydroimidazo[1,2-a]pyrazin-3-one), (Tokyo Kasei Kogyo Co. Ltd.) horseradish peroxidase (HRP), superoxide dismutase (SOD), Trolox, tryptophan, NaN_3 (Sigma-Aldrich, Germany) were used without further purification. $ClAlPcS_2$, $ZnPcS_2$ were synthesized by ing. Rakušan (VÚOS Rybitví).

Sodium hypochlorite (NaOCl) was prepared by reaction of NaOH with chlorine, which was generated by the oxidation of HCl by MnO_4. Its concentration was determined using iodometric titration. The stock solution was kept in small aliquots at −20°C until needed.

Chlorophyll *a* and *β*−carotene were isolated from *Hordeum vulgare* using thin layer chromatography. The pigment extract, which was prepared by the homogenisation of the plant material with $MgCO_3$ in 100% acetone, was purified by centrifugation for 5 min at 10.000 rpm. The supernatant was then separated using thin layer chromatography with mobile phase consisting of technical benzine, isoprophyalcohol and water in the ratio 100:10:0,25. After separation the zones of chlorophyll *a* and *β*-carotene were scratched out and silica gel with the pigment molecules was extracted to 96% ethanol. The suspension was separated by centrifugation and the ethanol extracts of chlorophyll *a* and *β*-carotene were used for further measurements. Both extracts were kept in the dark in the refrigerator.

CL was determined in Fluoroskan *Ascent* FL (Thermo*Labsystems*, Finland) at 25 °C and 37 °C in phosphate buffer pH 7.6. Chemiluminescent reaction was initiated by the injection of hydrogen peroxide in solution of CL compound in the presence of $CuSO_4$ or HRP. The solution of NaOCl is quite unstable at room temperature. In order to minimize the effect of NaOCl decomposition, the stock solution was defrosted immediately before priming the automatic dispenser and solution was kept on ice. The photodynamic effect was initiated by the irradiation of the solution of photosensitizer by a semiconductor laser (Lasotronic pocket therapy laser, power 50 mW, wavelength 670 nm). All manipulations with chlorophyll *a* and *β*-carotene were done in the dark room and the extracts were kept at −20 °C to minimize degradation processes. All measurements were performed in a volume of 200 μL.

RESULTS AND DISCUSSION

Luminol, CLA and MCLA chemiluminescence were studied in three ROS generating chemical systems *in vitro* (HRP - H_2O_2 system, $CuSO_4$ - H_2O_2 system, and NaOCl - H_2O_2 system). Superoxide dismutase and sodium azide were used as specific scavenger of superoxide anion radical and quencher of singlet oxygen, respectively. Their effect was compared with that of Trolox (water soluble analogue of vitamin E), which should scavenge all reactive species present in the reaction mixture, and tryptophan. As the photosensitizers we used phthalocyanines and chlorophyll *a*.

For a quantitative description of the quenching effect of various compounds the comparison of maximal CL intensities was used. The inhibition was determined as $(I_0 - I_Q)/I_0$, where I_0 is an average value (of three independent measurements) of the maximal CL intensities without quencher and I_Q is CL intensity in the presence of quencher. The dependence of the rate of inhibition on the quencher concentration was created by least-square approximation by a cubic polynomial using MATLAB. The value of this polynomial function corresponding to x% inhibition was designated IC_x. Analogically, the enhancement of the CL intensity to x% of the initial intensity was designated SC_x. The results for the effect of SOD, NaN_3 and tryptophan are shown in Table 1.

Table 1. Summary of the quenching effects of SOD, NaN_3 and tryptophan. IC_x is the concentration of quencher corresponding to x% inhibition of the initial intensity. SC_x is the concentration of quencher corresponding to x% inhibition of the initial intensity

	SOD	NaN_3	Tryptophan
Luminol (HRP - H_2O_2)	*enhancement* SC_{120}=28 U	*inhibition* IC_{20}=7.0 mM	*inhibition* IC_{120}=0.07 mM
CLA (HRP - H_2O_2)	*enhancement* SC_{150}=18 U	*enhancement* SC_{150}=4.4 mM	*inhibition* IC_{120}=2.5 mM
MCLA (HRP - H_2O_2)	*enhancement* SC_{150}=16 U	*enhancement* SC_{200}=0.48 mM	*inhibition* IC_{120}=1.9 mM
Luminol (CuSO$_4$ - H_2O_2)	*inhibition* SC_{20}=38 U	*enhancement* SC_{150}=4.8 mM	*inhibition* IC_{120}=0.13 mM
CLA (CuSO$_4$ - H_2O_2)	*change of kinetic*	*enhancement* SC_{150}=1.6 mM	*inhibition* IC_{120}=0.17 mM
MCLA (CuSO$_4$ - H_2O_2)	*change of kinetic*	*enhancement* SC_{150}=0.34 mM	*inhibition* IC_{120}=0.21 mM
Luminol (NaOCl - H_2O_2)	*inhibition* SC_{20}=60 U	*inhibition* IC_{120}=15.0 mM	*inhibition* IC_{120}=0.04 mM
CLA (NaOCl - H_2O_2)	*inhibition* SC_{15}=20 U	*inhibition* IC_{120}=0.08 mM	*inhibition* IC_{120}=0.38 mM
MCLA (NaOCl - H_2O_2)	*inhibition* SC_{15}=87 U	*inhibition* IC_{120}=0.08 mM	*inhibition* IC_{120}=0.37 mM

The HRP - H_2O_2 system: sodium azide acted as a week quencher of luminol CL and the quenching effect was depended on its concentration. It caused significant enhancement of CLA CL as well as of MCLA chemiluminescence.

SOD had only a small influence on luminol CL emission. The addition of SOD led to an expected increase of CLA and MCLA CL intensity.

The CuSO₄ - H₂O₂ system: the presence of NaN_3 led to strong increase of luminol CL, enhanced the CL of CLA and MCLA. The addition of SOD suppressed luminol CL and it did not significantly influence the CL intensity of CLA and MCLA.

The NaOCl - H₂O₂ system: NaN_3 caused the decrease of luminol CL intensity, in case of CLA and MCLA the CL intensity decreased. The presence of SOD led to slight suppression of luminol CL intensity (20%).

The photodynamic effect: the Cl intensity of both *Cypridina* analogues in the presence of irradiated phthalocyanines was very low. The quenching effect of NaN_3 (10 mM) was bigger than quenching effect of SOD (30 U). In case of luminol we were not able to detect CL emission.

CONCLUSIONS

The declared specifity of *Cypridina* luciferin analogues chemiluminescence based on superoxide and singlet oxygen and the suitability of this prove was proved. Both MCLA and CLA are able to visualize the photodynamic effect of irradiated extract of *chlorophyll a*. *β*-carotene suppressed the chemiluminescence of MCLA in the presence *of chlorophyll a* in both cases with and without laser irradiation. It can act as a quencher of singlet oxygen and excited triplet chlorophyll, but it can also react with ROS (especially singlet oxygen) forming xanthophylls [1,5]. All these effects combine together and result in strong antioxidant effect of *β*-carotene and carotenoids in general. It is not clear whether the chemiluminescence of CLA and MCLA is initiated by the reaction with singlet oxygen or superoxide or other reactive oxygen species formed by the photodynamic effect.

ACKNOWLEDGMENTS

This research was supported by the grant from Ministry of Education MSM 153100008.

REFERENCES

1. Bensasson RV, Land EJ, Truscott TG. Excited states and free radicals in biology and medicine, Oxford: Oxford University Press 1993.
2. Nakano M. Determination of superoxide radical and singlet oxygen based of Cypridina luciferin analogs. Methods Enzymol 1990; 186:585-94.
3. Nakano M.: Detection of active oxygen species in biological systems. Cell Mol Neurobiol 1998;18:565-79.
4. Elstner EF, Osswald W. Mechanism of oxygen activation during plant stress. Proc Roy Soc Ediniburgh, 1994;102B:131-54.
5. Montenegro MA, Nazareno MA, Durantini EN, Borsarelli DC. Singlet molecular oxygen quenching ability of carotenoids in a reverse micelle membrane mimetic system. Photochem Photobiol 2002;75:353-61.

DEVELOPMENT OF THE CHEMISTRY OF
THE IMIDAZOPYRAZIONONE-BIOLUMINESCENCE SYSTEM:
FROM THE BIO- AND CHEMILUMINESCENCE MECHANISM
TO A DESIGN OF SENSOR MOLECULES

T HIRANO, S NAKAI, S SEKIGUCHI, S FUJIO, S MAKI, H NIWA

*Dept of Applied Physics and Chemistry, The University of Electro-Communications,
Chofu, Tokyo 182-8585, Japan
Email: hirano@pc.uec.ac.jp*

INTRODUCTION

The imidazo[1,2-a]pyrazin-3(7*H*)-one (imidazopyrazinone) ring system is a core structure of the luminescent substrates isolated from marine bioluminescent organisms, such as the jellyfish *Aequorea* and the crustacean *Cypridina* (*Vargula*). To develop the chemistry of the imidazopyrazinone-bioluminescence system, we have systematically investigated the bio- and chemiluminescent properties of imidazopyrazinone derivatives as well as their physical properties. As the results of our studies, we could clarify the unique π-electronic character of the imidazopyrazinone π-system.[1] In this paper, we will explain the fundamental chemistry of imidazopyrazinone derivatives and will explore the problem of molecular recognition in bioluminescent processes and the problem of the chemiluminescence reaction. In addition to these studies, we found that the π-electronic character of imidazopyrazinone derivatives was sensitive to interactions with molecular environments, such as a hydrogen-bonding interaction and a Lewis acid/base interaction.[1, 2] These interactions with molecular environments caused the continuous spectral change of the imidazopyrazinone derivatives. We will also show that imidazopyrazinone derivatives are useful as sensor molecules for determining the hydrogen-bond donor strength of a solvent and the Lewis acidity of a metal ion.

METHODS

All new compounds were fully characterized by spectroscopic data. UV-visible absorption spectra were measured with a Varian Cary 50 spectrophotometer. Spectroscopic measurements were done by using spectral grade solvents at 25 °C. Semi-empirical MO calculations were carried out with the AM1-COSMO method in the MOPAC package (MOPAC2000 ver. 1.0, Fujitsu Ltd, Tokyo, Japan, 1999).

RESULTS AND DISCUSSION

The basic character of the imidazopyrazinone π-system. The physical properties of imidazopyrazinone derivatives **1–4** have been established by X-ray crystallography, UV/vis absorption spectroscopy, NMR, and a MO calculation. Since the imidazopyrazinone π-system contains an anti-aromatic 1,4-dihydropyrazine ring as a partial structure, imidazopyrazinones show many attractive characteristics. The imidazopyrazinone π-system has a planar ring structure and the weakened carbonyl character of the C3-O10 bond indicates that the imidazopyrazinone π-system has a zwitter-ionic resonance structure (**II**) possessing an aromatic 10 π-electron ring (Fig. 1). Imidazopyrazinone derivatives also underwent hydrogen-bonding interactions in solution. The observed structural characteristics were evaluated with the results of the AM1-COSMO calculations. The MO calculations showed that imidazopyrazinones have a localized negative charge on the O10 (net atomic charge $\approx$ –0.6) and a large dipole moment ($\mu \approx$ 12-15 D). The calculated HOMO levels of imidazopyrazinones are ca. 1.5 eV higher than that of a typical 10 π aromatic compound such as naphthalene. These fundamental properties of imidazopyrazinones closely relate to the bio- and chemiluminescent and the spectroscopic properties.

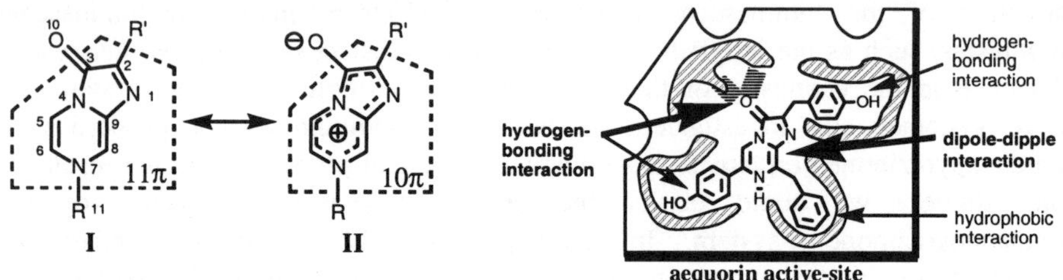

Figure 1. Resonance structures **I** and **II** of the imidazopyrazinone π-system and a supramolecular model of the aequorin active site.

Bio- and chemiluminescence mechanism. Coelenterazine (Cz) and *Cypridina* luciferin (CLn) have been well studied as bioluminescent imidazopyrazinones. Apoaequorin and *Cypridina* luciferase precisely recognize the appendages of these substrates and promote the luminescent reactions with O_2 as enzymatic processes. In addition, we propose that the imidazopyrazinone rings themselves in Cz and CLn play an important role for molecular recognition. The MO calculations indicate that Cz and CLn maintain the typical characteristics of the imidazopyrazinone π-system: Cz and CLn have zwitter-ionic character (**II**) and the ability to make hydrogen-bonding interactions. For the molecular recognitions by apoaequorin and *Cypridina* luciferase, dipole-dipole interactions and hydrogen-bonding interactions work as important attractive interactions as shown in Fig. 1.

For the chemiluminescence reactivity, the imidazopyrazinone π-system has a high HOMO level sufficient to react with triplet molecular oxygen. In the chemiluminescence experiment, an alcohol is employed as a solvent for stock solutions of imidazopyrazinone. The stock solution is mixed with an aprotic solvent such as DMSO containing base under O_2, and this initiates the chemiluminescence reaction. A reason to use the alcohol solvent for stock solutions is explained by stabilization of imidazopyrazinone molecule and its anion due to hydrogen-bonding interactions. Therefore, imidazopyrazinone becomes inert in alcohol solutions and it is easy to handle the stock solution. When the stock solution is mixed with the aprotic solvent containing base, the stabilization effect is reduced, resulting in generation of the reactive naked anion species for the chemiluminescence reaction.

Solvatochromic property – design of an indicator for the hydrogen-bond donor strength of a solvent. Imidazopyrazinones **1–4** show various solution colors from yellow to red. To establish the solvatochromic character, the absorption spectra of **1–4** in various solvents were systematically investigated (Fig. 2A). The wavenumbers E_a of the lowest energy bands correlated with the Kamlet-Taft's α values (Fig. 2B). Thus, it is clarified that the origin of solvatochromism is hydrogen-bonding interactions with solvent molecules. Solvent molecules (**D-H**) with the hydrogen-bond donor part interact with imidazopyrazinone molecules and alter the frontier orbital levels of the imidazopyrazinone π-system. Since the π-system has the zwitterionic character (**II**) possessing a localized negative charge on the O10, the O10 acts as the hydrogen-bond acceptor for hydrogen-bond donor molecules as illustrated in **III**. From these results, we showed that the imidazopyrazinones are applicable as an indicator of the hydrogen-bond donor strength of a solvent.

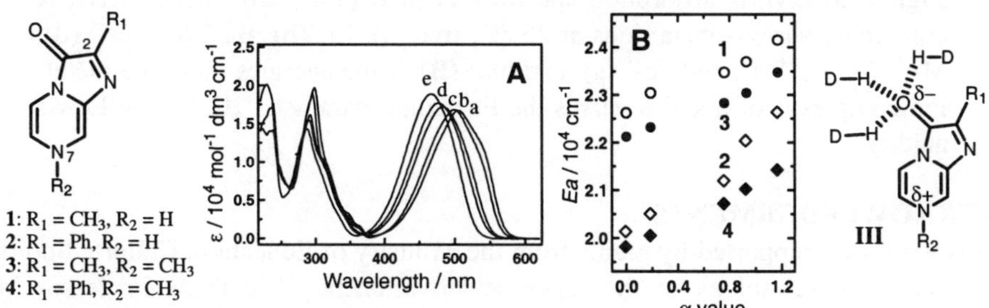

Figure 2. UV/vis absorption spectra (**A**) of **4** in various solvents; (a) DMSO, (b) CH_3CN, (c) 2-propanol, (d) CH_3OH, and (e) H_2O, and plots (**B**) of wavenumbers Ea of the lowest energy bands for **1–4** against the Kamlet-Taft's α value.

Metal-ion complexation – design of an indicator for the Lewis acidity of metal ions. Complexation of **5** and **6** with various metal ions (Li^+, Mg^{2+}, Ca^{2+}, Ba^{2+}, Sc^{3+}, and La^{3+}) was confirmed by spectral measurements in acetonitrile (Fig. 3A). We found the linear relationships between the energies E_a (eV) of the lowest energy absorption bands for the complexes and the Fukuzumi parameter ΔE (eV) for the Lewis acidity of the metal ions (Fig. 3B). With increasing Lewis acidity of the metal ion, the lowest energy bands exhibited a blue shift. It is clear that **5** and **6** act as

Lewis bases toward the metal ions. The O10 must be the actual center of the Lewis base and coordinates to the metal ion, because of the zwitter-ionic character (**II**) of an imidazopyrazinone. An enhancement of the Lewis acid/base interaction in the metal-ion complexes causes changes in the π-electronic character and increases the aromatic imidazopyrazine property, resulting in the blue shift of the lowest energy bands. Therefore, imidazopyrainone derivatives are potentially applicable as indicators for the Lewis acidity of metal ions. The design of 1,2- and 1,3-bis(2-phenylimidazopyzinon-7-ylmethyl)benzene derivatives demonstrated that complexation with the metal ion is enhanced by chelate effects.

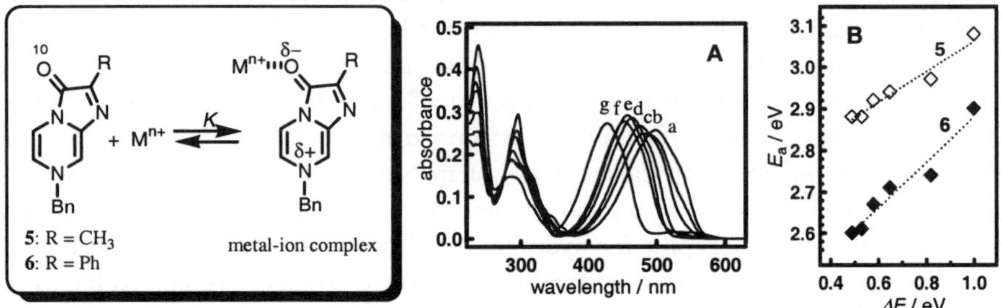

Figure 3. UV/vis absorption spectra (**A**) of **6** (1.4×10^{-5} M) in CH_3CN containing various metal ions at 25 °C; free (a), Li^+ (b), Ba^{2+} (c), Ca^{2+} (d), Mg^{2+} (e), La^{3+} (f), and Sc^{3+} (g), and plot (**B**) of the energies Ea of the metal-ion complexes of **5** and **6** versus the Fukuzumi parameter ΔE for the Lewis acidity.

ACKNOWLEDGEMENTS

This work was supported by grants from the Ministry of Education, Culture, Sports, Science and Technology of the Japanese Government. We thank Professor M. Ohashi (Kanagawa University), Dr. H. Ikeda (Tohoku University), and Professor T. Miyashi (Tohoku University) for kind discussions.

REFERENCES

1. Nakai S, Yasui M, Nakazato M, Iwasaki F, Maki S, Niwa H, Ohashi M, Hirano T. Fundamental studies on the structures and spectroscopic properties of imidazo[1,2-*a*]pyrazin-3(7*H*)-one derivatives. Bull Chem Soc Jpn 2003; 76: 2361-87.

2. Sekiguchi T, Maki S, Niwa H, Ikeda H, Hirano T. Metal-ion complexation of imidazo[1,2-*a*]pyrazin-3(7*H*)-ones: continuous changes in absorption spectra of complexes depending on the Lewis acidity of a metal ion. Tetrahedron Lett 2004; 45: 1065-9.

BIOSYNTHESIS OF CYPRIDINA LUCIFERIN FROM FREE AMINO ACIDS IN *CYPRIDINA (VARGULA) HILGENDORFII*

S KATO[1], Y OBA[1], M OJIKA[1], S INOUYE[2]

[1]Graduate School of Bioagricultural Sciences, Nagoya University, Chikusa-ku, Nagoya 464-8601, Japan
[2]Yokohama Research Center, Chisso Co., 5-1 Okawa, Kanazawa-ku, Yokohama 236-8605, Japan
Email: i033005d@mbox.nagoya-u.ac.jp

INTRODUCTION

The luminous marine ostracod *Cypridina hilgendorfii* (presently *Vargula hilgendorfii*) lives in the Japanese coast. The luminescence system of *V. hilgendorfii* has been investigated extensively, since Harvey reported the luciferin-luciferase reaction in 1917.[1] When the specimen is stimulated physically or electronically, it expels Cypridina luciferin and luciferase directly into the seawater to produce a brilliant bluish luminescence (λ_{max} = 460 nm). The isolation, structural determination and total synthesis of Cypridina luciferin have been achieved by Kishi et al.,[2,3] and they proposed that Cypridina luciferin may be biosynthesized from three amino acids or their equivalents: arginine, isoleucine and tryptophan (or tryptamine).[3,4] However the biosynthetic pathway of Cypridina luciferin has not been proven. Recently, we identified that L-tryptophan is a synthetic component of Cypridina luciferin by feeding experiments using deuterium labeled L-tryptophan.[5] In this study, other possible amino acids were examined and we concluded that Cypridina luciferin is biosynthesized from L-tryptophan, L-arginine, L-isoleucine, but not tryptamine.

MATERIALS AND METHODS

Animals. The specimens of *V. hilgendorfii* were collected at night using porcine liver as bait at Mukaishima, Hiroshima in Japan on 27 Sept. 2001, 20 Dec. 2001 and 7 Apr. 2004.

Labeled amino acides. [D_5]-L-Tryptophan and [D_5]-tryptamin were prepared by the deuterium-exchange method as previously described.[5] [$^{13}C_6$]-L-Isoleucine and [$^{15}N_2$]-L-arginine were obtained from Cambridge Isotope Laboratories and Spectra Stable Isotopes, respectively.

Feeding experiment. Feeding procedures were essentially same as previously reported.[5] Briefly, the stable isotope labeled amino acids and tryptamine were gelled by agarose (Type VII: Sigma) and the gel was fed to the specimens in a small dish. After feeding over 10 days, Cypridina luciferin was extracted from 4 animals with ethanol and the extracts were served to LC/ESI-TOF-MS analyses.

LC/ESI-TOF-MS. LC/ESI-TOF-MS was performed with an Agilent 1100 HPLC system (Hewlett-Packard) connected to a Mariner Biospectrometry (Applied Biosystems). HPLC condition: column, Cadenza CD-C18 (2.0 x 75 mm, Intakt); mobile phase, 25-65% MeOH (containing 0.1% formic acid) in 20 min; flow rate 0.2 mL/min, split ratio 40:1 (5 µl/min to MS); monitor at 280 nm; ESI-TOF-MS positive mode.

RESULTS AND DISCUSSION

ESI-TOF-MS analyses for natural and synthetic Cypridina luciferin. For natural and synthetic Cypridina luciferins, the monovalent and divalent ions peaks were observed.   Also the ion peaks of luciferinol[6] and luciferyl methyl ether[6] were detected (Table 1). They were generated with solvent during the measurement by ESI-TOF-MS analysis.

Table 1. Relative intensity of mass peaks in natural
and synthetic Cypridina luciferin by ESI-TOF-MS

Compounds (Mass ions)	Ion state	m/z (relative intensity, %)		Calculated
		Natural (found)	Synthetic (found)	
(monovalent)				
Cypridina luciferin	$[M]^+$	405.227 (100)	405.226 (100)	405.228 (100)
	$[M+1]^+$	406.233 (39.6)	406.227 (51.6)	406.230 (27.4)
Luciferinol	$[M+H]^+$	422.230 (18.3)	422.228 (29.4)	422.230
Luciferyl methyl ether	$[M+H]^+$	436.242 (45.0)	436.245 (30.8)	436.246
(divalent)				
Cypridina luciferin	$[M+2H]^{2+}$	203.619 (100)	203.614 (100)	203.621 (100)
	$[M+1+2H]^{2+}$	204.120 (27.5)	204.120 (29.8)	204.123 (27.4)
Luciferinol	$[M+2H]^{2+}$	211.615 (27.1)	211.616 (13.3)	211.619
Luciferyl methyl ether	$[M+2H]^{2+}$	218.623 (11.5)	-	218.626

The mass value of monovalent ion corresponding to Cypridina luciferin was mainly observed at m/z 405 as $[M]^+$ in the positive mode. The intensities of isotopic signals $[M+1]^+$ for natural and synthetic Cypridina luciferin were inconsistent with the calculated mass value. On the other hand, the divalent ion was observed at m/z 203.6 as $[M+2H]^{2+}$ and the intensities of isotopic signals were good agreement with that of the calculated mass value. The similar signal pattern was observed by MALDI-TOF-MS (data not shown) and FD-MS.[7] Thus, the divalent mass value for Cypridina luciferin was used for detecting Cypridina luciferin.

ESI-TOF-MS analyses for the stable iostope labeled compounds. To confirm the isotopic purity of the labeled compounds, $[D_5]$-L-tryptophan, $[D_5]$-tryptamine, $[^{13}C_6]$-L-isoleucine and $[^{15}N_2]$-L-arginine were analyzed by ESI-TOF-MS (Table 2).

Table 2. Relative intensity of mass peaks in stable
isotope labeled compounds by ESI-TOF-MS

Numbers of	Relative intensity (%)			
stable isotope	$[D_5]$-Trp	$[^{15}N_2]$-Arg	$[^{13}C_6]$-Ile	$[D_5]$-Tryptamine
+0				
+1				
+2	0.9	100.0		
+3	5.8	10.2		3.3
+4	36.1			21.6
+5	100.0		6.6	100.0
+6	13.8		100.0	9.4
+7			1.1	

LC/ESI-TOF-MS analysis of Cypridina luciferin extracted from feeding animals. After feeding over 10 days, 4 specimens were extracted by ethanol and the incorporation of the stable isotopes from amino acids was identified by LC/ESI-TOF-MS. The retention time of Cypridina luciferin on HPLC was confirmed using *dl*-synthetic Cypridina luciferin.[7] The mass spectral analyses indicated that $[D_5]$-L-tryptophan, $[^{13}C_6]$-L-isoleucine and $[^{15}N_2]$-L-arginine were all incorporated into Cypridina luciferin as a synthetic component. On the other hand, $[D_5]$-tryptamine did not incorporate into Cypridina luciferin at all. Thus, Cypridina luciferin is biosynthesized from L-tryptophan, L-arginine, L-isoleucine, but not tryptamine in living animals. The labeling efficiencies in Cypridina luciferin for $[D_5]$-L-tryptophan, $[^{15}N_2]$-L-arginine and $[^{13}C_6]$-L-isoleucine and in living animal were estimated approximately 19.1, 7.2 and 10.1%, respectively.

Table 3. Relative intensity of mass peaks in Cypridina luciferin by LC/ESI-TOF-MS

m/z	Relative intensity (%)				
$[M+2H]^{2+}$	Natural[a]	$[D_5]$-Trp[b]	$[^{15}N_2]$-Arg[b]	$[^{13}C_6]$-Ile[c]	$[D_5]$-Tryptamine[d]
203.6 (0)[e]	100.0	100.0	100.0	100.0	100.0
204.1 (+1)	27.5	25.1	28.5	25.5	30.1
204.6 (+2)	6.8	6.9	16.2	7.0	9.8
205.1 (+3)		2.9			
205.6 (+4)		6.3			
206.1 (+5)		16.3			
206.6 (+6)		5.6		11.5	
207.1 (+7)				3.4	

Underline shows the significant signal after feeding. [a] Without incorporation study. [b] Feeding for 10 days. [c] Feeding for 15 days. [d] Feeding for 11 days. [e] Values in parenthesis indicate the number of stable isotopic atoms.

                                                          Kato S et al.

In summary, after feeding the stable isotope labeled compounds to *V. hilgendorfii*, LC/ESI-TOF-MS analyses of Cypridina luciferin extracted from the specimens strongly suggested that three amino acids of L-tryptophan, L-arginine and L-isoleucine participate in biosynthesis of Cypridina luciferin (Scheme 1).

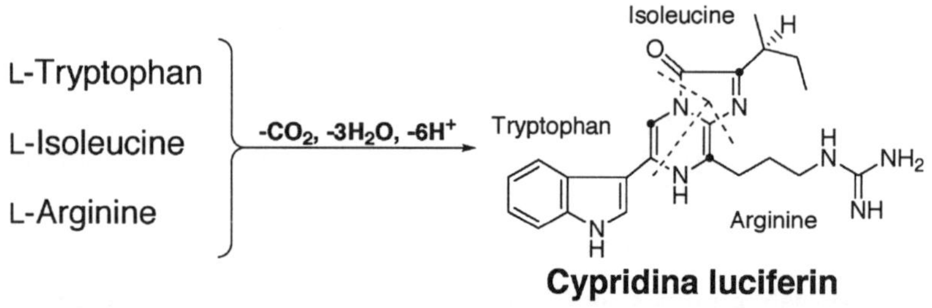

Scheme 1. Biosynthesis of Cypridina luciferin from three amino acids

ACKNOWLEDGEMENTS

We thank Drs. H. Michibata and T. Ueki, Mukaishima Marine Biological Laboratory, Hiroshima University, Japan for helping to collect animals. S.K. thanks JSPS Research Fellowship for Young Scientists.

REFERENCES

1. Harvey EN. Studies on bioluminescence. IV. The chemistry of light production in a Japanese ostracod crutacean, *Cypridina hilgendorfii*, Müller. Am J Physiol. 1917; 42: 318-41.
2. Kishi Y, Goto T, Hirata Y, Shimomura O, Johnson FH. Cypridina bioluminescence I. structure of Cypridina luciferin. Tetrahedron Lett 1966; 7: 3427-36.
3. Kishi Y, Goto T, Inoue S, Sugiura S, Kishimoto H. Cypridina bioluminescence III. total synthesis of Cypridina luciferin. Tetrahedron Lett 1966; 7: 3445-50.
4. McCapra F. Roth M. Cyclisation of a dehydropeptide derivative: a model for Cypridina luciferin. J C S Chem Commun 1972; 894-5.
5. Oba Y, Kato S, Ojika M, Inouye S. Biosynthesis of luciferin in the sea firefly, *Cypridina hilgendorfii*: L-tryptophan is a component in Cypridina luciferin. Tetrahedron Lett 2002; 43: 2389-92.
6. Toya Y, Nakatsuka S, Goto T. Structure of Cypridina luciferinol, "Reversibly oxidized Cypridina luciferin". Tetrahedron Lett 1983; 51: 5753-6.
7. Nakamura H, Aizawa M, Takeuchi D, Murai A, Shimomura O. Convergent and short-step syntheses of *dl*-Cypridina luciferin and its analogues based on Pd-mediated cross couplings. Tetrahedron Lett 2000; 41: 2185-8.

STUDIES ON THE CHEMILUMINESCENCE MECHANISM OF *CYPRIDINA* LUCIFERIN ANALOGUES: DISSOCIATION CONSTANTS OF THE SINGLET-EXCITED *CYPRIDINA* OXYLUCIFERIN ANALOGUES

R SAITO, E IWASA, A KATOH

[1]*Dept of Applied Chemistry, Seikei University, Musashino 180-8633, Japan*
Email: saito@ch.seikei.ac.jp

INTRODUCTION

The chemiluminescent reaction of *Cypridina* luciferin analogue (CLA), 2-methyl-6-phenyl-imidazo[1,2-*a*]pyrazin-3(7*H*)-one, with molecular oxygen or other reactive oxygen species in aqueous media gives two light emitting species, the singlet-excited state of 2-acetamido-5-phenylpyrazine, *Cypridina* oxyluciferin analogue (OCLA), [1](4c)*, and its conjugate base, [1](4c⁻)*, as shown in Scheme 1.[1] The ratio of these two species is affected by the medium pH.[1] Also a substituent at the 6-position of the imidazopyrazinone ring influences the ratio.[2] As the molecular mechanism for the formation of these emitting species, it has long been believed that the anionic [1](4c⁻)* is formed primarily after decomposition of the dioxetane intermediate, and then it is protonated to form the neutral [1](4c)* (**A** in Scheme 1).[1,3] Recently, this mechanism has become disputable due to the suggestion about the other possible protonation steps (**B** or **C** in Scheme 1).[4,5] To elucidate whether the protonation occurs in the excited states or not, pK_a value for the N-H dissociation in the singlet-excited state (pK_a*) of **4c** should be required. However, there has been no attempt to estimate the pK_a* value so far. The aim of the present study was to evaluate the plausibility of the protonation in the excited state. We report here syntheses and pK_a* values of **4c** analogues. Substituent effect on the pK_a* values will be also discussed for more quantitative understanding of the photophysical natures of the excited species responsible for the luminescence of the imidazopyrazinones.

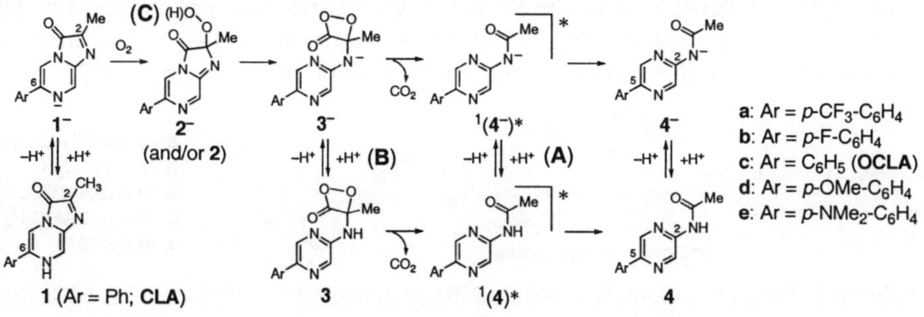

Scheme 1. Postulated chemiluminescence mechanism for *Cypridina* luciferin analogue (CLA) and its derivatives. The parentheses donate possible protonations.

MATERIALS AND METHODS

All chemicals except synthetic materials were commercially available and used as it was. UV-visible spectra were measured with a JASCO V-530 spectrophotometer. Fluorescence spectra were recorded on a JASCO FP-777 fluorescence spectrophotometer. Photometric titration was performed by measuring absorption spectra of **4a-e** in the Britton-Robinson buffers at various pH at 20 °C. The samples were prepared by mixing a 1.0 mM solution of **4** in DMSO (100 µL) and the Britton-Robinson buffer (1.9 mL).

2-Acetamido-5-(4-trifluoromethylphenyl)pyrazine (4a): mp 229-230 °C; ^{1}H-NMR (400 MHz, CDCl$_3$) δ/ppm 2.29 (3H, s), 7.75 (2H, d, J = 8.6 Hz), 7.96 (1H, broad s), 8.11 (2H, d, J = 8.6 Hz), 8.71 (1H, s), and 9.81 (1H, s). Anal Calcd for $C_{13}H_{10}N_3O$: C, 55.52; H, 3.58; N, 14.94. Found: C, 55.42; H, 3.30; N, 14.81.

2-Acetamido-5-(4-fluorophenyl)pyrazine (4b): mp 199-201 °C; ^{1}H-NMR (400 MHz, CDCl$_3$) δ/ppm 2.28 (3H, s), 7.18 (2H, t, J = 9.0 Hz), 7.88 (1H, broad s), 7.98 (2H, dd, J = 5.4 and 9.0Hz), 8.63 (1H, s), and 9.55 (1H, s). Anal Calcd for $C_{12}H_{10}FN_3O$: C, 62.33; H, 4.36; N, 18.17. Found: C, 62.29; H, 4.19; N, 18.05.

2-Acetamido-5-phenylpyrazine (4c): mp 164-166 °C; ^{1}H-NMR (400 MHz, CDCl$_3$) δ/ppm 2.28 (3H, s), 7.44-7.52 (3H, m), 7.89 (1H, broad s), 7.98 (2H, d, J = 7.1 Hz), 8.67 (1H, s), and 9.57 (1H, s). Anal Calcd for $C_{12}H_{11}N_3O$: C, 67.59; H, 5.20; N, 19.71. Found: C, 67.34; H, 5.13; N, 19.52.

2-Acetamido-5-(4-methoxyphenyl)pyrazine (4d): mp 201-203.5 °C; ^{1}H-NMR (400 MHz, CDCl$_3$) δ/ppm 2.27 (3H, s), 3.87 (3H, s), 7.02 (2H, d, J = 8.9 Hz), 7.85 (1H, broad s), 7.93 (2H, d, J = 8.9 Hz), 8.61 (1H, s), and 9.52 (1H, s). Anal Calcd for $C_{13}H_{13}N_3O_2$: C, 64.19; H, 5.39; N, 17.27. Found: C, 64.11; H, 5.34; N, 17.18.

2-Acetamido-5-(4-dimethylaminophenyl)pyrazine (4e): mp 246-247 °C; ^{1}H-NMR (400 MHz, CDCl$_3$) δ/ppm 2.26 (3H, s), 3.04 (6H, s), 6.80 (2H, d, J = 8.9 Hz), 7.79 (1H, broad s), 7.89 (2H, d, J = 8.9 Hz), 8.58 (1H, s), and 9.47 (1H, s). Anal Calcd for $C_{14}H_{16}N_4O$: C, 65.61; H, 6.29; N, 21.86. Found: C, 65.47; H, 6.28; N, 21.64.

RESULTS AND DISCUSSION

Synthesis. 2-Acetamido-5-arylpyrazines (**4a-e**) were synthesized by acetylation of 2-amino-pyrazines (**8a-e**) as shown in Scheme 2. Compounds **8a-e** were prepared by the palladium catalyzed cross-coupling of phenylboronic acids (**7a-e**) and 2-amino-5-bromopyrazine (**6**), which was prepared from the bromination of 2-aminopyrazine (**5**).

Scheme 2. *Reagents and conditions*: i) Bu$_4$NBr$_3$, pyridine, CHCl$_3$, 0 °C-r.t., 11 h, 62%; ii) PdCl$_2$(PPh$_3$)$_2$, PPh$_3$, 2 M Na$_2$CO$_3$, dioxane, reflux, 4-22 h, 55-93%; iii) acetyl chloride, pyridine, CHCl$_3$, 0 °C, 30 min, 25-68%.

Dissociation constants of 4a-e in the ground and excited states. The dissociation constants of **4a-e** in the ground states (pK_a) were measured by means of the photometric titration, and the results are listed in Table 1. The pK_a values were obtained in a range of 12.36-13.09 and

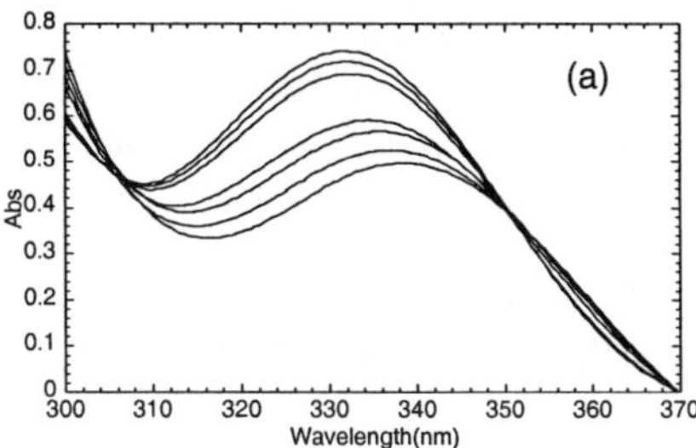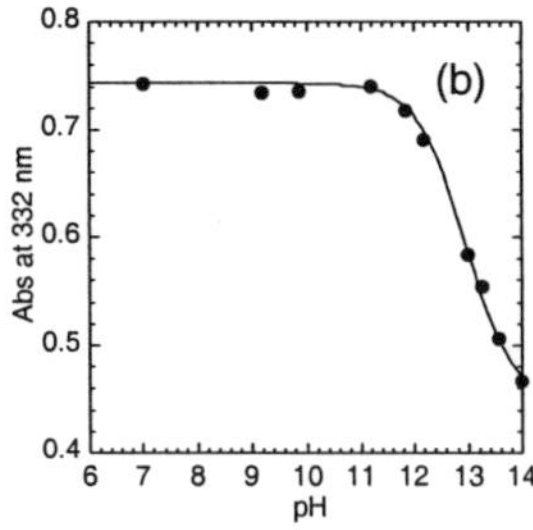

Figure 1. (a) Selected absorption spectra of **4d** in the Britton-Robinson buffers with various pH, and (b) selected plots of the absorbance at 332 nm against pH for **4d**.

increased with increasing the electron-donating nature of the *p*-substituent (R) on the 5-phenyl.

Generally, the dissociation constants of organic molecules in the excited states (pK_a^*) are deduced from the Förster cycle based on absorption and fluorescence data.[6,7] Thus, the pK_a^* values for the N-H dissociation in **4a-e** were estimated by using the following equation,

$$pK_a^* = pK_a - \frac{N_A hc\,(\Delta E_n - \Delta E_{an})}{2.303RT}$$

where N_A is the Avogadro number, h is the Planck's constant, c is the velocity of light, ΔE_n and ΔE_{an} are the 0-0 transition energy of protonated and deprotonated forms, respectively. The mean of the wavenumbers of absorption and fluorescence maxima was taken as the 0-0 energy. The results are given in Table 1, together with the 0-0 transition energies for both fully protonated and deprotonated forms. The pK_a^* values were estimated in a range of 11.90-12.86 and increased with increasing the electron-donating character of R as seen in pK_a. Owing to the non-fluorescent property in aqueous media,[8] the pK_a^* value for **4e** could not be determined. According to the Förster's theory, pK_a^* is lower than pK_a as seen in aromatic alcohols.[9,10] As expected, each compound showed the pK_a^* value smaller than pK_a. It is noteworthy that the difference between pK_a and pK_a^*, ΔpK_a, for **4a** (0.46) was much larger than that for **4d** (0.03). It was reported that 2-acetamidopyrazines possessing the electron-donating R, such as OMe or NMe_2, have CT character in the excited states such that they show specific solvatochromic fluorescence, while the analogues having the electron-withdrawing R, such as CF_3 or F, do not.[8] This apparent difference in the excited-state polarity seems to have much to do with the observed ΔpK_a variation.

In the chemiluminescent reaction of CLA in aqueous media, light emission arises from fully protonated ^{1}(**4c**)* at pH 7.0, while at higher pH over 8.5 the light emitter is deprotonated ^{1}(**4c**⁻)*.[1] If this reaction follows the proposed mechanism, i.e. the protonation occurs at the step A in Scheme 1, the pK_a^* of **4c** is supposed to be around 8.5. However, the actual value is 12.22 and much larger than the predicted value. Therefore, it is conceivable that the proton acceptor in the reaction is a less basic anion such as hydroperoxide anion.

CONCLUSION

We have demonstrated the N-H dissociation constants of 2-acetamido-5-arylpyrazines in the

                                                                    Saito R et al.

ground and singlet-excited states. Their pK_a^* were obtained in a range of 11.90-12.86. Judging from this result, it seems to be reasonable that the protonation does not occur mainly to the singlet-excited state amide anion. The protonation equilibrium may take place in the earlier stage such as step B in Scheme 1 or the initial stage of the reaction.

Table 1. Dissociation constants in the ground (pK_a) and singlet-excited states (pK_a^*) and the 0-0 energies of **4a-e** in Britton-Robinson buffers at 20 °C.

Compound (R)	pK_a	0-0 transition energy $/10^3 cm^{-1}$		pK_a^*
		Neutral species	Anion species	
4a (CF$_3$)	12.36	28.86	26.30	11.90
4b (F)	12.45	28.23	26.15	12.08
4c (H)	12.61	28.39	26.18	12.22
4d (OMe)	12.89	26.04	25.85	12.86
4e (NMe$_2$)	13.09	$-^{a)}$	$-^{a)}$	$-^{a)}$

a) not determined

REFERENCES

1. Fujimori K, Nakajima H, Akutsu K, Mitani M. Chemiluminescence of *Cypridina* luciferin analogues part 1. Effect of pH on rates of spontaneous autoxidation of CLA in aqueous buffer solutions. J Chem Soc Perkin Trans 2 1993: 2405-9.
2. Saito R, Hirano T, Niwa H, Ohashi M, unpublished data.
3. Goto T, Inoue S, Sugiura S, Nishikawa K, Isobe M, Abe Y. *Cypridina* bioluminescence V. Structure of emitting species in the luminescence of *Cypridina* luciferin and its related compounds. Tetrahedron Lett 1968; 37: 4035-8.
4. Usami K, Isobe M. Low-temperature photooxygenation of coelenterate luciferin analog synthesis and proof of 1,2-dioxetanone as luminescence intermediate. Tetrahedron Lett 1996; 52: 12061-90.
5. Teranishi K, Hisamatsu M, Yamada T. Synthesis and chemiluminescence properties of the peroxy acid compound as an intermediate of coelenterate luciferin luminescence 1997; 38: 2689-92.
6. Förster T. Elektrolytische dissoziation angeregter molecule. Z Electrochem 1950; 54: 42-6.
7. Weller A. Quantitative untersuchungen der fluoreszenzumwandlung bei naphtholen. Z Electrochem 1952; 56: 662-8.
8. Saito R, Hirano T, Niwa H, Ohashi M. Solvent and substituent effect on the fluorescent properties of coelenteramide analogues. J Chem Soc Perkin Trans 2 1997: 1711-6.
9. Stryer L. Excited state proton-transfer reactions. A deuterium isotope effect on fluorescence. J Am Chem Soc 1966; 88: 5708-12.
10. Wolfbeis O S, Koller E, Hochmuth P. The unusually strong effect of a 4-cyano group upon electronic spectra and dissociation constants of 3-substituted 7-hydroxycoumarins. Bull Chem Soc Jpn 1985; 58: 731-4.

BIOSYNTHESIS OF *VARGULA HILGENDORFII* LUCIFERIN, ARISEN FROM L-ARGININE, L-TRYPTOPHAN, AND L-ISOLEUCINE

Y TOYA

*Laboratory of Organic Chemistry, Aichi University of Education,
Kariya 448-8542, Japan
Email: ytoya@auecc.aichi-edu.ac.jp*

INTRODUCTION

The bioluminescence of the small marine ostracod crustacean, *Vargula* (formerly *Cypridina*) *hilgendorfii* found around the coast of Japan, has been investigated so far by many workers since Harvey confirmed its luciferin (substrate)-luciferase (enzyme) reaction in 1917.[1] Vargula luciferin (**1**) was isolated as crystals in 1957 and its structure was finally determined by total synthesis in 1966.[2] The luciferase was a single polypeptide chain with 555 amino acids, deduced by cloning its cDNA in 1989.[3] The luciferin has been assumed to arise from three amino acid components or their equivalents, i.e. tryptamine (tryptophan, Trp), isoleucine (Ile), arginine (Arg), or their tripeptide without experimental evidences for a long time (Fig. 1).[2,4] Recently Oba *et al.*[5] have first revealed that the D-labeled L-Trp fed to Vargula individuals was incorporated into the luciferin, by using LC/ESI-TOFMS. However, their result has been remained some questions because of the use of a mixture of labeled L-Trp (100:40 indole-D_5/indole-D_4) and low incorporation (9.2%).

We have developed a new methodology for the feeding of the labeled amino acids to *V. hilgendorfii*. In this report we describe that L-Arg, L-Trp, and L-Ile were effectively incorporated into Vargula luciferin, therefore, it was clarified to be biosynthesized from these amino acids.

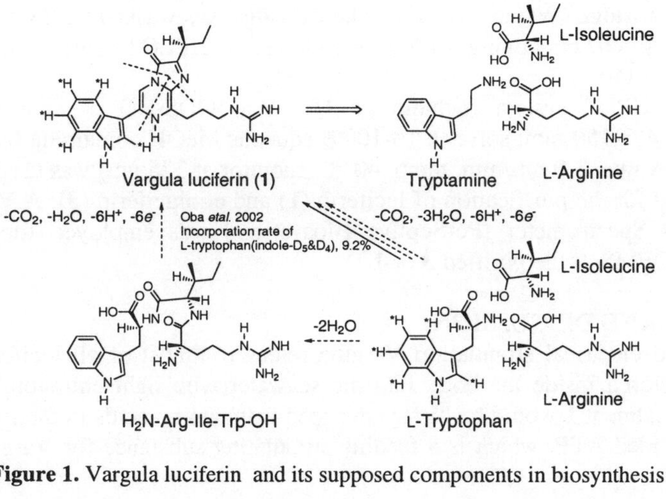

Figure 1. Vargula luciferin and its supposed components in biosynthesis

METHODS

Animals

Vargula individuals were collected at Shima County (Pacific coast of Mie Prefecture, Japan) and were kept in aquariums before the incorporation experiment.

Materials

L-Arg (guanido-$^{15}N_2$, ^{15}N, >98%) hydrochloride, L-Trp (indole-D_5, D, 98%), and L-Ile (Uniform-$^{13}C_6$, ^{13}C, 98%) were purchased from Cambridge Isotope Laboratories, Inc. The enrichment of these compounds (>95%) was reconfirmed by MALDI-TOFMS analysis. All other chemicals were of highest grade commercially available.

Complete release of luciferin (1) from Vargula by electric stimulation

The individuals were put into 10 mL of artificial seawater in a Petri dish, which was equipped with two carbon electrodes (spare lead). In order to release the luciferin completely, AC 30 V was charged to the electrodes for one second repeatedly (20-30 times), until no bioluminescence was observed after the pulse stimulation.

Preparation of the formula bait

The formula bait for the incorporation experiment of Vargula was designed by referring to the components of a feed for prawn shrimp. The bait was composed with ATP (a feeding stimulant for Vargula[6]), amino acids, sugars, vitamins, etc., and no proteins and the mixture was gelled in 3% agar. For the incorporation experiment, one amino acid in the bait was replaced with labeled L-Trp, L-Arg, or L-Ile.

Feeding experiment

After the electrical stimulation, Vargula individuals were cultured by feeding small portion of the bait every day for 14-15 d. In the case of labeled L-Trp or L-Arg feeding, 10 individuals were fed together in 10 mL of artificial seawater in a Petri dish. For L-Ile incorporation, 24-well plates were used to culture each Vargula individually in 1 mL of artificial seawater.

Extraction and purification of etioluciferin (3) from the spent seawater

The luminescence-spent seawater after the electric stimulation was loaded on a CEP-PAK C18 cartridge (Waters, 39545). The cartridge was washed with water and then eluted with MeOH containing 0.1% TFA followed by HPLC to give etioluciferin (3).

Instrumentation

A JASCO HPLC system including a Develosil ODS-HG 5 column (Nomura Chemicals; 4.6x150 mm; solvent, 10-100% aqueous MeOH containing 0.1% TFA in 30 min; flow rate, 1.0 mL/min; temp, 40 °C; monitor at 325 nm) was used for HPLC analysis and for the purification of luciferin (1) and etioluciferin (3). A Voyager-DE PRO Mass Spectrometer (PerSeptive Biosystems) was employed for measuring MALDI-TOFMS of the purified 3 or 1.

RESULTS AND DISCUSSION

By repeated electrical stimulation Vargula released almost of all luciferin (1) and luciferase stored inside the body into the seawater with light emission. Therefore, after this treatment 1 would be bio-synthesized from amino acids in the formula bait. Since we added ATP, which is a feeding stimulating substance for Vargula,[6] to the bait, its feeding was promoted and amino acid should be incorporated into **1** with high efficiency. To prevent from dilution of the labeled amino acids, Vargula was fed the bait, containing no proteins, and cultured independently for avoiding of the cannibalization. After 14-15 days' feeding Vargula recovered its bioluminescent ability, then the electrical stimulation was subjected again and the oxidized product was extracted from the luminescence-spent seawater.

As shown in Fig. 2 Vargula luciferin (**1**) was transformed to Vargula oxyluciferin (**2**) with emission. However, in our experiment only Vargula etioluciferin (**3**) was detected in the seawater extract by HPLC analysis. When [D_5]-L-Trp was fed to Vargula purified **3** gave its molecular ion peaks at *m/z* 310 (M+H)$^+$ and *m/z* 315 (labeled), while [$^{15}N_2$]-L-Arg gave the peaks at *m/z* 310 and *m/z* 312 (labeled), respectively. Therefore, the Incorporation rate of labeled L-Trp and L-Arg into **3** was determined to be 27% and 35%, respectively. By pulse stimulation some other digestive enzyme might be released, then **2** could be hydrolyzed to 3.

As the incorporation of L-Ile could not be proven by analysis of **3**, **1** was isolated from the freeze-dried Vargula whole bodies (13-21 individuals), and analysed by MS. When [$^{13}C_6$]-L-Ile was fed, **1** showed its molecular ion peaks at *m/z* 406 (M+H)$^+$ and *m/z* 412 (labeled) and the peak intensity ratio of 412/406 indicate very high incorporation rate of 42% (Fig. 3).

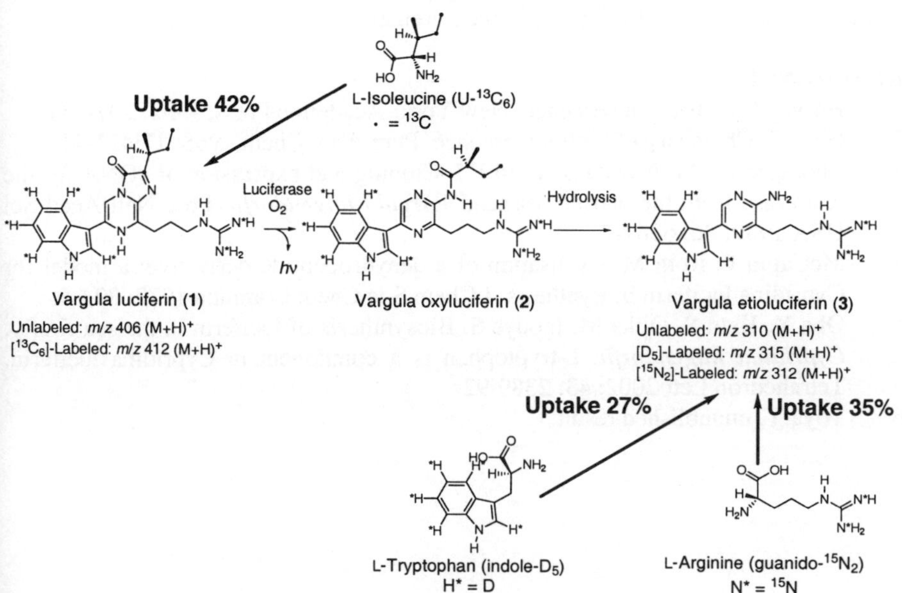

Figure 2. Results of the feeding experiment

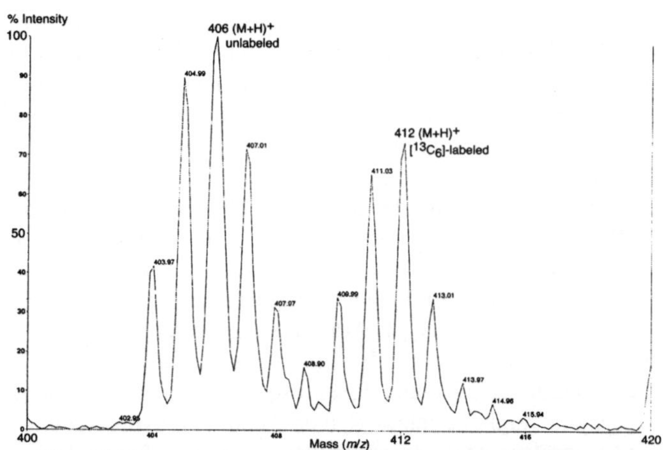

Figure 3. The mass spectrum of **1** obtained after the feeding experiment

In conclusion, we clarified that **1** was bio-synthesized from orally uptaken L-Trp, L-Arg and L-Ile by using an efficient feeding methodology. Such a high incorporation rate indicates that in Vargula body luciferin might be *de novo* synthesized from amino acids without any salvage cycle.

ACKNOWLEDGEMENTS
We greatly thank Prof. T. Kondo of Nagoya University for the supply of labeled amino acids and for MALDI-TOFMS measurement.

REFERENCES

1.   Harvey E N. Bioluminescence. New York: Academic Press, 1952: 297-331.
2.   Goto T. Chemistry of bioluminescence. Pure Appl Chem 1968; 17:421-41.
3.   Thompson E M, Nagata S, Tsuji F I. Cloning and expression of cDNA for the luciferase from the marine ostracod *Vargula hilgendorfii*. Proc Natl Acad Sci USA 1989; 86: 6567-71.
4.   McCapra F, Roth M. Cyclisation of a dehydropeptide derivative: a model for Cypridina luciferin biosynthesis. J Chem Soc Chem Commun 1972: 894-5.
5.   Oba Y, Kato S, Ojika M, Inouye S. Biosynthesis of luciferin in the sea firefly, *Cypridina hilgendorfii*: L-tryptophan is a component in Cypridina luciferin. Tetrahedron Lett 2002; 43: 2389-92.
6.   Toya Y. unpublished result.

PART 5

CHEMILUMINESCENCE

ON THE ROLE OF THE SINGLET-OXYGEN DIMOL CHEMILUMINESCENCE IN DIOXIRANE REACTIONS

W ADAM[1], VP KAZAKOV[2], DV KAZAKOV[*2], RR LATYPOVA[2],
GY MAISTRENKO[2], DV MAL'ZEV[2], FE SAFAROV[2]

[1]*Institute of Organic Chemistry, University of Wuerzburg, Am Hubland,
D-97074 Wuerzburg, Germany
E-mail: wadam@chemie.uni-wuerzburg.de
Department of Chemistry, Facundo Bueso 110, University of Puerto Rico,
Rio Piedras, PR 00931, USA*
[2]*Institute of Organic Chemistry, Ufa Scientific Center of the RAS,
71 Prospect Oktyabrya, 450054 Ufa, Russia
E-mail: chemlum@ufanet.ru*

INTRODUCTION

Dioxiranes, three-membered-ring cyclic peroxides, are known as highly efficient and selective oxidants, capable of performing a variety of transformations for synthetic purposes.[1-3] It is known[4-7] that some reactions of these peroxides are accompanied by chemiluminescence due to the release of singlet oxygen. For instance, infra-red chemiluminescence (IR-CL) of $^1O_2$ at λ 1270 nm is emitted in the reaction[4,5] of tertiary amines and N-oxides with dimethyldioxirane (DMD) and methyl(trifluoromethyl)dioxirane (TFD), as well as during the anion-catalyzed[6] breakdown of the dioxiranes. Furthermore, IR-CL emission is produced in the ketone-catalyzed decomposition of the monoperoxysulfate ion HSO_5^- through the intermediary dioxirane.[7]

We report here the chemiluminescence of these reactions observed in the visible spectral region (Vis-CL), along with the IR-CL emission of $^1O_2$. A spectral analysis of the emission as well as the influence of the solvent on the Vis-CL intensity revealed that the singlet-oxygen dimol is one of the emitters of the chemiluminescence:

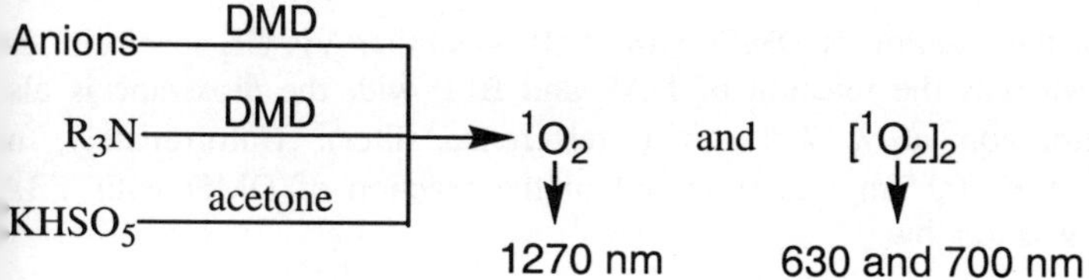

EXPERIMENTAL SECTION

The dioxirane solutions in the parent ketone and ketone-free in CCl_4 or CH_2Cl_2 were prepared as described in the literature.[1-4] Triple salt $2KHSO_5{\cdot}KHSO_4{\cdot}K_2SO_4$ (Curox) tetrabutylammonium salts, KO_2, NaCl, 4-dimethylaminopyridin (DAP), tribenzylamine (TBA), 1-benzyl-4piperidone (BPP), and 1,4-diazabicyclo[2.2.2]octane (DABCO) were used without further purification. All solvents were dried and purified prior to use. The solvent influence on the Vis-CL intensity was determined by means of a red-sensitive photomultiplier, cooled by liquid nitrogen. For wavelength selection, interference filter with a transmission maximum at λ 700 nm and 626 nm and or cut-off filters were employed.

RESULTS AND DISCUSSION

Reaction of DMD with tertiary amines

In the reaction of DMD with DAP, TBA and BPP, the chemiluminescence in the IR ($^1O_2$) and visible spectral regions are emitted. As an example, Figure 1 shows a

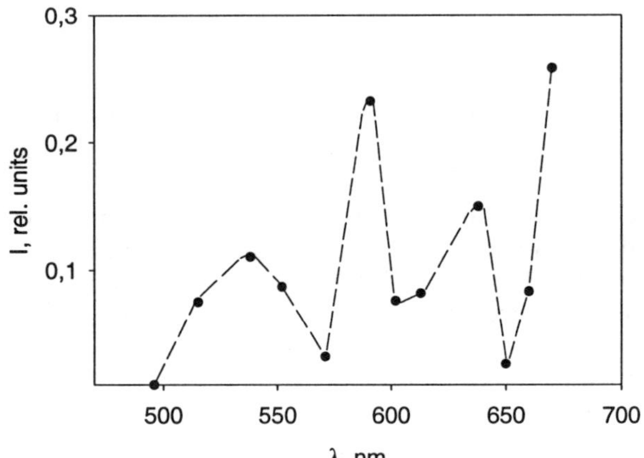

Figure 1. Vis-CL spectrum of the reaction of DMD with
4-dimethylaminopyridine (20 °C, CCl_4, [DAP] = $1{\times}10^{-2}$ M, [DMD] = $7{\times}10^{-2}$ M).

Vis-CL spectrum for the reaction of DMD with DAP. A similar Vis-CL spectrum was recorded for BPP, whereas the reaction of DAP and BPP with the dioxirane is also accompanied by emission at λ 700 nm (interference filter). Unfortunately, no chemiluminescence at λ 700 nm was observed in the reaction of DMD with TBA because its intensity was too low.

The emission at $\lambda > 600$ nm is assigned to the singlet-oxygen dimol species, which is known[8] to have characteristic bands at λ 633 and 703 nm. The CL intensity at λ 626 and 700 nm for the reaction of DMD with amines BPP and DAP is more then twice in acetone-deuteroacetone (1:1) and acetone-deuterochloroform(1:1) than in pure acetone. Clearly, this increase in the Vis-CL intensity is a consequence of the prolonged $^1O_2$ lifetime in deuterated solvents.

Reaction of DMD with anions[6]

The singlet-oxygen dimol emission at $\lambda > 600$ nm was also detected in the reaction of DMD with the quaternary ammonium halides $n\text{-Bu}_4\text{N}^+\text{Cl}^-$, $n\text{-Bu}_4\text{N}^+\text{Br}^-$ and $n\text{-Bu}_4\text{N}^+\text{I}^-$. A more then twofold decrease in the Vis-CL intensity at λ 700 nm and 626 nm was observed for the reaction of the dioxirane with the chloride and bromide anions in CCl_4-acetone (1:1) or CCl_4-CH_2Cl_2 (1:1) mixtures compared to pure carbon tetrachloride. Moreover, when the reaction of **DMD** with Na^+Cl^- was carried out in a 1.0 : 3.5 mixture of water and acetone, the Vis-CL emission intensity at $\lambda > 602$ nm was more then twenty times lower compared to the reaction of **DMD** with $n\text{-Bu}_4\text{N}^+\text{Cl}^-$ in a 1.0 : 1.0 mixture of CCl_4 and CH_2Cl_2. This decrease in the Vis-CL intensity is due to the lower lifetime of the $^1O_2$ in the aqueous acetone versus the nonaqueous (CH_2Cl_2, CCl_4) media. Likewise, the diminished singlet-oxygen lifetime in acetone and methylene chloride relative to carbon tetrachloride is responsible for the reduced Vis-CL intensity in the CCl_4-CH_2Cl_2 or CCl_4-acetone solvent mixtures (1:1).

Acetone-catalyzed decomposition of $KHSO_5$

We have found that in addition to the IR-CL of $^1O_2$, reported previously by Lange and Brauer[7], decomposition of $KHSO_5$ by acetone leads to the Vis-CL emission, whose spectrum consist of two emission bands at λ 580-610 nm and 610-645 nm. Furthermore, also the CL at λ 700 nm (interference filter) was detected. The emission intensities at λ 626 nm and 700 nm is increased more then three times in D_2O compared to H_2O, which implies that the singlet-oxygen dimol species intervenes.

Effect of DABCO on the singlet oxygen dimol emission

DABCO is known[9] to enhance the dimol emission at λ 633 nm and 703 nm, generated by the thermal decomposition of disodium 3,3'-(1,4-naphthylidene)dipropionate endoperoxide, as well as by the ClO^-/H_2O_2 system. In our reactions described above, however, a diminution of the singlet-oxygen dimol emission by DABCO was observed. In fact, a control experiment revealed that the singlet-oxygen dimol chemiluminescence intensity, generated in the thermolysis of 1,4-naphthalene endoperoxide in either CCl_4 or CH_2Cl_2, is also reduced by DABCO. In contrast, the previously reported[9] enhancement of the dimol chemiluminescence intensity by DABCO in the reaction of ClO^- with alkaline hydrogen peroxide was confirmed.

In conclusion, we have shown that reactions of DMD with amines and anions, as well as with $KHSO_5$ in acetone is accompanied by CL, in which the singlet-oxygen dimol species intervenes. Our results also caution that DABCO cannot be regarded as a reliable probe for the detection of $(^1O_2)_2$ in chemiluminescent systems.

ACKNOWLEDGEMENTS

The research in Ufa was supported by the RFFI (02-03-32515a), PNSh (grant No 591.2003.3) and OKHIM (160603-687). DVK is grateful to the Alexander-von-Humboldt Foundation for the Return Fellowship, as well as to the Presidium of RAS (6th Young Scientists Projects Competition, grant No 126). WA thanks the Deutsche Forschungsgemeinschaft (DFG) and the Fonds der Chemische Industrie for generous funding.

REFERENCES

1. Adam W, Curci R, Edwards JO. Dioxiranes - a new class of powerful oxidants. Acc Chem Res 1989; 22: 205-11.
2. Kazakov VP, Voloshin AI, Kazakov DV. Dioxiranes: from oxidative transformations to chemiluminescence. Russ Chem Rev 1999; 68: 253-86.
3. Adam W, Saha-Müller CR, Zhao C. Dioxirane epoxidation of alkenes. Org React 2002; 61: 219-516.
4. Adam W, Briviba K, Duschek F, Golsch D, Kiefer W, Sies H. Formation of singlet oxygen in the deoxygenation of heteroarene N-oxides by dimethyldioxirane. J Chem Soc, Chem Commun 1995; 1831-32.
5. Ferrer M, Sánchez-Baeza F, Messeguer A, Adam W, Golsch D, Görth F, Kiefer W, Nagel V. The release of singlet oxygen in the reaction of dioxiranes with amine N-oxides. Eur J Org Chem 1998; 2527-32.
6. Adam W, Kazakov DV, Kazakov VP, Kiefer W, Latypova RR, Schlücker S. Singlet-oxygen generation in the catalytic reaction of dioxiranes with nucleophilic anions. Photochem Photobiol Sci 2004; 3: 182-8.
7. Lange A, Brauer HD. On the formation of dioxiranes and of singlet oxygen by the ketone-catalysed decomposition of Caro's acid. J Chem Soc, Perkin Trans 2 1996; 805-11.
8. Khan A.U. The discovery of the chemical evolution of singlet oxygen. Some current chemical, photochemical, and biological applications. Int J Quant Chem 1991; 39: 251-67.
9. Mascio PD, Sies H. Quantification of singlet oxygen generated by thermolysis of 3,3'-(1,4-naphthylidene)dipropionate- monomol and dimol photoemission and the effects of 1,4-diazabicyclo[2.2.2]octane. J Am Chem Soc 1989; 111: 2909-14.

SOLVENT EFFECTS ON THE CHEMILUMINESCENCE OF TCPO IN PRESENCE OF 7-AMINO-4-TRIFLUOROMETHYLCOUMARIN

MJ CHAICHI[1], M SHAMSIPUR[2], A KARAMI[3], K ALIZADEH[3], O NAZARI[1]

[1]Dept of Chemistry, Mazandaran University, Babolsar, Iran
[2]Dept of Chemistry, Razi University, Kermanshah, Iran
[3]Dept of Chemistry, Tarbiat Modaress University, Tehran, Iran
Email:jchaichi@yahoo.com

INTRODUCTION

Among different coumarin derivatives used, 7-Amino-4-trifluoromethylcoumarin (ATFMC) revealed the most promising characteristics as an efficient fluorescent emitter.[1] AFTMC is used in the synthesis of a substrate for fluorimetric assay of proteolytic enzymes[2] and for use as a laser dye.[3] We have recently investigated the chemiluminescence reactions of some peroxyoxalate esters, hydrogen peroxide and AFTMC.[1,4-6] In this paper we report the solvent effects on the kinetics of the chemiluminescence process of the peroxyoxalate chemiluminescence in the presence of AFTMC.

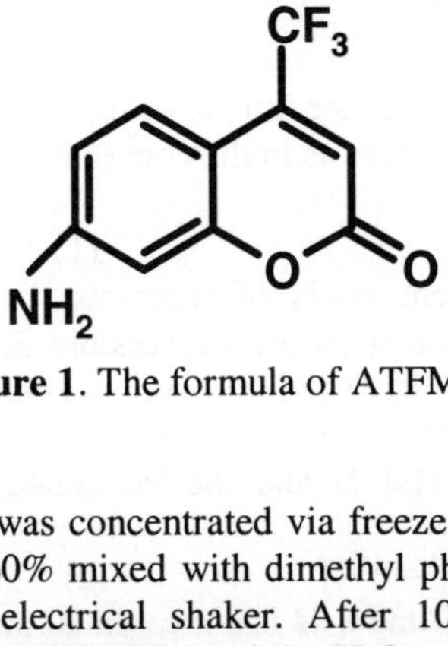

Figure 1. The formula of ATFMC

METHODS

Hydrogen peroxide (30%) was concentrated via freeze drying (using a model FD-1 Eyela freeze dryer) up to 60% mixed with dimethyl phthalate in a 1:1 v/v portions and shacken well on an electrical shaker. After 10 h, the organic phase was separated, dried on anhydrous Na_2SO_4 and the H_2O_2 concentration was determined by a standard potassium permanganate solution. Then a standard stock solution of hydrogen peroxide(1.5 M in 80:20 v:v dimethylphthalate:*tert*-butylalcohol containing 5.0×10^{-3} M sodium salicylate) was prepared from this solution.

Effect of solvent on chemiluminescence

The cell containing 1.0 mL solvent, 100 μL ATFMC(0.01 M in EtOAc) and 250 μL TCPO(0.01 M in EtOAc). Light intensity decay curve were obtained by introducing the 100 μL standard stock solution of hydrogen peroxide. The experiment was done with shaking.

Chemiluminescence detection was carried out with a homemade apparatus equipped with a model BPY47 photocell (Leybold, Huerth, Germany). The apparatus was connected to a personal computer via a suitable interface (Micropars, Tehran, Iran). Experiments were carried out with magnetic stirring (500 rpm) in a light-tight flattened bottom glass cell of 15 mm diameter at room temperature. All fluorescence and chemiluminescence spectra were recorded on a Model LS-50B Perkin Elmer instrument.

RESULTS AND DISCUSSION

Our experiments revealed that the addition of hydrogen peroxide to a colorless aprotic solvents such as ethyl acetate, containing ATFMC results very intense blue light. In protic solvents such as methanol, the light color shifted to green and the light intensity was greatly reduced. The strong emission of aminocoumarin dyes derives from the polar character of low-laying excited states. The Stokes shift and yield of fluorescence or chemiluminescence influenced by the maintenance of a large excited state dipole moment. Excitation leads to a polar, planar excited state of the dye which is moderately stabilized by interaction with solvent. Solvent to solute hydrogen bonding by protic solvents to the carbonyl oxygen stabilizes the electronic excited states and thus causes λ_{max} of fluorescence or chemiluminescence shift to lower energy i.e. longer wavelength. The stronger hydrogen bonds to the more negatively charged carbonyl oxygen and by the more positively charged aminegroup led to shifts to lower energy.[7,8]

Analysis of rate data

A simplified model for evaluating the kinetic of the CL reaction has been developed in terms of three pool of substances as in the consecutive first-order reactions:[9]

$$\overset{r}{A} \rightarrow \overset{f}{B} \rightarrow C \tag{1}$$

where **A**, **B**, and **C** represent pools of reactants, intermediates, and products respectively, and both reaction steps are irreversible first order reactions. Since **B** will rise with rate constant r (=rise) and fall with rate constant f (=fall), only a single pulse of light can be described. The chemiluminescence signal is proportional to the concentration of intermediate(s) **B** and the integrated rate equation for the CL intensity versus time is:

$$I_t = [Mr / (f - r)][e^{-rt} - e^{-ft}] \tag{2}$$

where I_t is a light intensity at time t, M is a theoretical maximum level of intensity if the reactants were entirely converted to a chemiluminescence-generating material and r and f are, respectively, the first order rate constants for the rise and fall of the burst of CL. The model permits an estimate of intensity at the maximum (J), time of maximum intensity (τ_{max}) and the total light yield (Y) from the reaction after the collection of only a part of the emitted light. These parameters were given by the following formulas:

$$J = M \, (f/r)^{(f/(r-f))} \tag{3}$$

$$\tau_{max} = \{\ln(f/r)\}/(f-r) \qquad\qquad (4)$$

$$Y = \int_0^\infty I_t \, dt = M/f \qquad\qquad (5)$$

The rate constants r and f and other parameters consist of M, J, τ_{max} and Y evaluated by fitting equation (2) to a non-linear least-squares curve fitting program KINFIT.[10]

Table 1. The kinetic parameters evaluated for effect of solvent on CL system

Solvent	r 1/mi	f 1/min	M μv	J μv	I_t μv	τ_{max} min	T_{exp} min	Y μv.min	$T_{1/2}$ min
Benzene	7.7	0.6	557	44	459	0.36	0.35	913	0.92
MeOH	34.6	0.1	18	18	22	0.16	0.57	138	1.00
DMF	9.8	1.6	140	98	103	0.22	0.32	88	0.74
DMSO	31.2	0.8	58	52	58	0.12	0.11	72	0.90
Iodobenzene	71.3	0.4	36	35	37	0.07	0.11	79	1.54
Bromobenzene	7.7	0.7	776	609	643	0.33	0.29	1085	1.34
Butyraldehyde	15.1	0.1	46	44	54	0.32	0.53	348	2.16
Nitroethane	11.4	0.3	58	52	55	0.33	0.13	195	2.11
Dichloromethane	13.8	0.9	607	498	487	0.21	0.22	643	1.09
Chloroform	6.6	0.6	729	570	576	0.39	0.41	116	1.73
EtOAc	16.7	1.1	782	648	709	0.18	0.16	736	0.76
Benzylchloride	3.1	1.5	66	33	37	0.45	0.41	43	1.36
Hexane	6.0	0.9	875	626	659	0.37	0.37	975	1.22
Aceton	10.7	0.4	171	151	150	0.32	0.40	459	2.18
1-Propanol	11.0	0.3	186	168	170	0.33	0.37	613	2.74
Diethylether	11.1	0.9	850	678	736	0.24	0.19	910	1.00
Toluene	8.0	0.6	398	320	337	0.34	0.28	622	1.48
m-Xylene	7.2	0.5	316	258	275	0.39	0.28	613	1.63
Nitromethane	3.5	0.6	52	36	43	0.60	0.43	85	2.17
Acethophenone	23.7	0.7	284	256	238	0.15	0.23	419	1.41
EtOH	25.2	0.1	72	71	77	0.23	0.17	846	4.40
Triethylphosphate	0.5	0.5	652	240	248	1.90	1.40	1240	5.35
Tribromomethane	27.9	0.4	34	32	35	0.16	0.47	96	1.37

The very low amounts of f is led to high amount of Y and in these situations we cannot use Y for comparable with other solvents. The light intensity in the protic solvents are lower than that of aprotic solvents. On the other hand the wavelength of maximum intensity of FL and CL in protic solvents are higher rather than that of aprotic solvents.

Table 2. Wavelengths of maximum intensity for the fluorescence (λ_{max}^{FL}) and chemiluminescence (λ_{max}^{CL}) for ATFMC with different solvents

Solvent	λ_{max}^{FL} (nm)	λ_{max}^{CL} (nm)
Benzene	434.8	460.9
1-Propanol	482.4	474.3
DMF	476.6	472.1
EtOAc	454.3	455.7
Diethylether	450.7	450.8
EtOH	487.4	476.5
MeOH	478.0	475.8
Toluene	433.7	461.6

REFERENCES

1. Chaichi MJ, Karami AR, Shockravi, A, Shamsipur M. Chemiluminescence characteristics of coumarin derivatives as blue fluorescers in peroxyoxalate-hydrogen peroxide system Spectrochim Acta Part A 2003; 59:1145-50.

2. Smith RE, Bissell, ER, Mitchell, AR, Pearson KW. Direct photometric or fluorometric assay of proteinase using substrates containing 7-Amino-4-trifluoromethylcoumarin. Thromb Res 1980; 17:393-402.

3. Fletcher AN. Laser dye stability part 3. Bicyclic dyes in ethanol. Appl Phys 1977; 14:295-302.

4. Chaichi MJ, Shamsipur M. A study of chemiluminescence from reactions of peroxyoxalate esters, hydrogen peroxide and 7-amino-4-trifluoromethylcoumarin. In:Stanley PE, Kricka LJ. Editors. Bioluminescence & Chemiluminescence: Progress & Current Applications. World Scientific Publishing Co. Pte. Ltd. 2002:141-4.

5. Shamsipur M, Chaichi MJ. Quenching effect of DL(±)α-methylbenzylamine on peroxyoxalate chemiluminescence of 7-amino-4-trifluoromethylcoumarin. J Photochem & Photobiol A: Chemistry 2003; 155:69-72.

6. Shamsipur M, Chaichi MJ. Quenching effect of triethylamine on peroxyoxalate chemiluminescence in presence of 7-amino-4-trifluoromethylcoumarin. Spectrochim Acta Part A 2001; 57:2355-8.

7. Jones G, Jackson WR, Kanoktanaporn S, Halpern AM. Solvent effects on photophysical parameters for coumarin laser dyes. Opt Commun 1980; 33:315-20.

8. Kamlet MJ, Dickinson C, Taft RW. Linear solvation energy relationships. Solvent effects on some fluorescence probes. Chem Phys Lett 1981; 77:69-72.

9. Hadd AG, Seeber A, Birks JW. Kinetics of two pathways in peroxyoxalate chemiluminescence. J Org Chem 2000; 65: 2675-83.

10. Dye JL, Nicely VA. A general purpose curve fitting program for class and research use. J Chem Edu 1971; 48: 443-8.

CHEMILUMINESCENCE IN THE REACTIONS OF URANIUM AND LANTHANIDES

VP KAZAKOV, SS OSTAKHOV, DV KAZAKOV[*], AV MAMIKIN,
VA ANTIPIN, SN KLIMINA, LN KHAZIMULLINA, OA KOCHNEVA
*Institute of Organic Chemistry, Ufa Scientific Center of the RAS,
71 Prospect Oktyabrya, 450054 Ufa, Russia
E-mail:chemlum@ufanet.ru*

INTRODUCTION

Since the discovery of electrochemiluminescence of uranyl ion (UO_2^{2+}) and lanthanide ions,[1,2] these metals have found numerous applications as an activators of chemiluminescence (CL) arising in oxidation of organic and inorganic compounds or decomposition of dioxetanes.[3,4] Moreover, CL has been also observed[4] during the oxidation of uranium by various oxidants.

This paper summarizes our recent achievements in the field of CL of uranium compounds and lanthanide chelates occurring during their oxidation by XeO_3, XeF_2, O_3, $S_2O_8^{2-}$ and dioxiranes as well as during decomposition of 1,2-dioxetanes.

RESULTS AND DISCUSSION

Chemiluminescence of the uranium compounds[4]

Chemiluminescence of uranium may be classified on two general types: (a) uranyl ion accepts energy from the excited species formed in chemiluminescent reaction; (b) uranyl ion participates in the reductive-oxidative transformations:

Scheme 1.

Type (a):

$$S + O_3 \longrightarrow SO^* ; \quad SO^* + UO_2^{2+} \longrightarrow [SO \cdots UO_2^{*2+}] \longrightarrow h\nu$$

Type (b):

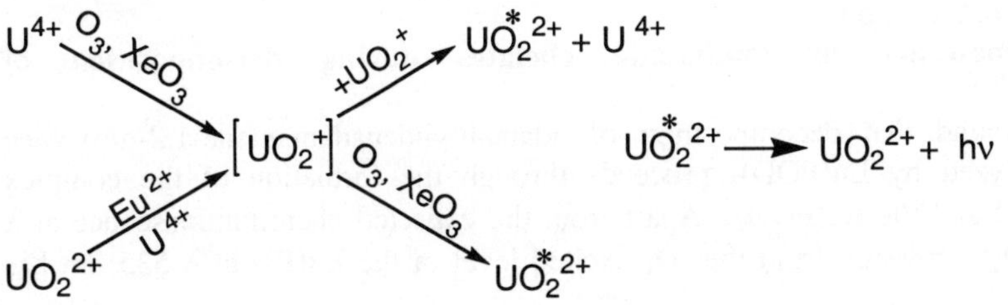

Type (a) is illustrated by the reaction of sulfur with ozone (Scheme 1), in which the first step is the excitation of the sulfur oxide SO*. The latter transfer energy to the uranyl ion inside its coordination sphere. The yield of UO_2^{2+} excitation in this reaction is close to unity.

Type (b) is CL which is observed in oxidation of the tetravalent uranium or during reduction of the UO_2^{2+} (Scheme 1). The most interesting example is oxidation of the uranium (IV) by the XeO_3. This reaction is accompanied by very high yields of excited states formation, that allows one to detect extremely low concentrations (ca. 10^{-12} M) of the uranium by means of non-sophisticated chemiluminescent equipment.

The other example of the type (b) is reduction of the UO_2^{2+} by the europium (II). The key chemiluminescent step here is disproportionation of the uranium (V).

Chemiluminescence of the uranium is observed not only in solution but also in the solid phase. For instance, solid-phase decomposition of the uranyl or europium (III) persulfate leads to the formation of UO_2^{2+} in excited state by energy transfer mechanism, whereas electron transfer is responsible for the uranyl ion excitation (through the intermediary uranium (V)) in the oxidation of $U(SO_4)_2$ by XeF_2.

Chemiluminescence in oxidation of europium by dimethyldioxirane (DMD)[5]

We have revealed a new type of lanthanide CL, when light emission arises due to the oxidation of organic ligands rather than as a result of outersphere energy transfer, as shown in Scheme 2.

Scheme 2.

$$Eu(III)L_3 \; + \; \text{(dimethyldioxirane)} \longrightarrow Eu(III)P^* \longrightarrow Eu^*(III)P \longrightarrow hn \; (570\text{-}650 \; nm)$$

L - heptafluorodimethyloctanedione (FOD) or thenoyltrifluoroacetone (TTA)
P- product of L oxidation by DMD

These results testify that the β-diketonates of europium are not always passive and may significantly contribute to the production of CL as a result of their oxidation by peroxides. This circumstance should be taken into account when lanthanide complexes are used as «inert» activators for studying of chemi- or bioluminescent reactions.

Chemiluminescence of lanthanide chelates during decomposition of dioxetanes [6-8]

We have found that decomposition of adamantylidenadamantane-1,2-dioxetane (DO), catalyzed by $Eu(FOD)_3$ proceeds through the formation of the complex between DO and the $Eu(FOD)_3$. Apart from the expected chemiluminescence at λ 613 nm ($^5D_0$), emission from the $^5D_1$ excited level of the Eu(III) at λ 535 nm has been observed during the catalytic decomposition. This phenomenon is explained by the luminescence from the excited non-equilibrium complexes of the europium (III) with the Ad=O - product of dioxetane decomposition (Scheme 3).

Scheme 3.

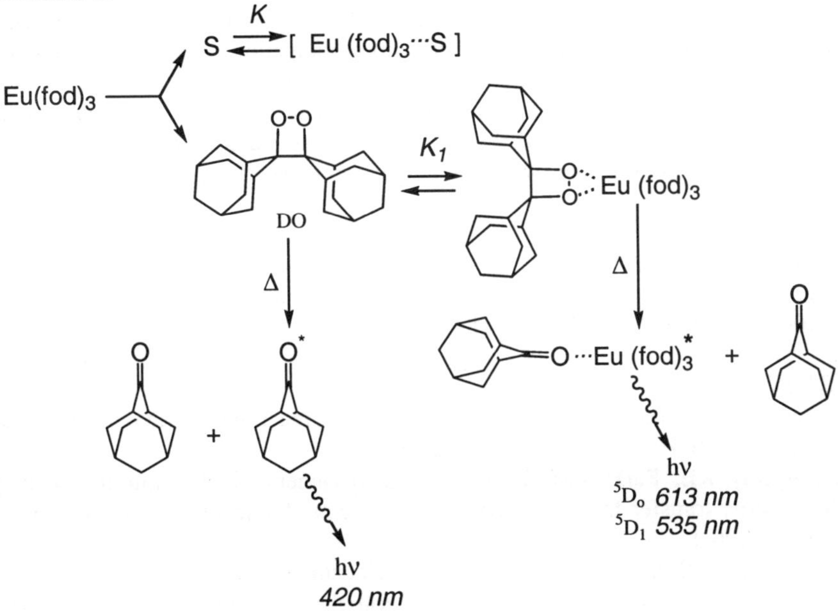

S: sulfoxides, sulfones, ketones and amines

Chemiluminescence intensity is affected when dioxetane is replaced from the coordination sphere of the Eu(III) by the other ligand, such as sulfoxides, sulfones or ketones (Scheme 3). This effect allowed us to develop a chemiluminescent procedure to study equilibriums in solutions. If we know stability constant (K_1) for the Eu(fod)$_3$-DO complex, we may obtain an equilibrium constant (K) for the complex of Eu(fod)$_3$ with a organic substrates.

Moreover, by means of the method of luminescent-kinetic spectroscopy[7] we have shown that Eu(III) ion in excited state forms more stable complex (up to two orders of magnitude) with sulfoxides, sulfones, ketones and amines then that in ground state. Since it is known that excitation of the europium is caused by electron transitions in the inner metal-centered 4f-states, our findings testify that f-electrons participate in the chemical bonding.

We have also revealed that europium-catalyzed decomposition of DO results in quantum chain reactions with energetic branching. Complex formation between Eu(fod)$_3$ and DO (or Ad=O) plays an important role in this process since it is complexing that initiates chain reaction. The key step in this chain process is transfer of energy from the excited europium on the vibrational levels of dioxetane, what leads to the DO decomposition with regeneration of the europium excitation. Branching of the chain is caused by the formation of Eu(III) ions in two excited $^5D_1$ and $^5D_0$ states.

In summary, we have reported a new class of intense chemiluminescent reactions (in liquid and solid phase) with uranium participation; We have shown that decomposition of adamantylidenadamantane-1,2-dioxetane (DO), catalyzed by Eu(FOD)$_3$, results in a quantum chain reactions with energetic branching; We have revealed that Eu(III) ion in excited state forms more stable complex with organic ligands then that in ground state, which is a strong evidence for participation of the f-electrons in chemical bonding.

ACKNOWLEDGEMENTS

The research was supported by the RFFI (02-03-32515a, 02-03-32406), PNSh (grant No 591.2003.3) and OKHIM (160603-687).

REFERENCES

1. Kazakov VP. On the chemiluminescence of some reactions in the concentrated sulfuric acid. Zh Phys Khim 1965; 39: 2936-41.
2. Hemingway RE, Park S-M, Bard AJ. Electrogenerated chemiluminescence. XXI. Energy transfer from an exciplex to a rare earth chelate. J Am Chem Soc 1975; 97: 200-1.
3. Elbanowski M, Makowska B, Staninski K, Kaczmarek M. Chemiluminescence of systems containing lanthanide ions. J Photochem Photobiol, A-Chem 2000; 130: 75-81.
4. Kazakov VP. Chemiluminescence of uranyl, lanthanides and d-elements, Moscow: Nauka, 1980.
5. Kazakov DV, Mainstrenko GYa, Kotchneva OA, Latypova RR, Kazakov VP. Chemiluminescence in the oxidation of europium β-diketonates by dimethyldioxirane. Mendeleev Commun 2001: 188-90.
6. Kazakov VP, Voloshin AI, Ostakhov SS. Quantum chain reactions with energetic branching: catalytic decomposition of dioxetanes. Kinet Catal 1999; 40: 180-93.
7. Kazakov VP, Ostakhov SS, Voloshin AI, Alyab'ev AS. Effect of the excited 4f level of Eu(fod)3 (Hfod-heptafluorodimethyloctanedione) on the kinetics and thermodynamics of complex formation in solution. Participation of f electrons in coordination bonds. Russ J Coord Chem 2001; 27: 138-44.
8. Kazakov VP, Voloshin AI, Shavaleev NM. Chemiluminescence in visible and infrared spectral regions and quantum chain reactions upon thermal and photochemical decomposition of adamantylidenadamantane-1,2-dioxetane in presence of chelates of Pr(dpm)$_3$ and Pr(fod)$_3$. J Photochem Photobiol, A-Chem 1998; 119: 117-23.

GREEN LUMINESCENCE EMITTED FROM ADSORBED OXYGEN — AURORA HAS APPEARED ON THE SURFACE OF MATERIALS?

J-I KIMURA

CL Advisor, Tohoku Electronic Industrial Co., Ltd.
6-6-6 Shirakashidai, Rifu, Miyagi, 981-0134, Japan
jkr@ym.catv.ne.jp

INTRODUCTION

Anomalous initial peaks in chemiluminescence (CL) intensity-time curves have been reported by many investigators for different materials, i.e., salad oil,[1] polymers,[2] and nitrocellulose.[3] The characteristic CL peak disappeared when a sample was reheated after being cooled down to ambient temperature in nitrogen. The most authors have attributed the CL peaks to the decomposition of hydroperoxides formed by autoxidation during storage in air. Activation energy reported[2] for the CL peaks is apparently lower than expected values for the decomposition of organic hydroperoxides. We found that cellulose filters continued to emit weak luminescence for longer than 1 week in flowing nitrogen when they were heated at around 100 °C.

The mechanism of light emission for the characteristic CL peaks is still debated. In photoluminescence studies of organic compounds such as filter paper and milk powder, we found green light emission spectra and then succeeded in taking colour pictures similar to green aurora with high sensitive colour CCD camera. The observed green light emission may imply that significant amount of oxygen couls exist on the surface of organic materials having oxygen-containing functional groups. The reported low activation energies for CL peaks of organic materials will be interpreted as the bond dissociation energy of oxygen bonds (weak covalent bonds) between oxygen molecules and neighbouring oxygen atoms of oxygen-containing functional groups in organic materials.

MATERIALS AND METHODS

Materials

Filter Papers used in this study were obtained from Advantec Co. (Japan) and the milk powder is the product of Morinaga Milk Industry Co. (Japan).

Apparatus and Procedure

Luminous intensity-time curves and spectra of test samples stimulated thermally and photochemically were measured with a chemiluminescence spectrometer CLA-FS1. Colour photos of the photoluminescence of samples placed in a cell with a diameter of 50 mm were taken with the Cube-2DI, which is employing an ultra high sensitive colour CCD camera of Bitran Co., Japan. Both types of the apparatus are available from the Tohoku Electronic Industrial Co., Japan. The black light has power of 240 $\mu W\ cm^{-2}$ and wavelength of 365 nm. In order to obtain a spectrum for steep decay photoluminescence curves, a sample was irradiated with the NUV light for 5 sec

"

outside the apparatus and then the sample chamber was set back into the apparatus. Luminescence measurement was started normally 5 sec after the end of irradiation under a predetermined cutoff filter. The latest model of chemiluminescence spectrometer MLA-GOLDS (Tohoku Electronic Industrial Co.) makes the spectrum measurement a simple process.

RESULTS AND DISCUSSIONS

A filter paper was heated in temperature ranging from 90 °C to 120 °C in a flowing nitrogen of 60 mL min^{-1}. The anomalous initial peak in the CL curves can be seen in Figure 1. The initial peak disappeared when it was heated again after being cooled down to room temperature. The luminous intensities decayed exponentially after the peaks. Thus, kinetic analysis of the luminescence decay curves was conducted as follows; (1) decay curves after the peak were plotted in logarithmic fashion against time, (2) the slopes of the straight lines were plotted against reciprocal absolute temperatures as depicted in the right figure in Figure 1.

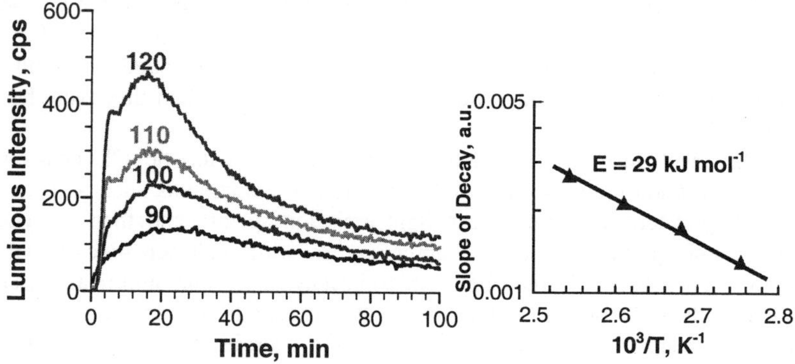

Figure 1. (Left) Typical CL curves for filter paper heated at different temperatures in nitrogen. (Right)The Arrehenius plot of the slope after the CL peak against reciprocal temperatures provided activation energy of 29 kJ mol^{-}

The Arrhenius plot gave activation energy of 29 kJ mol^{-1} for the initial decay parts. Activation energy for steady-state luminescence of cellulose was obtained to be 50 kJ mol^{-1}. Those value are significantly low compared to the reported activation energies for the thermal decomposition of organic hydroperoxides which is ranging from 96 kJ mol^{-1} to 116 kJ mol^{-1}.[4]

In order to clarify the CL curves of filter paper observed in nitrogen atmosphere we also adopted photoluminescence. A photoluminescence curve of filter paper measured in nitrogen is depicted in Figure 2. The photoluminescence curve exhibited exponential decay having a radiative lifetime of about 3 sec. The luminescence spectrum is also plotted in Figure 2 where logarithm of luminous intensities were plotted against wavelength. The emission spectrum showed

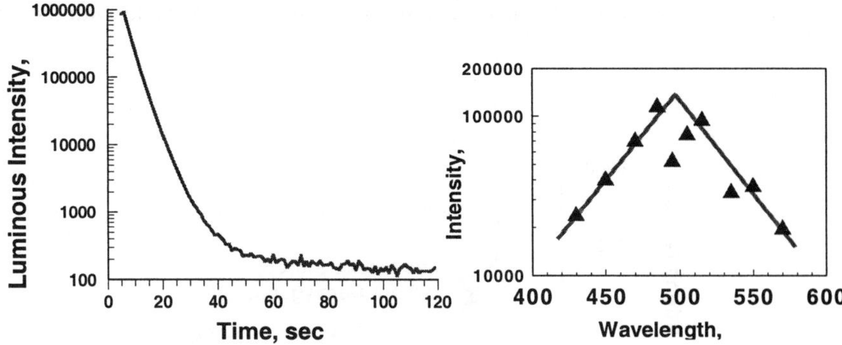

Figure 2. (Left) Photoluminescence curve of filter paper irradiated with UV. (Right) Emission spectrum of filter paper indicating green luminescence.

characteristic triangle having a peak at near 500 nm. The emission band of 480 nm has been assigned to transition of electronically excited oxygen molecules to the ground state as follows ($[^1\Delta_g + {}^1\Sigma_g^+] \rightarrow 2\ {}^3\Sigma_g^-$). This type of oxygen emission is known as one-photon two-molecule process, where two electronically-excited oxygen molecules emit light in the collision process. Observed band spectra shown in upper right figure of Figure 2 suggest that the color of luminescence from filter paper will be green. Evidently, bluish green photo was taken with a high sensitive colour CCD camera Cube-2DI. It will be natural to draw an anolgy between the green light emission from the surface of paper and green aurora caused by atomic oxygen.[5]

As described above two different values of activation energy (29 and 50 kJ mol^{-1}) were observed. These experimental facts may suggest that there might exhist two different types of adsorved oxygen: one will be weakly adsorved oxygen molecules between O_2 and O_2 above monomolecular layer and the other will be strongly adsorbed oxygen between O_2 and OH groups on the surface of cellulose. When filter paper was heated, weakly adsorbed oxygen molecules could be liberated firsty followed by strongly adsorbed oxygen molecules. These two process could produce the initial CL peaks shown in Figure 1.

Fig. 3 shows the photoluminescence curves of milk powders: (1) control is fresh milk powder taken out from a bottle, (2) milk powder stored for 2 weeks in a refregirator after being opened, and (3) milk powder stored in the refregirator for 2 weeks containing a package of deoxygenating agent. Emission decay curve for fresh milk powder stored for 2 weeks was almost similar to that for the fresh milk powder. The initial luminous intensity for milk powder containing a deoxygenating agent was lowered to almost 2 % of that for control. This is a direct evidence that oxygen molecules are the light emitter in the photoluminescence of milk powder because deoxygenated agents can absorb mainly oxygen from materials.

The fresh powder milk also was also taken bluish green photo with the color CCD camera.

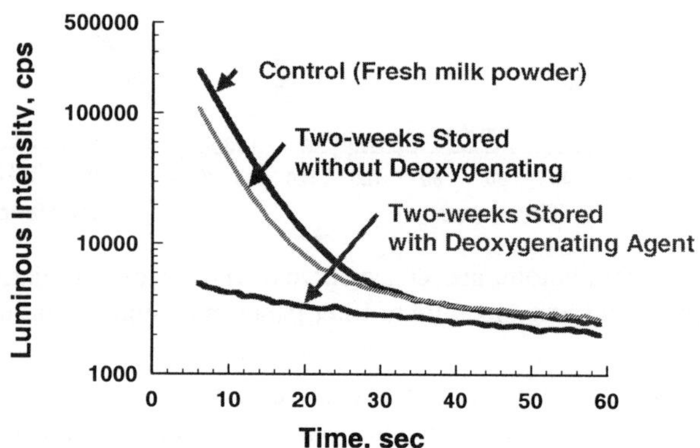

Figure 3.   Photoluminescence curves for a fresh milk powder (Control) irradiated with NUV light. Two glass bottles of milk powder were stored in a refrigerator for two weeks: one bottle contained deoxygenating agent and the other none.

CONCLUSIONS

Adsorbed oxygen molecules on the surface of organic materials could play vital role in chemiluminescence and photoluminescence.   The surface-bound oxygen molecules must be considered as significant in any autoxidation of organic materials at the ambient conditions because desorbed oxygen molecules must be energetic.

REFERENCES

1.   Usuki R, Kaneda T, Yamagishi A, Takyu C, Inaba H. Estimation of oxidative deterioration of oils and foods by measuring of ultra-weak chemiluminescence. J Food Sci 1979; 44: 1573-6.
2.   Suzuki T, Sunose T, Amasaki I, Ozawa T. Chemiluminescence of epoxy resin. Polym Degrad Stabil 2002; 77: 87-91.
3.   Kimura J-I, Salamone JC. eds. Nitrocellulose, Polymeric Materials Encyclopedia, Boca Raton, FL: CRC Press, 1996: 4582-7.
4.   George GA, Grassie N. eds. Use of chemiluminescence to study the kinetics of oxidation of solid polymers. Developments, Polymer degradation-3, London: Applied Science, 1981: 173-99.
5.   Slanger TG, Copeland RA. Energetic oxygen in the upper atmosphere and the laboratory. Chem Rev 2003; 103:4731-38.

THERMO- AND PHOTO-LUMINESCENCE FROM A PHOTOCATALYST TITANIUM DIOXIDE (TiO_2)

J-I KIMURA

CL Advisor, Tohoku Electronic Industrial Co., Ltd.
6-6-6 Shirakashidai, Rifu, Miyagi, 981-0134, Japan
jkr@ym.catv.ne.jp

INTRODUCTION

Spectroscopic techniques such as UV-Visible absorption/emission, FTIR, Raman, XPS and EPR have been extensively employed to investigate the photocatalytic activity of semiconductors as well as excited-state dynamics.[1] Molecular oxygen is often assumed as the oxidizing agent in photocatalytic oxidation reactions in solutions. The observed oxygenation has been believed to occur by generation of active specie or species at the photoexcited catalyst surface which desorbs into solution where reaction occurs.[2] Adsorbed oxygen onto the surface of TiO_2 serves as a trap for the photogenerated conduction band electron in many heterogeneous photocatalytic reactions in solution.[3] It has been found that photocatalytic activity is nearly completely suppressed in the absence of oxygen and the steady-state concentration of oxygen has a significant effect on the relative rate of photocatalyzed oxidation of organic molecules at TiO_2 particles in aerated water.[4]

The absence of completely acceptable TiO_2 photocatalytic oxidation mechanisms under atmospheric conditions might be explained by oxygen molecules adsorbed on the surface of TiO_2. Recently we have found that adsorbed oxygen molecules on the surface of organic materials can emit green luminescence by irradiation of NUV- and blue-light. The surface-bound oxygen molecules must be considered as significant in oxidation of organic materials at the ambient air and temperature. By analogy, it is natural to think that metal oxides could adsorb molecular oxygen on the surface because they are oxygen rich-compounds. Both anatase and rutile TiO_2 exhibited emission of weak light by stimulation of heat as well as light. We wish to demonstrate our luminescence techniques in TiO_2 research.

MATERIALS AND METHODS
Material
Both anatase and rutile titanium oxide powder were of chemical reagent grade provided by Kanto Chemical Co. , Japan.
Apparatus and Procedure
A luminescence spectrometer type CLA-FS1 was employed to measure both thermoluminescence and photoluminescence of TiO_2. Colour photographs of the photoluminescence for TiO_2 were taken with a high sensitive camera Cube-2DI equipped with a sensitive colour CCD (a product of Bitran Co., Japan). Both types of the apparatus are available from the Tohoku Electronic Industrial Co., Japan.

RESULTS AND DISCUSSION

Figure 1 (left) shows specially designed sample chamber equipped with two LEDs on the opposite sides of the chamber wall . The photograph in Figure 1 (right) is a image of photoluminescence of anatase TiO_2 (a-TiO_2) irradiated with the UV lamp in air. This photo was taken with Cube-2DI. There are many dark spots possibly due to apparent large particles (a few mm) formed by caking of the fine TiO_2 powders.

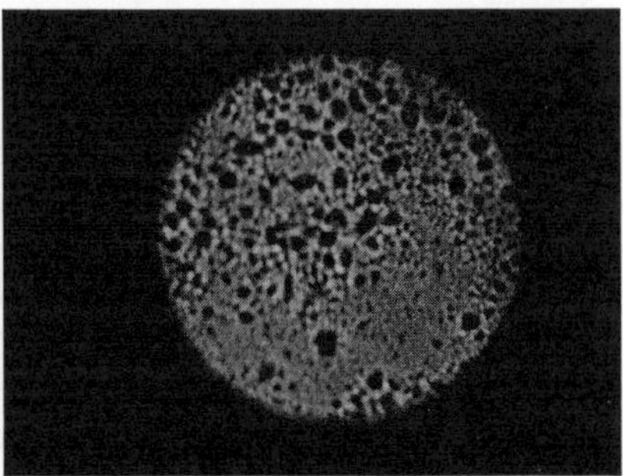

Figure 1. (Left) Experimental setup for photoluminescence spectrometer equipped with NUV (375 nm)- and RGB-LED. (Right) Photo of anatase TiO_2 irradiated with a NUV light in air.

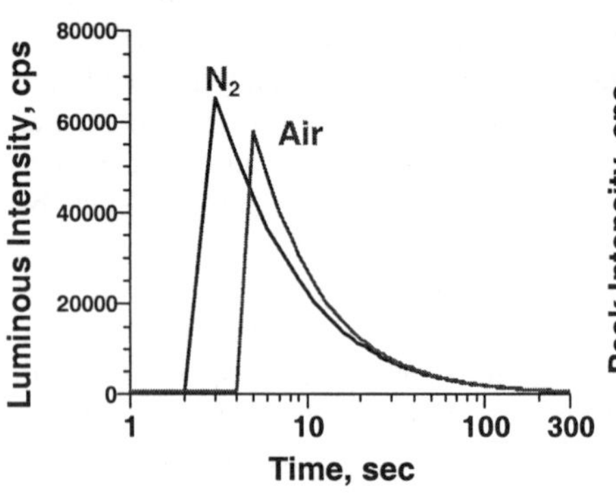

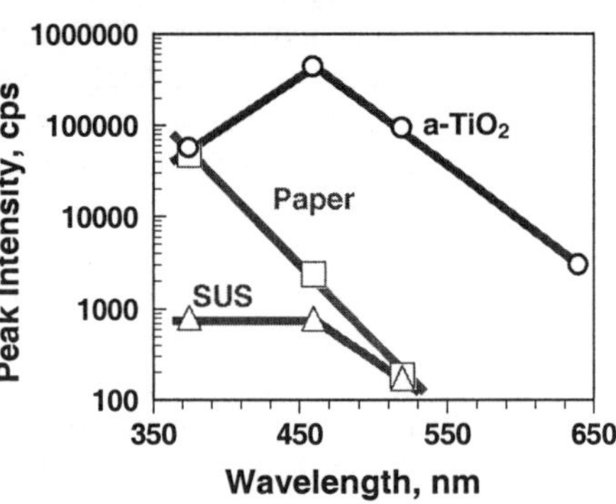

Figure 2. Photoluminescence curves for anatase TiO_2 powder irradiated with NUV-LED in air and nitrogen

Figure 3. Effect of wavelength on the peak luminous intensity for anatase TiO_2 powder, stainless steel cell and paper in air.

Figure 2 shows luminescence decay curves for the a-TiO_2 powder irradiated with NUV-LED both in air and in nitrogen. The peak luminous intensities measured in different atmospheres were the same within experimental error. One can see two steps in the decay curves. The half-life time of the first step was a few seconds which are the same as that for filter paper and stainless steel sample pan. The half-life time for the second step was as long as 300-400 sec. Figure 3 depicts the wavelength dependence of the photoluminescence peak intensity for a-TiO$_2$ powder, a stainless steel, and a filter paper. Surprisingly the a-TiO$_2$ exhibited the highest intensity at 460 nm (blue-LED). Moreover, the a-TiO$_2$ emitted considerable luminescence when it was irradiated with a red-LED.

Figure 4 shows luminescence spectra for the a-TiO$_2$ irradiated with NUV-, blue- and green-LED in air, where logarithm of luminous intensity was plotted against wavelength. All the spectra obtained for different LEDs were basically the same. Many peaks can be seen in visible light region and several in near NUV region below 400 nm. Several strong peaks appeared in the wavelength ranging from 600 nm to 700 nm. These spectra may suggest that the colour of the a-TiO$_2$ should be reddish white in accordance with the colour of the photograph shown in Figure 1.

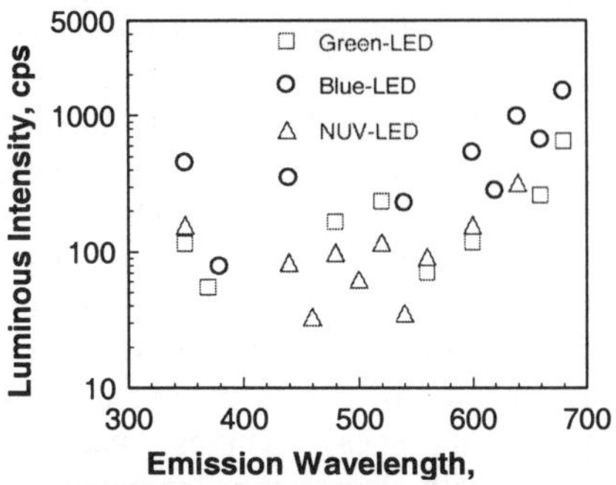

Figure 4. Photoluminescence spectra for a-TiO$_2$ powder irradiated with UV- and RGB-LED

Figure 5 shows the thermoluminescence curve of the a-TiO$_2$ heated to temperatures from 100 to 160 °C in air. Activation energy was obtained to be 93 kJ mol^{-1} (22 kcal mol^{-1}) as depicted in Figure 5. Activation energy for the a-TiO$_2$ heated in nitrogen was also obtained to be 118 kJ mol^{-1} (28 kcal mol^{-1}). These activation energies could be those for desorption of oxygen molecules from the surface of a-TiO$_2$ particles. Figure 6 shows the thermoluminescence spectrum of TiO$_2$ heated at a temperature of 160 °C in air. The thermoluminescence spectrum is similar to the photoluminescence spectrum shown in Figure 4.

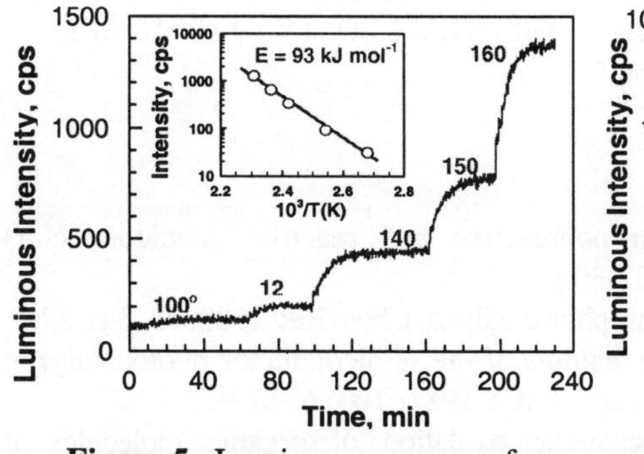

Figure 5. Luminescence curve for anatase TiO$_2$ powder heated in air.

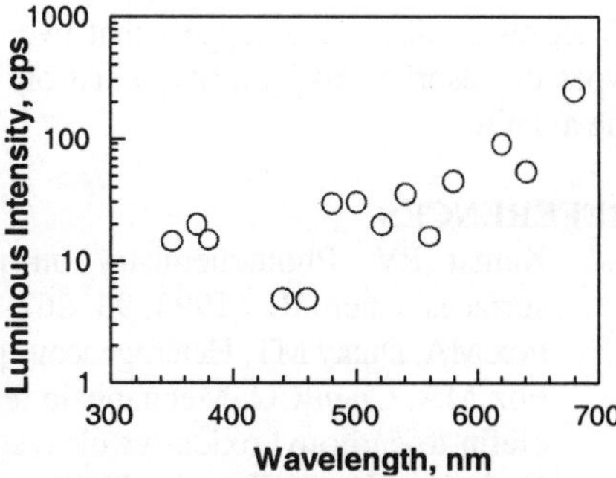

Figure 6. Luminescence spectra for anatase TiO$_2$ heated at 160 °C in air.

This agreement suggests that the light emitters of the a-TiO$_2$ would be the same for thermoluminescence and photoluminescence, which could be singlet oxygen molecules.   Extensive work has shown that organic waste materials can be decomposed on irradiated TiO$_2$.   Active species of the oxidative power of irradiated TiO$_2$ have been reported to be hydroxyl radical ($\cdot$OH) and superoxide anion (O$_2\cdot$-).[2] From the observed spectra (see Figures 4 and 6), several emission bands such as 350-380 nm, 480 nm, and 630 nm can be assigned to singlet oxygen.   It should be pointed out that ratio of electronically excited oxygen to thermally excited oxygen might be quite small.   Virtually energetic molecular oxygen (i.e., hot oxygen) may play predominant role in TiO$_2$-photocatalyzed oxidative reactions.

In order to confirm the adsorption of oxygen molecules on the a-TiO$_2$ powder simultaneous photo- and heat-activated luminescence measurements were conducted with NUV-LED as a light sauce.   Figure 7 summarized the results of luminescence decay curves of the a-TiO$_2$ measured at temperatures ranging from 15 to 90 °C in air. Observed decay curves exhibited initial luminous intensity and the slope of the exponential decay curves became smaller with increasing temperature. This fact can be attributed to the quenching by evolved molecular oxygen because the evolution rate of oxygen increased with increasing temperature as depicted in Figure 5. Accordingly reverse slope of the decay curves would be proportional to the quenching rate.   Arrhenius plot of the reversed slope gave a straight line and the activation energy was calculated to be of 42 kJ mol^{-1}.   Thermally excited luminescence gave the activation energy to be 93 kJ mol^{-1} (Figure 5). These two values may suggest that two types of adsorbed oxygen may exist on the a-TiO$_2$.

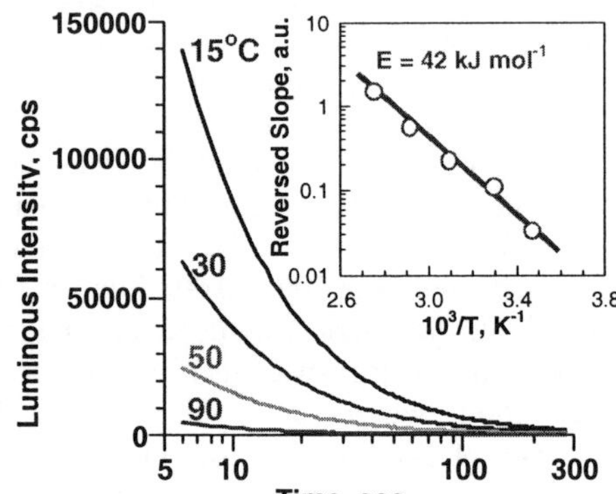

Figure 7. Simultaneous photo- and heat-luminescence of a-TiO$_2$ irradiated with NUV-LED at RT to 90 °C in air.

REFERENCES

1.   Kamat PV. Photochemistry on nonreactive and reactive (semiconductor) surfaces, Chem Rev 1993; 93: 267-300.
2.   Fox MA, Dulay MT. Heterogeneous photocatalysis, Chem Rev 1993; 93: 341-57.
3.   Fox MA, Chen CC. Mechanistic features of the semiconductor photocatalyzed olefin-to-carbonyl oxidative cleavage, JACS 1981; 103: 6757-9.
4.   Gerischer H, Heller A. Photocatalytic oxidation of organic molecules at TiO$_2$particles by sunlight in aerated water, J Electrochem Soc 1992; 139: 113-8.

RELATIONSHIP BETWEEN HEAT OF REACTION AND CHEMILUMINESCENCE EFFICIENCY OF CHEMILUMINESCENT REACTIONS

M KIMURA[1], H IGA[1], H ARAKI[1], M MATSUMOTO[2]

[1] *Dept of Chemistry, Okayama University, Okayama, 700-8530, Japan*
[2] *Dept of Chemistry, Kanagawa University, Hiratsuka, 259-1205, Japan*
kimuram@cc.okayama-u.ac.jp

INTRODUCTION

Primitive organisms had to excrete oxygen due to its toxic nature. We think that one of the reasons for toxicity is heat generation by oxidation, which may damage susceptible organisms. A luminescent organ probably exists as a device for oxygen treatment. McCapra noted that concerted [2+2] cleavage of a four member dioxetane should be a key step for changing stored energy to light.[1] A dioxetane structure should be a crucial intermediate for oxygen treatment, because of its ability to change energy into light as opposed to heat. In order to confirm the energy transformation, we investigated the relationship between heat of reaction and chemiluminescence efficiency. We selected three typical stable dioxetanes: dioxetane **1a** (chemiluminescence quantum yield = Φcl = 0.25), dioxetane **1b** (Φcl = 0.012), and dioxetane **2**(Φcl = 0.23), with the expectation that that a dioxetane with higher efficiency in chemiluminescence might generate a smaller amount of heat.[2-4] We have now measured calorimetrically the heat of reaction of [2+2] cleavage of **1a**, **1b**, and **2** in the solid state and in solution.

1a: R = Si(CH$_3$)$_2$t-Bu
1b: R = CH$_3$

2: R = Si(CH$_3$)$_2$t-Bu

155

MATERIALS AND METHODS

We prepared an aluminum capsule and two types of quartz cell. One is a plain transparent cell with an aluminum cap with a small hole for filling with a solution. The other is coated with Chinese ink made of carbon for absorbing thermal energy formed in the reaction of the dioxetanes (Figure 1).

  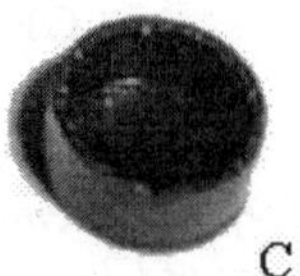

Figure 1. A: An aluminum capsule   B: A transparent cell   C: A coated cell

Thermolysis of solid **1a, 1b**: 1) in an aluminum capsule, 1 mg of **1a** or **1b** were sealed in a cylindrical, air-tight, pressure-resistant, aluminum sample-holder (volume 30 mm^3) and were then decomposed in a DSC calorimeter at a constant heating rate (10 °C/min).

Reaction of dioxetane **1a** and **2** initiated by addition of TBAF/DMSO: In a representative case, 0.3 mg of **1a** was dissolved in ca. 10 µL of DMSO and the reaction initiated by addition of 10 µL of 1 M TBAF/DMSO solution at room temperature.

RESULTS

Figure 2 shows the change in heat flow with temperature; the area under the curve is a measure of heat evolved. $Q1 = -72.9{\pm}1.1$ kcal/mol for solid **1a**, $Q2 = 76.0 \pm 1.23$ kcal/mol for liquid **1b** as a result of three measurements. In the two types of quartz cell, $Q3 = -69.8$ kcal/mol for solid **1a** in a coated quartz cell and $Q4 = -71.5 \pm 1.25$ kcal/mol for solid **1a** in a transparent quartz cell (result of three measurements). Under these conditions, **1a** has a melting point of ca. 134 °C and the heat of melting can be estimated as ~5 kcal/mol. Since the measurements were made at constant volume, Q (reaction heat) is proportional to the internal energy ΔU. We did not recalculate the enthalpy ΔH because part of heat of melting and sublimation (appeared from 170 to 210 °C, ~15 kcal/mol) of adamantanone are included, for which correction causes a loss of reliability of the analysis.

Q5 had a value of -56 kcal/mol for **1a** and Q6 had a value of -47 kcal/mol for **2** in coated cells and Q7= -49kcal/mol for **1a**, Q8 = -42 kcal/mol for **2** in transparent cells (result of three measurements). The reason these Q's are lower than those in the case of solid samples is under investigation although we have as yet no satisfactory explanation.

From the retro [2 + 2] reaction of **1a**, **1b** and **2** in the solid state into ketones and a light emitting counter parts, thus 72 to 76 kcal/mol are liberated [see equation

1)]. Several calculations indicate that between 60 and 100 kcal/mol should be liberated from the [2 + 2] ring opening reaction of dioxetanes.[5] Thermolysis of tetramethyl-1, 2-dioxetane by Lechtken and Hoehne gave a reaction enthalpy of 71 kcal/mol with an activation energy of 25 kcal/mol, which is just enough to produce singlet or triplet acetone.[6,7] In the case of solid **1a** and **2**, a significant difference in heat evolution was not observed between thermolysis in a coated quartz cell and in a transparent cell. The quantum yield of **1a** was 0.0013 measured based on the value for **1b**, which is very low considering the high efficiency ($\Phi cl = 0.25$) obtained by initiation with tetrabutylammonium fluoride (TBSF/DMSO. This DSC method probably can not be applied to measuring such small difference (> 6%) in the heat of reaction.

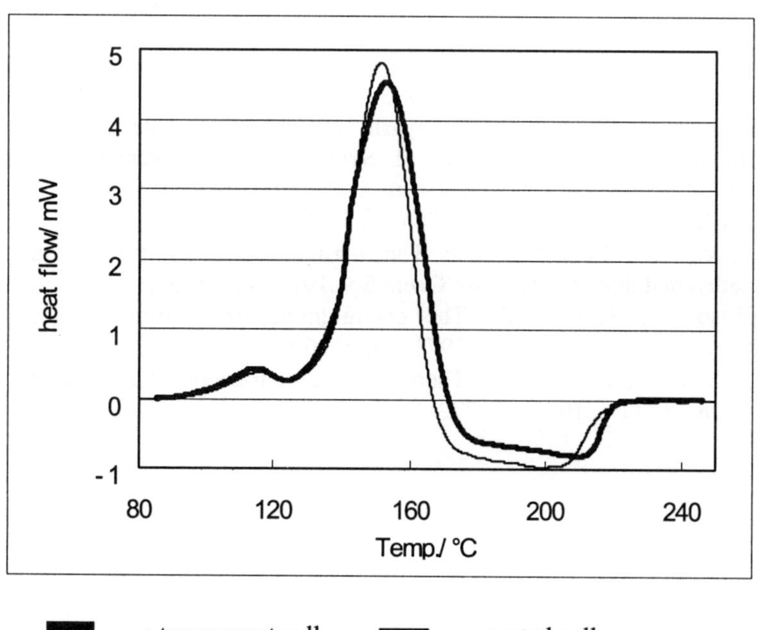

a transparent cell   a coated cell

Figure 2.   DSC thermograms of dioxetane **1a**

In the case of the reaction of dioxetanes **1a** and **2** initiated by addition of TBAF/DMSO, significant differences in heat evolution were observed in the comparison between reactions in a coated quartz cell and in a transparent cell. The quantum yield of chemiluminescence measured by Schaap et al., for **1a** is 0.25 which is high enough to measure difference in reaction heat between reactions in a coated quartz cell (Q5, Q6) and a transparent cell (Q7, Q8).[2] The reaction heat differences measured for the coated and transparent cells by DSC method, $\Delta E1 = Q5 - Q6 = 6.6$ kcal/mol and $\Delta E2 = Q6 - Q7 = 5.5$ kcal/mol represented the amount of reaction heat converted to light. Considering the transparent area of the coated cell is 50% of the whole surface area, ca. 23% of the reaction heat is transferred to emitted light energy, which is consistent with the chemiluminescence efficiencies of **1a** and **2**. These results confirm that stored energy in dioxetanes is transferred into harmless visible light calorimetrically.

REFERENCES

1. McCapra F. Charge transfer dioxetanes–simple rationalization. Tetrahedron Lett 1993: 6941-4.
2. Schaap AP, Chen TS, Handley RS, DeSilva R, Giri BP.Chemical and enzymatic triggering of 1,2-dioxetanes. 2: Fluoride-induced chemiluminescence from tert-butyldimethylsiloxy-substituted dioxetanes. Tetrahedron Lett 1987; 28: 1155-8.
3. We thank Prof. K. Fujimori in the Tsukuba University for the information on the quantum yield of **1b**.
4. Matsumoto , Watanabe N, Kobayashi H, Matsubara J, Kitano Y, Suganuma H, Matsubara J, Kitano Y, Ikawa H. Synthesis of 3-alkoxymethyl-4-aryl-3-tret-butyl-4-methoxy-1,2-dioxetanes as a chemiluminescent substrate with short half-life emission. JCS Chem Commun 1995: 431-2.
5. O'Neal HE, Richardson WH. The thermochemistry of 1,2-dioxetane and its methylated derivatives. J Am Chem Soc.1970; 92: 6553-6.
6. Wilson T, Schaap AP. The chemiluminescence from cis-diethoxy-1, 2 dioxetane. J Am Chem Soc 1971; 93: 4126-6.
7. Lechtken P, Hoehne G. Thermolysis of tetramethyl-1,2-dioxetane. Angew Chem   Intl Ed. 1973; 12: 772-3.

THE HIGH ENERGY KEY INTERMEDIATES IN THE PEROXYOXALATE CHEMILUMINESCENCE OF 2,4,6-TRICHLOROPHENYL *N*-ARYL-*N*-TOSYLOXAMATES

R KOIKE, J MOTOYOSHIYA, H AOYAMA

Department of Chemistry, Faculty of Textile Science & Technology,
Shinshu University, Ueda, Nagano, 386-8567, Japan
Email: koikery@pmac103.shinshu-u.ac.jp

INTRODUCTION

There has been a growing interest in the mechanism of the peroxyoxalate chemiluminescence. A CIEEL mechanism[1] can be applied to this chemiluminescence, and many efforts have been made to determine the high energy key intermediate that interacts with the fluorophores. In the general peroxyoxalate chemiluminescence, the most likely key intermediate are dioxetanedione (**I**)[2] or dioxetanone (**II**),[3] the latter of which still bears an eliminating group. Therefore, it is of significance to elucidate which is the key intermediate. In the course of our study of peroxyoxalate chemiluminescence,[4] we report here the results of the mechanistic study using the reaction of 2,4,6-trichlorophenyl *N*-aryl-*N*-tosyloxamates[5] with the hybrid structure of well-known TCPO [bis(2,4,6-trichlorophenyl)oxalate] and the oxamides.

RESULTS AND DISCUSSIONS

The reaction of the oxamates (**1**) and aqueous hydrogen peroxide in THF provided chemiluminescence in the presence of the fluorophores, and whose emission spectra were in good agreement with the fluorescence spectra of the fluorophores. However, the bisoxamide PhNTsCOCONTsPh (**2**) is unreactive to an aqueous hydrogen peroxide, and the reaction of EtOCOCOOTCP (**3**) with hydrogen peroxide under the same conditions formed ethyl monoxalate EtOCOCOOH and TCPOH without light emission. Therefore, in the reaction of the oxamates the initial nucleophilic acyl substitution takes place at the carbonyl of the TCP site and then gives tosylanilides and CO_2 *via* the cyclic peroxides. The reaction of the oxamates with hydrogen peroxide was monitored by ^{1}H NMR spectrum under the pseudo-first order conditions in THF-d_8. As shown in Figure 1(a), a Hammett relationship between the σ-values and the elimination kinetic constants was established (ρ = + 1.75) except the case of *p*-dimethylamino derivative (**1g**), in which dimethylaminophenyl-*N*-tosylanilide was accompanied. A good Hammett relationship between the initial maximum intensities (I_0) and the σ-values in the presence of DPA (9,10-diphenylanthracene) was also observed under neutral or basic conditions. Assuming that I_0 is proportional to the concentration of the key intermediate, it can be regarded as a criterion for the generation rate of the high-energy intermediates. While the ρ-value was estimated to be +2.66 under neutral conditions, it decreased to + 1.20 under basic conditions using Na_2CO_3 as shown in Figure. 1(b), (c). Considering the Bender's study of the amide hydrolysis, these results show that the high energy key intermediate that interacts with the fluorophores will be dioxetanone (**II**) still bearing the eliminating group rather than dioxetanedione (**I**). The ρ-value of the amide hydrolysis was estimated to be + 0.1 by Bender *et al.*[6], in which a proton transfer generating a quaternary amino group is involved before an elimination of the amide. If dioxetanedione (**I**) is the key

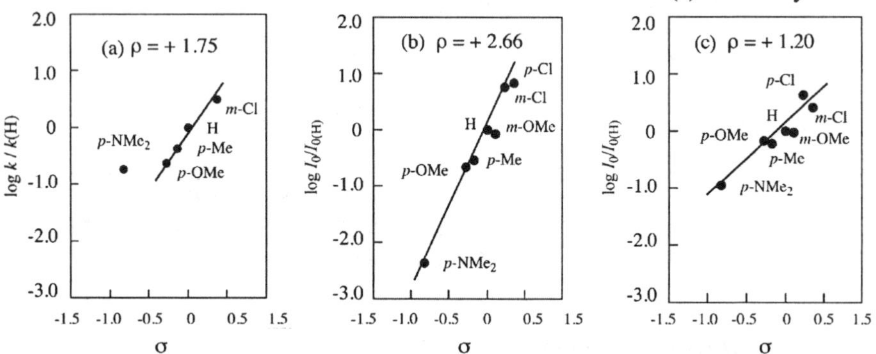

Figure 1. The Hammett relationship in the chemiluminescence reaction of oxamates (**1**). (a) Product formation. (b) Under neutral conditions. (c) Under basic conditions.

intermediate, the ρ-value should be small similarly to the amide hydrolysis, and furthermore, the ρ-value should be large under basic conditions because of stability of eliminated *N*-tosylanilide anions. Thus, the interaction of the crucial intermediate with the fluorophores takes place before the liberation of the tosylanilides.

On the other hand, when the oxamate **1h** having a fluorescent naphthyltosylamide group reacted with hydrogen peroxide, only a feeble chemiluminescence was observed in the absence of a fluorophore. But when fluorescent *N*-2-naphthyl-*N*-tosylanilide (**4**) was externally added, an enhanced light emission was observed, whose chemiluminescence spectrum was in good agreement with the fluorescence spectrum of **4** having the maximum intensity at 402 nm. The double reciprocal plot of Φ_{CL} vs. the concentration of **4** was found to be a straight line as shown in Figure 2, establishing a bimolecular process between **4** and the key intermediate. This relation holds even when the concentration of **4** is lower than that of **1h**. A similar bimolecular process was shown in the reaction of other oxalates having the fluorescencent phenol groups in the presence of DPA. Therefore, dioxetanone (**II**) is strongly suggested as the key intermediate. If dioxetanedione (**I**) is the key intermediate, chemiluminescence should be observed in the absence of externally added **4**, and Φ_{CL} should not depend on the concentration of **4**.

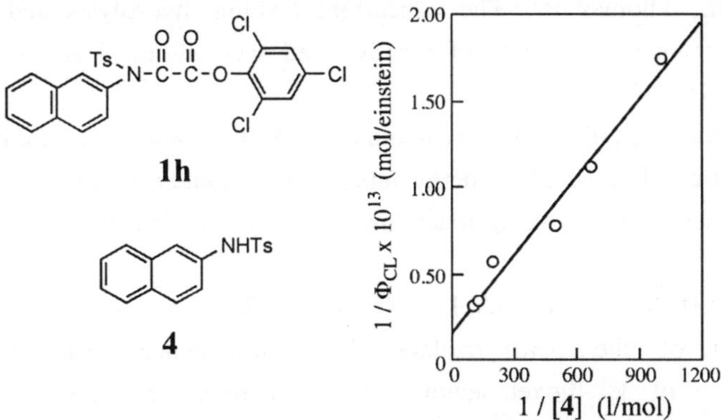

Figure 2. Double reciprocal plot of Φ_{CL} vs. [**4**]

From these results, the most likely key intermediate that interacts with fluorophores is dioxetanone (**II**) still bearing the eliminating group rather than dioxetanedione (**I**) in this system. Recently, the six-membered ring was suggested by Lee *et al.*,[7] and B. Rechard *et al.*[8] reported that the key intermediate was a dioxetanedione in the certain peroxyoxalate chemiluminescence reactions. However, we believe that the results obtained in the present study can be applied to the general peroxyoxalate chemiluminescence.

REFERENCES

1. Schuster GB. Chemiluminescence of organic peroxide. Conversion of ground-state reactants to excited-states products by the chemically initiated electron-exchange luminescence mechanism. Acc Chem Res 1979; 12: 366-73.

2. Stevani CV, Lima DF, Toscano VG, Baader WJ. Kinetic studies on the peroxyoxalate chemiluminescence reaction: imidazole as a nucleophilic catalyst. J Chem Soc Perkin Trans 2 1996: 989-95.

3. Catherall CLR, Palmer TF. Chemiluminescence from reactions of bis(pentachlorophenyl)oxalate, hydrogen peroxide and fluorescent compounds. J Chem Soc Faraday Trans 2 1984; 80: 837-49.

4. Motoyoshiya J, Sakai N, Imai M, Yamaguchi Y, Koike R, Takaguchi Y, Aoyama H. Peroxyoxalate chemiluminescence of *N,N'*-bistosyl-1H,4H-quinoxaline-2,3-dione (TsQD) and related compounds. Dependence on electronic nature of fluorophores. J Org Chem 2002; 67: 7314-8.

5. Koike R, Motoyoshiya J, Takaguchi Y, Aoyama H. The key intermediates that interact with the fluorophores in the peroxyoxalate chemiluminescence reaction of 2,4,6-trichlorophenyl *N*-aryl-*N*-tosyloxamates. Chem Commun 2003: 794-5.

6. Bender ML, Thomas RJ. The concurrent alkaline hydrolysis and isotopic oxygen exchange of a series of *p*-substituted acetanilides. J Am Chem Soc 1961; 83: 4183-8.

7. Lee JH, Rock JC, Park SB, Schlautman MA, Carraway ER. Study of the characteristics of three high-energy intermediates generated in peroxyoxalate chemiluminescence (PO-CL) reactions. J Chem Soc Perkin Trans 2 2002: 802-9.

8. Bos R, Barnett NW, Dyson GA, Lim KF, Russell RA, Watson SP. Studies on the mechanism of the peroxyoxalate chemiluminescence reaction part 1 confirmation of 1,2-dioxetanedione as an intermediate using ^{13}C nuclear magnetic resonance spectroscopy. Anal Chem Acta. 2004; 502: 141-7.

CHEMILUMINESCENCE STUDIES ON THE PHOTOCHEMICAL PRODUCTION OF HYDROGEN PEROXIDE FROM PORPHYRINS AND THEIR AGGREGATES

K KOMAGOE, S OSADA, T SHINDO, K TAMAGAKE

Faculty of Pharmaceutical Sciences, Okayama University
1-1-1 Tsushima-naka, Okayama 700-8530, Japan

INTRODUCTION

Photodynamic therapy (PDT) for tumor treatment is one of the major topics in porphyrin chemistry as well as in photobiology in recent years.[1] Although singlet oxygen produced by energy transfer is widely accepted as a key intermediate, an electron transfer mechanism which may cause a direct formation of superoxide or hydrogen peroxide should not be overlooked. The difficulty to distinguish these mechanisms lies on the fact that these active intermediates are mutually convertible.[2] From a theoretical point of view, however, we expect that an aggregation effect would be suppressive or promotive depending if an energy transfer or electron transfer operates. The purpose of this work is to find a clear experimental evidence for the promotion of electron transfer by aggregation of porphyrins.

EXPERIMENTAL

5-50 $\times 10^{-6}$M aqueous solutions of tetrakis(N-methyl-4-pyridyl)porphin (TMPyP) and/or tetrakis(4-sulfonatophenyl)porphin (TSPP) were placed under the light from a 1kW slide projector for 5-10 min and put into an plastic cuvette, then 5×10^{-5}M luminol in 0.1M Na_2CO_3 buffer together with hemine or FeTMPyP as a catalyzer were injected to it. The emission intensity from the cell was recorded for 2-10 minutes immediately after the mixing. The amounts of photoproduced hydrogen peroxide were determined by comparing the observed intensities with the standard ones at various concentrations of pure hydrogen peroxide. Absorption spectra of the solutions were also measured in order to check the status of porphyrins and the possible photodegradation. The photoactivities of uroporphyrin (UP) and coproporphyrin (CP) were also examined at various pH in the same way.

RESULTS

As shown in Fig.1, either TMPyP or TSPP showed almost no emission when they are used alone whereas the emission from the one-to-one mixture of those solutions was quite strong. The time profiles of the emission were quite similar to those obtained with pure hydrogen peroxide. Addition of catalase suppressed the emission completely. Those observations assured us that the chemiluminescence is solely due to the hydrogen peroxide produced by the light irradiation.

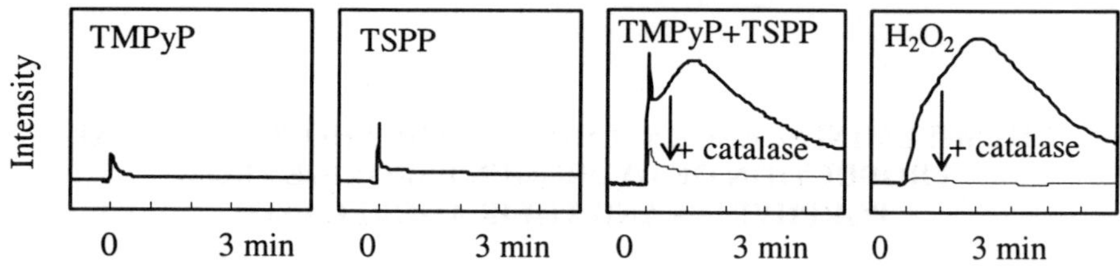

Figure 1. Luminol chemiluminescence stimulated by light irradiated porphyrin solutions and pure hydrogen perpxide

The mixtures showed broadened absorption spectra compared to those of individual spectra indicating the formation of aggregate due to the electrostatic attractions between the cationic and anionic porphyrins. The size of the aggregate were expected larger than 5×10^{-7}m since the filtrated solution from a membrane filter showed neither absorption nor photoactivity. Use of deuterium oxide or heavy water as solvent had no effect on the emission intensities suggesting involvement of singlet oxygen is unlikely in this process.

UP which has eight ionizable carboxylic groups also showed strong emission at pH>7 where no aggregates are expected because of the highly negative charge of the ionized UP (Fig.2).

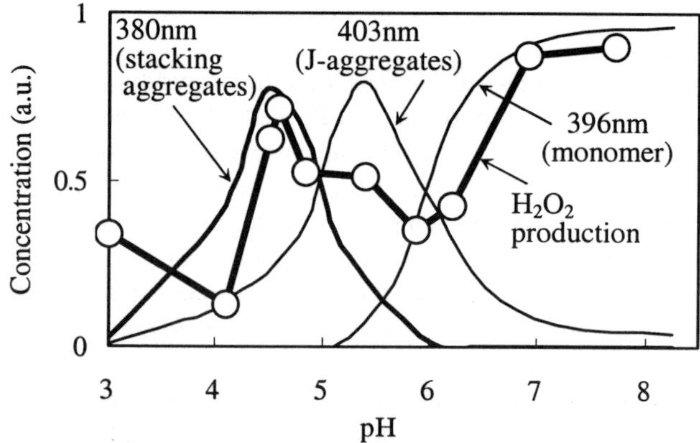

Figure 2. Change of photoproduction rate for hydrogen peroxide and aggregation types of porphyrin with pH

Lowering the pH to 6 caused a decrease in the emission intensity reflecting the partial decrease of charge. Further lowering of pH caused an increase of the intensity again giving a maximum at pH4.5 where aggregation was confirmed spectroscopically. Under pH4, a sharp drop took place corresponding to the complete loss of the solubility of UP. Although UP is already known as a photosensitizer for the production of hydrogen peroxide,[3] this is the first work which suggests that the production mechanism is bimodal. The high efficiency at higher pH should arise from the highly negative charge of the parent molecule which increases the electron releasing ability to O_2. The drop of the efficiency around pH 6 should be due to decrease of the negative charge by partial protonation. The second maximum at pH 4-5 seems to correlate to the case of TMPyP-TSPP aggregate.

Absorption spectra of UP are also pH dependent strongly. The monomeric band at 396 nm (pH>7) shifts to 403 nm at pH 5.4 and then to 380 nm at pH 4.5 (Fig.3).

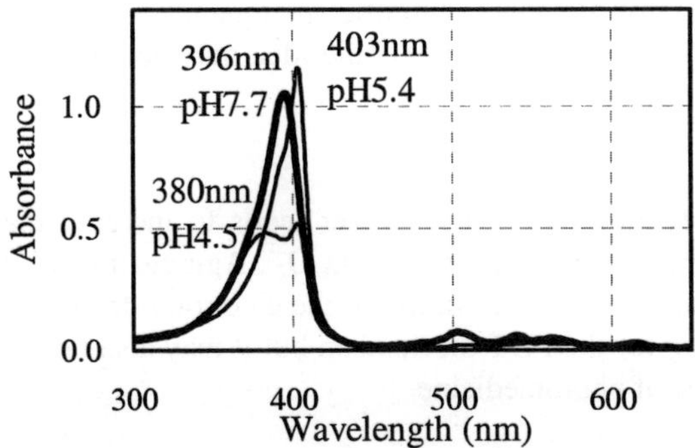

Figure 3. Absorption spectra of UP at various pH

Based on the characteristic features of these bands, we assigned them to monomer, J-like aggregates and pi-pi stacking aggregates. The change of the H_2O_2 production curve is well correlated to the status of the porphyrin in the solution as shown in Fig.2. CP showed quite similar result as UP.

DISCUSSIONS

Photochemistry of aggregated compounds has not drawn much attention because it usually tends to shorten the lifetime of excite states and reduces the quantum yield of photoproducts.[4] However, it may act as a favorable factor in charge transfer systems. Our observation of the enhanced photoproductivities of hydrogen peroxide by both TMPyP-TSPP complex and UP aggregates clearly supported such an idea. Among two types of aggregates, only pi-pi stacking aggregates (380 nm) showed the enhancement. According to semiempirical MO calculations, the other type of aggregation, a 'J-aggregate', was speculated to form a coplanar aggregation linked by three strong hydrogen bonds between partially protonated carboxylic groups (COOH)-(COO⁻) which are stretched out from the two porphyrins. Pi-pi interaction is unlikely in such aggregates. These considerations lead to a plausible conclusion that the electrons released from porphyrin to oxygen are pi electrons and only aggregates which make pi-pi interaction possible are responsible for the promotion of electron release. The high efficiency observed at pH>7 in UP should be understood as due to the elevation of the potential for the pi electrons in the porphyrin ring by complete ionization of the surrounding carboxylic groups.

For the PDT mechanism, there have been many reports about involvement of singlet oxygen produced by energy transfer from excited state of porphyrins or other photosensitizers.[5] Our intention is not to deny such mechanism however the electron transfer mechanism which leads to production of O_2^- or hydrogen peroxide by the following scheme

$$Por + light \rightarrow Por^*, \quad Por^* + O_2 \rightarrow Por^+ + O_2^-, \quad 2O_2^- + 2H^+ \rightarrow H_2O_2 + O_2$$

should not be overlooked since accumulated porphyrins in tumor cells may aggregate to some extent because of the poor solubilities in aqueous environment. Even if such scheme is not the case for the actual PDT, the idea may help to develop a new type of photomedicine.

CONCLUSIONS

We found that aggregation of photosensitizer tends to increase the photochemical production of hydrogen peroxide probably via O_2^-. Aggregation which induces pi-pi interaction is responsible for the activation of the electron release. The importance of this finding is not only for the PDT mechanism but it may also be applied to develop an idea for a new type of photomedicine.

REFERENCES

1. Lane N. New light on medicine. Sci Amer 2003;288:38-45.
2. Hoebeke M, Schuitmaker HJ, Jannink LE, Dubbelman TMAR, Jakobs A, Vorst AV. Electron spin resonance evidence of the generation of superoxide anion, hydroxyl radical and singlet oxygen during the photohemolysis of human erythrocytes with bacteriochlorin a. Photochem Photobiol 1997;66:502-8.
3. Menon IA, Becker MAC, Persad SD, Haberman HF. Quantitation of hydrogen peroxide formed during UV-visible irradiation of protoporphyrin, coproporphyrin and uroporphyrin. Clin Chim Acta 1989;186:375-81.
4. Damoiseau X, Tfibel F, Hoebeke M, Fontaine-Aupart MP. Effect of aggregation on bacteriochlorin a triplet-state formation : a laser flash photolysis study. Photochem Photobiol 2002;76(5):480-5.
5. Nyman ES, Hynninen PH. Research advances in the use of tetrapyrrolic photosensitizers for photodynamic therapy. J Photochem Photobiol B:Biology 2004;73:1-28.

EXCITED STATES OF DIOXINS AS STUDIED BY AB INITIO QUANTUM CHEMICAL COMPUTATIONS: ANOMALOUS LUMINESCENCE CHARACTERISTICS

T KOTO, K TOYOTA, K SATO, D SHIOMI, T TAKUI

*Departments of Chemistry and Materials Science, Graduate School of Science,
Osaka City University, 3-3-138 Sugimoto, Sumiyosi-ku, Osaka, 558-8585, Japan
Email: t-koto@sci.osaka-cu.ac.jp*

INTRODUCTION

The luminescence of dioxins, notorious environmental pollutants, has been attracting interest from both the experimental and theoretical perspective. Ryzhikov et al. observed the luminescence quantum yields of unsubstituted dibenzo-*p*-dioxin (DD) and the related molecules, and they attributed the very low quantum yield of fluorescence (ϕ_{fl}) in DD to fast intersystem crossings to the low-lying triplet $n\pi^*$ states.[1] Phosphorescence spectra of dioxins have unique vibronic structures in each congener and are very useful for their detection and identification.[2,3] However, theoretical interpretation on the luminescence spectra is not straightforward due to insufficient information on energy levels of the low-lying excited states. Semiempirical methods, density functional theory (DFT), and configuration interaction with single excitations (CIS) method have been used in calculating the singlet and triplet excited states.[2-5] However, the predicted energies were not always satisfactory for direct comparison with experiment. Obviously, more accurate calculations are desired for correct understanding of the low-lying excited states and luminescence characters of dioxins. We present a new theoretical interpretation on the low fluorescence quantum yields of dioxins, based on high-level ab initio calculations. In the recent study,[6] we have studied the theoretical spectra of dipole-allowed singlet excited states and detailed assignment of the observed UV spectra of DD,[1] 2,3,7,8-tetrachlorodibenzo-*p*-dioxin (TCDD), and octachlorodibenzo-*p*-dioxin (OCDD),[7] obtained by the symmetry adapted cluster configuration interaction (SAC-CI) calculations,[8] reproducing the observed bands within an error of 0.3 eV.

COMPUTATIONAL METHODS

The equilibrium structures of DD, TCDD, and OCDD were obtained by DFT B3LYP/6-31G* calculations. In the SAC-CI calculations, the Huzinaga-Dunning double zeta plus polarization (DZP) basis set[9] was selected for chlorine atoms and DZP + diffuse basis of Chipman[10] for hydrogen, carbon, and oxygen atoms. Vertical excitation energies and oscillator strengths were calculated for six excited states for all symmetries. We chose the *z*-axis to be perpendicular to the molecular plane, and the *y*-axis to be along two oxygen atoms. To discuss the difference for ϕ_{fl} between DD and 9,10-dihydroanthracene (DHA), six excited states for all dipole allowed and forbidden symmetries were calculated. We carried out the present calculations by using Gaussian 03 suit of programs.[11]

RESULTS

Molecular structure

We obtained planar D_{2h} structures for DD, TCDD, and OCDD in good agreement with X-ray diffraction data. The equilibrium structure for DHA is bent along an axis through two methylene groups, and in the C_{2v} point group rather than D_{2h}. The SCF energies of DD, TCDD, and OCDD are -609.925861, -2444.469648, and -4279.978835 in hartree, respectively.

Assignment of UV spectra of DD, TCDD, and OCDD

Recently, we have calculated excited singlet states of dioxins for the dipole-allowed symmetries (B_{1u}, B_{2u}, and B_{3u}) in which electronic absorption spectra between ca. 150 and 300 nm are involved, presenting the comprehensive theoretical spectra of DD, TCDD and OCDD by SAC-CI.[6] All the B_{2u} and B_{3u} states in the observed region are of $\pi\pi^*$ character. Transitions to the B_{1u} states are of very small oscillator strength due to the direction of the transition moment perpendicular to the molecular plane. The long-wavelength band (A) located around 300 nm (4.13 eV) and the strongest band (B) in the short-wavelength is attributed to the $1B_{3u}$ and $2B_{3u}$ state, respectively. The short-wavelength band includes another electronic state $1B_{2u}$, which probably contributes to the band C seen as either a maximum or a shoulder in experiment. The band D near the edge of the observation range in TCDD and OCDD is assigned the transition to the $2B_{2u}$ state. Several bands that have not been documented so far are predicted between 150-200 nm.

Theoretical interpretation of the large difference in the quantum yield of luminescence between DD and DHA

Ryzhikov et al. reported the quantum yield of fluorescence (ϕ_{fl}) of DD and its analogue DHA is 0.003 and 0.26, respectively, which are quite different from each other.[1] Their proposed mechanism is fast intersystem crossing to the low-lying triplet $n\pi^*$ states for the low value of ϕ_{fl} in DD in terms of the El-Sayed rule.[12]

In addition to the above-mentioned optically allowed excited singlet states, we have also calculated the low-lying excited states of the forbidden A_g, B_{1g}, B_{2g}, B_{3g}, and A_u symmetries to search for all low-lying electronic states at the SAC-CI level of theory for DD. Because DHA is of C_{2v} symmetry, we carried out calculations for DHA having D_{2h} symmetry as well as C_{2v} to consider two effects: a bending effect as well as an effect of substitution in 9 and 10 positions. The present calculation showed that any $n\pi^*$ states involving lone pairs of oxygen atoms do not appear in the observed range up to 200 nm. This shows the above-mentioned assumption of the presence of the low-lying $n\pi^*$ states is not likely.

The lowest allowed excited singlet state of DD is the $1B_{3u}$ state at 4.26 eV, which is responsible for the band A. This $1B_{3u}$ state was found to be S_3, below which two forbidden states S_1 ($1B_{1g}$) and S_2 ($1A_g$) were found at 4.14 and 4.18 eV, respectively. In DHA the S_1 state was found to be $1B_1$ (4.83 eV). The calculation clearly shows that the S_1-S_0 transition is optically allowed in DHA but forbidden in DD. Thus, according to the Kasha's rule, ϕ_{fl} should be lower in DD than that in DHA.

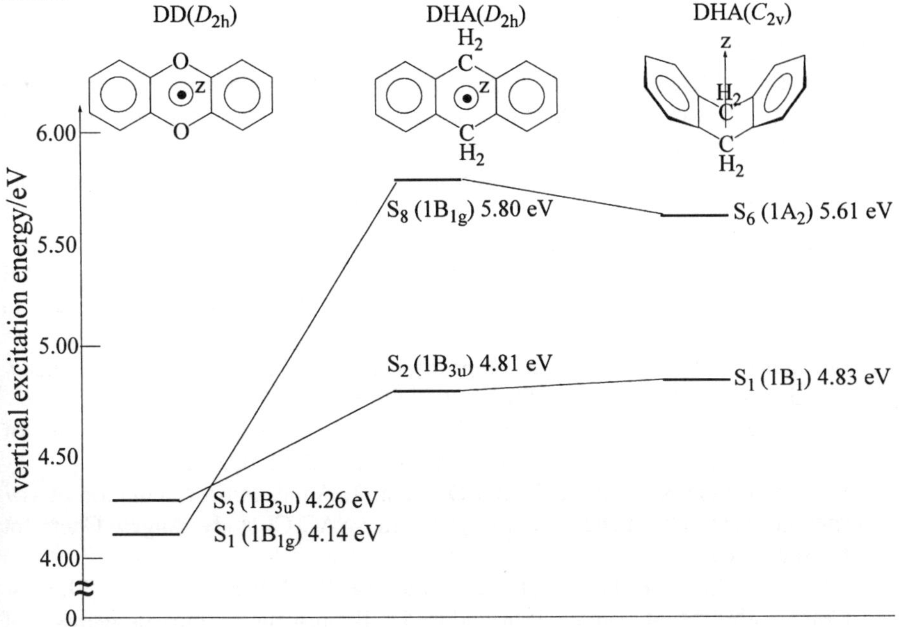

Figure 1. Energy diagram of excited singlet states of DD, DHA (D_{2h}), DHA (C_{2v})

Fig. 1 illustrates the energy levels of excited singlet states with respect to structural changes from DHA to DD. The $1B_{3u}$ (S_3) state in DD corresponds to the $1B_1$ (S_1) state in DHA, and the $1B_{1g}$ (S_1) state in DD corresponds to the $1A_2$ (S_6) state in DHA, according to the calculated electronic structures. Clearly, the remarkable energy lowering of the $1B_{1g}$ state makes the S_1-S_0 transition forbidden in DD. In Fig.1 the energy levels were shown for the structure of DHA optimized under the D_{2h} point group as well as the C_{2v} equilibrium structure to see the substitution and geometry effects separately. The change in energy is larger for oxygen-substitution than for the symmetry change, indicating that the energy of the state is highly sensitive to substitution at these positions.

This large energy shift in the forbidden $1B_{1g}$ ($1A_2$ in DHA) state may be explained as a rough estimation in terms of orbital energies between $3b_{1u}$ ($3a_1$) (HOMO) and $2a_u$ ($12a_2$). HOMO is destabilized by 0.53 eV in the substitution of oxygen atoms for methylene groups, and the overall destabilization including the bending effect is 0.72 eV. This destabilization of HOMO is mainly caused by anti-bonding interaction between oxygen and carbon atoms, while the change of orbital energy of the virtual $2a_u$ ($12a_2$) orbital is relatively small since the 9 and 10 position is node for the a_u symmetry. Thus, very low ϕ_{fl} can be explained by the effect of oxygen atoms on the π orbitals.

REFERENCES

1. Ryzhikov MB, Rodionov AN, Stepanov AN. Spectral luminescence characteristics of the dihetero derivatives of dihydroanthracene with group VI elements. Russ J Phys Chem 1989;63:1378-80.
2. Gastilovich EA, Klimenko VG, Korol'kova NV, Rauhut G. Excited electronic states and effect of vibronic-spin-orbit coupling on the radiative deactivation of the lowest triplet states of dioxin. Chem Phys 2001;270:41-54.
3. Gastilovich EA, Klimenko VG, Korol'kova NV, Nurmukhametov RN. Optical spectra and photophysical properties of polychlorinated dibenzo-p-dioxin derivatives. Russ Chem Rev 2000;69:1037-56.
4. Okamoto Y, A new dioxin decomposition process based on a hybrid density functional calculation. Chem Phys Lett 1999;310:355-60.
5. Hirokawa S, Imasaka T, Urakami Y. MO study on the $S_1 \leftarrow S_0$ transitions of polychlorinated dibenzo-*p*-dioxins. THEOCHEM 2003;622:229-37.
6. Koto T, Toyota K, Sato K, Shiomi D, Takui T. An ab initio calculation of UV spectra of polychlorinated dibenzo-*p*-dioxins: SAC-CI study. Angew Chem Int Ed submitted.
7. Funk DJ, Oldenborg RC, Dayton D-P, Lacosse JP, Draves JA, Logan TJ. Gas-phase absorption and laser-induced fluorescence measurements of representative polychlorinated dibenzo-*p*-dioxins, polychlorinated dibenzofurans, and a polycyclic aromatic hydrocarbon. Appl Spectrosc 1995;49:105-13.
8. Nakatsuji H. Cluster expansion of the wavefuncton. Excited states. Chem Phys Lett 1978;59:362-4.
9. Dunning TH Jr. Gaussian basis functions for use in molecular calculations. I. Contraction of ($9s5p$) atomic basis sets for the first-row atoms. J Chem Phys 1970;53:2823-33.
10. Chipman D. Gaussian basis sets for calculations of spin densities in first-row atoms. Theor Chim Acta 1989;76:73-84.
11. Frisch MJ, et al. Gaussian 03. Revision B.01. Pittsburgh PA:Gaussian, Inc.;2003.
12. El-Sayed MA. Spin-orbit coupling and the radiationless processes in nitrogen heterocyclics. J Chem Phys 1963;38:2834-8.

CHEMILUMINESCENCE REACTION OF 4-STYRYLPHTHALHYDRAZIDES. REMARKABLE SUBSTITUENT EFFECT ON THE EMITTING SPECIES AND CHEMILUMINESCENCE EFFICIENCY

J MOTOYOSHIYA, K YOKOTA, M HOTTA, Y NISHII, H AOYAMA

Dept of Chemistry, Faculty of Textile Science and Technology, Shinshu University, Ueda, Nagano, 386-8567, Japan

Email: jmotoyo@giptc.shinshu-u.ac.jp

INTRODUCTION

Of the artificial chemiluminescent compounds, luminol (5-amino-2,3-dihydro-1,4-phthalazinedione) is the most popular, and has been applied not only in analytical chemistry but also in other fields. The luminol chemiluminescence is based on the light emission from an excited 3-aminophthalate ion [1] generated by oxidation with hydrogen peroxide or atmospheric oxygen in the presence of bases and catalysts. This peculiar chemiluminescence property and its industrial value have attracted continuous interest and prompted many chemists to investigate its reaction in detail [2,3,4] and to exploit various chemiluminescent phthalhydrazides. [5,6] Among them 4-styrylphthalhydrazide (**1a**) was involved in a very unique reaction, photochemistry without light, [7] in which the *cis-trans* isomerization occurred, in spite of a low efficiency, during the oxidation. Such a diversity in the phthalhydrazide chemiluminescence encouraged us to investigate the chemiluminescence reaction of 4-styrylphthalhydrazides (**1a-g**) whose fluorescent and chemical properties would vary based on the electronic nature of the substituents.

RESULTS AND DISCUSSION

The 4-styrylphthalhydrazides (**1a-g**) were prepared by the Horner-Wadsworth-Emmons reaction of dimethyl 4-dimethylphosphonomethylphthalate and 4-substituted benzaldehydes followed by the reaction with hydrazine. This synthetic sequence provides a convenient route for various chemiluminescent phthalhydrazide derivatives. Especially, phthalhydrazide (**1g**), having a distyrylbenzene moiety, was designed to increase both solvent solubility and fluorescence intensity, and it might be superior to luminol. Alternatively, the potassium phthalates (**2a-g**) were prepared by saponification of the corresponding dimethyl phthalates, because they were the most likely candidates as emitters in the chemiluminescence reactions of the phthalhydrazides. The fluorescence spectral data of **2a-g** are shown in Table 1. A substantial red-shift as well as an increase in the fluorescence quantum yield (Φ_F) with respect to **2d** and **2e** were observed, which are due to the introduction of a highly electron-donating substituent and an extended conjugation; the Φ_F's for **2f** and **2g** are greater by 15-35 times than that for **2a** along with the red-shifts by more than 80 nm. A Hammett relationship was established between Φ_F and the σ-values, the substituent constants estimated depending their electronic nature, as shown in Fig. 1.

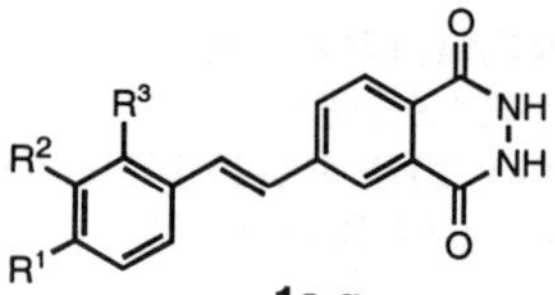

1a-g

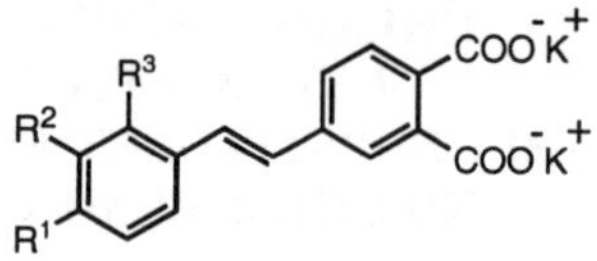

2a-g

a R^1=H , R^2= H, R^3= H
b R^1=Cl., R^2= H, R^3= H
c R^1= H, R^2= OMe, R^3= H
d R^1= OMe, R^2= H, R^3= H
e R^1= OMe, R^2= H, R^3= OMe
f R^1= NMe_2, R^2= H, R^3= H
g R^1=

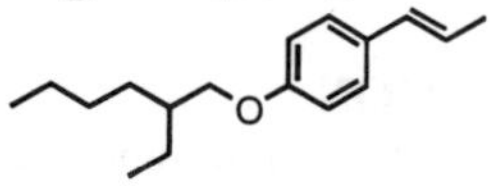

R^2 = H, R^3 = H

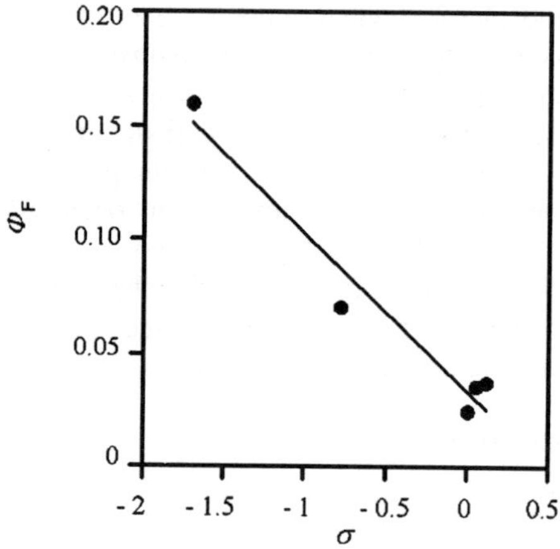

Figure 1. Relationship between Φ_F and σ of dipotassium phathlates (**2a-g**)

The chemiluminescence reactions of the phthalhydrazides (**1a-g**) were carried out in the aerobic DMSO solution in the presence of ^{t}BuOK. The chemiluminescence intensities relative to the luminol chemiluminescence are also described in Table 1. As expected from their strong fluorescence, **1f** and **1g** produced much stronger chemiluminescence than the others. Since their chemiluminescence spectra agreed well with the fluorescence spectra of the corresponding potassium phthalates, the emitters are the phthalate ions (**2f** and **2g**) similar to the luminol chemiluminescence.

Table 1. Fluorescence Spectra Data of Dipotassium Phthalates (**2a-g**) and Chemilumines--cence Quantum Yield (rel. Φ_{CL}) of the Reactions of Phthalhydrazides (**1a-g**)

	substituent			Fluorescence potassium phthalates (**2**)		Chemiluminescence phthalhydrazide (**1**)	
	R^1	R^2	R^3	λ_{max} [a]	Φ_F [b]	$\Phi_{CL} \times 10^3$ [c]	rel. Φ_{CL} [d]
a	H	H	H	363	0.025	0.07	0.012
b	Cl	H	H	370	0.038	0.05	0.008
c	H	OMe	H	368	0.036	0.10	0.02
d	OMe	H	H	377	0.071	0.22	0.04
e	OMe	H	OMe	397	0.096	1.2	0.21
f	NMe_2	H	H	452	0.16	2.0	0.34
g	Ar	H	H	442	0.30	23	3.8

$Ar = $ —⟨ ⟩—CH=CH—⟨ ⟩—$OCH_2CH(CH_2CH_3)CH_2CH_2CH_2CH_3$

[a] Mesured in DMSO in the presence of ^{t}BuOK : fluorescence was recorded by irradiati--on at the absorption maximum. [b] Determined by comparison with 9,10-diphenylanthracene. [c] Determined by comparing with luminol chemiluminescence : photons were counted during 512 sec after the reaction were started. [d] Relative to luminol chemiluminescence.

Notably, the emission efficiency of **1g** was superior to that of luminol, whereas only a faint light emission was observed in the reactions of **1a**, **1b** and **1c** under similar conditions. The less chemiluminescent **1a-c** would behave just like phthalhydrazide, the parent compound of luminol but lacking an amino group, namely, an energy transfer chemiluminescence would take place, because the product, 4-styrylphthalate ions is much weakly fluorescent. The chemiluminescence spectra exhibited the very weak emission at 528 and 508 nm for **1a** and **1b**, respectively, but they are not for the fluorescence of **2a** and **2b**. Interestingly, **1c** and **1d** showed the emission at around 500 nm accompanied by the emission matching to the fluorescence of the corresponding phthalate ions, **2c** and **2d**. Applying to the known relationship [7] in the fluorescence and chemiluminescence spectra of the parent phthalhydrazide monoanion giving peaks at 435 nm and 526 nm, respectively, the peaks around 500 nm observed in the reactions of **1a-d** might be the emission from their excited monoanions. As the fluorescence intensity and the reactivity of **1c** and **1d** would take an intermediate position among the 4-styrylphthalhydrazides, its chemiluminescence would be provided by both of the excited fluorescent substrates. Such an energy transfer chemiluminescence in **1a-c** is trivial because it was reported [8] that even during the reaction of luminol, the energy transfer chemiluminescence occurs in the presence of the very highly fluorescent dye such as fluorescein.

 Motoyoshiya J et al.

In summary, control of the chemiluminescence efficiency as well as the emitter by the terminal substituents was demonstrated for the chemiluminescence reactions of 4-styrylphthalhydrazides. The strongly electron-donating character of the terminal substituents is coincident with the increase in the Φ_F values of the corresponding phthalate ions by a contribution of an electronic pull-push system, which provides an efficient chemiluminescence with the excited phthalate ions being the emitters. On the other hand, the energy transfer chemiluminescence takes place when the fluorescence of the phthalate ions is weak.

REFERENCES
1. White E H, Zafiriou O C, Kagi H H, Hill H M. Chemiluminescence of luminol: The chemical reaction. J Am Chem Soc 1964; 86:940-1.
2. White E H, Bursey M M. Chemiluminescence of luminol and related hydrazides: The light emission step. J Am Chem Soc 1964; 86: 941-2.
3. White E H, Roswell D F. The chemiluminescence of organic hydrazides. Acc Chem Res 1970; 3: 54-62.
4. Merenyi G, Lind J S. Role of a peroxide intermediate in the chemiluminescence of luminol. A mechanistic study. J Am Chem Soc 1980; 102: 5830-5.
5. Ishida J, Takada M, Hara S, Sasamoto K, Kina K, Yamaguchi M. Development of a novel chemiluminescent probe, 4-(5',6'-dimethoxybenzothiazolyl) phthalhydrazide. Anal Chim Acta 1995; 309: 211-9.
6. White E H, Wiecko J, Roswell D F. Photochemistry without light. J Am Chem Soc 1969; 91: 5194-6.
7. White E H, Roswell D F, Zafiriou O C, The anomalous chemiluminescence of phthalhydrazide. J Am Chem Soc 1969; 34: 2462-8.
8. Voicescu M, Vasilescu M, Constantinescu T, Meghea A. On the luminescence of luminol in DMSO in the presence of potassium superoxide-8-crown-6-ether and fluorescein. Luminescence 2002; 97: 60-7.

FLUORESCENCE AND CHEMILUMINESCENCE CHARACTERISTICS OF BISINDOLES

M NAKAZONO, M ASECHI, K ZAITSU

Graduate School of Pharmaceutical Sciences, Kyushu University
3-1-1 Maidashi, Higashi-ku, Fukuoka 812-8582, Japan
Email: zaitsu@phar.kyushu-u.ac.jp

INTRODUCTION

Indole derivatives have been used for fluorescence (FL), chemiluminescence (CL) and bioluminescence assays. Various indole derivatives were synthesized and their CL characteristics were investigated.[1] However, indole derivatives have not been frequently used compared to the typical FL and CL reagents such as dansyl chloride, fluorescein, luminol and acridinium esters in terms of emission wavelength and intensity. For the selective and highly sensitive FL and CL assays, indole derivatives, which have a long emission wavelength of more than 600 nm and have strong CL intensity, should be developed.

We observed changes in the FL and CL of indole derivatives by changing the number of indole moieties. In this study, we synthesized bisindoles, which have two indole moieties in the structure such as 1,2-bis(1*H*-indole-3-yloxoacetyl)-ethylenediamine (**I**) and 3,4-bis(3-indolyl)-1*H*-pyrrole-2,5-dione (**II**) (Fig. 1), and then measured their FL and CL.

METHODS

Apparatus

A Hitachi F2000 (Tokyo, Japan) was used to measure the FL with a quartz cuvette (30 x 10 x 10 mm). Lumat LB 9501 (Berthold, Wildbad, Germany) was used to measure the CL with a round-bottom glass tube (75 x 12 mm i.d.).

Synthesis

1,2-bis(1*H*-indole-3-yloxoacetyl)ethylenediamine (**I**) and 3,4-bis(3-indolyl)-1*H*-pyrrole-2,5-dione (**II**) were synthesized by previously reported methods.[2,3]

CL measurements of indole, I and II

To 200 µL of a 100 µmol/L indole, **I** or **II** in dimethylformamide (DMF) (indole and **II** were also dissolved in CH_3CN) was added 100 µL of 1-200 mmol/L NaOH. After standing for 25 s, the CL reaction was initiated by the addition of 100 µL of 1-1000 mmol/L H_2O_2 using the automatic injection system in the luminometer. The CL emission was measured for 5 min, and the integral photon counts were used for estimating the CL intensities.

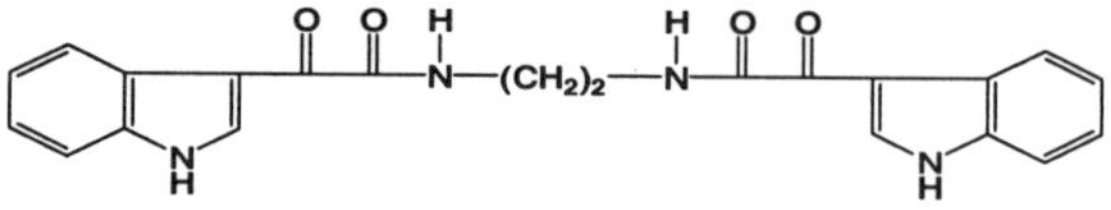

Figure 1. Structures of **I** and **II**.

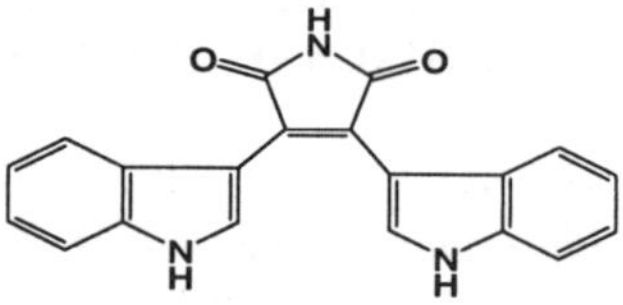

RESULTS

Fluorescence

The FL intensities of 10 μmol/L indole, **I** and **II** were measured in DMF, CH_3CN, CH_3OH and C_2H_5OH.   There was no relationship between the FL intensities and molar extinction coefficient of the indole derivatives.   The FL intensities of **I** and **II** were lower than that of indole in DMF.   Their excitation (ex) and emission (em) maxima wavelengths were as follows: indole (ex. 280 nm, em. 326 nm), **I** (ex. 325 nm, em. 495 nm) and **II** (ex. 366 nm, em. 571 nm) (Table 1).   The emission maxima wavelengths of **I** and **II** were longer than that of indole, and their Stokes' shifts observed for **I** and **II** were 170 and 205 nm, respectively.   The FL intensity of **II** linearly increased with the increasing concentration of **II** in the range of 0.1-10 μmol/L.   However, the FL intensities of **II** decreased in the presence of protic solvents such as CH_3OH, C_2H_5OH and H_2O. The hydrogen bond formation between the oxygen of the carbonyl group in the maleimide structure of **II** and the hydrogen of the hydroxyl group in CH_3OH, C_2H_5OH and H_2O was supposed to decrease the FL intensities of **II**.   **II** should be used for selective FL assays which need detection at emission maxima wavelengths greater than 550 nm in aprotic solvents.

Chemiluminescence

The CL intensities of 100 μmol/L indole, **I** and **II** were measured in DMF, CH_3OH and CH_3CN.   Indole and **II** did not emit light in the presence of H_2O_2 and NaOH in CH_3OH.   The CL intensities of **I** and **II** in DMF were 3- and 45-fold stronger than that of indole in DMF, respectively.   The CL intensities of indole and **II** in CH_3CN were 23- and 63-fold stronger than those of indole and **II** in DMF (Table 2).   The CL intensities of the indole derivatives in CH_3CN were strong when compared to that of the indole derivatives in DMF or CH_3OH. This indicated that the production of singlet oxygen in the presence of H_2O_2 and NaOH oxidized the indole derivatives in the CL reaction.[4,5]   The CL intensity of **II** was 124-fold stronger than that of indole in CH_3CN. The reason why **II** had a strong CL intensity was postulated as follows: singlet oxygen produced in CH_3CN reacted with **II**, the dioxetane structure was formed as the intermediate and strong light was produced *via* decomposition of the dioxetane (Fig. 2). **II** can be used for the highly sensitive CL assay of singlet oxygen.

Table 1.　Relative FL intensities of indole, **I** and **II**.

Compound [a]	Solvent	ε / $M^{-1}cm^{-1}$ (nm)	Max Ex (nm)	Max Em (nm)	Relative FL intensities [b]	
Indole	DMF	5720 (280)	280	326	1	1
	90 % DMF-H_2O	8040 (280)	280	327		1.6
	CH_3OH	7550 (271)	271	331		1.4
I	DMF	22290 (325)	325	495	0.006	1
	90 % DMF-H_2O	21880 (325)	325	522		1.4
II	DMF	5760 (366)	366	571	0.21	1
	90 % DMF-H_2O	4910 (368)	368	591		0.09
	CH_3CN	4990 (364)	364	574		0.7
	90 % CH_3CN-H_2O	8190 (368)	368	538		0.05
	CH_3OH	5620 (372)	372	603		0.002
	C_2H_5OH	5760 (372)	372	604		0.005

a: 10 μM, b: FL intensities of indole, I and II in DMF were taken as 1.

Table 2.　Relative CL intensities of indole, **I** and **II**.

Compound [a]	Solvent	NaOH / mM	H_2O_2 / mM	Relative CL intensities [b]		
Indole	DMF	25	500	1	1	
	CH_3CN	5	100		23	1
I	DMF	50	5	3		
II	DMF	20	500	45	1	
	CH_3CN	20	250		63	124

a: 100 μM, b: CL intensities of indole, I and II in DMF were taken as 1.

We observed strong CL intensity changes of the indole derivatives by changing the number of indole moieties. It was postulated that bisindole has selectivity in the FL assay and high sensitivity in the CL assay.

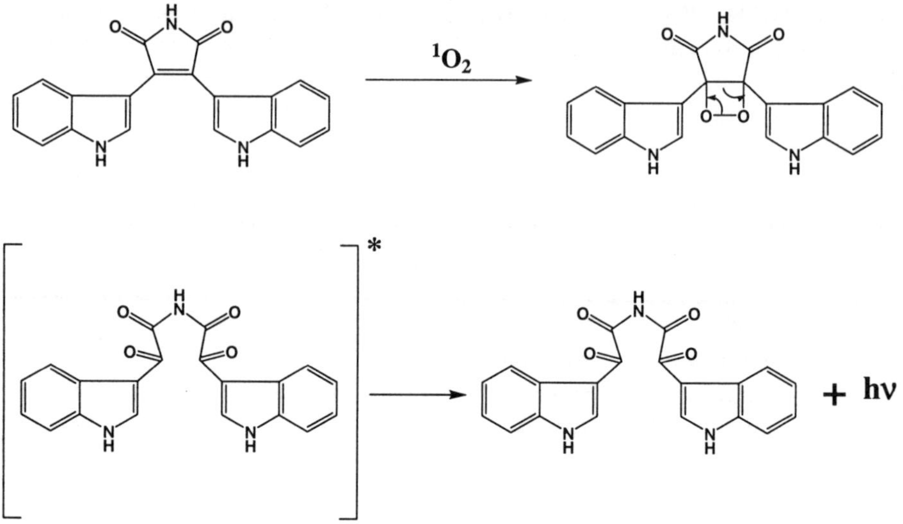

Figure 2. Possible CL mechanism of **II**.

REFERENCES

1. Sugiyama N, Akutagawa M, Gasha T, Saiga Y, Yamamoto H. The chemiluminescence of indole derivatives. I. Bull Chem Soc Jpn 1967; 40: 347-50.
2. Nakazono M, Sho Y, Zaitsu K. Lasting chemiluminescence of 3-indoleglyoxylyl chloride and its enhancement. Anal Sci 2003; 19: 123-7.
3. Zhu G, Conner SE, Zhou X, Shih C, Li T, Anderson BD, Brooks HB, Campbell RM, Considine E, Dempsey JA, Faul MM, Ogg C, Patel B, Schultz RM, Spencer CD, Teicher B, Watkins SA. Synthesis, structure-activity relationship, and biological studies of indocarbazoles as potent cyclin D1-CDK4 inhibitors. J Med Chem 2003; 46: 2027-30.
4. Wiberg KB. The mechanisms of hydrogen peroxide reactions. I. The conversion of benzonitrile to benzamide. J Am Chem Soc 1953; 75: 3961-4.
5. Mckeown E, Waters WA. Chemiluminescence as a diagnostic feature of heterolytic reactions which produce oxygen. Nature 1964; 203: 1063.

PHOTOSENSITIVE LUMINOL RELEASING COMPOUND, LUMINOL-*O*-4,5-DIMETHOXY-2-NITROBENZYLATE

M NAKAZONO, K ZAITSU

Graduate School of Pharmaceutical Sciences, Kyushu University
3-1-1 Maidashi, Higashi-ku, Fukuoka 812-8582, Japan
Email: zaitsu@phar.kyushu-u.ac.jp

INTRODUCTION

Luminol derivatives are used for the simple and highly sensitive chemiluminescent assays of DNAs and enzyme activities. Luminol produces 3-aminophthalic acid and emits light in the presence of peroxide in an alkaline medium.

Photosensitive compounds, which have a 2-nitrobenzyl group in their structures, have been utilized for developing novel caged compounds such as a caged ATP. [1,2] E. H. White reported that the luminol derivative, which has a methyl group at the carbonyl oxygen of luminol, was non-chemiluminescent. [3] We synthesized luminol-*O*-2-nitrobenzylate (**I**) which has a 2-nitrobenzyl group at the carbonyl oxygen of luminol. We found that **I** released luminol upon light irradiation at 366 nm, and used **I** for light power measurements. The released luminol emitted light in the presence of H_2O_2 and NaOH. In the range of 0.01-1 μmol/L **I**, the CL intensities of the released luminol increased with the increasing concentration of **I** with a light irradiation time of 5 min. **I** (1 μmol/L in DMF) was photoirradiated for 10 s with 366 nm light using an ultrahigh-pressure mercury lamp. The CL intensity of **I** after the light irradiation linearly increased with the light power in the range of 450-650 mW/cm^2. [4]

It was reported that the 4,5-dimethoxy-2-nitrobenzyl esters of cyclic AMP and cyclic GMP were more preferable than the simple 2-nitrobenzyl ester from the standpoint of the efficiency of the light-induced release of cyclic nucleotides. [5] Thus, we synthesized luminol-*O*-4,5-dimethoxy-2-nitrobenzylate (**II**) for comparison with **I** as a CL reagent for light power measurement (Fig. 1).

METHODS

Apparatus

The irradiation system consisted of an Optical Module X (model SX-UI500 MQQ) and a power supply unit (model BA-H500) (360 nm cutoff, 70-450 mW/cm^2, and the distances between the collimator lens and the quartz cuvette were 40-160 cm) was obtained from USIO Electronics (Tokyo, Japan). The light power meter was purchased from Advantest (Tokyo, Japan). A Lumat LB 9501 (Berthold, Wildbad, Germany) was used to measure the CL with a round-bottom glass tube (75 x 12 mm i.d.).

Synthesis

Luminol-*O*-4,5-dimethoxy-2-nitrobenzylate **II**.   To stirred DMF (100 mL) was added luminol (0.53 g, 3 mmol) and sodium hydride (60 % in oil, 0.12 g, 3 mmol). This solution was then stirred at ambient temperature for 10 min.   4,5-Dimethoxy-2-nitrobenzyl bromide (0.83 g, 3 mmol) was added to the solution. The mixture was stirred at ambient temperature for 8 h. H_2O (250 mL) was added to the solution and the organic layer was extracted with ethyl acetate (600 mL).   To the organic layer was added CH_3OH (200 mL) and this solution was dried with magnesium sulfate and the filtrate was concentrated.   The resulting compound was washed with $CHCl_3$ then recrystallized from CH_3OH to give **II** as a yellow powder (0.08 g, 7.2 % yield, mp 258 °C). ^{1}H-NMR (($CD_3)_2$S=O): 3.88 (s, 6H, -OCH_3), 5.58 (s, 2H, Benzyl H), 6.91 (m, 2H, ArH), 7.48 (m, 1H, ArH), 7.72 (s, 1H, ArH), 11.54 (s, 1H, -CO-NH-). FAB MS: 373.25 [M+H]$^+$,   Anal. Calcd. for $C_{17}H_{16}O_6N_4$: C, 54.84; H, 4.33; N, 15.05. Found: C, 54.18; H, 4.31; N, 14.84.

Light power measurement using II

Three mL of 1 µmol/L **II** in DMF was added to a quartz cuvette.   The solution was photoirradiated for 10 s with a 360 nm cutoff using the ultrahigh-pressure mercury lamp (70-450 mW/cm^2).   To 200 µL of the photoirradiated solution was added 100 µL of 10 mmol/L NaOH.   After standing for 25 s, to the mixture was added 100 µL of 500 mmol/L H_2O_2 using the automatic injection system in the luminometer.   The CL emission was measured for 5 min, and the integral photon counts were used for estimating CL intensities.

RESULTS

The photosensitive compound, which has a 2-nitrobenzyl group in the structure, produces the precursor and 2-nitrosobenyl derivative with light irradiation *via* a photorearrangement.   It is postulated that **II** produces luminol and 4,5-dimethoxy-2-nitrosobenzaldehyde by light irradiation (Fig. 1).   **II** released luminol by light irradiation at 366 nm using the ultrahigh-pressure mercury lamp.   The total photon count reached a maximum with the light irradiation time of 2 min (Fig. 2).   **II** released luminol 2.5-fold times faster than **I**.   Therefore, the light irradiation time of 2 min was used for evaluating the maximum CL conditions of the released luminol. This indicated that **II** could be used for the measurement of light power.   In the concentration range of 0.01-1 µmol/L **II**, the CL intensities of the released luminol increased in proportion to the increase in the concentration of **II** with a light irradiation (450 mW/cm^2) time of 2 min.   The CL intensities increased in proportion to the light power in the range of 70-450 mW/cm^2 (Fig. 3).

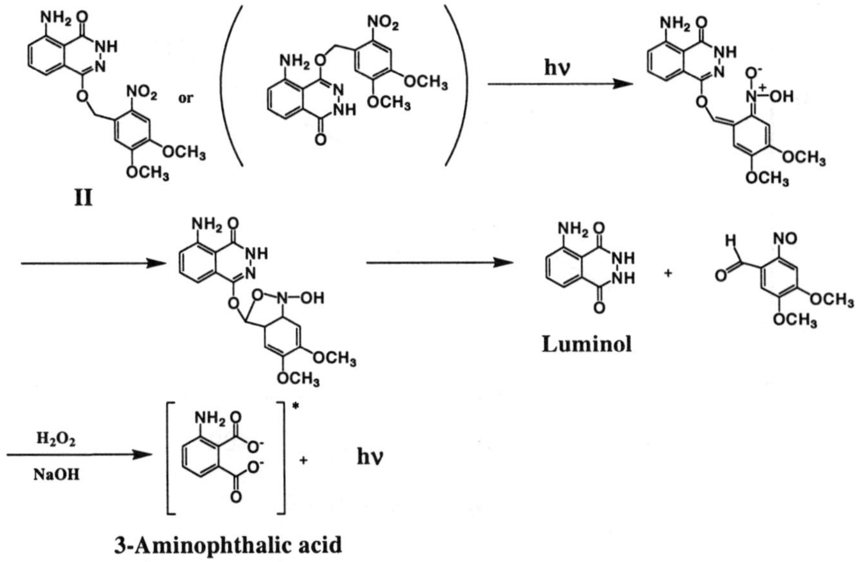

Figure 1. Possible mechanism for the photosensitive release of luminol.

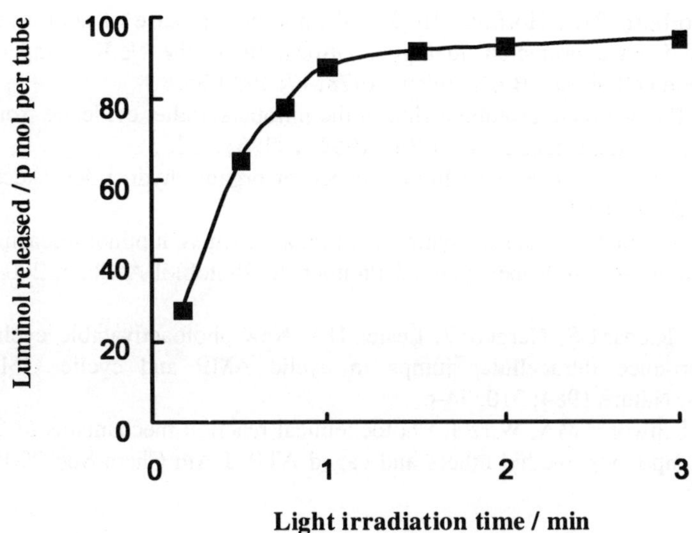

Figure 2. Effect of light irradiation time on the luminol release from **II**.
[**II**] = 1 μmol/L, [NaOH] = 10 mmol/L, [H_2O_2] = 500 mmol/L.

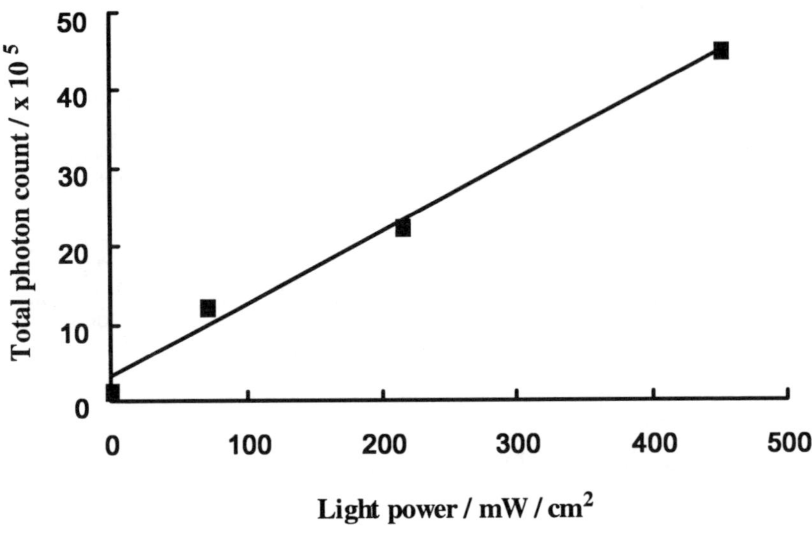

Figure 3. Measurement of light power.
[**II**] = 1 μmol/L, [NaOH] = 10 mmol/L, [H_2O_2] = 500 mmol/L.

REFERENCES

1. Kaplan JH, Forbush BIII, Hoffman JF. Rapid photolytic release of adenosine 5'-triphosphate from a protected analogue: Utilization by the Na:K pump of human red blood cell ghosts. Biochemistry 1978; 17: 1929-35.
2. Mitchison TJ. Polewards microtubule flux in the mitotic spindle: evidence from photoactivation of fluorescence. J Cell Biol 1989; 109: 637-52.
3. White EH, Roswell DF. The chemiluminescence of organic hydrazides. Accts Chem Res 1970; 3: 54-62.
4. Nakazono M, Asechi M, Zaitsu K. Synthesis of photosensitive luminol releasing compound, luminol-*O*-2-nitrobenzylate. J Photochem Photobiol A Chem 2004; 163: 149-52.
5. Nerbonne JM, Richard S, Nargeot J, Lester HA. New photoactivatable cyclic nucleotides produce intracellular jumps in cyclic AMP and cyclic GMP concentrations. Nature 1984; 310: 74-6.
6. Ll'ichev YV, Schwörer MA, Wirz J. Photochemical reaction mechanisms of 2-nitrobenzyl compounds: methyl ethers and caged ATP. J Am Chem Soc 2004; 126: 4581-95.

PREPARATION OF A CROWN-ETHER-MODIFIED ISOLUMINOL DERIVATIVE AND ITS CHEMILUMINESCENCE PROPERTIES IN AN ORGANIC MEDIUM

H OKAMOTO, M KIMURA

Department of Chemistry, Graduate School of Natural Science and Technology, and Department of Chemistry, Faculty of Science, Okayama University, Okayama 700-8530, Japan
E-mail: hokamoto@cc.okayama-u.ac.jp

INTRODUCTION

Molecular photodevices, whose properties can be controlled by specific additives, have been extensively developed[1-3] since the pioneering studies of Vögtle on chromoionophores.[3] For such conventional chromophores, absorption and photoluminescent properties have been modified through interactions between the host dyes and specific guests. Since chemiluminescence provides potent highly sensitive analytical probes,[4,5] it would be of interest to construct a chemiluminophore having a host function which displays a change in its chemiluminescence properties on addition of a specific guest. Such a chemiluminophore may serve as a novel luminescent chemosensor.[6,7] Herein, we describe preparation of an isoluminol derivative having an aza-15-crown-5 ionophore **1** and its chemiluminescence behavior in acetonitrile in the presence of alkali-metal salts.

MATERIALS AND METHOD

The crowned isoluminol **1** has been prepared by the route shown in Scheme 1 and confirmed by NMR, IR, UV-VIS spectra as well as HRMS. The chemiluminescence spectra were recorded on a multi-channel photodiode array detector (Hamamatsu Photonics).

RESULTS AND DISCUSSION

The synthetic route of the title isoluminol derivative **1** is shown in Scheme 1. 4-Amino-*N*-methylphthalimide was treated with sodium hydride and then reacted with penta(ethylene glycol) ditosylate to afford a crowned phthalimide **3**. In this reaction, an α,ω-bichromophoric podand **4** was also obtained. Subsequent reaction of the phthalimide **3** with hydrazine provided the desired crowned isoluminol **1**,

Scheme 1.

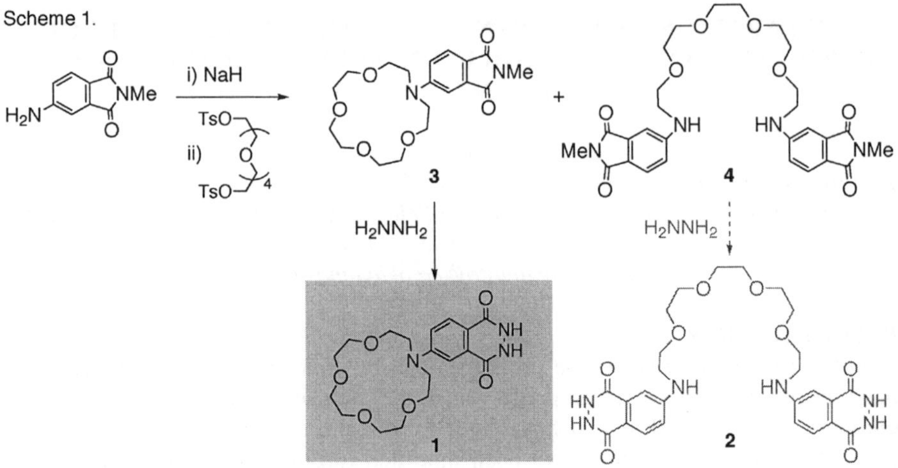

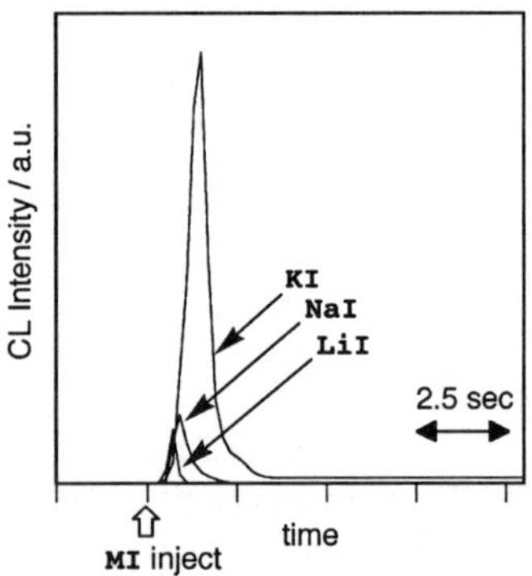

which was the first luminol analogue possessing an aza-crown ionophore. Preparation of a bifunctional podand **2** from the precursor **4** is currently also under examination.

The crowned isoluminol **1** displayed quite weak chemiluminescence (CL) in acetonitrile in the presence of hydrogen peroxide and tetrabutyl ammonium hydroxide (TBAOH). On the other hand, addition of alkali-metal salts (as iodide or perchlorates) to this mixture triggered off intense, blue CL emission (Fig. 1). The CL

Figure 1. Time course of the chemiluminescence of the crowned isoluminol **1** in (0.8 mM) MeCN in the presence of H_2O_2 (133 mM), TBAOH (8 mM) detected at 450 nm upon addition of alkali-metal iodides (MI) (20 mM).

emission was short-lived and ceased within a few second at room temperature (Fig.. 1). Whereas, in aqueous alkaline hydrogen peroxide solution, such CL emission was not detected by addition of alkali-metal iodide.

The CL spectra of the isoluminol **1** observed are shown in Fig. 2. The maximum

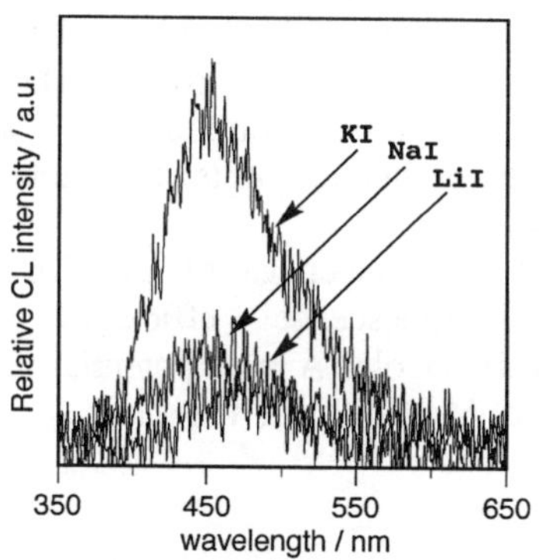

Figure 2. Chemiluminescence spectra of the crowned isoluminol **1** (0.72 mM) in MeCN in the presence of H_2O_2 (90 mM) and TBAOH (6.7 mM) detected upon addition of alkali-metal iodides (8 mM).

of the CL spectrum (λ_{CL}) was affected little by the alkali-metal cation used (Li^+, Na^+, K^+, λ_{CL} 450-470 nm), whereas, the intensity of the CL emission changed depending on the cation; the relative intensity of the CL emission increased for $Li^+ < Na^+ < K^+$ ($Li^+ : Na^+ : K^+ = 1 : 1.8 : 7.5$).

In the case of *N,N*-dimethylamino derivative of isoluminol, which possesses no ionophore function, CL emission was also observed under the same CL conditions as used for the crowned isoluminol **1**. However, the CL intensity was not changed remarkably by the metal cation. Therefore, the aza-crown host function of the isoluminol **1** might play a significant role in the CL modulation under the present conditions.

ACKNOWLEDGEMENTS

The present work was supported by the Grant-in-Aide (No. 13740398) of the Ministry of Education, Culture, Sports, Science and Technology of Japan.

REFERENCES

1. Lehn JM. Supramolecular Chemistry. Concepts and Perspectives. Weinheim: WCH: 1995.
2. de Silva AP, Gnaratne HQN, Gunnlaugsson T, Huxley AJM, McCoy CP, Rademacher JD, Rice TE. Signalling recognition events with fluorescent sensors and switches. Chem Rev 1997; 97: 1515–66.
3. Löhr HG, Vögtle F. Chromo- and fluoroionophores. A new class of dye reagents. Acc Chem Res 1985; 18: 65–72.

4.   Dodeigne C, Thunus L, Lejeune R. Chemiluminescence as a diagnostic tool. A review. Talanta 2000; 51: 415-39.
5.   Imai K. ed. Bioluminescence and Chemiluminescence. Basics and Applications. Tokyo: Hirokawa Publishing Co., 1989.
6.   Okamoto H, Owari M, Kimura M, Satake, K. Preparation of a crown-ether-modified lophine peroxide as a guest-sensitive novel chemiluminophore and modulation of its chemiluminescence by metal cations. Tetrahedron Lett 2001; 42: 7453-5.
7.   Kimura M, Morioka M, Tsunenaga M, Hu ZZ. Effect of conformational change on chemiluminescence efficiency of 2-(Ar)-4,5-diphenyl-4H-hydroperoxy-4H-imidazole: A new potential signalling system. ITE Lett Batteries New Technol Med 2000; 1: 418-21.

CHEMILUMINESCENT STUDY ON OXIDATION OF MONO-, DI-, AND POLY-SACCHARIDES

T TAMEFUSA[1], J KIMURA[2], R ITO[1], K INOUE[1],
Y YOSHIMURA[1], H NAKAZAWA[1]

[1.]Department of Analytical Chemistry, Hoshi University, Tokyo 142-8501, Japan
[2.]Tohoku Electronic Industrial Co, Ltd, Miyagi 981-0134, Japan

INTRODUCTION

Recently, ultra weak chemiluminescence (CL) which is undetectable with the naked eye can be measured with high sensitivity photo-detectors. In addition, it was clarified that various substances emit ultra weak chemiluminescence in oxidation reactions. The chemiluminescence method has been applied to the assessment of oxidation in different fields, such as macromolecules, foods, living bodies, and environments.[1-3]

Cellulose is used as a food additive. Since it has the outstanding flowability and granulation properties, it is widely used as a binding agent in drugs. Cellulose has been reported to emit strong chemiluminescence upon heating or exposure to ultraviolet (UV) irradiation. Analysis of the surface physical properties of cellulose and elucidation of the chemiluminescence mechanism was undertaken to evaluate the stability of foods and drugs containing cellulose. We also investigated the surface properties of cellulose and the chemiluminescent mechanism.

Mechanism of autoxidation

Heating and UV irradiation of cellulose forms an active site for reaction with oxygen. Heat or UV light causes chemical reaction of many substances in an oxygen atmosphere. Oxygen is one of the most important factors in autoxidation, and the reaction is remarkably influenced by the concentration and diffusion of oxygen. In the autoxidation process, oxygen-containing compounds are generated including hydroperoxide, aldehyde, ketone, carboxylic acid, ester, and alcohol. Oxygen is adsorbed in these products and the adsorbed oxygen can emit chemiluminescence as a result of heating or UV irradiation. Luminous intensity is proportional to the quantity of the oxygen-containing products.

METHODS

Materials

Analytical grade glucose, fructose, galactose, ribose, maltose, lactose and sucrose were obtained from Sigma Aldrich Japan Co. (Tokyo, Japan). Analytical grade cellobiose was purchased from Kanto Chemical Co. (Tokyo, Japan). Cellulose was imported from Scientific Polymer Products(NY, USA).

Instruments

CL intensities and spectra were measured using a CLA-FS1 (Tohoku Electronic Industrial Co, Japan). Enthalpy change was calculated by CS Chem3D (Cambridge Soft Corporation, USA) and by MOPAC (Fujitsu Limited, Japan).

Experiments

CL intensities of all the saccharides were measured at elevated temperatures below 100 °C in both a nitrogen and oxygen atmosphere. The average CL intensities for different temperatures were plotted in the Arrhenius plot. Activation energies were calculated from the slope of straight lines in the plots. In order to examine temperature effect on CL emission of cellulose in an inert atmosphere, powdered cellulose was stored for two weeks in air at ca. 25 °C and ca. 10 °C.

All the saccharides were irradiated with near ultraviolet (375 nm) for 6 s and then CL measurement started after waiting for 1 s. CL intensities were measured using different wavelength filters, and the spectrum was calculated.

RESULTS

Activation energies of saccharides by heating

The activation energies of saccharides in both a nitrogen and oxygen atmosphere are shown in Fig. 1. The observed activation energies of all the saccharides varied from 4 kcal/mol to 20 kcal/mol in both a nitrogen and oxygen atmosphere. It should be noted that the values were low compared to those for normal chemical reactions.

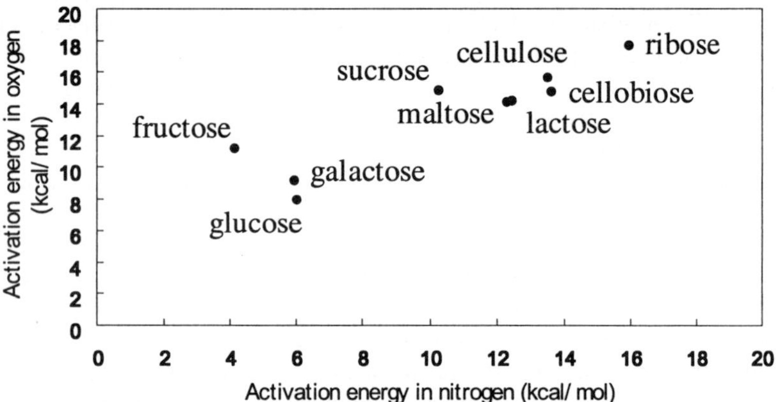

Figure 1. Activation energy of saccharides
in both a nitrogen and oxygen atmosphere

Measurement of CL spectra (heated samples)

Fig. 2 illustrates CL spectra of sucrose heated at 80 °C in both a nitrogen and oxygen atmosphere. All the saccharides produced a CL peak in the range of 350-370 nm probably due to excited carbonyl compounds and at 620-640 nm due to singlet oxygen. Therefore, it is suggested that the chemiluminescence occurred by the same mechanism in both a nitrogen and oxygen atmosphere.

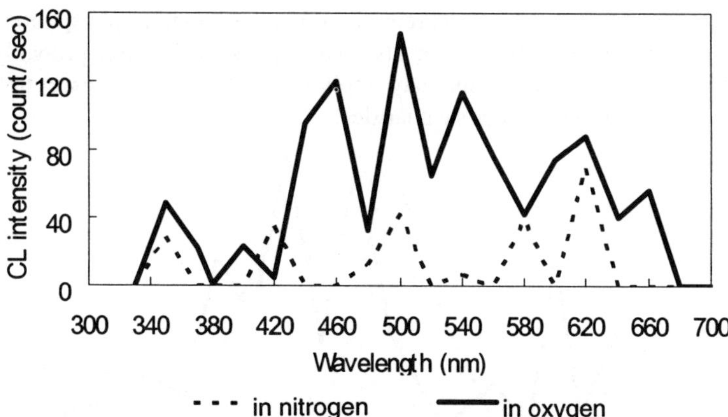

Figure 2. CL spectra of sucrose at 80 °C in both nitrogen and oxygen atmosphere

Effect of the preservation temperature on chemiluminescence
CL intensities of the powdered cellulose are shown in Fig. 3. The powdered cellulose stored at 10 °C for two weeks produced a higher CL intensity than that stored at 25 °C for the same period. Therefore, it was suggested that adsorbed oxygen is the light emitter but not unstable intermediates such as hydroperoxides and dioxetanes.

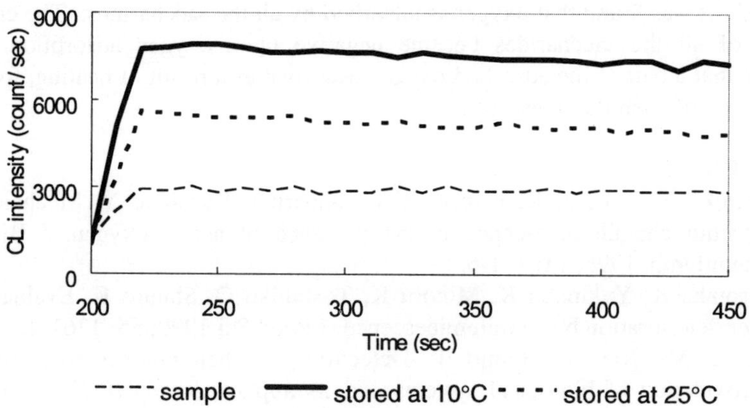

Figure 3. CL intensities of the powder cellulose

Measurement of CL spectra (UV irradiated samples)
Fig. 4 illustrates the CL spectra of ribose and fructose irradiated with a UV lamp.
All the saccharides exhibited green chemiluminescence. The broad bands observed
between 460-560 nm (blue green to green) could be attributed to phosphorescence of
adsorbed oxygen. Only ribose gave also a CL peak at 360 nm probably due to
excited carbonyl compounds. This may imply that only ribose is susceptible to UV
irradiation compared to the other saccharides.

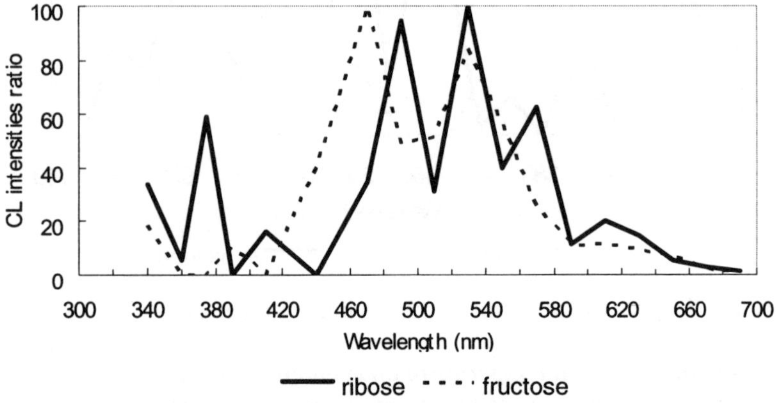

Figure 4. CL spectra by UV irradiation

Calculation of the enthalpy change using MOPAC
The energies of adsorption for oxygen and saccharides were examined by using
MOPAC. It was found that oxygen is adsorbed by all the saccharides. The enthalpy
change of all the saccharides became negative upon oxygen adsorption. This
suggests that a part of the adsorbed oxygen desorbed as a result of heating, followed
by emission of chemiluminescence.

REFERENCES

1.   Yumiko Y, Takashi K, Kazuyoshi O, Kiharu I, Kazuhiko Y. Mechanism of
      catechin chemiluminescence in the presence of active oxygen. J Biolumin
      Chemilumin 1996; 11: 131-6.
2.   Hirotaka K, Yukinobu K, Minoru K, Toshihiko O, Shunro K. Evaluation of
      beer deterioration by chimiluminescence. J Food Sci 1990; 55: 1361-4.
3.   Teruo M, Rie S, Humio I. Detection of chemiluminescence in lipid
      peroxidation of biological systems and its application to HPLC. J Biolumin
      Chemilumin 1989; 4: 475-8.

SOLID SURFACE ENHANCEMENT EFFECTS ON CHEMILUMINESCENCE: INVESTIGATION OF HIGH PERFORMANCE SOLID MEDIA AND ITS APPLICATION TO ANALYTICAL CHEMISTRY

T YOSHINAGA[1], T ICHIMURA[2], H HIRATSUKA[3]

[1] *Dept of Applied Chemistry, Kyushu Institute of Technology, Sensuicho, Tobata, Fukukoka 804-8550, Japan*
[2] *Dept of Chemistry, Tokyo Institute of Technology, Ookayama, Meguro-ku, Tokyo152-8552, Japan*
[3] *Dept of Chemistry, Gunma University, Tenjincho, Kiryu, Gunma 376-8515,Japan*
Email: yosinaga@che.kyutech.ac.jp

INTRODUCTION

Chemiluminescence (CL) has been widely used in many fields, such as nutrition industries, forensic, biological, agricultural, and medical fields,[1-3] and furthermore, it also has been utilized for light sources in the case of natural disasters, and interior decorations. We have studied the chemiluminescent characteristics[4-6] on solid surfaces, using diaryloxalate (as a representative of CL reagents) and many kinds of solid media expected to be used as solid surface enhancers for detectors or to control CL characteristics. We have found the CL surface enhancement effects using powdered silica and other inorganic compounds. Solid surface enhancement effects on CL have been studied using many solid media, such as filters, organic polymers (synthetic and natural), inorganic materials (oxides and other compounds) and other materials. We have found that most of the solid media used were classified into four or five types using three basic indices of "relative intensity", "relative lifetime", and "relative CL energy". In consequence, we have applied these CL enhancement effects to the analysis of hydrogen peroxide and obtained good results.

EXPERIMENTAL

Reagents: the constituents of (a) humidity regulators and (b) chemiluminescent reagents are described in detail elsewhere.[4-6] (c) Solid media. Filters [cellulose, glass fiber, active carbon etc.], polymers (natural, synthetic powder or film etc.), inorganic material [oxides, other compounds, ceramics etc.], and others [organic compounds, metallic salts etc.] were used. **Procedure:** the stick is bent and the inner glass tube is broken up to mix the A and B solutions to initiate the CL reaction, then, the CL reagent is extracted and dropped on the solid medium. The CL intensity is measured using a light power meter or an appropriate detector at regulated intervals.[4-6]

RESULTS AND DISCUSSION
Filter materials
Several filter materials were tested as solid media that could interact with a CL reagent. These materials were made from cellulose-, active carbon-, glass fiber-, silica fiber-, and membrane-filters (nitrocellulose, PTFE etc.). A glass plate (hollow slide glass for microscope observation) was chosen as the standard blank solid, since it has a smooth surface and is inactive to the CL reagent. Among these filters, glass fiber filters have shown the most interesting solid surface enhancement effects i.e., it has given the highest relative CL intensities (about 10 times: R= 10) to the blank one and the active carbon filter was inhibiting (R=0.04) vs . blank solid (R =1.00).
Polymer materials
Natural, semi-natural, or synthetic polymer powders such as cellulose, chitin/chitosan, acetylcellulose, nylon, PVC, and PVA powders were used as solid media. PVA (polyvinyl alcohol) gave the highest intensity (R= 7.2). while, chitosan and nylon gave relatively high intensities. However, when they are prepared as film with smooth surface, then the relative CL intensities became weak like a solid of glass plate.
Inorganic materials
Many inorganic materials including oxides, carbonate, and ceramics were investigated to find out if the solid media show solid surface enhancement effects or some interesting properties.   They were classified as follows; (a) higher intensities and shorter lifetime group, (b) lower intensities and longer lifetime group, (c) higher intensities and longer intensity group, (d) lower intensities and shorter lifetime group, and in addition, (e) higher energy group than blank solid.
Metallic salts
We have so far used solvent insoluble solid media. However, we have found that if the volume of CL reagent used is relatively small compared to the solid media, we could measure the CL intensities even if the solid media are solvent soluble. Sodium salicylate has been used as a superior catalyst in CL reactions. Therefore, we tested the metallic salts of salicylic acid, other organic acids, and many other inorganic salts. Mainly alkaline and alkaline earth metal ions were studied. Sodium and potassium ions gave the highest intensities among many cationic species, although sodium ion had the greater effect. We also have found that CL intensity further increased as the number of sodium ions in the compound increased. Fig. 1 shows the influence of cationic species on CL intensities.

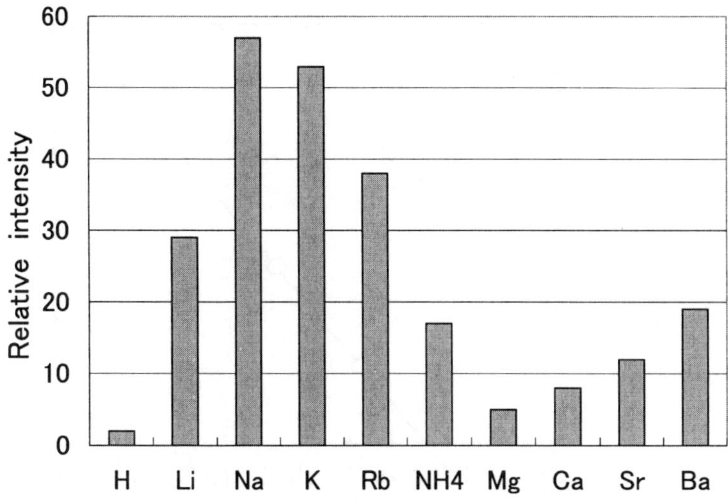

Figure 1. The relationship between CL intensities and
cationic species in salicylates as solid media

Application to H_2O_2 analysis

We have applied these solid surface enhancement effects to the measurement of
hydrogen peroxide (H_2O_2). In Fig. 2, the relationship between relative CL intensity
and absolute amount (mole) of peroxide is shown. As Fig. 2 shows, the lowest
detection limit of ca.10^{-11} mol was obtained when sodium salicylate was used as an
active solid medium and the Anritsu light power meter was used as an detector which
can detect photons of 10^{-9} W (=J/s), whereas a detection limit of 5×10^{-18} mol was
obtained when a higher sensitivity detector (Lumicounter-2500: Microtech Nichion,
JAPAN) was used.

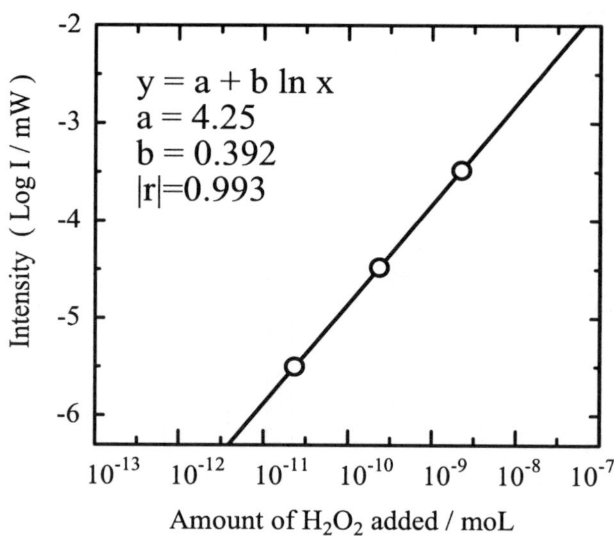

Figure 2. The relationship between relative CL intensities
and amount of H_2O_2 added

REFERENCES

1.   Usuki R, Ogawa M, Kaneda T. Estimation of soybean lipoxygenase activity by measurement of chemiluminescence. J Jpn Soc Food Sci 1987; 34: 229-33.

2.   Alapont AG, Zamora LL Calatyd JM. Indirect determination of paracetamol in pharmaceutical formulations by inhibition of the system luminol-H_2O_2- $Fe(CN)_6^{3-}$ chemiluminescence. J Pharm Biomed 1999; 21:311-7.

3.   Quickenden TI, Cooper PD. Increasing the specificity of the forensic luminol test for blood. Luminescence 2001; 16: 251-3.

4.   Yoshinaga T, Tanaka Y, Ichimura T, Hiratsuka H, Kobayashi M, Hoshi T. Solid surface enhancement effects on chemiluminescence: diaryloxalate and polymers as media solids. J Luminescence 1998; 78: 221-9.

5.   Yoshinaga T, Tanaka Y, Ichimura T, Hiratsuka H, Hasegawa M, Kobayashi M, Hoshi T. Solid surface enhancement effects on chemiluminescence: diaryloxalate as chemiluminescent reagent and inorganic materials as media solids. Bull Chem Soc Jpn 2001; 74: 1507-16.

6.   Yoshinaga T, Akimoto S, Takemura S, Hiratsuka H, Hasegawa M, Kobayashi M, Hoshi T. Solid surface enhancement effects on chemiluminescence influence of cationic species in solid media. Chem Lett 2003; 32: 102-3.

PART 6

1,2-DIOXETANES

ON THE CIEEL MECHANISM OF TRIGGERABLE DIOXETANES: DOES THE ELECTRON JUMP OR IS IT CHARGE TRANSFER?

W ADAM, AV TROFIMOV

[1]*Department of Chemistry, University of Puerto Rico, Rio Piedras, PR 00931, USA and Institute of Organic Chemistry, University of Wuerzburg, D-97074, Germany*
[2]*Institute of Biochemical Physics, Russian Academy of Sciences, 119991 Moscow, Russia*

INTRODUCTION

Chemically initiated electron-exchange luminescence (CIEEL) constitutes a general phenomenon, the important example of which is the firefly bioluminescence. A long-standing mechanistic dichotomy on the CIEEL process concerns concerted *versus* stepwise cleavage of the dioxetane ring. As it is shown in Scheme 1 (*on the left*), the

Scheme 1

concerted cleavage may operate with concomitant excited-state generation through *partial charge transfer* (CT) from the electron donor (ED) to the peroxide. Alternatively, a stepwise process may apply (Scheme 1, *on the right*), in which initially one-electron transfer to the peroxide bond takes place and causes dioxetane cleavage into a radical-ion pair, followed by excited-state generation through electron back-transfer (BET). Herein, we consider this mechanistic query as applied to the CIEEL-active dioxetanes with the phenolate functionality as the electron-donating group (ED in Scheme 1); these are of particular interest for chemiluminescent bioassays.

METHODS

The chemiluminescence (CIEEL) emission was measured as reported.[1,2] For the semiempirical calculations, the PM3 and AM1 methods were used, as implemented in the MOPAC 6.00 and VAMP 5.0 software packages.

RESULTS AND DISCUSSION

To assess the role of *electron transfer* in the dioxetane decomposition, a comparison of the reaction pathways for the neutral dioxetane and its negatively charged ion is necessary (Fig. 1). The energy profiles for the cleavage of a 1,2-dioxetane and its dioxetane radical anion as a function of stretching the O-O bond, as calculated by the PM3 method, are displayed in Fig. 1a. These energy profiles disclose two significant mechanistic features of the CIEEL process: The electron transfer to the dioxetane ring is initiated by elongation of the O-O bond, which lowers substantially the

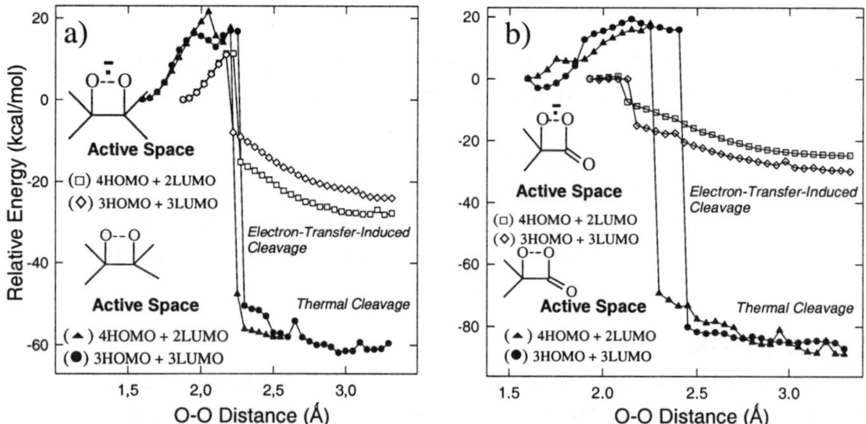

Figure 1. (a) Energy profiles for the cleavage of the 1,2-dioxetane (filled symbols) and its anion (open symbols), as calculated by the PM3/MECI method with the active space of six molecular orbitals; (b) the same for the corresponding α-peroxy lactone (filled symbols) and its anion (open symbols).

activation energy for the dioxetane cleavage, but a small energy barrier still remains. This induced electron transfer is even more pronounced for the α-peroxy lactones, as manifested by the negligible cleavage barrier (Fig. 1b). Such unusual features of the electron-transfer-induced decomposition require mechanistic rationalization.

First, the original mechanism of the CIEEL process involves, as a key tenet, the chemically *activated* transfer of the electron from the donor (activator) to the acceptor (peroxide) functionality. Indeed, as estimated from electrochemical data, the electron transfer from an activator is endothermic at the equilibrium geometry of the peroxide bond. Thus, the chemical activation process involves stretching of the oxygen-oxygen bond to accommodate the transferred electron. Consistent with the semiempirical computations, the O-O bond in the radical anion is markedly elongated relative to the neutral dioxetane (*cf.* Fig. 1).

Second, whereas the cleavage barrier for the dioxetane is reduced on electron transfer (Fig. 1a), for the α-peroxy lactone it is negligible (Fig. 1b); accordingly, the O-O bond is *irreversibly* cleaved in the α-peroxy-lactone radical anion. These findings are consistent with the well-known fact that the α-peroxy lactones are considerably more efficient in the electron-transfer-induced chemiluminescence than the corresponding dioxetanes.

The data displayed in Fig. 1 support the *full*-electron-transfer mechanism. To understand the process of the excited-state generation, the nature of the CIEEL emitter and the chemiexcitation mechanism need to be established. Fig. 2a displays the CIEEL spectra of the monocyclic (*m*-**1a**) and bicyclic (*m*-**1b**) dioxetanes and the fluorescence spectra of the *meta*-oxybenzoate ions derived from the electron-

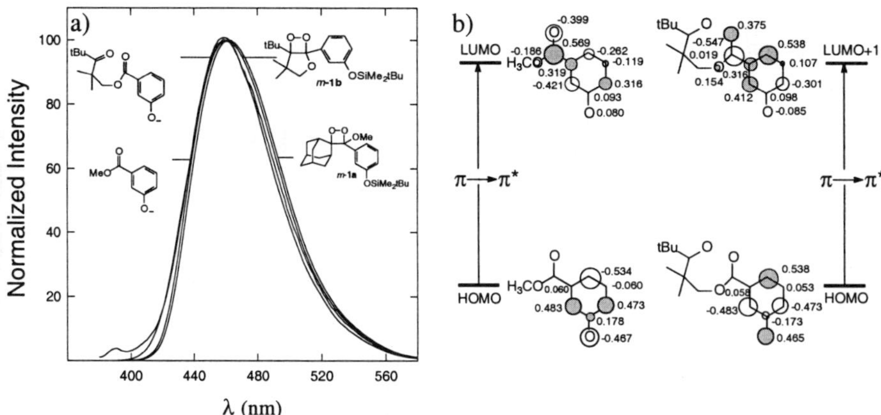

Figure 2. (a) Normalized spectra of the CIEEL intensity in the fluoride-ion-triggered decomposition of the dioxetanes *m*-**1a** and *m*-**1b** and the fluorescence intensity (λ_{ex} at 330 nm) of the *m*-oxybenzoate ions in MeCN. (b) Pertinent molecular orbitals for the electronic excitation of the oxyanions derived from the *m*-**1a** (*on the left*) and the *m*-**1b** (*on the right*) dioxetanes, as calculated by the AM1 method.

transfer-induced cleavage of these dioxetanes. The coincidence of the fluorescence and the CIEEL spectra (Fig. 2a) shows that in both cases the same chromophore is responsible for the CIEEL emission. This is substantiated by the computational results in Fig. 2b, which show that the pertinent molecular orbitals for the excitation of the CIEEL emitter look very similar. The light emission derives *exclusively* from the $\pi \rightarrow \pi^*$ excitation of the oxybenzoate ion.

The chemiexcitation efficiency depends decisively on the location (*meta versus para* position) of the triggerable phenolate functionality: The excitation yield for the *meta* regioisomer is *ca.* 200 times higher than for the *para*-substituted derivative. What is the origin of this dramatic difference? In Scheme 2, is summarized the electronic analysis of the mechanistic alternatives for the CIEEL generation of the regioisomeric dioxetane phenolates *m*-**2** and *p*-**2**, to explain this observation. The CT path constitutes the concerted chemiexcitation without intermediates, whereas for the BET channel the two paths *a* and *b*, should be considered. Scheme 2 reveals a significant structural difference in the two emitters, *i.e.*, *m*-**3** is a crossed-conjugated and *p*-**3** an extended-conjugated system. The consequence in regard to regioselection may be found in path *b*. The extended-conjugated, *para*-patterned anion-diradical intermediate *p*-**D** leads on spin annihilation directly to the resonance-stabilized, mesomeric structure *p*-**D'**, which is equivalent to the *ground state* of the methyl oxybenzoate ion *p*-**3**. In contrast, for the crossed-conjugated, *meta*-patterned anion-diradical intermediate *m*-**D**, the high-energy, spin-coupled *m*-**D'** structure connects with the *m*-**3** ground state. The *intramolecular* BET process in path *b*

Scheme 2

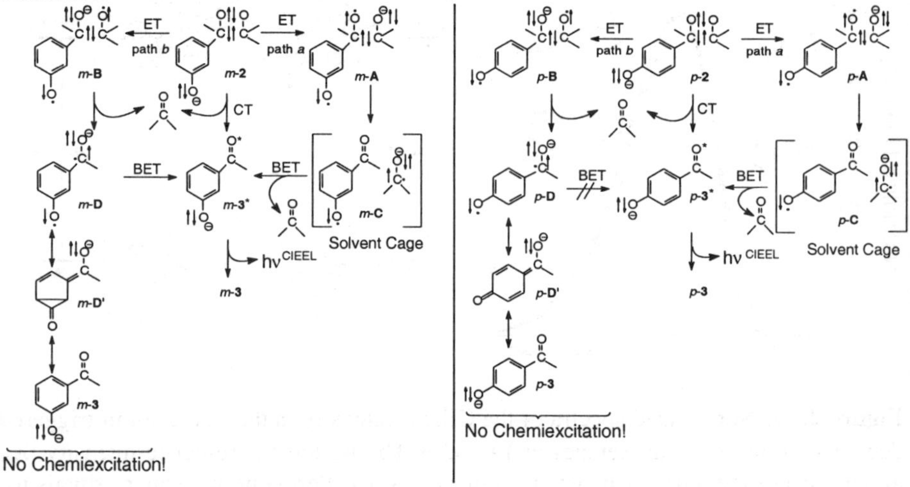

would be expected to compete efficiently with the deactivation of *m*-**D** to its ground state *m*-**3** and afford the electronically excited *m*-**3*** emitter. As a consequence, the chemiexcitation should be more effective for the *m*-**D** than the *para* regioisomer. This mechanistic rationale would account for the higher CIEEL efficiency observed for the *meta- versus* the *para*-substituted dioxetane; however, the observed viscosity dependence of the chemiexcitation yield (*i.e.*, a 2.5-fold enhancement for a 4-fold viscosity increase) supports the BET process in path *a*. For this *intermolecular* BET channel, the difference in the experimental chemiexcitation efficiencies of the *meta* and *para* regioisomers may be explained in terms of energy considerations: The excited state of the crossed-conjugated *meta*-oxybenzoate ion *m*-**3*** is by *ca.* 12 kcal/mol lower in energy than its extended-conjugated *para* regioisomer *p*-**3***,[2] and, thus, the intermolecular BET process is more efficient for the meta regioisomer.

In conclusion, the present analysis supports the full-electron-transfer rather than partial-charge-transfer mechanism for the CIEEL generation.

REFERENCES

1. Adam W, Bronstein I, Trofimov AV, Vasil'ev RF. Solvent-cage effect (viscosity dependence) as a diagnostic probe for the mechanism of the intramolecular chemically initiated electron-exchange luminescence (CIEEL) triggered from a spiroadamantyl-substituted dioxetane. J Am Chem Soc 1999; 121: 958-61.

2. Adam W, Trofimov AV. The effect of meta versus para substitution on the efficiency of chemiexcitation in the chemically triggered electron-transfer-initiated decomposition of spiroadamantyl dioxetanes. J Org Chem 2000; 65: 6474-8.

CHEMILUMINESCENCE INVOLVING THE PHOSPHORUS CHEMISTRY. PHOSPHA-1,2-DIOXETANES AS THE MOST LIKELY HIGH-ENERGY INTERMEDIATES IN AUTOXIDATION OF PHOSPHONATE CARBANION

J MOTOYOSHIYA, H AOYAMA

Department of Chemistry, Faculty of Textile Science & Technology,
Shinshu University, Ueda, Nagano 386-8567, Japan
Email: jmotoyo@giptc.shinshu-u.ac.jp

INTRODUCTION

Phospha-1,2-dioxetanes (**1**) have been proposed to be the most likely intermediates in the oxygenation of phosphonium ylides sometimes called the oxy-Wittig reaction, which generates the corresponding carbonyl compounds and phosphine oxides.[1] For a structural similarity to 1,2-dioxetanes (**2**), well known as chemiluminescent species, the phospha-1,2-dioxetanes are also expected to provide chemiluminescence along with decomposition. Indeed, a light emission has been observed during the singlet oxygenation of the phosphonium ylides and phosphazines.[2] Chemiluminescence was also observed during the oxygenation of phosphonate carbanions by this group.[3] These chemiluminescent reactions are believed to involve phospha-1,2-dioxetanes as the most likely chemiluminescent species. Elucidation of

$$
\begin{array}{ccc}
\underset{O-O}{\overset{|\quad\;|}{\underset{|}{>}\!P\!-\!C\!-}} & \quad & \underset{O-O}{\overset{|\quad\;|}{-C\!-\!C\!-}} \\[4pt]
\mathbf{1} & & \mathbf{2}
\end{array}
$$

such intermediates will be very important for the further understanding of the related reactions, the Wittig type of reactions, for example, the Horner-Wadsworth-Emmons (HWE) reaction,[4] a very important olefination reaction, whose intermediates have been not fully determined.[7] In this study we investigated the chemiluminescent autoxidation of the phosphonate carbanions from the viewpoint of chemiluminescence, thus providing strong support for the phospha-1,2-dioxetanes.[5]

RESULTS AND DISCUSSION

The autoxidation of 9-phosphono-9,10-dihydroacridanes (**3**) provided chemiluminescence which lasted enough to be spectroscopically detected, and whose emission spectra were in completely agreement with the fluorescence spectrum of the acridone anion generated from **5** under basic conditions. The formation of the weakly fluorescent 9-phosphonoacridine (**7**) accompanies these chemiluminescent oxidation reactions.

3, X = H
4, X = Me

5, X = H
6, X = Me

7

a : R=OMe

b : R=OEt

c : R=OCH$_2$CF$_3$

d : R=

e : R=

Similar reactions with **4** also displayed chemiluminescence due to fluorescence of the *N*-methylacridone (**6**). There is a conspicuously much larger Φs value for **3** than that for **4**, in spite of the by-product formation in the reactions of **3**. Such the large difference in Φ_{CL} is significant and explains the intramolecular CIEEL (chemically initiated electron exchange luminescence) mechanism as proposed for the firefly luciferin bioluminescence,[6] in which strong support for the CIEEL mechanism was due to the drastic decrease in the emission efficiency when the phenolic proton of the firefly luciferin was replaced by a methyl group. Similar circumstances are furnished in the present chemiluminescence, because an amide anion in **3** would more easily release an electron to the acceptor than the methylated nitrogen of **4**. An increase of Φs by an electron attracting 2,2,2-trifluoroethyl (TFE) group was observed, which is probably due to the promotion of ring closure (vide infra) as well as the CIEEL process.

Of interest was the drastic change in the emission profiles. The maximum intensity appeared at about 60s for **3a**, **3b**, and **3e**, while that for **3d** and **3c** occurred faster, especially, the maximum of **3d** happend within 1s. The curves of the time course of the emission could be matched by an equation of the combined exponential contribution of the two parameters, k_a and k_b, which is adapted to a sequential reaction process. Thus, the emission intensity I at any time t is expressed by the following equation,

$$I = M \{\exp(-k_a t) - \exp(-k_b t)\}$$

where M is proportional to the relative maximum intensity (rel. I_{max}.). Such a sequential reaction can been seen in the peroxyoxalate chemiluminescence,[7] in which the key species are presumed to be the cyclic peroxides such as 1,2-dioxetanones. In the present system, if either the kinetic parameter, k_a or k_b, corresponds to a ring-closing step producing a phospha-1,2-dioxetane, the structure of the phosphonate substituents should affect the rates as has been found in the HWE reaction.[8] Thus, an enhancement of the rates is expected for **3c**, **3d**, **4c** and **4d**, because the formation of the phospha-1,2-dioxetane ring would be promoted by the electronegative substituents or the strained 5-membered cyclic moieties.[9] According to the technique used for the kinetic analysis of the peroxyoxalate chemiluminescence reaction,[7] the computational work fitting the emission curves allowed us to estimate the relative values of both parameters k_a and k_b in each reaction. As expected, a drastic increase in k_a by some hundred times compared to the others was found in the reaction of **3d**. An appreciable acceleration in **4c** and **4d** was also found, but that for **3c** was negligible. Since such a large enhancement of the rate in **3d** was also observed when the reactants and base concentrations were changed, k_b does depend on the initial carbanion formation or the sequential oxygenation, but on the valence transformation of the phosphorus atoms. This suggests that the light emission synchronizes with the ring formation, namely, the high-energy species providing chemiluminescence should be the phospha-1,2-dioxetanes (**8**). Therefore, we could understand the difference in the rates of the ring formation depending on the phosphonate substituents.

8   **9**   **10**

The autoxidation of other phosphonate carbanions derived from diethyl diphenylmethylphosphonate (**9**) and diethyl fluorenylphosphonate (**10**) showed that DBA (9,10-dibromoanthracene, a triplet energy aacceptor) enhanced the chemiluminescence in spite of the lower energy for the excited triplet benzophenone (68-69 kcal/mol) and fluorenone (53 kcal/mol) than that for singlet DBA (71 kcal/mol). The Stern-Volmer plot of the double reciprocal of the DBA concentration and the chemiluminescence quantum yields established a bimolecular process with the fluorophor and the excited species in these chemiluminescence reactions. The emission quantum yields at the infinitive DBA concentration were calculated to be 4.1×10^{-8} for **9** and 1.7×10^{-8} for **10**. The detection of no remarkable difference in the emission quantum yields between **9** and **10** is contrary to the finding of the preferential formation of the singlet excited fluorenone by decomposition of the corresponding 1,2-dioxetane, which is probably due to the different counter part for fragmentation. The possibility of the formation of the excited phosphate ion is readily excluded, because there is no overlapping region in the absorption spectra of the phosphate ion and DBA. These results show that chemiluminescence is not a significant reaction for the acridanyl phosphonates but a general event in the autoxidation of the phosphonate carbanions if the reaction conditions are adjusted.

Consequently, phospha-1,2-dioxetanes are the only reasonable intermediates that produce the excited carbonyl fragments and simultaneously satisfy the experimental results in the present study. In this study, the substituent effect on the ring closure step in the oxy-Wittig type reaction was revealed by chemiluminescence decay. The promotion of the ring formation results in the acceleration of light emission during oxygenation of the phosphonate carbanions and results in the preferential Z-olefin formation during the HWE reaction.

REFERENCES

1. Johnson AW. Phosphonium Ylids. In: Ylid chemistry. New York: Academic Press, 1966: Chap. 3, 16-125.
2. Akasaka T, Sato R, Ando W. Oxidation of phosphazine by singlet oxygen. High-field ^{31}P NMR spectroscopic studies of 3-phospha-1,2-dioxa-4,5-diazine and phospha-1,2-dioxetane. J Am Chem Soc 1985; 107: 5539-40.
3. Motoyoshiya J, Isono Y, Hayashi S, Kanzaki Y, Hayashi S, Chemiluminescent oxidation of phosphonates: Phospha-1,2-dioxetanes as possible intermediates. Tetrahedron Lett 1994; 35: 5875-78.
4. Maryanoff BE, Reitz AB. The Wittig olefination reaction and modifications involving phosphoryl-stabilized carbanions. Stereochemistry, mechanism, and selected synthetic aspects. Chem Rev 1989; 89: 863-927.
5. Motoyoshiya J, Ikeda T, Tsuboi S, Kusaura T, Takeuchi Y, Hayashi S, Yoshioka S, Takaguchi Y, Aoyama H. Chemiluminescence in autoxidation of phosphonate carbanions. Phospha-1,2-dioxetanes as the most likely high-energy intermediates. J Org Chem 2003; 68: 5950-5.
6. Koo J–Y, Schmidt SP, Schuster DB. Bioluminescence of the firefly: key steps in the formation of the electronically excited state for model systems. Proc Natl Acad Sci USA 1978; 75: 30-3.
7. Hadd AG, Robinson AL, Rowlen KL, Birks JW. Stopped-flow kinetics investigation of the imidazole-catalyzed peroxyoxalate chemiluminescence reaction. J Org Chem 1998; 63: 3023-31.
8. Motoyoshiya J, Kusaura T, Kokin K, Yokoya S, Takaguchi Y, Narita S, Aoyama, H. The Horner-Wadsworth-Emmons reaction of mixed phosphonoacetates and aromatic aldehydes: geometrical selectivity and computational investigation. Tetrahedron 2001; 57: 1715-21.
9. Breuer E, Bannet D. M. The preparation of some cyclic phosphonates and their use in olefin synthesis. Tetrahedron 1978; 34: 997-1002.

CHEMILUMINESCENCE OF UNSUBSTITUTED AND PHENOXIDE SUBSTITUTED 1,2-DIOXETANES

J TANAKA, C TANAKA, M MATSUMOTO

Dept of Chemistry, Kanagawa University, Hiratsuka, 259-1293, Japan
Email: tanaka-hadano@tbb.t-com.ne.jp

In the chemiluminescence from unimolecular decomposition, the product molecule is excited by the energy produced from the fission of the chemical bond and some activation energy supplied from the surroundings. The electron in the HOMO of the parent molecule is promoted to the LUMO of the product molecule and emits light.

The change of charge distribution during the reaction is important to clarify the intramolecular electronic process. In this paper we will focus on the change of HOMOs and the charge density distribution of unsubstituted 1,2-dioxetane (DO) and phenoxide substituted 1,2-dioxetane (PHOD) during the reaction.

Bond breaking in unsubstituted 1,2-dioxetane (DO)

In Fig.1 the potential energy curve for chemiluminescence of DO and the change of HOMO and NHOMO are illustrated for the initial state **A**, the transition state **B** and the transition state **Q** to the emitting triplet state. The molecule at **B** is more twisted than the molecule at **A** ; the torsional angle $\angle$O4-C2-C1-O3 is $12°$ in **A** and $28°$ in **B**.[1] The increase of torsion angle induces changes of HOMO and NHOMO in a way that the anti-bonding character of O3-O4 bond is mixed in NHOMO. The O3-O4 bond distance is 2.00 Å at **B** and the bond is opened further by going down the potential curve. The singlet to the triplet intersystem crossing occurs in the region **D** enclosed by the dotted lines. Appearance of isoenergetic $n^2\pi$ and $n\pi^2$ states in the region **D** is a reason why dioxetane ring is involved in many chemiluminescent processes. The HOMO and NHOMO in **Q** suggest that the triplet $n\text{-}\pi^*$ state will be formed on the C2-O4 side. The activation energy is required to cross over **Q** for the C1-C2 bond fission. Finally DO decompose and the triplet state, which can emit light, is formed on formaldehyde.

Charge and electron transfer

The charge transfer and the electron transfer are widely used concepts, but they are sometimes used without distinction. In the chemically initiated intramolecular electron transfer luminescence (CIEEL), the term "electron transfer" is appropriate since an electron from the electron pair in the hydroxyl group formed by deprotonation is moved to the dioxetane group. A biradical is formed by electron transfer from the HOMO of the phenoxide anion to the LUMO of the dioxetane group. Accompanying with this electron transfer, the charge density redistributes in the underlying MOs, therefore the charge

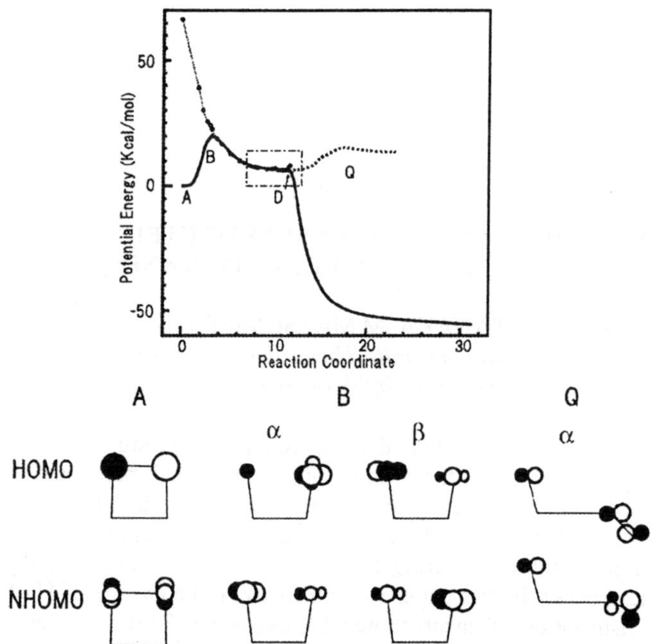

Figure 1. (Top) Potential energy curve for chemiluminescence and decomposition
of 1,2-dioxetane. **A** is starting point. **B** is the transition state and the box
enclosed by **D** is the region of intersystem crossing. **Q** is the transition state
to the triplet excited state. (Bottom) HOMO and NHOMO at **A** and **B. Q**.

density on each atom are given by the fraction of the charge summed over the filled
MOs. The charge transfer is a useful concept to show the charge densities of atoms and
groups are influenced by interacting with other atoms or groups or even molecules.
Charge density in PHOD anion
The atomic charge densities for *m*-hydroxybenzaldehyde and formaldehyde groups of
syn-m-hydroxyphenyl-1, 2-dioxetane anion are calculated by using the uB3LYP method
of Gaussian 98 program.   At point **A**, the charge transfer from the
hydroxybenzaldehyde anion to the formaldehyde group is only $0.20e$. On the transition
state at **B**, it reaches to $0.5e$. Near the point **C**, a fraction of charge density of $0.13e$ is
returned from O4 to O3 of the carbonyl group shown in Fig.3.   At **D** the C1-C2 bond
begin to elongate and a small amount of charge density $0.1e$ comes back to O4 from the

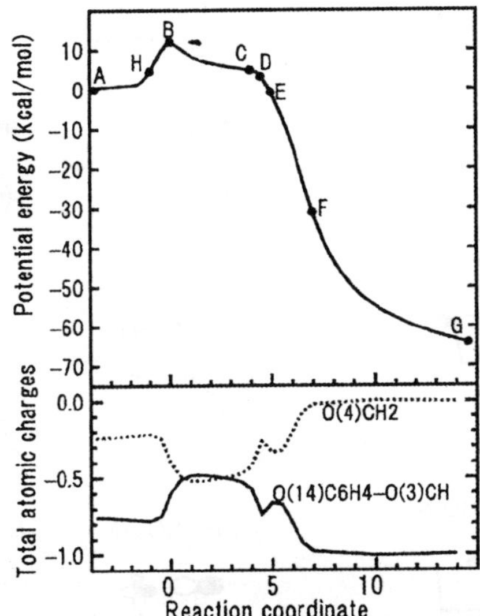

Figure 2. (Top) Potential energy curve. (Bottom) Charge densities for the benzaldehyde group and formaldehyde group during decomposition of oxyanion of 3-(3-ydroxyphenyl)-1.2-dioxetane (*syn* form)

benzene ring on the way from **D** to **E**. After passing **E** the C1-C2 bond opens to bond length of 2.04 Å at **F** and extra charge on formaldehyde group disappears.

HOMO suggests nature of excited state

Immediately after the deprotonation from *syn-m*-hydroxyphenyl-1, 2-dioxetane, the charge on $O(14)$-C_6H_4 is $0.53e$ and $O(3)$-CH is $0.27e$.. The HOMO at **A** in Fig.3 shows that both the α-HOMO and β-HOMO are on the phenoxide group. When the reaction proceeds to the transition state **B**, an electron is transferred to the O3-O4 σ^* orbital (anti-bonding) as shown in β-HOMO of **B** in Fig.3. Going down the potential curve the O3-O4 bond is broken at **C**, where the O3-O4 bond length is 2.36 Å. The extra charge accumulated on O4 is partly back to O3 and both the α-HOMO and β-HOMO are concentrated on O3 and they are symmetric and anti-symmetric combination of O3 n and O3 π type orbitals. This result implies that the excited state (S_1) of the $n\pi^*$ type which is symmetric combination of these orbitals, coexists near the ground state (S_0) at **C**. The reaction proceeds to the point **D**, the β-MO and the α-MO show that an electron transfer occurrs from the phenoxide group to the C1-O3 π^* orbital. Furthermore at the point **E** the α-MO shows involvement of the excitation at O4. This

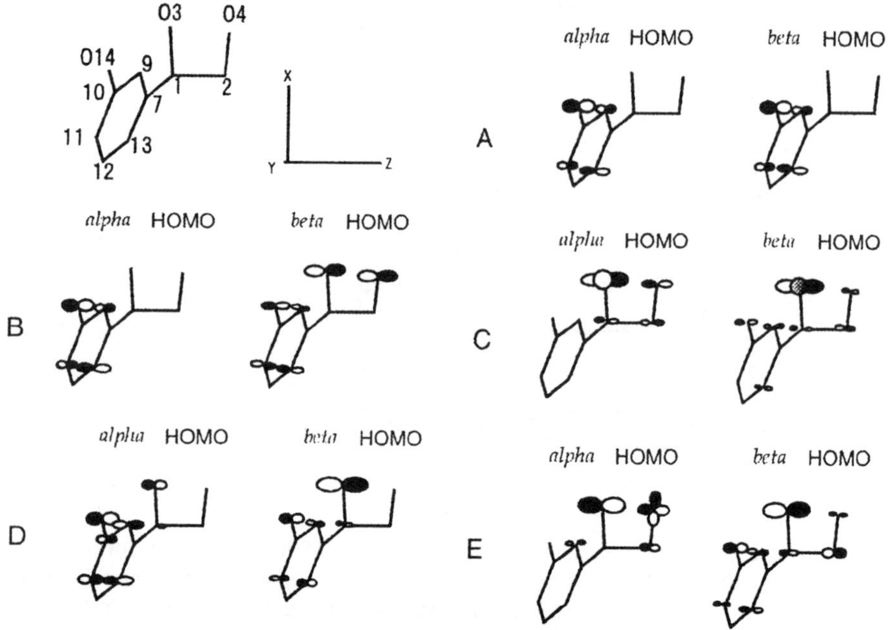

Figure 3. α-HOMO and β-HOMO for oxyanion of
3-(3-hydroxyphenyl)-1,2-dioxetane (*syn* form) during the decomposition
reaction. **A, B, C.D**.and **E** are the same shown in Fig.2.

may be due to the back electron transfer of $0.1e$ mentioned in the previous section. Involvement of several excited configurations is envisaged by viewing at the HOMO of the ground state calculated by the uB3LYP method. This method deals with the ground state only, but the HOMO obtained revealed that the S_1 state is close to the S_0 state. This is most important condition for chemiluminescence to be produced.

ACKNOWLEDGEMENTS
We thank to the Research Center for Computational Science in Okazaki for the use of computers.

REFERENCES
1.   Tanaka C. Tanaka. J. Ab Initio Molecular Orbital Studies on the Chemiluminescence of 1,2-Dioxetanes, J. Phys. Chem. A; 2000, 104: 2078-2090.
2.   Frisch M.J. et al. Gaussian 98 and 03, Gaussian Inc.Pittsburgh, 1998, 2003.

INTRAMOLECULAR CIEEL MECHANISM OF CHEMILUMINESCENCE OF PHENOXIDE SUBSTITUTED 1,2-DIOXETANES

C. TANAKA, J. TANAKA, M. MATSUMOTO

Dept of Chemistry, Kanagawa University, Hiratsuka, 259-1293, Japan
Email :tanaka-hadano@tbb.t-com.ne.jp

More than twenty years ago, Schaap et al [1] discovered chemiluminescence from phenoxide substituted dioxetanes (PHOD) as a model for firefly bioluminescence. McCapra[2] had suggested a charge or an electron transfer is involved for such process, but confirmative evidences had not been presented at that time. Recently successive synthetic efforts on phenoxide.substituted dioxetanes produced a series of efficient chemiluminescent molecules.[3]

1,2-dioxetanes with no electron-donating groups show chemiluminescence from the triplet state via a biradical mechanism.[4-6]   On PHOD a biradical will be formed in the transition state, since an electron is transferred from the phenoxide group to the O-O antibonding orbital. In this article we present a theoretical study on chemiluminescence mechanism by the ab initio MO calculation on three isomers of PHOD to show the detail of the chemically initiated intramolecular electron transfer (CIEEL) process. All calculations are performed with Gaussian programs[7] using the computers in Research Center for Computational Science in Okazaki.

Fig.1 shows the optimized geometries of *syn-m* isomer. Figs. 2 and 3 show the potential energy curves along the intrinsic reaction coordinate (IRC) for the ground states S_0 of *syn-m-* and   *p-* isomers, respectively. The roman characters, **A**, **H**, **B**, **C**, etc are characteristic points of the reaction.   In Fig.1, the geometry at the starting point **A** shows that the O3-O4 bond is under the plane of the phenoxide ring.   The O3-O4 bond length is 1.50Å in A. From **A** to **H** the electron configuration is closed shell. The **B** is the transition state and the O3-O4 bond is crossing the plane of the phenoxide ring. The node of the O3-O4 σ * orbital is conformity with the π orbital of the phenoxide ring. Accordingly, the electron transfer is feasible in this conformation, and a charge density is increased by $0.3e$ on O3-O4.   The bond length increases to 1.85Å . The activation energies of the reaction are calculated as 12 kcal/mol and 7 kcal/mol for *syn-m-* and   *p-* isomers, respectively.   These values are in agreement with experiments and much smaller compared to unsubstituted and tetramethyl-1, 2-dioxetanes.   After crossing B the biradical state (open shell electronic structure) continues to **F**. The atomic charge density on O4 is $0.5e$ up to the point **C**. The O3-O4 distance is 2.5Å indicating   that

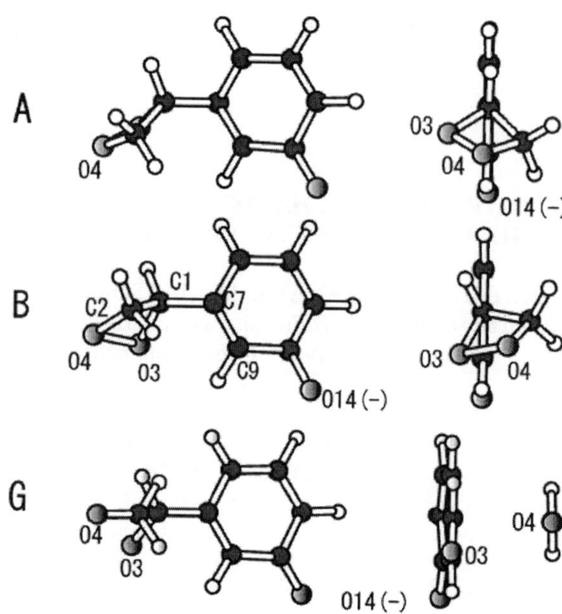

Figure 1. Optimized geometries for oxyanion of 3-(3-hydroxyphenyl)-1,2-dioxetane (*syn* form) in the initial state (A), the transition state (B) and the product (G).

the bond is opened.   A part of the accumulated charge on O4 is returned to O3 atom. At this point the excited state S_1 is isoenergetic with S_0 where the S_0 is $n^2\pi$ and S_1 is $n\pi^2$ configuration on non-bonding orbital and π orbital of O3 .   This is found by the MCSCF calculation.

In the region from **C** to **D**, the S_1 and S_0 states are isoenergetic.   At **D** the character of the excited state is completely changed to the $\pi\pi^*$ type of   intramolecular electron transfer from the phenoxide HOMO to the C1-O3 π^* orbital. At the point **E** the C1-C2 bond elongates to 1.67Å . Small amounts of charge migrate   from the benzaldehyde group to O4 and the α -HOMO involves the O4 π^* orbital.   It may mean that the $n\pi^*$ excited state of the C2-O4 is involved in the intermediate state of chemiluminescence.[8]

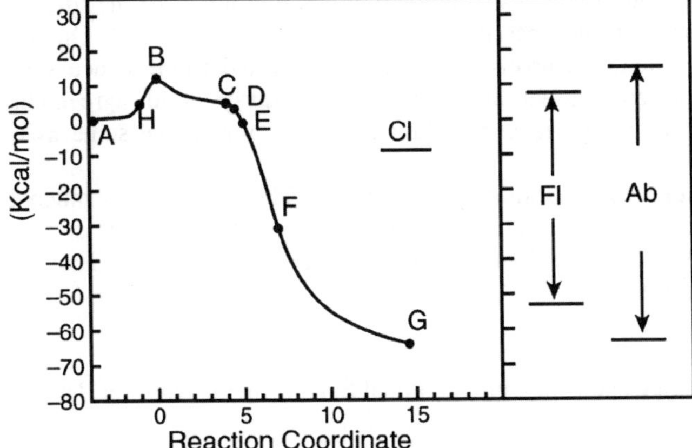

Figure 2.   Potential energy curve for the decomposition of oxyanion of 3-(3-hydroxyphenyl)-1, 2-dioxetane (*syn* form) (Left)   Absorption and fluorescence energy levels of   3-hydroxybenzaldehyde anion (Right).

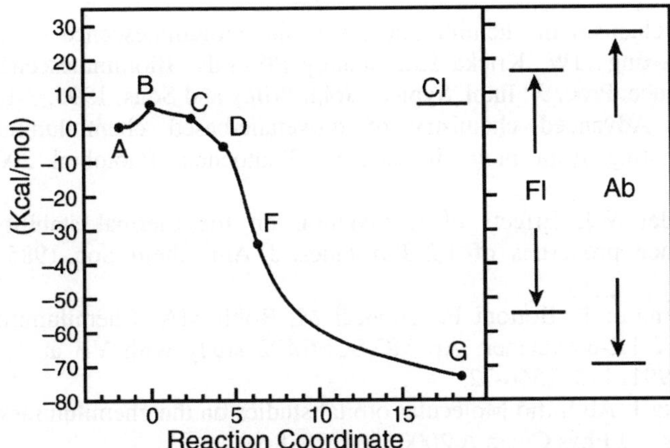

Figure 3.   Potential energy curve for the decomposition of oxyanion of 3-(4-hydroxyphenyl)- 1,2-dioxetane (Left)   Absorption and fluorescence energy level of   4-hydroxybenzaldehyde anion (Right) .

At **F** the formaldehyde group separates from the oxy-anion group.   At the final point **G** the estimated level for chemiluminescence (CL) is illustrated in Fig.2. On the right column of Fig. 2, the calculated levels for the absorption and fluorescence spectra of *m*-hydroxybenzaldehyde are shown and these are in fairly good agreement with experimental values. The energy level for fluorescence is almost the same as the CL level.

In Fig. 3 the results for *p*-isomer are illustrated. The CL level is much higher compared to *syn-m-* isomer, therefore the activation energy to the CL level is larger than for *syn-m-* isomer.   The low yield of chemiluminescence of *p*-isomer is reasonably explained by these calculations.

ACKNOWLEDGEMENTS

We thank to the Research Center for Computational Science in Okazaki for the use of SGI2000 computers.

REFERENCES

1.   Schaap AP, Gagnon SD. Chemiluminescence from a phenoxide-substituted 1,2-dioxetane: A model for firefly bioluminescence, J Am Chem Soc; 1982; 104: 3504-6.
2.   McCapra F. Mechanism in chemiluminescence and bioluminescence unfinished business. In: Hastings JW, Kricka LJ., Stanley PE. Eds. Bioluminescence and Chemiluminescence. Proc. 9th Intnl. Symp.   John Wiley and Sons, 1996; 7-15.
3.   Matsumoto M. Advanced chemistry of dioxetane-based chemiuluminescent substrates originating from bioluminescence.   Photochem Photobiol 2004; 5: 27-53.
4.   Adam W, Baader WJ, Effects of methylation on the thermal stability and chemiluminescence properties of 1,2-dioxetanes. J Am Chem Soc 1985; 107: 410-6.
5.   Reguero M, Bernardi F, Bottoni F, Olivucci M, Robb MA. Chemiluminescent decomposition of 1,2-dioxetanes: An MC-SCF/MP2 study with VB analysis J Am Chem Soc 1991; 113: 1566-72.
6.   Tanaka C, Tanaka J. Ab Initio Molecular orbital studies on the chemiluminescence of 1,2-dioxetanes.   J Phys Chem A 2000; 104: 2078-90.
7.   Frisch MJ et al. Gaussian 98 and 03, Gaussian Inc. Pittsburgh, PA. 1998, 2003.
8.   Fujimori K. Private communication.

PART 7

INSTRUMENTATION & DEVICES

SINGLE-MOLECULE IMAGING OF PROTEIN IN LIVING CELLS BY PIN-FIBER VIDEO-MICROSCOPY

Y HIRAKAWA[1], T HASEGAWA[1], T MASUJIMA[1], M TOKUNAGA[2],
N TSUYAMA[3], M KAWANO[3]

*[1] Dept of Frontier Medical Science, Hiroshima University, 1-2-3 Kasumi,
Hiroshima 734-8551, Japan*
*[2]Structural Biology Center, National Institute of Genetics, Mishima,
Shizuoka 411-8540, Japan,*
*Research Center for Allergy and Immunology, RIKEN, Yokohama,
Kanagawa 230-0045, Japan*
*[3]Dept of Bio-Signal Analysis, Yamaguchi University, 1-1-1 Minami-kogushi, Ube,
Yamaguchi 755-8505, Japan*
Email: yhirakaw@hiroshima-u.ac.jp

INTRODUCTION

A single molecule imaging technique is an attractive method which can reveal micro-kinetics of biological molecules. By using this technique, remarkable results have so far been reported.[1-3] In these results, the technique of near-field illuminations, which utilize total internal reflection or a microscopic optical aperture, was adopted. Near field methods need an interface to generate evanescent light illuminating the samples. As the result, the region which can be observed by this technique is sometimes restricted. In order to overcome this point, our group have proposed a novel microscope system of "pin-fiber video-microscope" and reported its applications.[4-6] This system has a unique illumination source consisted of a single optical fiber. With this single fiber system, a selectable region, intensity and area of irradiation in a sample on a microscope are possible by changing the position of the fiber output. In this report, we present single molecular imaging of proteins in a living cell by pin-fiber video-microscope. Kinetic behaviour of proteins caused by extracellular stimulation was visualized. It was found that by analyzing the single molecule images of protein, this new video-microscope has a potential to reveal kinetics of intercellular proteins.

METHODS

In the experiments, signal transducer and activator of transcription 1 (STAT1), in HeLa cells was selected for the visualized target. The experimental setup is depicted in Fig. 1. The most characteristic point of pin-fiber video-microscope is that its light source needs no interface, because it does not utilize evanescent light.

Instrumentation

A coherent radiation from a laser (473 nm, 5 mW, Crystal Laser, Model BCL-005-M) was injected into a single optical fiber (Thorlabs, SM500) with a coupling lens (Thorlabs, A390TM-A). The illuminated region and area were controlled by a micro-manipulator (Narishige, NHN-21 and NHW-4) holding the fiber output. A

                                                                    Hirakawa Y et al.

fluorescent image from a sample on the microscope (Zeiss, Axioplan) was optically filtered by passing through an interference filter and a notch filter. For elimination of unnecessary fluorescent light from unfocused planes, total optical system of the microscope was changed to a confocal system by setting an additional optical system which consisted of two convex lenses and one pinhole in front of the detector. The noise-reduced fluorescent signal was monitored by a high sensitive CCD camera (Hamamatsu, C7190) and recorded on a digital video recorder (Sony, DSR-30). The recorded digital images were processed and analyzed by computer software (Apple, iMovie; National Institute of Health, NIH images; and Adobe, Photoshop).

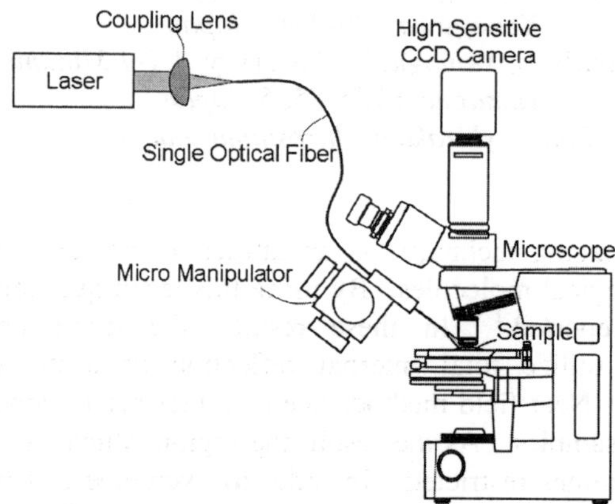

Figure 1. Experimental setup

Sample preparations

STAT1 is one of transcription factor proteins and it is activated by stimulation of interferon (IFN) γ. For visualizing this protein, cDNA encoding STAT1 was fused to green fluorescent protein (GFP) gene by recombination, and STAT1-GFP expression plasmid was transfected into living HeLa cells. In order to suppress background level of images for single molecular imaging, the degree of expression of STAT1-GFP fusion protein was reduced to the level that no fluorescent signals were detected by normal fluorescent microscope mode with a normal CCD camera instead of the high-sensitive CCD camera.

RESULTS

By confirming single step bleaching of fluorescent signals, condition of single molecular imaging was established. With focusing on cell membrane, STAT1-GFPs were recognized as small bright spots when they were recruited to IFN γ receptors after IFN γ stimulation. STAT1 is phosphorylated by tyrosine kinase complex with IFN γ receptor and forms a homodimer. A great number of bright fluorescent spots appeared and rapidly disappeared one after another. Fig. 2 indicates one of the typical still images of this phenomenon. The fact that wide area can be visualized under single molecular imaging condition is one of the merits of pin-fiber video-microscope.

By focusing on a particular STAT1-GFP spot and analyzing the time course of the recorded images, it was found that different behavior types of STAT1-GFP image intensity existed. The results are shown in Fig. 3. One type was that after appearance of single STAT1 spot, spot brightness was enhanced by a factor of 2 (Fig. 3 (a)). In this case, it might be visualized that one STAT1-GFP came to a receptor first, and after a few hundred ms, another STAT1-GFP was recruited to the same receptor, then these two STAT1-GFPs were dimerized together. The other type was that the STAT1-GFP spot intensity was rapidly increased to this twice brightness (Fig. 3 (b)). This might mean that two STAT1-GFPs were recruited simultaneously to a receptor and they were dimerized. The weak intensity curve labelled with "Single" in Fig. 3 might correspond to STAT1-GFP which did not dimerize or dimerized with non-fluorescent endogenous STAT1. Although this different type of temporal variation of STAT1-GFP fluorescence needs to be investigated further, these preliminary results suggest that pin-fiber video-microscope has a potential to analyze kinetics of intracellular molecules including proteins by using single molecular images.

In conclusion, single molecular observation of protein inside a living cell was tried by pin-fiber video-microscope. Under single-molecule imaging conditions tested, appearance and disappearance of STAT1-GFP spots was clearly observed probably due to phosphorylation mediated activation of STAT1 that resulted in its nuclear translocation from cytoplasm through the cell membrane. Because the two-fold increase in brightness might indicate dimerization of STAT1, this fact suggest that the kinetics of STAT1 dimerization could be observed by this novel method.

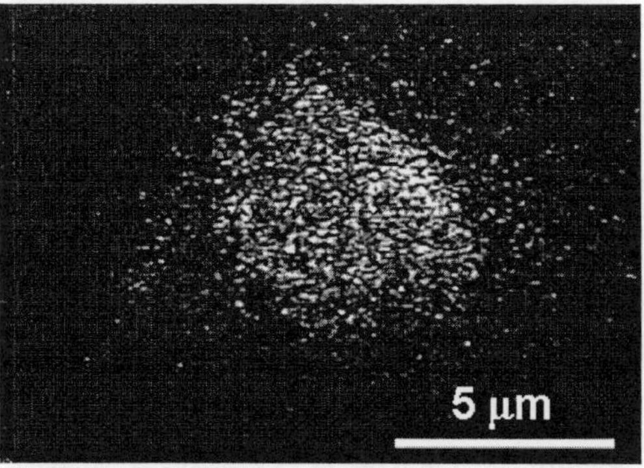

Figure 2. STAT1-GFP single molecule spots visualized by wide area illumination

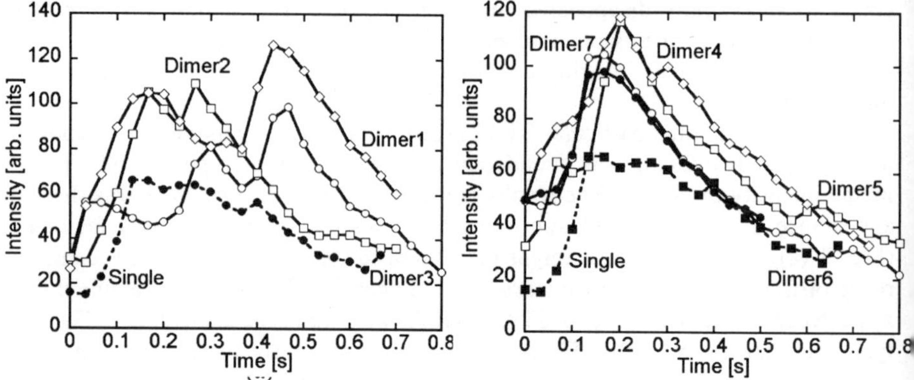

Figure 3. Two-types of temporal variations of STAT1 fluorescent spots intensities

(a)   **(b)**

ACKNOWLEDGEMENTS

This research was partially supported by the Japan Ministry of Education, Culture, Sports, Science and Technology, Grant-in-Aid for Scientific Research.

REFERENCES

1   Funatsu T, Harada Y, Tokunaga M, Saito K, Yanagida T. Imaging of single fluorescent molecules and individual ATP turnovers by single myosin molecules in aqueous solution. Nature 1995: 374; 555-9.
2   Weiss S. Fluorescence spectroscopy of single biomolecules. Science 1999: 283; 1676-83.
3   Xu X, Yeung ES. Direct Measurement of single-molecule diffusion and photodecomposition in free solution. Science 1997: 275; 1106; Chemical characterization of single cells and single molecules. Trends in analytical Sciences 1997: 1; 173-81 (http://neo.pharm.hiroshima-u.ac.jp/tals/).
4   Hirakawa Y, Suzutoh M, Ohnishi H, Shingaki T, Eyring EM, Tokunaga M, Masujima T. Analysis of the nano-kinetic movement of a single DNA by a pin-fiber video scope. Anal Sci 2002: 18; 1293-4.
5   Suzuto M, Hirakawa Y, Ohnishi H, Tachino S, Shingaki T, Eyring EM, Masujima T. Nano-kinetics of probe-particles in solution visualized by a pin-fiber video scope. Anal Sci 2003: 19; 43-7.
6   Hirakawa Y, Suzuto M, Ohnishi H, Shingaki T, Eyring EM, Tokunaga M, Masujima T. Observation and analysis of single DNA nano-kinetics by Pin-Fiber Video Scope. Analyst 2003: 128; 676-80.

SCANNING NEAR FIELD OPTICAL/ATOMIC FORCE MICROSCOPY (SNOW/AFM)—NOVEL IMAGING TECHNIQUE IN NANO-METER SCALE AND DNA-NANOFISH METHOD

T OHTANI[1], JM KIM[2], T YOSHINO[1], H NAKAO[1], M SASOU[1],
S SUGIYAMA[1], H MURAMATSU[2]

[1]*Instrumentation Engineering Lab., Nano-biotechnology Group,
National Food Research Institute, Kanondai 2-1-12, Tsukuba, Ibaraki, 305-8642 Japan*
[2]*School of Bionics, Tokyo University of Technology, 1404 Katakura, Hachioji,
Tokyo 192-098 Japan*
Email: ohtani@affrc.go.jp

INTRODUCTION

The fluorescence *in situ* hybridization (FISH) method is applied to the detection of the location of a specific gene on the DNA. Because the resolution of the conventional FISH method by light microscopy is practically limited to the half micro-meter level, a novel technique is required that would enable the visualization of a specific gene on DNA at the nano-meter scale level.

In this study, we describe the development of the "DNA nano-FISH" method that directly defines the location of a specific gene on the DNA fiber with a high resolution using a scanning near-field optical/atomic force microscope (SNOM/AFM). Theoretically, this method is expected to exceed the limited resolution of conventional FISH. We describe a principle of the SNOM/AFM and show the high-resolution fluorescent images of DNA using a SNOM/AFM.

METHODS

Scanning near-field optical/atomic force microscope (SNOM/AFM)

Figure 1 shows the principle of scanning near-field microscopy (SNOM) as compared with conventional fluorescence microscopy[1]. When different dyes at intervals less than 500 nm are located on a transparent substrate such as a cover slip, all dyes will be excited at once because the visible light wave length is typically about 500 nm in length (Fig. 1a). However, by SNOM, an optical fiber having a sharp tip with a small aperture (50-100nm) is used for excitation of the dye molecule. If an optical fiber probe with a 100 nm aperture is used, individual dyes can be excited with the near-field light from the aperture, as if illuminated with a spotlight of 100 nm radius (Fig. 1b). Thus, the resolution of the SNOM method surpasses the optical limitation. The range of the near-field from the probe tip is nearly equal to the diameter of the tip aperture.

The SNOM/AFM (SNOAMTM, Seiko Instruments Inc., Chiba) [2] used in this SNOM/AFM study was mounted on a conventional inverted light microscope; a bent optical

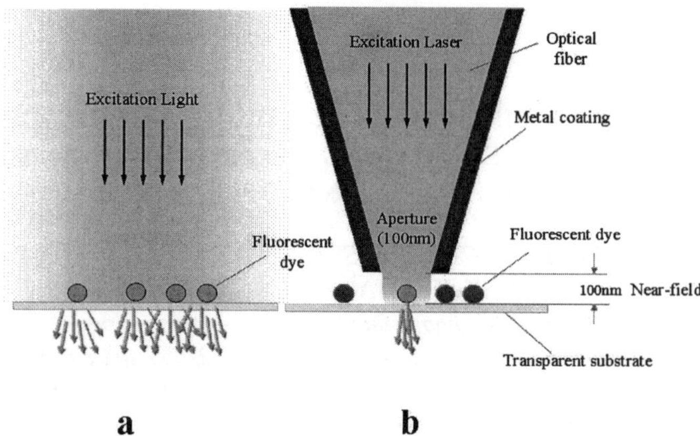

Figure 1 Principle of the SNOM/AFM (b) as compared with that of the standard fluorescence microscopy (a).

a. Principle of standard inverted light microscopy. Four different dyes are excited at once because the wavelength of the visible light is more than about 500 nm.

b. Principle of scanning near-field optical microscopy. Since individual dyes can be exited with the near-field light which is made by the tip aperture of about 100 nm in diameter, thus, the individual dyes are discriminated by this method.

fiber probe was used for this SNOM/AFM instead of a silicon AFM probe and controlled by a dynamic force mode. The excitation light is coupled with the end of the optical probe and transmitted to the tip of the probe, where the near-field light is produced. The excited far-field light from the labeled fluorescent dye on the DNA fiber was collected by a photo-detector through the objective lens of the original light microscope. A resonant frequency of an optical fiber probe was controlled around 17 kHz (SPM 3800N, Seiko Instruments Inc.). We used a set of three piezo scanners with a scan range of 20 μm (Seiko Instruments Inc.) and an excitation Ar ion laser (488 nm wavelength). The scan speed ranged from 0.14 to 0.19 Hz, yielding an imaging time for each frame of about 50 minutes. The amplitude reference (set point) was adjusted as low as possible, between -0.032 and -0.048, so as not to damage the chromosome samples.

DNA solution and stretching method

Double-stranded lambda-phage DNA (48.5 kbp, 570 μg/mL) was supplied from Wako Pure Chemicals Inc., in TE buffer (10 mM Tris, 1 mM EDTA, pH 8). The DNA solution was diluted to a final concentration of 5.7×10^{-3} ng with the TE buffer including 1mM DMSO. To visualize the DNA, we prestained it with dimeric cyanine dye YOYO-1 (Molecular Probes, Inc., adsorption 491 nm, emission 509 nm), at a ratio of 1 dye/5 base pairs. Mica sheets of 0.01 mm thickness modified with 3-(aminopropyl)triethoxysilne or methyltrimethoxysilane[3] were attached on the cover slip with an adhesive glue[4]. Just before the experiment, the mica was freshly cleaved. The samples were fixed on a steel ring and observed by SNOM/AFM.

The DNA was stretched on the mica with a spin stretching method and suck up method[5].

Preparation of FISH Sample

A peptide nucleic acid (PNA) 15-mer probe was designed for hybridizing to the top part of the ea47 gene of lambda-phage DNA (48.5 kb, 16.5 μm). The ea47 gene locates to almost the center position of the DNA. Alexa 532 pigments (A532, Molecular Probes, Excitation maximum 532 nm/ Emission maximum 554 nm) was conjugated to 5' end of the PNA probe. The PNA probe was hybridized to the DNA strands in Tris-HCl (pH. 7.4, 10% DMSO) for 90 s at reaction temperatures of 65 or 70°C. For high efficiency labeling, the PNA probe was

RESULTS

DNA stretching on modified mica surface

The high-resolution image of the methyltrimathoxysilane-coupled mica substrate[3] can be achieved using AFM. The surface looked uniform and had no large surface structures. Aggregates or debris that hindered visualization and manipulation of DNA were not found by AFM observation. The height of DNA was much higher than the roughness of the surface structures, and the topography of the straightened DNA was clearly observed. The surface roughness on the DNA stretched mica was essentially the same as those on the methyltrimethoxysilane-coupled mica. The surface root-mean-square (RMS) roughness of the silanized mica substrates swere under 0.09 nm. The value suggested that the coupled silane on the mica surfaces was very thin, and might be considered a single molecular layer.

Detection of PNA probe on DNA molecule

After the several fundamental investigations, we finally detected the PNA probe hybridized on the lambda phage DNA molecule.

Figure 2 shows an over lapped fluorescence image of Alexa 532 labeled PNA probe and the YOYO-1 stained lambda phage DNA using SNOM/AFM. Figure 2 indicated that the SNOM/AFM system in this study could clearly detect the fluorescence dye conjugated with PNA. In our experiment, one or two molecules were conjugated; because the

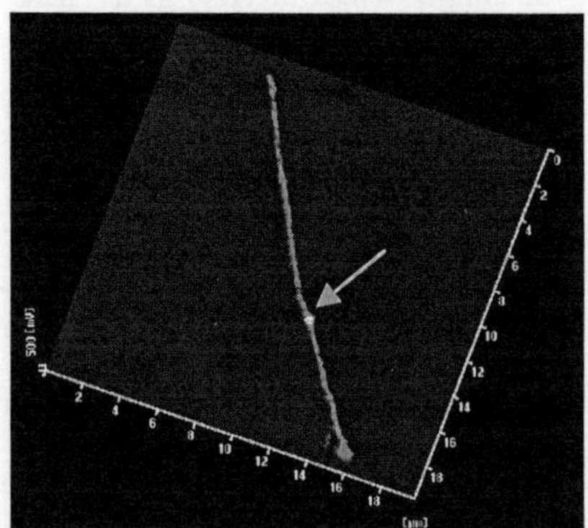

Figure 2 Fluorescence image of PNA probe (arrow) hybridized on the lambda phage DNA molecule using SNOM/AFM.

photon counts from the fluorescence area exhibited one or two-step decrease (data not shown) during continuous imaging of the PNA probe. We believe that the Fig. 2 showed the direct identification of the PNA probe on the DNA molecule in nano-mater scale.

Line profile of the enlarged image of the PNA fluorescence spot in the different experiment of Fig. 2 indicated that the optical resolution of about 20 nm for the DNA length direction was achieved. (data not shown). It corresponded to the best optical resolution ever reported at room temperature. Additionally, in this sample, a consecutive scan to confirm the number of the excited fluorescence molecules showed one-step photo bleaching (data not shown). In this experiment, the fluorescence resolution was confirmed by control imaging for a single dye molecule scattered surface.

CONCLUSIONS

DNA molecule can be stretched on an appropriate substrate, such as hydrophobic mica surface, by spin stretching method or a suck up method. Using the SNOM/AFM technique, a labeled DNA molecule and a hybridized PNA probe on DNA molecule can be detected at the nano-meter scale. "DNA-nanoFISH" method with SNOM/AFM has the ability to become one of the conventional tools in the scientific and medical field. However, further investigation of pre-treatment of samples and hardware of SNOM/AFM will be necessary to improve the resolution and establish the practical DNA-nanoFISH method.

ACKNOWLEDGMENTS

This study was performed through Special Coordination Funds of BRAIN (Bio-oriented Technology Research Advancement Institution) and the Ministry of Education, Culture, Sports, Science and Technology of the Japanese Government.

REFERENCES

1.   Ohtani T, Shichiri M, Fukushi D, Sugiyama S, Yoshino T, Kobori T, Hagiwara S, Ushiki T. Imaging of chromosomes at nano-meter scale resolution using scanning near-field optical/atomic force microscopy. Arch Histol Cytol 2002; 65: 425-34.

2.   Muramatsu H, Homma K, Yamamoto N, Wang J, Sakata-Sogawa K, Shimamoto N. Imaging of DNA molecules by scanning near-field microscope. Materials Sci Eng C. 2000; 12: 29-32.

3.   Sasou M, Sugiyama S, Yoshino T, Ohtani T. Molecular flat mica surface silanized with methyltrimethoxysilane for fixing and straightening DNA. Langmuir 2003; 19: 9845-9.

4.   Muramatsu H, Chiba N, Homma K, Nakajima K, Ataka T, Ohta S, Kusumi A, Fujihira M. Near-field optical microscopy in liquids. Appl Phys Lett 1995; 24: 3245-7.

5.   Nakao H, Hayashi H, Yoshino T, Sugiyama S, Otobe K, Ohtani T. Development of novel polymer-coated substrates for straightening and fixing DNA. Nano Lett 2002; 2: 475- 9.

A NEW INSTRUMENT FOR AUTOMATED LUMINESCENT ASSAYS

WH SYMONDS, DJ SQUIRRELL, RS JACKSON.
Dstl Porton Down, Salisbury, Wiltshire, SP4 0JQ, UK

INTRODUCTION

The development of automated luminometers is focused primarily on devices that achieve high sample throughput rates, typically through the use of 96 and 384 well microtitre plates. Whilst suited for many applications, these systems require a skilled operator and often do not provide rapid results when the time taken to prepare the microtitre plate is included in the assay time. An instrument that can perform on demand, automated, near real time analysis using a variety of luminescent assay protocols has been developed. The instrument has been designed to automate various luminescent assays including adenylate kinase[1] (AK) assays and those that use magnetic separation steps in conjunction with bioluminescence[2].

INSTRUMENTATION DEVELOPMENT

The instrument is based on integrated fluidics and comprises a single acrylic fluidic block (Fig. 1) in which are fabricated channels, wells and reagent handling structures (fabricated by Carville Ltd, UK). This "integrated fluidic device" is, in essence, an automated pipettor under computer control. The computer control provides the flexibility and potential to perform complicated fluid handling steps and to build sequences of steps into a protocol or assay. The fluidic block was designed such that assay reagents can be moved from storage vessels to a common reaction chamber without cross-contamination between assay components. Multiple valve banks and a syringe stepper motor driven pump (Lee Products Ltd., UK) are used to move fluids around the device. Sample can be introduced through one of the reagent reservoir positions or from an internal flow-through reservoir. The internal reservoir can be connected to a sampler such as a cyclone aerosol collector and facilitates automated batch testing. A stepper motor driven stirrer and two stepper motor coupled linear worm drives were incorporated to enable the device to perform magnetic bead manipulations and thus to allow specific, as well as generic, assays.

National Instruments™ (Texas, US) hardware was used to control the valves and stepper motors. A 12 V DC power supply (RS, UK) was used to power the valves. Software to control the valves and stepper pumps was written using National Instruments™ LabVIEW™ 5.1. A photomultiplier tube (PMT) was supplied by Biotrace International (Bridgend, UK) and was mounted so that the light measurements could be made directly from the reaction chamber. As a developmental iteration the National Instruments™ control hardware was substituted for electronic control components integrated into a light tight box together with the fluidic block. A compact controlling computer was built in order to make the device portable.

Figure 1. The integrated fluidic device showing; A, reagent and sample reservoirs;

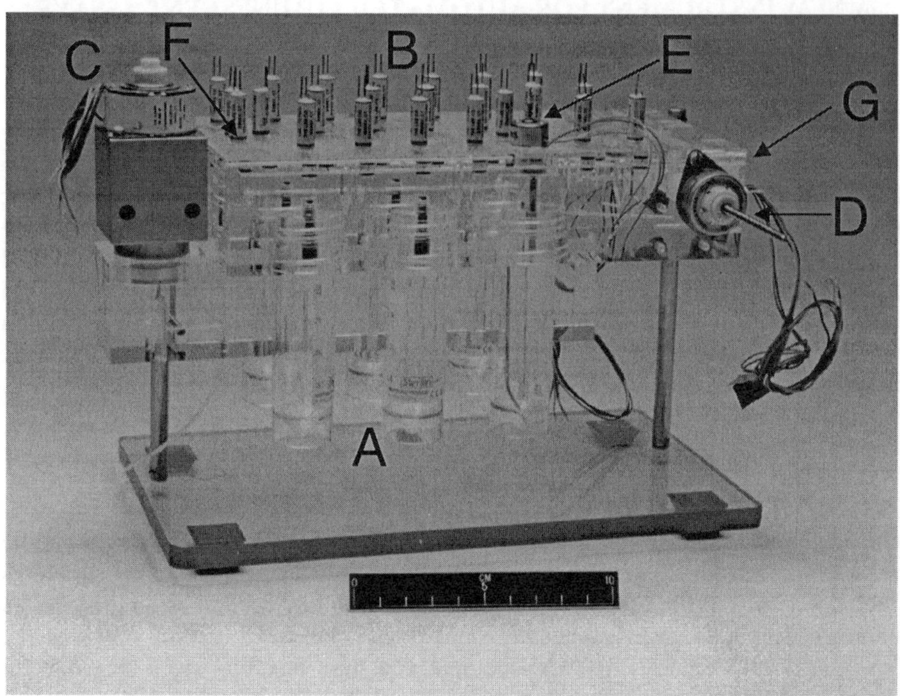

B, manifold-mounted valves arranged in banks of 3; C, integrated 750 μL stepper motor driven syringe pump; D, linear worm drive; E, stirrer stepper motor; F, pressure release and pump check valves; G, external sample port; and 10 cm scale.

METHODS AND MATERIALS
Instrumentation
The integrated fluidic device was tested in comparison with a manual bioluminescent assay, which was performed in 3.5 mL polypropylene assay cuvettes where light measurements were made using a Multi-Lite™ luminometer obtained from Biotrace International. Bacteria were cultured in a shaking incubator (Stuart Scientific, UK).
Reagents
Erwinia herbicola NCIMB 12126 was obtained from the National Collection of Industrial and Marine Bacteria (Aberdeen, UK), HEPES buffer sachets and magnesium acetate were obtained from Sigma (Poole, UK), adenylate kinase assay kits were obtained from Acolyte Biomedica (Salisbury, UK), sterile tissue culture grade distilled water was obtained from Gibco (Paisley, UK), L-broth and tryptone soya agar plates were obtained from Oxoid (Basingstoke, UK).

Methods

Cultures of *E. herbicola* were grown overnight at 37 °C in L-broth with shaking at 150 rpm. 100 mmol/L stock Buffer pH7.4 was prepared from sachets of HEPES powder. 10 fold serial dilutions of *E. herbicola* from overnight cultures were made in 10 mmol/L HEPES buffer. *E. herbicola* cell concentrations were measured using triplicate tryptone soya agar plates incubated for 24 hours at 37 °C. Each dilution was tested in triplicate in AK assays: 100 μL sample plus 100 μL 15 mmol/L magnesium acetate and 100 μL MAKAR (microbial adenylate kinase assay reagent), incubation for 5 min at room temperature, and then addition of 100 μL bioluminescent reagent. The light from the reaction chamber in the integrated fluidic device was measured for 10 s and averaged to allow comparison with the Multi-Lite™, which integrates the light output for 10 s.

RESULTS

Results from serial dilutions of *E. herbicola* assayed by AK in the integrated fluidic device in comparison with the manual luminometer are shown in Fig. 2. Results from dilution buffer blanks plus 2 standard deviations are shown to allow comparison of the backgrounds in the two systems.

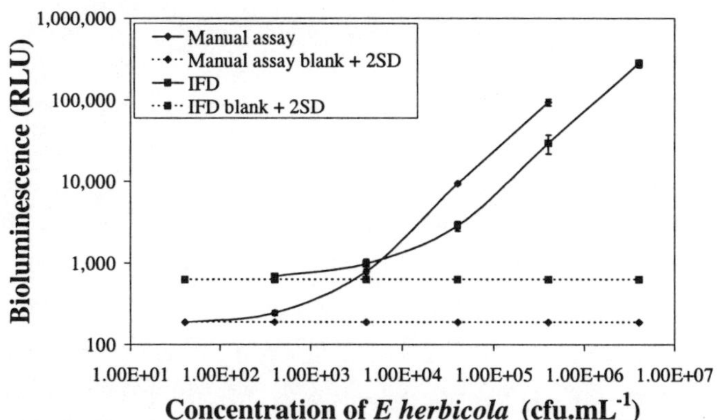

Figure 2. Comparison of a manual AK assay with an automated AK assay performed in the integrated fluidic device using *E. herbicola*.

DISCUSSION

An automated device, which allows for performing bioluminescent assays has been developed.

Results from the current integrated fluidic device indicate analytical sensitivities similar to manual assays, but limits of detection 10 times worse as a consequence of higher backgrounds and relatively inefficient light collection. Development is needed to improve the device through optimisation of the optical coupling between the PMT and reaction chamber. In other testing the sensitivity of the integrated fluidic device was compared to the sensitivity of an automated flow injection luminometer and was found to be very similar (results not shown). Possible applications of the device could include water testing and environmental monitoring.

Protocols that enable the device to perform immuno-magnetic bead assays using AK bioluminescence as an endpoint detection system have also been developed, and are currently being evaluated.

REFERENCES

1.   Squirrell DJ, Murphy MJ. Adenylate kinase as cell marker in bioluminescent assays. In: Campbell AK, Kricka LJ, Stanley PE, Eds. Bioluminescence and Chemiluminescence: Fundamentals and Applied Aspects. Chichester: John Wiley and Sons, 1994: 468-9
2.   Squirrell DJ, Price RL, Murphy MJ. Rapid and specific detection of bacteria using bioluminescence. Anal Chim Acta 2002; 457: 109-14

PART 8

APPLICATIONS OF LUMINESCENCE

CONSTRUCTION OF A NOVEL BIOLUMINESCENT BACTERIAL BIOSENSOR FOR REAL-TIME MONITORING OF CYTOTOXIC DRUGS ACTIVITY

HM ALLOUSH[1], JE ANGELL[1], MA SMITH[1], PJ HILL[2], VC SALISBURY[1]
[1]Faculty of Applied Sciences, University of the West of England, Bristol, UK
[2]School of Biosciences, University of Nottingham, Loughborough, Leicester, UK

INTRODUCTION

Cytosine arabinoside (Ara-C), a synthetic pyrimidine nucleoside analogue, is the mainstay of treatment for acute myeloid leukaemia (AML) and routinely figures in therapeutic protocols.[1] The prime determinants of Ara-C cytotoxicity depends on its uptake and subsequent phosphorylayion by deoxycytidine kinase (dCK) into its active metabolite Ara-CTP as shown in Fig. 1.[2] Factors influencing these steps include rates of intracellular anabolism and catabolism of the drug, the presence of competing deoxycytidine triphosphate pools, and incorporation and retention into DNA. Hence, resistance to chemotherapy is common and represents a major obstacle to effective treatment of patients with AML.[3] The prognosis of patients with AML remains disappointing as 30% of newly diagnosed patients fail to achieve remission. *In vitro* assessment of Ara-C efficacy has traditionally involved measurement of cell death or S phase activity in treated cells. These take days to perform and give no indication of drug's uptake as a factor contributing to sensitivity.[4]

Bioluminescence has been used as an sensitive real-time reporter of bacterial survival within human cell lines and to monitor bacterial sensitivity to antimicrobial agents.[5,6] In this study, we report the construction of a constitutively bioluminescent *E. coli* strain which is sensitive to Ara-C that can be used as a potential intracellular reporter of the drug uptake and efflux in human cells.

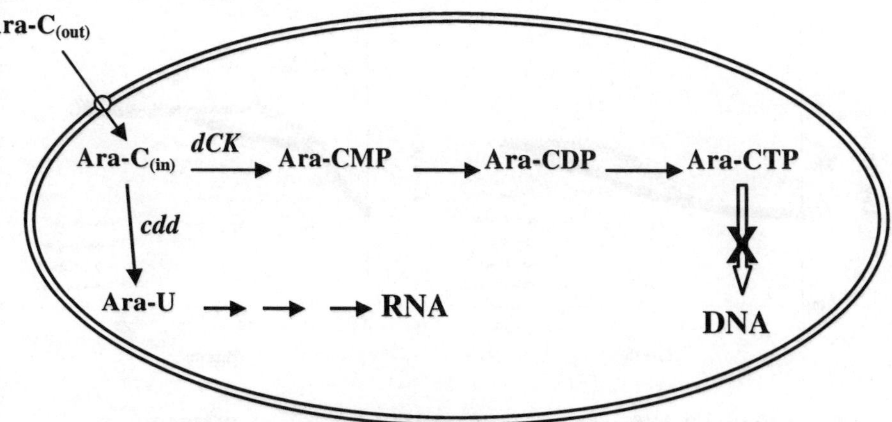

Figure 1. Transport and metabolism of cytosine arabinoside in humans

MATERIALS AND METHODS

Bacterial strains and growth media

Ara-C has no effect on *E. coli* since it lacks dCK gene and has the ability to deaminate Ara-C into Ara-U by cytidine deaminase (cdd). An *E. coli* strain (SØ5218) was rendered sensitive to Ara-C, by transforming a *cdd*-deficient strain (SØ5110) with the IPTG-inducible pTrc99-A plasmid carrying the human dCK gene.[7] Luria-Bertani (broth and agar) medium was used for routine culturing of *E. coli*. Growth inhibition experiments were performed in minimal salt medium with 0.2% glucose as a carbon source and thiamine (1μg/mL) and leucine (50 μg/mL) as nutritional requirements. Vitamin-free casamino acids (CA) was added at 0.2%. The antibiotics kanamycin and ampicillin were added at 10 and 100 μg/mL, respectively.

Transformation of *E. coli* SØ5218 with the *lux* operon

The *luxCDABE* cassette from *Photorabdus luminescens* was subcloned as an *Eco*RI fragment into the broad host range vector pBBR1MCS-2.[8] The resulting vector, pMCS2-LITE, was used to transform *E. coli* SØ5218 by electroporation and transformants were selected on kanamycin-plates. Luminescent colonies (*lux*+) were picked up using an ICCD 225 photon counting camera (Photek Ltd., UK).

Measurement of Ara-C activity

An overnight culture of *E. coli* SØ5218 *lux*[+], grown in minimal medium with antibiotics, was diluted into a pre-warmed fresh minimal medium containing 1 mM IPTG to achieve an initial optical density (OD_{600}) of 0.02. The culture was divided into samples of 200μL in a 96-well microtitre plate. Control samples were set up in which IPTG was omitted. Ara-C was added at 0, 25, 50 and 100 μM which is equivalent to the maximum plasma concentration during treatment. Growth (OD_{600}) and light output (Relative Light Units, RLU) were monitored over 24 h at 37 °C in a multimode microplate reader (GENios Pro, Tecan).

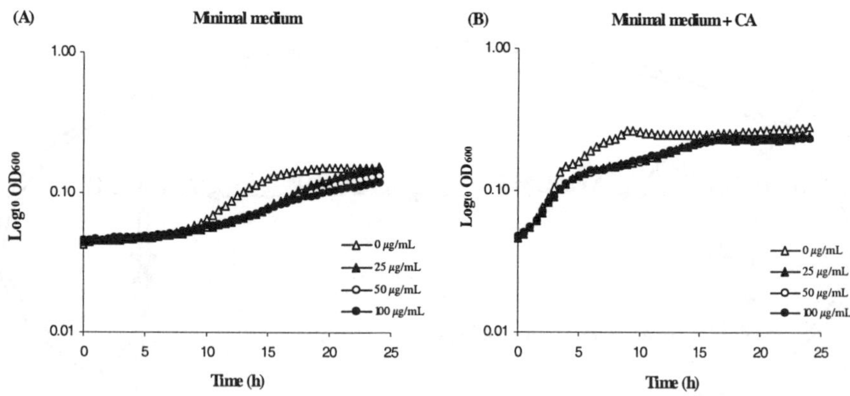

Figure 2. Effect of Ara-C on the growth of *E. coli* SØ5218 *lux*[+]

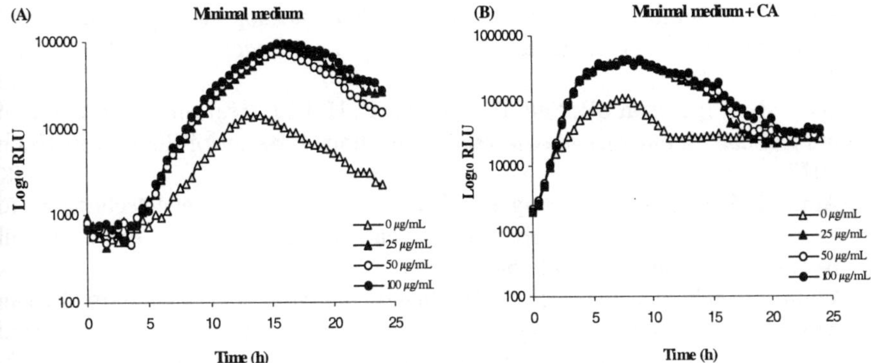

Figure 3. The effect of Ara-C on the bioluminescence of *E. coli* SØ5218 *lux*$^+$

RESULTS AND DISCUSSION

The effect of Ara-C on the growth of *E. coli* SØ5218 *lux*$^+$ is shown in Fig. 2. As previously reported[7], Ara-C caused a significant reduction in the growth rate of *E. coli* SØ5218 *lux*$^+$ at all concentrations used. This effect was more significant in the minimal medium without casamino acids and is completely abolished when assayed in the rich medium LB broth (data not shown). Fig. 2 shows the activity of Ara-C towards *E. coli* SØ5218 *lux*$^+$ by monitoring its bioluminescence. Unexpectedly, Ara-C caused a significant increase in light output of the drug-treated cultures. One possible explanation is that the DNA damage caused by Ara-CTP induces bioluminescence that could be used in DNA repair by a photo reactivation process.[9] Nevertheless, the bioluminescent derivative of *E. coli* SØ5218 is able to detect Ara-CTP and, therefore, has the potential for use as a biosensor within human AML cells as a rapid and non-invasive tool for screening of cancer cell sensitivity to nucleoside analogues.

ACKNOWLEDGEMENTS

The authors thank Prof. Staffan Eriksson (University of Copenhagen, Denmark) for providing *E. coli* strains SØ5110 and SØ5218. We are grateful for Claudia Marques for her invaluable help.

REFERENCES

1. Hiddemann W, Buchner T. Current status and perspectives of therapy for acute myeloid leukaemia. Semin Hematol 2001; 38:3-9.
2. Daher GC, Harris BE, Diasio RB. Metabolism of pyrimidine analogues and their nucleosides. Pharmacol Ther 1990; 48:189-222.
3. Galmarini CM, Mackey JR, Dumontet C. Nucleoside analogues: mechanisms of drug resistance and reversal strategies. Leukemia 2001; 15:875-90.

4.   Smith MA, Smith JG, Pallister CJ, Singer CRJ. Haemopoietic growth, the cell cycle and sensitivity of AML cells to Ara-C. Leuk Lymphoma 1996; 23:467-72.

5.   Qazi SNA, Harrison SE, Self T, Williams P, Hill PJ. Real-time monitoring of intracellular *Staphylococcus aureus* replication. J Bacteriol 2004; 186:1065-1077.

6.   Arain TM, Resconi AE, Sing DC, Stover CK. Reporter gene technology to assess activity of antimycobacterial agents in macrophages. Antimicrobial Agents Chemother 1996; 40:1542-4.

7.   Wang J, Neuhard J, Eriksson S. An *Escherichia coli* system expressing human deoxyribonucleoside salvage enzymes for evaluation of potential antiproliferative nucleoside analogues. Antimicrobial Agents Chemother 1998; 42:2620-5.

8.   Kovach ME, Elzer PH, Hill DS, Robertson GT, Farris MA, Roop RM, Peterson KM. Four new derivatives of the broad-host-range cloning vector pBBR1MCS, carrying different antibiotic-resistance cassettes. Gene 1994; 166: 800-2.

9.   Czyż A, Wróbel B, Węgrzyn G. *Vibrio harveyi* bioluminescence plays a role in stimulation of DNA repair. Microbiology 2000; 146:283-8.

METHOD FOR IMPLEMENTING BIOLUMINESCENCE-BASED ANALYTICAL ASSAYS IN NANOLITER VOLUMES

DA BARTHOLOMEUSZ, RH DAVIES, TSM YANG, JD ANDRADE

Dept of Bioengineering, University of Utah, Salt Lake City UT 84107, USA

INTRODUCTION

Bioluminescence-based analytical assays were used to measure various analytes in nanoliter sample volumes. Nanoliter volumes of multiple bioluminescent analytical assays were deposited in an array format and lyophilized. ATP-firefly luciferase (FFL) and NADH-bacterial luciferase (BL) platform reactions were compared. We achieved parallel sample delivery via sample-hydrated membranes. A CCD camera measured the luminescent kinetics for each assay. These miniaturized assays and instruments can be prepared as micro-analytical systems to operate in point-of-care (POC) diagnostic devices.

METHODS
ChemChip fabrication

We built arrays of clear bottom reaction wells, or ChemChips, consisting of 5x5 arrays of 1 mm diameter holes spaced 2 mm apart. The holes were cut in 15 mm squares out of 0.180 mm thick adhesive backed vinyl film with a knife plotter. The array patterns were sealed to 15 mm square glass cover slips after manually removing the cut holes. The glass cover slips became the clear bottom for the 140 nL wells (Fig. 1A).

Reagents and samples

ATP, NADH, Lactate, and Galactose assays were formulated according to Table 1. Mixed analyte samples were made at various concentrations in 50 mM Trizma buffer (pH 8.0). Lactate and Galactose assays were measured using samples without ATP and NADH since they interfere with their respective competition and production reactions.

Table 1. Analytical assay recipes

Reagents	ATP	Galactose	Reagents	NADH	Lactate
FFL -*Promega*	5 µM	3.3 µM	FMN -*Sigma*	200 µM	100 µM
Luciferin -*Biosynth*	0.1 mM	0.0 67mM	RCHO -*Sigma*	200 µM	100 µM
CoA -*Fluka*	100 µM	67 µM	BL -U of *Utah*	2 µM	1 µM
Mannitol -*Pfanstiel*	0.4 M	0.27 M	Oxidoreductase -*Roche*	2.4 u/mL	1.2 u/mL
40kD Dextran -*Pharmacia*	1mg/mL	0.67 mg/mL	Sucrose - *Pfanstiel*	0.5 M	0.5 M
8kD PEG -*Sigma*	5mg/mL	3.3 mg/mL	8kD PEG -*Sigma*	2 mg/mL	2 mg/mL
BSA -*Sigma*	30 µM	20 µM	BSA -*Sigma*	5 mg/mL	5 mg/mL
ATP -*Sigma*		5 µM	GPT -*Sigma*		125 µg/mL
Mg^{++} -*Sigma*		50 µM	Glutamate -*Sigma*		2 mM
Galactokinase -*Sigma*		0.2 u/mL	Lactate Dehydrogenase -*Sigma*		250 µg/mL
			NAD^+ -*Sigma*		1 mM

Figure 1. A) Empty ChemChip, B,C) Reagent deposition system

Reagent deposition

Individual ChemChip wells were filled with reagent cocktails via a computer controlled XYZ stage with a syringe pump and solenoid dispensing system (Fig. 1B,C). A miniature solenoid valve with a 0.002" nozzle (INKX0516350AA, The Lee Company) dispensed reagents in 10 ms pulses, pressured at 8 PSI. The drops were calibrated at 360±10 nL. A tray of 25 chips was cooled to less than -60 °C using dry ice (Fig. 1C shows the chips on the cold plate), allowing the reagent droplets to freeze within seconds of dispensing. This process prevented evaporation and maintained reagent stability prior to lyophilization.

Lyophilization

Lyophilization was performed in two stages in a VirTis Genesis 12 pilot plant lyophilizer. The chips were placed in the sample chamber of the lyophilizer, which had been previously cooled to at least –50 °C. Primary lyophilization began when the sample chamber was connected to the condenser chamber cooled to –70 °C with a system pressure below 100 mTorr. Primary lyophilization was performed for 48-72 hr. Secondary lyophilization was then performed for 12-24 hours after changing the sample chamber to 25 °C.

Simple sample delivery

25 µL samples were dispensed on the center of 14 mm diameter circles Whatman qualitative membrane filters clamped to the center of the ChemChips. Since the membranes hydrate uniformly, less than 1 µL of the 25 µL sample was delivered to each of the 25 wells. Given a 2 mm well spacing and a 0.18mm thick membrane, only 510 nL of sample was available to each well. The sample wicked along the membrane and into each well, whereupon the reagents rehydrated and bioluminescence reactions began. Since reagent drops were larger than the volume of the wells, a convex meniscus formed above each well. This convex structure, porous and hydrophilic in nature after lyophilization, facilitated drawing the sample from the membrane into each well without the risk of bubble formation.

Detection

An Andor DV-434 CCD was used to take a series of six 30-s exposures to record the bioluminescence activity for each assay (CCD temp = –50 °C, binning = 4×4 pixels). CCD images can be seen in Fig. 2. Although we used a sensitive CCD camera, some of the assays were bright enough to see with the human eye. These assays produced an estimated 10 nanoWatts/steradian/cm^2.[1] Such a signal produces a current signal of about 50 pA on the CCD. This is about 50 times greater than the dark current for the less expensive photodiode arrays (Hamamatsu S8593 and S8550) (assuming a collection angle of 1steradian, an area of 5.3 mm^2, and a photosensitivity of 0.3 A/W on the photodiode arrays) These arrays would enable the ChemChip to be implemented in a less expensive POC diagnostic device.

Data Analysis

The native Andor data files were opened in ImageJ (from NIH) where a macro integrated the CCD counts across the area of each well for each exposure. We used Matlab to sort the data by analalyte and sample concentration. Calibration curves for ATP, NADH, and Lactate assays were created by time integrating the CCD counts across each well. Time integrated CCD counts were then averaged across all 5 rows for each column of analyte. An average integration was also taken across multiple chips that were tested at each sample concentration. The calibration curve for the Galactose assay was based on the area integrated CCD counts at t=180 s, divided by the area integrated CCD counts at t=30 s (the brightest exposure period).

RESULTS

Fig 3 shows the kinetics for the ATP assays. Fig. 4 shows assay calibration curves. The estimated detection limit for each assay was as follows: 0.51 picomoles ATP ±19%, 5.1 picomoles NADH ±21%, 5.1 picomoles Galactose ±26%, and 51 picomoles Lactate ±22%.

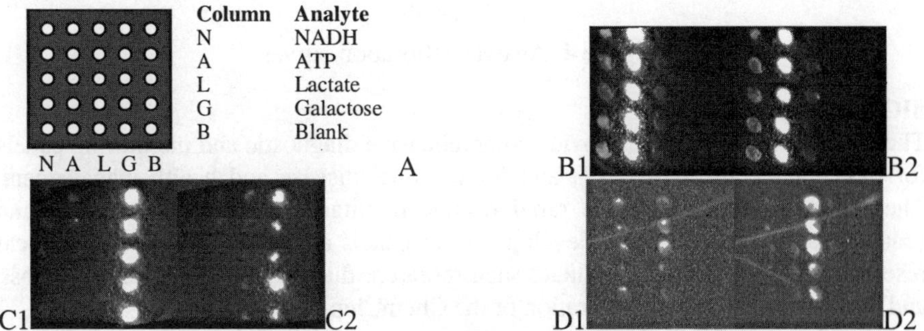

Figure 2. CCD images of luminescent ChemChip arrays
A) Bioluminescence assays were dispensed in separate columns for replicate data (5 rows per column). **B1)** NADH and ATP at 1 and 0.1 mM, respectively. **B2)** NADH and ATP at 0.01 and 0.001 mM, respectively. This is dimmer than B1 due to lower concentration of analytes. **C1)** Galactose assay (1 mM sample) at first 30 s exposure. **C2)** Galactose assay (1 mM sample) at sixth 30 s exposure. This competition luminescence dims with time. **D1)** Lactate assay (10 mM sample) at first 30 s exposure. (Streaks of light across are due to a cracked cover slip). **D2)** Lactate assay (10 mM sample) at sixth 30 s exposure. Being a production assay, the luminescence increases with time.

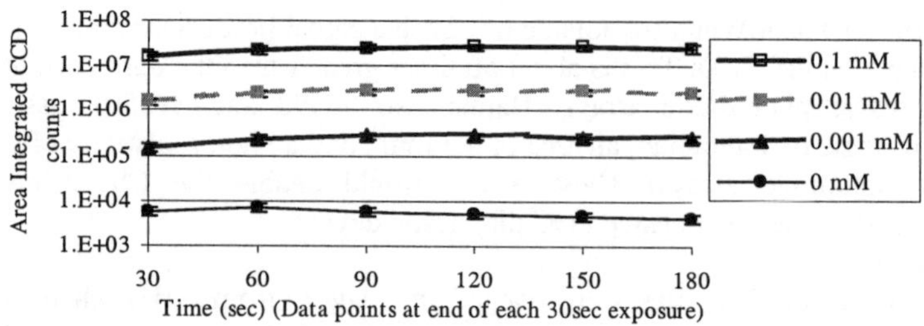

Figure 3. ATP kinetic data averaged over 3 chips, 5 rows/chip

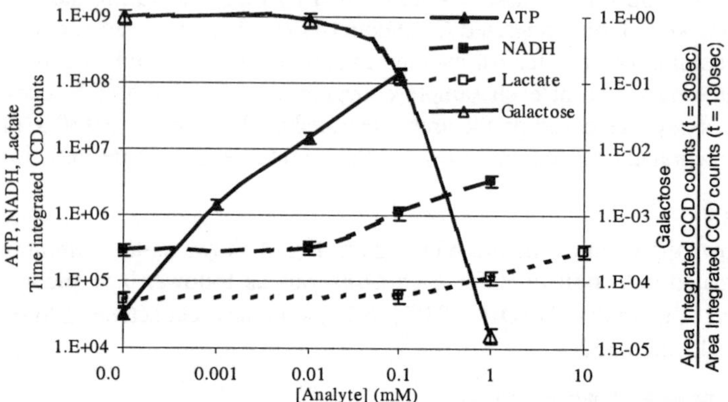

Figure 4. Analyte calibration curves

SIGNIFICANCE

The ChemChip system will provide comprehensive diagnostic and monitoring panels for basic and clinical research, and for personal disease and health management. These systems would provide rapid results, facilitate patient empowerment, and reduce health care costs. The development of panels appropriate to specific clinical research areas will greatly facilitate such research, due to the ease of use, low cost, and multi-parametric data generation of the ChemChip systems.

ACKNOWLEDGMENTS

We acknowledge the support of NIH RFP#PAR01-057 Project#1R21RR17329, Technology Development for Biomedical Applications Grant, and our industrial partners. We also thank Dr. J. Harris of the University of Utah for use of the CCD.

REFERENCE

1   Bartholomeusz DA, Andrade JD. Photodetector calibration method for reporting bioluminescence measurements in standardized units. Luminescence 2002; 17:77-115.

IMMOBILIZATION OF BIOLUMINESCENT SYSTEMS AND THEIR APPLICATIONS

EN ESIMBEKOVA[1], VA KRATASYUK[2]
[1]Institute of Biophysics SB RAS, Akademgorodok, 660036 Krasnoyarsk, Russia
[2]Krasnoyarsk State University, pr.Svobodnyi 79, 660041 Krasnoyarsk, Russia
E-mail: bpl@ibp.ru

INTRODUCTION

Bioluminescent enzyme systems based on bacterial and firefly luciferases offer a unique and general tool for analysis of the many analytes and enzymes in the environment, research and clinical laboratories and other fields.

The agents for bioluminescent assays are luminous bacteria, recombinant luminous organisms, luciferases and others enzymes for multienzymatic bioluminescent assays. The extremely high amplification of these luminescent systems allows rapid methods to be set up which can be applied to a very small amounts of biological samples.[1] The sensitivity of these methods is often at the nanomolar level, on the border between conventional enzymatic and immunological methods. Moreover they are applicable to analytes present at very low concentration and when high sensitivity is not required, analysis time can be reduced to few seconds.[2]

However, native bioluminescent enzymes are generally subject to inactivation in vitro, and hence not suitable for routine analytical use. Immobilized enzymes and whole bacteria largely solve this instability problem, and hence enable the routine use of bioluminescent analysis with high speed, specificity, simplicity, sensitivity and accuracy. Immobilization also enables the development of automated luminescent biosensors.

METHODS

A variety of procedures are available for coupling proteins to insoluble solid supports. There are three categories of immobilization techniques - chemical modification, physical absorption and gel entrapment. Experimental procedures, characteristics and peculiarities of different immobilized systems are tabulated and discussed in review.[3]

RESULTS

Immobilized reagents for bioluminescent analysis have their peculiarities. For example, immobilization reagent kit should contain all the reaction components, and the procedure must provide opportunities for coimmobilizing luciferase with other enzymes and their substrates. In addition, methods for production of immobilized bioluminescent reagent must also meet the following requirements.

- Immobilization should not involve active sites important for catalysis, i.e. the activity of the final product (immobilized bioluminescent reagent) must approach 100 %.
- Immobilized luciferase and other enzymes must retain substrate specificity, with their kinetic constants unaltered.
- Optimal environment for enzyme stability and extended storage life.
- The immobilized reagent should be stable both during use and storage.

Chemical methods of immobilization give better yields of active immobilized luciferase than physical ones, and agarose, collagen, epoxy methacrylate and nylon have proved to be the most effective of the different solid supports which have been investigated.[3] But chemical procedures involve covalent coupling, and usually lead to some protein inactivation. Gel entrapment technique has the advantages of better protein stability and ease of processing. An important general feature of these immobilized enzymes can be incorporated into flow cells, used for multiple assays, recycled and reused in automated devices.

Entrapment of enzymes and cells, especially in polysaccharide gels is very popular, because it has the advantage of high enzyme stability and ease of preparation. Firefly and bacterial luciferases are very labile, and hence agarose is mostly used for its low gelling temperature (26-30 °C).

We have developed a disc shape biosensor based on bacterial NADH-FMN oxidoreductase and luciferase immobilized into a starch gel.[4-6] Properties of the resulting luciferase biosensor depends on the preparation conditions. Best results were obtained when 50-100 μL of gel per disc were used. Drying of discs also affected enzyme activity and stability, and optimum drying time was 2-3 hours. With shorter drying time, the discs are still wet and rapidly disintegrate, while longer drying times make the discs fragile. In either case, enzyme activity decays rapidly during reuse.

The biosensor had following characteristics: 0.1 mL preparation of luciferase immobilized into starch gel disc, diameter 7-8 mm, width 50 μm, dry weight 9±0.5 mg. The enzyme activity in the disc increases with increasing the activity of soluble luciferase preparations used for immobilization.[5]

Entrapment of luciferase from *Photobacterium leiognathi* in starch gels increases it's K_m for dodecanal and tetradecanal to 1/3, but the change is insignificant for decanal. Also, K_m of aldehydes with different chain length is smaller for immobilized luciferase than for the soluble enzyme.[5]

The characteristics of the immobilized luciferase depend on the time of drying, amount of gel and gel concentration, the nature of lining used for drying and on the characteristics of the initial enzyme activity.[3,5] During immobilization it is important to preserve the activity of functionally important groups of enzymes and high specificity of luciferase to aldehydes. The method of immobilization into gel enables the coimmobilization of luciferase and other enzymes with their substrates. The method of preparing the reagent will depend upon the type of analysis required. For example, the reagent for analysis of NADH must consist of NADH:FMN oxidoreductase, luciferase, FMN and aldehyde.

Analytical usefulness of immobilized bioluminescent assays depends on properties of their immobilized enzymes. The most popular application of immobilized bioluminescent systems is for analysis and monitoring of chemical and biochemical analytes and environmental pollutants. The wide range of analytes measured and monitored by immobilized bioluminescent systems has been reviewed.[3] Stability, sensitivity, precision, and effects of interfering substances and the microenvironment are also discussed.

Bacterial luciferase coimmobilized with NAD(P)H: FMN oxidoreductase on starch gel has been used for bioluminescent assay of aldehydes.[4,6] Co-immobilization of bacterial luciferase, NAD(P)H:FMN oxidoreductase and their substrates is referred to as "multifunctional immobilized biosensor" and is a new trend for use of bioluminescent analysis, e.g. toxicity biotest and bioassay. The main principle of this luciferase biotest is the correlation between toxicity of the sample being studied and changes in bioluminescence parameters *in vitro*. Toxicity of the sample is measured by the changes in bioluminescence intensity compared with that of a control. Multifunctional immobilized biosensors based on luciferase have been used for the following bioassays.

- Control of toxicity of waste water, water and air.[7-9]
- Assay of degree of corn and bread infection by fungi.[10]
- Rapid continuous control of physical loading for prognosis of astronauts health and destructive influence of physical load and stress.[11]
- Salts of platinum acid control of skin purity.[12]
- Bioluminescent assay of endotoxicosis in the clinics.[13-14]

The effect of blood serum, lymph and other biological liquids on bioluminescent reactions has been studied. The immobilized luciferase shows lower sensitivity than the soluble enzyme in human clinical tests, and in analysis of corn and bread infection by fungi.[13] But this problem is overcome by using larger samples, and hence this new biosensor can be successfully used for toxicity bioassays.

So, there is a great possibility of application of immobilized bioluminescent systems as biosensors and for different researches in the fields of biology, molecular biology, enzymology, biotechnology and others due to the properties of bioluminescent systems.

ACKNOWLEDGMENTS

This work was supported by the Ministry of Education of the Russian Federation (grant PD 02-1.4-316) and the U.S. Civilian Research and Development Foundation for the Independent States of the Former Soviet Union (grant KY-002-X1, Science Education Center "Yenisei", grant Y1-B-02-11 and grant Y1-B-02-12).

REFERENCES

1. Roda A, Girotti S, Ghini S. Continuous-flow determination of primary bile acids by bioluminescence with use of nylon-immobilized bacterial enzymes. Clin Chem 1984; 30: 206-10.
2. Girotti S, Roda A, Angelotti M. Bioluminescence flow system determination of branched-chain L-amino acids in serum and urine. Anal Chim Acta 1988; 205: 229-37.
3. Kratasyuk V, Esimbekova E. Polymer Immobilized Bioluminescent Systems for Biosensors and Bioinvestigations. In: Arshady R (Ed), Polymeric Biomaterials, The PBM Series, Citus Books, London 2003: 301-43.
4. Kim N, Kratasyuk V. Luciferase biosensors for the analysis of aldehydes. In: Jezowska-Trzebiatowska B, Kochel B, Slawinski J, Strek W (Eds) Biological luminescence, World Scientific, Singapore 1990: 564-72.
5. Kratasyuk V, Abakumova V, Kim N. A gel model for the functioning of luciferase in the cell. Biochemistry (Russian) 1994; 59: 761-5.
6. Abakumova V, Kratasyuk V. Bioluminescent immobilized sensors. In: Poncelet D (Ed), International Workshop on Bioencapsulation 1996: Talk 10.
7. Kuznetsov A, Tyulkova N, Kratasyuk V, Abakumova V, Rodicheva E. The investigation of properties of chemicals for bioluminescent bioassays. Siberian ecological journal (Russian) 1997; 5: 459-65.
8. Kudryasheva N, Kratasyuk V, Esimbekova E, Vetrova E, Kudinova I. Development of the bioluminescent bioindicators for analysis of environmental pollution. Field Anal Chem Tech 1998; 2: 277-80.
9. Kratasyuk V, Vetrova E, Kudryasheva N. Bioluminescent water quality monitoring of salt lake Shira. Luminescence 1999; 14: 193-5.
10. Kratasyuk V, Egorova O, Esimbekova E, Kudryasheva N, Orlova N, L'vova L. A biological luciferase test for the bioluminescent assay of wheat grain infection with Fusarium. Appl Biochem Microbiol (Russian)1998; 34: 688-91.
11. Kratasyuk V. Principle of luciferase biotesting. In: Jezowska-Trzebiatowska B, Kochel B, Slawinski J, Strek W, eds. Biological luminescence, World Scientific, Singapore, 1990: 550-8.
12. Kratasyuk V, Kim N. The catalitic characteristics of luciferase biosensors. In: Akhapkin Yu, Bartsev S, Vsevolodov N, eds. Biotechnics – a novel strategy of computerization, Moscow: Nauka 1990: 82-7.
13. Esimbekova E, Kratasyuk V, Abakumova V. Bioluminescent method non-specific endotoxicosis in therapy. Luminescence 1999; 14: 197-8.
14. Voevodina T, Kovalevskii A, Kratasyuk V, Schultz V, Nifantyev O. Bioluminescent technique to analyse degree of endotoxication, In: Jezowska-Trzebiatowska B, Kochel B, Slawinski J, Strek W, eds. Biological luminescence, World Scientific, Singapore 1990: 573- 81.

DETERMINATION OF BASIC COMPOUNDS WITH PEROXYOXALATE CHEMILUMINESCENCE DETECTION

H KAWANISHI, M TSUMURA, T FUKUSHIMA,
M KATO, T TOYO'OKA

*School of Pharmaceutical Sciences, University of Shizuoka,
52-1 Yada, Shizuoka 422-8526, Japan*

INTRODUCTION

Chemiluminescence (CL) reactions require no excitation light source, and thus avoid interfere by light scattering. Since CL permits attainment of large signal-to-noise ratio, determination of wide range analytes can be assayed. Luminol and its analogues are most famous CL reagents, and directly produce emission of light due to oxidation reactions with oxygen species such as hydrogen peroxide (H_2O_2). CL reactions involving an energy transfer reaction is also well known. Oxalates react with H_2O_2 to yield intermediate peroxides, which produce light by energy transfer to a co-existing fluorophore.[1] Therefore, combination of oxalate with a fluorophore is important to yield intense emission. A large number of fluorophores are detected by the CL reaction with aryloxalates and H_2O_2. Among aryloxalates, bis(2,4,6-trichlorophenyl)oxalate (TCPO) is most popular for the CL reaction. Fluorophores derived from target compounds and H_2O_2 in samples have been determined in this CL detection system. Since the CL reaction using TCPO and H_2O_2 proceeds in the presence of base catalyst, the determination of basic compounds may also be possible.

In this paper, the determination of basic compounds including drugs was carried out by flow injection analysis (FIA) using TCPO and H_2O_2 as the CL reagents, and DNS-amino acid was selected as the fluorophore. The concentration on these reagents, which affect the CL intensity, was optimised for FIA. The CL intensity of various compounds possessing imidazole ring and related structures was determined under a CL reaction conditions. The detection of histamine in micro titre plates with a multi-label counter was also studied.

MATERIALS AND METHODS
Chemicals

Imidazole, 1-methylimidazole, L-histidine, ethylenediamine, caffeine and teophylline were purchased from Kanto Chemicals. Cimetidine, famotidine, omeprazole, thioperamide and dansyl-L-phenylalanine (DNS-Phe) were from Sigma. Histamine (Nacalai Tesque), 3-methylhistamine dihydrochloride (Calbiochem), bis(2,4,6-trichlorophenyl)oxalate (TCPO; Tokyo Kasei) and hydrogen peroxide (31% H2O2; Mitsubishi Gas Chemicals) were used as received. Triazolam, estazolam, midazolam and alprazolam were generously supplied from pharmaceutical companies. Special reagent grade of acetonitrile (CH₃CN) and de-ionized and distilled water was used throughout the experiments.

Flow injection analysis (FIA)

Figure 1 shows schematic flow diagram for the HPLC-CL detection system. It consists of three LC-6A pumps, a SIL-6B auto injector and a CLD-10A CL monitor equipped with a 80-μL spiral flow cell (Shimadzu). The signals obtained from the CL monitor were recorded on a Shimadzu C-R7A Plus Chromatopak. Appropriate concentrations of chemicals were injected into a stream of a mobile phase (70% acetonitrile in water) through an auto injector. The reagent solution (I) consists of a mixture of fixed concentrations of TCPO and DNS-Phe in CH₃CN; whereas the reagent solution (II) is H₂O₂ in water. The reagent solutions were delivered with two different pumps. The mobile phase and the reagents were continually degassed with a DEGASYS DG-1310 (Uniflows). The flow rates of the mobile phase and the reagents (I) and (II) were 0.5mL/min.

Histamine analysis by multi-label counter

Histamine was detected with a multi-label counter (Wallac 1420 ARVOsx, Perkin Elmer). Forty-μL CH3CN solution containing 0.25 mM TCPO and 2.5 μM DNS-Phe were added to 20 μL histamine solutions diluted to appropriate concentrations. After addition of 20 μL water containing 200 mM H2O2 to each well, the luminescence produced was immediately measured. The calibration curve was conducted by CL intensity against added amounts of histamine (10 fmol~50 pmol).

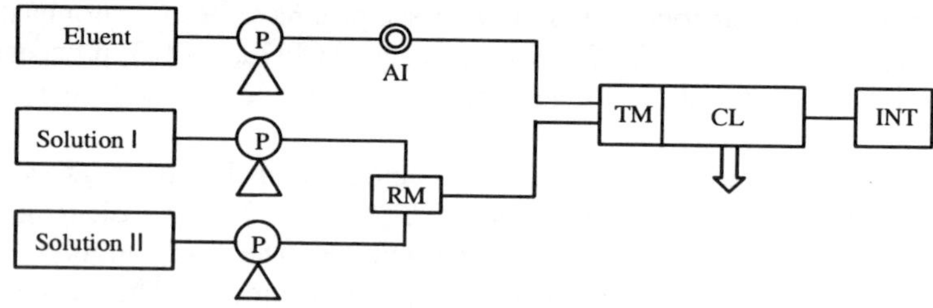

Figure 1. Schematic flow diagram of FIA-CLdetection
P, pump; AI, auto injector; RM, rotating mixing device; TM, T-type mixing device; CL, chemiluminescence detector; INT, integrator.
Eluent, water-acetonitrile (3:7);
Solution I , mixture of 5μ M DNS-Phe and 0.25mM TCPO in acetonitrile;
Solution II , 20mM H₂O₂ in water

RESULTS AND DISCUSSION

It is well known that imidazole buffer is an excellent base catalyst to increase the luminescence in a peroxyoxalate CL detection system.[2-4] The results suggest that the compounds possessing imidazole ring structure seem to be indirectly detected with the peroxyoxalate CL. Thus, various imidazole analogues including some drugs were screened with the CL system. The luminescence was generated constantly from the chemical reaction between TCPO/H₂O₂ and DNS-Phe. Although the strong CL intensity was obtained from higher concentrations, the background emission also increased. Thus, appropriate concentrations are required for sensitive detection and hence the concentrations of these reagents were optimised. Figure 2 shows the effect of the concentrations of TCPO and H_2O_2 on the CL intensity. Based upon the results, the final concentrations of DNS-phe, TCPO and H_2O_2 for FIA were 2.5μM, 0.125mM and 10mM, respectively. Among the compounds tested, several such as histamine efficiently produce luminescence under the proposed CL conditions (Fig. 2). In contrast, some compounds showed no light emission (e.g. alprazolam and triazolam). The results demonstrate that the emission is dependent upon the structure, complicated molecules are inferior. Judging from the results of imidazole and histamine, the effect on the CL intensity is independent of basicity of the compounds, but dependent on imidazole ring structure. However, the mechanism leading to the higher luminescence is currently unknown.

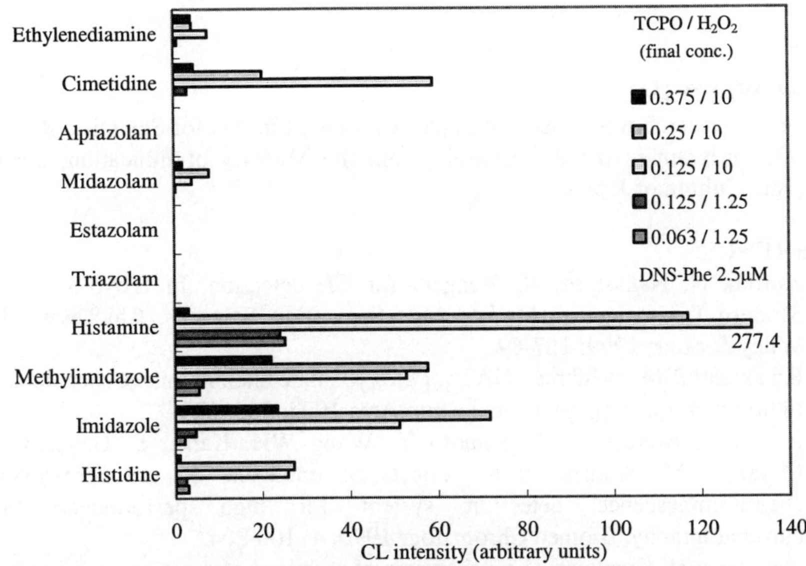

Figure 2. Effect of TCPO and H_2O_2 concentrations on CL intensity. The other conditions are shown in Fig 1.

Since strong emission was observed in histamine, detection with a multi-label counter was also tried. In this method, 0.125 mM TCPO, 2.5 μM DNS-Phe and 50 mM H_2O_2 were used. Figure 3 shows the calibration curve of histamine. Good linearity was observed in the range of 50 fmol~50 pmol. Under the conditions, the linearity was not obtained at concentration higher than 50 pmol. The detection limit of histamine was approximate 10 fmol. The method will be further optimised in our laboratory and applied to real specimens.

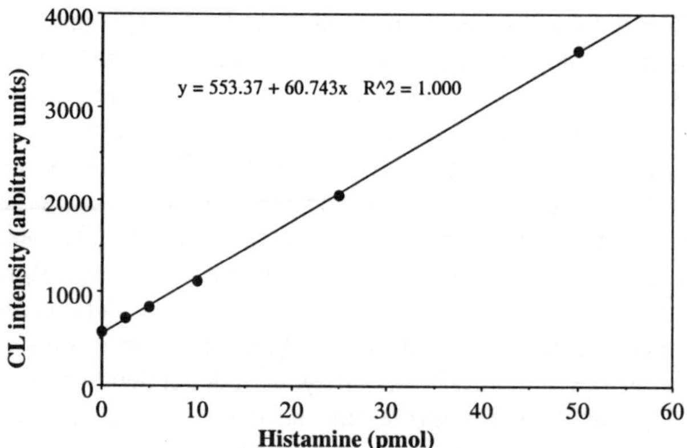

Figure 3. Calibration curve of histamine
Reagents (final concentration): DNS-Phe, 2.5uM; TCPO, 0.125mM; H2O2, 50mM.

ACKNOWLEDGMENT

The present research was supported in part by a Grant-in-Aid for Scientific Research and COE program in the 21st Century from the Ministry of Education, Science, Sports and Culture of Japan.

REFERENCES

1. Kuroda N, Nakashima K. Reagent for CL detection. In: Toyo'oka T. ed. Modern Derivatization Methods for Separation Sciences. Chichester: John Wiley & Sons, 1999: 167-89.
2. Kwakman PJM, Brinkman UATh. Peroxyoxalate chemiluminescence detection in liquid chromatography. Anal Chim Acta 1992; 266: 175-92.
3. Imai K, Nishitani A, Tsukamoto Y, Wang WH, Kanda S, Hayakawa K, Miyazaki M. Studies on the effects of imidazole on the peroxyoxalate chemiluminescence detection system for high performance liquid chromatography. Biomed Chromatogr 1990; 4: 100-204.
4. Neuvonen H. Kinetics and mechanisms of reactions of pyridines and imidazole with phenyl acetates and trifluoroacetates in aqueous acetonitrile with low content of water: nucleophilic and general base catalysis in ester hydrolysis. J Chem Soc Perkin Trans II 1987; 266: 159-67.

DETERMINATION OF ARTEMISININ BY HPLC WITH ON-LINE PHOTOREACTOR AND PEROXYOXALATE CHEMILUMINESCENCE DETECTION

N KURODA, A AMPONSAA-KARIKARI, N KISHIKAWA,
Y OHBA, K NAKASHIMA

*Graduate School of Biomedical Sciences, Course of Pharmaceutical Sciences,
Nagasaki University, 1-14 Bunkyo-machi, Nagasaki 852-8521, Japan*

INTRODUCTION

Artemisinin (Fig. 1), isolated in 1972 from Chinese medicinal plant *Artemisia annua L.*, is a novel antimalarial drug with a sesquiterpene lactone structure containing an internal endoperoxide linkage which is essential for the drug's activity. Artemisinin and its derivatitives form a series of antimalarial compounds with activity against chloroquine-resistant malaria parasites. In contrast to chloroquine, artemisinin penetrates the blood brain barrier, which makes it especially valuable for the treatment of cerebral malaria.[1]

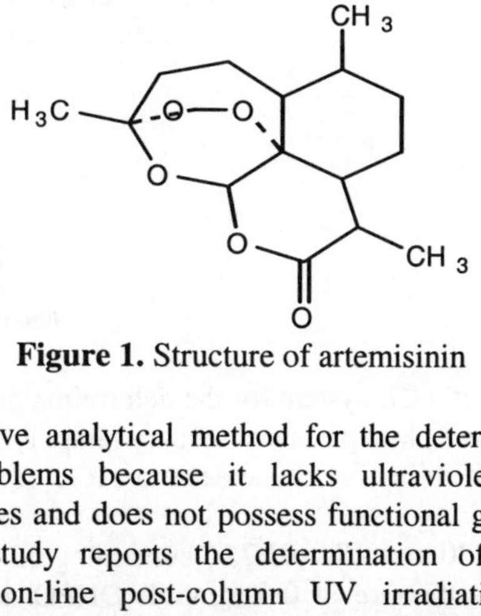

Figure 1. Structure of artemisinin

Development of selective analytical method for the determination of artemisinin poses challenging problems because it lacks ultraviolet (UV) absorption or fluorescent chromophores and does not possess functional groups with potential for derivatization. This study reports the determination of artemisinin by liquid chromatography with on-line post-column UV irradiation and peroxyoxalate chemiluminescence (PO-CL) detection. A similar method as previously reported by us for the determination of organic peroxide.[2] In this method, after artemisinin is eluted from the HPLC column, it is UV irradiated to generate hydrogen peroxide, which is determined by PO-CL detection.

EXPERIMENTAL

Materials and reagents

Artemisnin was obtained from Acros organics (New Jersey, USA). Solution of artemisinin was prepared in acetonitrile and diluted appropriately with mobile phase to obtain the working solutions. Bis (2,4–dinitrophenyl) oxalate (DNPO) and imidazole were obtained from Tokyo Chemical Industry (Tokyo, Japan); imidazole was recrystallized from acetonitrile before use. 2,4,6,8-Tetrathiomorpholino-pyrimidol [5,4–d] pyrimide (TMP) was synthesed in our laboratory.[3] Water was deionized by Autosill WG 220(Yamato Kagaku, Tokyo) and passed through Organo Puric-z (Organo, Tokyo) before used.

HPLC-PO-CL system

The HPLC-PO-CL system (Fig. 2) consisted of two LC 9A liquid chromatographic pumps (Shimadzu, Kyoto, Japan), a Rheodyne 7125 injector (Cotati, CA, USA) with a 20-μL sample loop, an ultraviolet lamp, Toshiba GL-10 (10 W, 254 nm), a Chemcosorb 5-ODS-UH column (150 x 4.6 mm I.D.), CLD-10A detector (Shimadzu), a Rikadenki R-61 recorder. PTFE tubing (6.0 m x 0.5 mm I.D., GL Sciences, Tokyo) coiled around the ultraviolet lamp as the on-line for the UV radiation reactor.

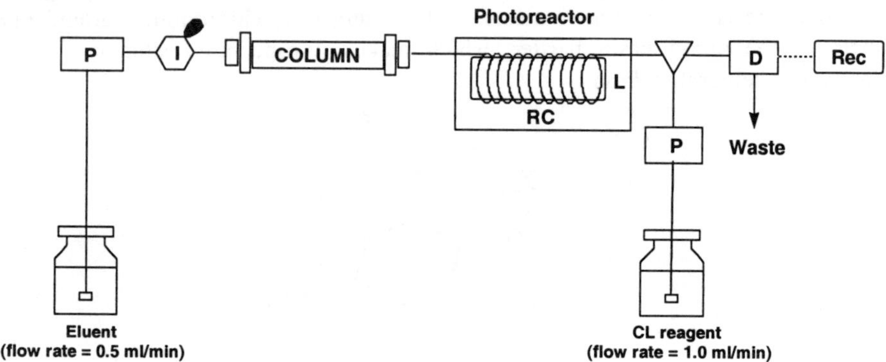

Figure 2. HPLC PO-CL system for the determination of artemisinin.
P, pump; I, injector; L, low-pressure mercury lamp; RC, reaction coil; D, chemiluminescence detector; Rec, recorder.

Imidazole-HNO_3 buffer (20 mmol/L, pH 7.5) containing 60 % acetonitrile was used as a mobile phase and a mixture of 0.5 mmol/L DNPO and 1.5 μmol/L TMP in acetonitrile as a post column CL reagent. The flow rates of the mobile phase and the CL reagent were set at 0.5 and 1.0 mL/min, respectively. Artemisinin injected into the system passed through the on-line photoreactor in 2.35 min.

RESULTS AND DISCUSSION

Optimization of UV irradiation

We optimized the detection conditions for the determination of artemisinin with the aim of maximizing the artemisinin peak height and decreasing noise level. Since hydrogen peroxide is generated as artemisinin passes through the on-line UV irradiation reactor, three types of UV lamps were evaluated; a Toshiba G-10 (10 W, 254 nm), Shigemi AL-ISH (15 W, 254 nm) and National FL-10 BL-B (10 W, 350 nm). National FL-10 BL-B, which emits relatively long wavelength, gave no CL signal. Signals were obtained for lamps at wavelength 254 nm. At 254 nm wavelengths, a 10 W powered lamp gave better results compared with a 15 W powered lamp.

The length of the reaction coil was also evaluated since it affects the CL intensity. The effect of coil lengths ranging from 1.0 m to 9.0 m on CL intensity was examined. On examination of the coil length, 6 m gave the best result.

Optimization of CL condition

On examination of a scope of buffers, imidazole-HNO$_3$ buffer gave the best result and the effect of different concentrations of imidazole were examined and the largest signal-to-noise ratio (S/N) was obtained at 20 mmol/L. The effect of the buffer pH was also examined, both peak height and S/N ratio increased with increasing pH up to 8.0, but taking the durability of the ODS column into consideration, pH value of 7.5 was chosen. The effects of the concentrations of CL reagents were also investigated; the increases in the CL intensities and S/N ratio considered, 0.5 mmol/L DNPO and 1.5 μmol/L TMP were chosen for further experiments.

Fig. 3 shows a typical chromatogram of artemisinin.

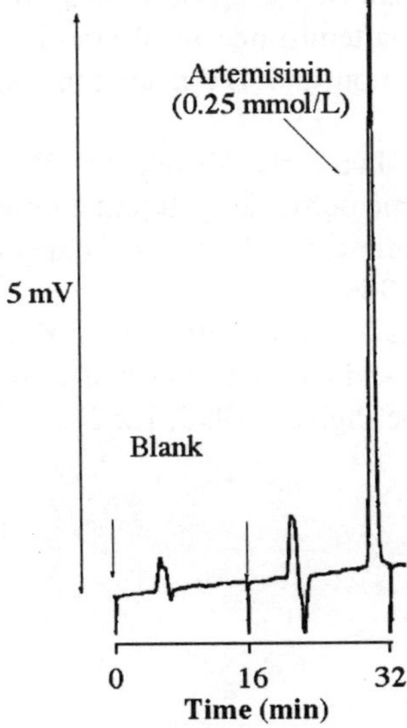

Figure 3. Chromatogram of artemisinin

Calibration curve, detection limit and reproducibility
Calibration curve showed good linear relationship (r=0.9998) between artemisinin concentration and CL intensity. The detection limit at S/N ratio of 3 was 5 μmol/L (100 pmol/injection). The reproducibility of the proposed method was determined using 1 mmol/L and 0.25 mmol/L artemisinin, the relative standard deviation for within-day (n=5) and between-day (n=3) analysed were < 3% and < 11%, respectively.

CONCLUSION

Artemisinin could be detected by the use of HPLC-PO-CL system with on-line UV irradiation. In the proposed method, artemisinin was UV irradiated to generate H_2O_2, which was determined by PO-CL detection and the resulting CL was proportional to the concentration of artemisinin. The detection limit of artemisinin obtained for the proposed method was 100 pmol/injection (S/N = 3).

The proposed method has good selectivity, high sensitivity, less time consuming and gives satisfactory reproducibility. This method should be applicable to the determination of artemisinin in biological fluid and this is currently being investigated.

REFERENCES

1   Edlund PO, Westerlund D, Carlqvist J, Wu BL, Jin YH. Determination of artesunate and dihydroartemisinine in plasma by liquid chromatography with post-column derivatization and UV-detection. Acta Pharma Suec 1984; 21: 223-34.
2.   Wada M, Inoue K, Ihara A, Kishikawa N, Nakashima K, Kuroda N. Determination of organic peroxide by liquid chromatography with on-line post-column ultraviolet irradiation and peroxyoxalate chemiluminescence detection. J Chromatogr A 2003; 987: 189-95.
3.   Nakashima K, Akiyama S, Tsukamoto S, Imai K. Synthesis of pyrimido [5,4-d] pyrimidine derivatives and their ultraviolet absorption and fluorescence spectral properties. Dye Pigment 1990; 12: 21.

PHOTINA™: AN IMPROVED Ca²⁺-SENSITIVE PHOTOPROTEIN

N MASTROIANNI, M FOTI, S BOVOLENTA, M STUCCHI,
A ROSSIGNOLI, S CORAZZA
Enabling Technologies, AXXAM, via Olgettina 58, Milan 20132, Italy
Email: nadia.mastroianni.nm@axxam.com

INTRODUCTION

Calcium-sensitive photoproteins are important tools for analyzing calcium-mediated signal transduction processes in mammalian cells.[1,2] The luminescent reaction is based on immediate photon release (flash luminescence) upon calcium binding to the coelenterazine-photoprotein complex.[3] We have created a very sensitive chimeric photoprotein, Photina™, that can be used in a variety of cell-based functional assays that utilize measurement of intracellular calcium to evaluate the activity of proteins, particularly G-protein coupled receptors (GPCRs) and plasma membrane ion channels. Although the large and rapid increase in intracellular calcium concentration following GPCR and ion channel stimulation can be detected by various reporters such as calcium-sensitive fluorescent dyes, the use of an extremely bright bioluminescent photoprotein is preferred as its background is virtually absent in contrast to fluorescent dyes. Moreover, calcium measurement with photoproteins, besides producing rapid signals, generates a high signal-to-noise ratio with a broad range of detection sensitivity. The use of cells which express both Photina™ (as a reporter system) and a receptor involved in the modulation of intracellular calcium provides a valid system for the screening of compounds for their effects on the release of intracellular calcium. The robust flash luminescence signal obtained with the Photina™ cell line as well as its high signal-to-noise ratio allow the use of small assay volumes, essential features for the set-up of high throughput screening assays in the pharmaceutical industry.

METHODS

In vitro transcription and translation

Translation of the photoproteins was carried out using the Wheat Germ Extract System TNT® T7 kit from Promega (Madison, WI), according to the manufacturer's instructions.

Tagged expression in mammalian cells and cell culture conditions

The sequence encoding the mitochondrial-targeting peptide from subunit VIII of human cytochrome c oxidase was fused in frame at the 5' end of the Photina™ gene and the resultant construct was cloned into the pcDNA3 vector lacking the neomycin resistance cassette, for expression in mammalian cells.

All reagents for cell culture were purchased from GIBCO (Carlsbad, CA). CHO-K1 cells were cultured in DMEM/F12 with Glutamax supplemented with 1.35

mM sodium pyruvate, 10% FBS, 11 mM Hepes, 0.2% sodium bicarbonate, 1% penicillin/streptomycin. Standard propagation conditions consisted of seeding 3.0×10^5 cells in a T75 flask twice a week, recovering about 9×10^6 cells/T75 flask.

Recombinant protein production and purification
The complete coding sequence of Photina™ was subcloned into the pET28a(+) vector for expression in *E. coli* BL21(DE3) competent cells. Photina™ synthesis was induced with 1 mM IPTG. As the recombinant protein accumulates within the host cells in the form of inclusion bodies, cells were disrupted by high pressure dispersion with a French press and then centrifuged. The final pellet was resuspended in 20 mM Tris-HCl pH 7.0, 6 M urea, 5 mM $CaCl_2$, 5 mM DTT and stirred overnight at 4 °C. The scheme used to obtain highly-purified Photina™ includes ion-exchange chromatography on DEAE-Sepharose Fast Flow resin in 6 M urea with salt gradient elution (0-0.5 M sodium acetate) followed by ion-exchange chromatography on a Mono P HR 5/5 column with salt gradient elution (0-0.5 M sodium acetate) in the absence of urea.

RESULTS AND DISCUSSION
In order to create a very sensitive photoprotein, the sequence and structural similarities as well as the unique features of different well-known photoproteins were carefully analyzed. Based on this study a chimeric photoprotein, named Photina™, was designed. Indeed, the region between the first two calcium binding sites of the obelin gene was replaced with the corresponding fragment from the photoprotein clytin, producing a novel photoprotein with improved luminescence.

In vitro transcription and translation of Photina™
In order to verify the activity of the chimeric product, in vitro transcription and translation experiments were performed. Active photoprotein formation was monitored by measuring luminescence from the translation mixture in the presence of coelenterazine after the addition of calcium ions. As shown in Fig. 1, Photina™ produces an intense luminescence signal in response to calcium stimulation (25 mM $CaCl_2$) which is generally higher than that observed with natural photoproteins.

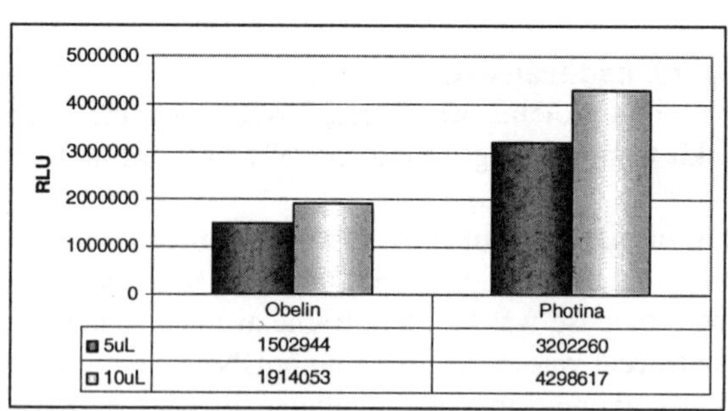

	Obelin	Photina
■ 5uL	1502944	3202260
□ 10uL	1914053	4298617

Figure 1. Calcium-stimulated luminescence of in vitro-translated Photina™

Photina™ expression in mammalian cells

The new photoprotein, characterized by an optimized codon usage for expression in mammalian cells and a reduced number of cysteine residues, was used as a reporter system in cell based assays. Moreover, a mitochondrial-targeted form of Photina™ was generated, allowing the measurement of calcium concentrations at subcellular level and a refined analysis of calcium-dependent signaling pathways in response to activation of cellular receptors.[4] The resulting reporter cell line, named CHO mito-Photina™, was transfected with a fusion construct of the histamine H3 receptor gene and the Gα16 gene sequences. In order to analyze the functional expression of the resulting stable cell line, cells were seeded in 384 MTP at different densities ranging from 250 to 1000 cells/well and tested for their response to the selective histamine H3 receptor agonist, imetit. Following a four hour incubation with 5 μM coelenterazine, ligand was injected at different concentrations and light release was measured using a CCD camera over a total integration time of 50 s. In Fig. 2 results are presented as kinetic curves, and the corresponding EC$_{50}$ value was determined to be 4 nM.

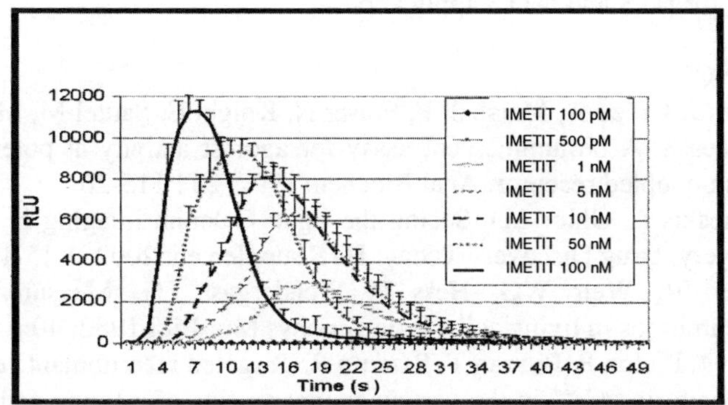

Figure 2. Dose-dependent light emission upon imetit stimulation of the histamine H3 receptor transfected into CHO mito-Photina™ cells

Recombinant Photina™ and measurement of light production

Functional characterization of the new photoprotein was carried out by biochemical analysis of purified recombinant protein. Different quantities of recombinant Photina™ ranging from 0.09 to 12.5 ng were charged with 10 μM coelenterazine for 4 h at 4 °C. After incubation, 100 μM CaCl$_2$ was applied and the total RLU (relative light units) was recorded for 10 s using a Berthold Luminometer. As shown in Fig. 3, recombinant Photina™ gives a consistent signal even when used at very low concentrations.

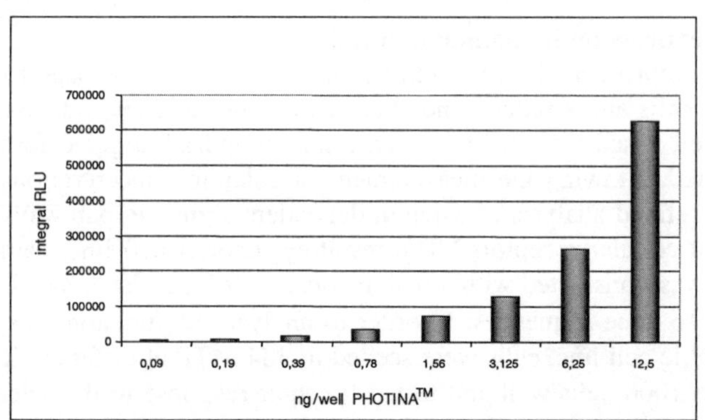

Figure 3. Light production of recombinant Photina^{TM} upon 100 µM CaCl$_2$ injection

In conclusion, the new chimeric photoprotein Photina^{TM} exhibits a robust flash luminescence signal in different expression systems, rendering this protein a reporter well-suited for HTS and uHTS applications.

REFERENCES

1. Stables J, Green A, Marshall F, Fraser N, Knight E, Sautel M, Milligan G, Lee M, Rees S. A bioluminescent assay for agonist activity at potentially any G-protein-coupled receptor. Anal Biochem 1997; 252:115-26.
2. Mattheakis L, Ohler LD. Seeing the light: Calcium imaging in cells for drug discovery. Drug Discovery Today: HTS supplement 2002; 1:15-19.
3. Blinks JR, Weir WG, Hess P, Prendergast FG. Measurement of Ca^{2+} concentrations in living cells. Prog Biophys Mol Biol 1982; 40:1-114.
4. Brini M, Pinton P, Pozzan T, Rizzuto R. Targeted recombinant aequorins: tools for monitoring Ca^{2+} in the various compartments of a living cell. Microsc Res Technique 1999; 46:380-9.

DEVELOPMENT OF FLUORESCENCE PROBES FOR BIOLOGICAL APPLICATIONS, BASED ON PHOTOINDUCED ELECTRON TRANSFER

T NAGANO

*Graduate School of Pharmaceutical Sciences, The University of Tokyo,
Tokyo 113-0033, Japan*
Email: *tlong@mol.f.u-tokyo.ac.jp*

INTRODUCTION

Human beings are highly receptive to an enormous amount and variety of information from the external environment, and more than 90% of it is thought to be visual. It is desirable that seeing into cells or cultured tissues is accomplished by using noninvasive techniques, without isolating of cellular constituents. Therefore, I believe that techniques to visualize physiological or pathopysiological changes in the cells or cultured tissues will become increasingly important in life sciences.

Fluorescence imaging is the most powerful technique currently available for continuous observation of the dynamic intracellular processes of living cells. Fluorescein is widely employed as the core of various fluorescence probes used in imaging important biological effectors. Despite the extensive use of fluorescein derivatives and the importance of the applications, the mechanism that controls the quantum yield of fluorescence has not been fully established. I report herein photoinduced electron transfer (PeT) mechanism that can control the fluorescence quantum yields of fluorescein and boron dipyrromethene (BODIPY) derivatives.

DESIGN OF FLUORESCENCE PROBES BASED ON PHOTOINDUCED ELECTRON TRANSFER

3-Aminofluorescein has a low quantum yield of 0.015, whereas its amide derivatives fluoresce strongly. To our knowledge, little more is known about the relationship between the chemical structures of fluorescein derivatives and their fluorescent properties. Our working hypothesis is that the fluorescence properties of fluorescein derivatives are controlled by PeT process from donor moiety (benzoic acid) to acceptor moiety (fluorophore) (Fig. 1a).

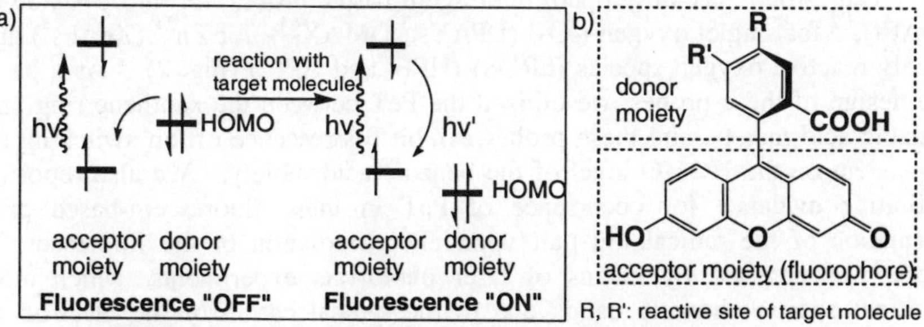

Figure 1. Photoinduced electron transfer mechanism

PeT is a widely accepted mechanism for fluorescence quenching, in which electron transfer from PeT donor to the excited fluorophore diminishes the fluorescence of the fluorophore. We consider that it is appropriate to divide the fluorescein structure into two parts, i.e., the benzoic acid moiety as the PeT donor and the donor (xanthene) moiety as the fluorophore, based on their spectral data and X-ray analysis (Fig. 1b). Our hypothesis is that if the highest occupied molecular orbital (HOMO) energy level of benzoic acid moiety is high enough for electron transfer to the excited xanthene ring, the quantum yield will be small. In other words, fluorescein derivatives with high quantum yields must have benzoic acid moieties with low HOMO energy levels. The HOMO energy levels of 3-aminobenzoic acid, 3-benzamidobenzoic acid, 9,10-diphenylanthracene-2-carboxylic acid (DPA-COOH) and 9,10-diphenylanthracene-9,10-endoperoxide-2-carboxylic acid (DPA-EP-COOH) were estimated by semiempirical (PM3) calculations. DPA-COOH and aminobenzoic acid, which are the benzoic acid moieties of weakly fluorescent fluorescein derivatives, have relatively higher HOMO levels than DPA-EP-COOH and amidobenzoic acid. These results are in accordance with our hypothesis. Further, to confirm the hypothesis, we synthesized 9-[2-(3-carboxy)naphthyl]-6-hydroxy-3H-xanthen-3-one (NX) and 9-[2-(3-carboxy)anthryl]-6-hydroxy-3H-xanthen-3-one (AX). The excitation maximum (Ex_{max}) and emission maximum (Em_{max}) of fluorescein, NX and AX were not much altered among these fluorescein derivatives. However, the quantum yields were greatly altered: NX is highly fluorescent, whereas AX is almost nonfluorescent. Thus, a small change in the size of conjugated aromatics, namely from naphthalene to anthracene, causes a great alteration of fluorescence properties. When HOMO levels of benzoic acid moieties were compared, we found that HOMO levels of benzoic acid and naphthoic acid, which are present in highly fluorescent fluorescein and NX, are lower than that of the xanthene ring, while the HOMO level of anthracenecarboxylic acid, which is present in the scarcely fluorescent AX, is higher than that of the xanthene ring. These results are consistent with the idea that a PeT process controls the fluorescence properties of fluorescein derivatives and that these properties can be predicted from the HOMO level of the benzoic acid moiety, with a threshold around -8.9 eV. This, in turn, provides a basis for developing novel fluorescence probes with fluorescein-derived structure.

Our group has developed fluorescein-based probes for nitric oxide (NO) (DAFs),[1,2] for singlet oxygen ($^1O_2$) (DPAXs,[3] DMAXs[4]), for Zn^{2+} (ZnAFs[5]) and for highly reactive oxygen species (hROS) (HPF[6] and APF[6]) (Fig. 2). As a basis for the design of these probes, we utilized the PeT between the xanthene ring and the benzoic acid moiety, and these probes exhibit fluorescence off/on switching that is dependent on the HOMO level of the benzoic acid moiety. We also reported the definitive evidence for occurrence of PeT in these fluorescein-based probes.[7] Formation of the radical ion pair upon photoirradiation of the fluorescein-based probe was detected by means of laser photolysis experiments, which afforded transient spectra showing bands due to the radical cation of the electron donor moiety and the xanthene radical anion. The rates of PeT and the back electron transfer were determined and analyzed in terms of the Marcus theory of electron transfer. The results provided a quantitative basis for rational design of fluorescein-based probes with high efficiency in fluorescence off/on switching.

DAF-2
(NO probe)

DPAX
($^1O_2$ probe)

DMAX
($^1O_2$ probe)

ZnAF-2
(Zn^{2+} probe)

HPF
(hROS)

APF
(hROS)

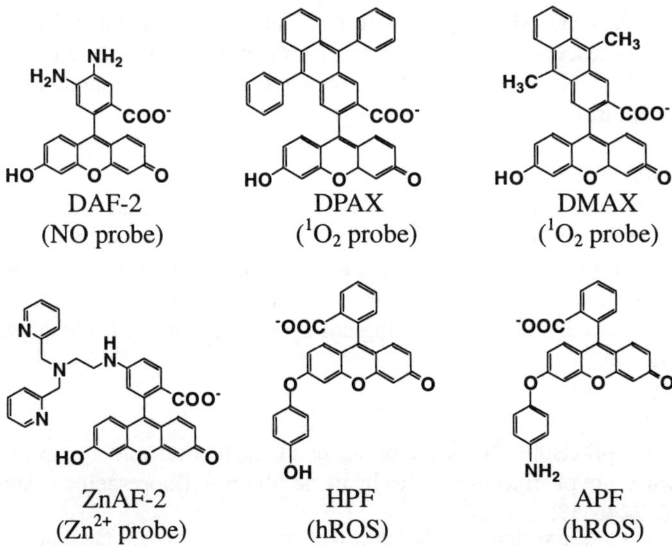

Figure 2. Novel fluorescein-based fluorescence probes

FLUORESCENCE PROBES BASED ON BORON DIPYRROMETHENE

BODIPYs are of interest as chromophores due to their desirable photophysical properties. It is also easy to modify BODIPY chemically for preparation of various derivatives. However, BODIPY-based functional probes are not yet available for

DAMBO-P^H
(NO probe)

ZnAB
(Zn^{2+} probe)

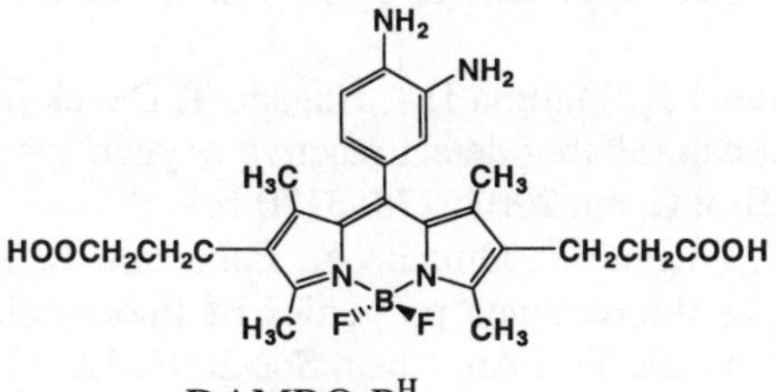

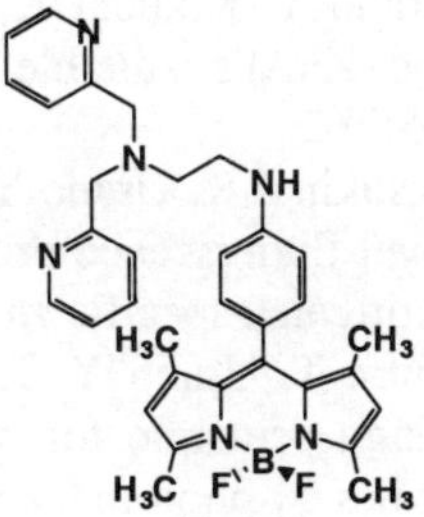

Figure 3. Novel BODIPY-based fluorescence probes

biological use.   The PeT-dependent fluorescence off/on switching mechanism is applicable to BODIPY fluorophore.

Novel fluorescence probes for nitric oxide (DAMBO-P^H)[8] and for Zn^{2+} (ZnAB) have been developed based on BODIPY structure (Fig. 3).   DAMBO-P^H is a pH-independent and more highly sensitive fluorescence probe for nitric oxide than DAF-2.   ZnAB has the advantages of less sensitivity to solvent polarity and pH than ZnAF-2, fluorescein-based Zn^{2+} probe, and is also not influenced by other cations such as Na$^+$, K$^+$, Ca^{2+} and Mg^{2+}, which exist at high concentration under physiological conditions.

CONCLUSION

The results obtained are consistent with our hypothesis that the fluorescence properties of fluorescein and BODIPY derivatives are determined by a PeT process from the benzoic acid moiety.   This provides a practical strategy for rational design of functional fluorescence probes to detect certain biomolecules and developed probes should be widely useful in biological systems from the point of sensitivity and specificity.

REFERENCES

1.   Kojima H, Nakatsubo N, Kikuchi K, et al. Detection and imaging of nitric oxide with novel fluorescent indicators: diaminofluoresceins. Anal Chem 1998; 70: 2446-53.
2.   Kojima H, Urano Y, Kikuchi K, Higuchi T, Nagano T. Fluorescent indicators for imaging nitric oxide production. Angew Chem Int Ed 1999; 38: 3209-12.
3.   Umezawa N, Tanaka K, Urano Y, Kikuchi K, Higuchi T, Nagano T. Novel fluorescent probes for singlet oxygen. Angew Chem Int Ed 1999; 38: 2899-2901.
4.   Tanaka K, Miura T, Umezawa N, et al. Rational design of fluorescein-based fluorescence probes. -Mechanism-based design of a maximum fluorescence probe for singlet oxygen-. J Am Chem Soc 2001; 123: 2530-6.
5.   Hirano T, Kikuchi K, Urano Y, Nagano T. Improved fluorescent probes for zinc, ZnAFs, suitable for biological applications. J Am Chem Soc 2002; 124: 6555-62.
6.   Setsukinai K, Urano Y, Kakinuma K, Majima H J, Nagano T. Development of novel fluorescence probes that can reliably detect reactive oxygen species and distinguish specific species. J Biol Chem 2003; 278: 3170-5.
7.   Miura T, Urano Y, Tanaka K, Nagano T, Ohkubo K, Fukuzumi S. Rational design principle for modulating fluorescence properties of fluorescein-based probes by photoinduced electron transfer. J Am Chem Soc 2003; 125: 8666-71.
8.   Gabe Y, Urano Y, Kikuchi K, Kojima H, Nagano T. Highly sensitive fluorescence probes for nitric oxide based on boron dipyrromethene chromophore -rational design of potentially useful bioimaging fluorescence probe-. J Am Chem Soc 2004; 126: 3357-67.

HPLC WITH FLUORESCENCE DETECTION OF MORPHINE IN RAT PLASMA USING 4-(4,5-DIPHENYL-1H-IMIDAZOL-2-YL)BENZOYL CHLORIDE AS A LABEL

K NAKASHIMA, Y OGATA, MN NAKASHIMA, M WADA

*Department of Clinical Pharmacy, Graduate School of Biomedical Sciences,
Nagasaki University, 1-14 Bunkyo-machi, Nagasaki 852-8521, Japan
Email: naka-ken@net.nagasaki-u.ac.jp*

INTRODUCTION

Morphine (MOR), a potent opioid analgesic, has been used for short-term treatment of postoperative and traumatic pain as well as for long-term treatment of severe pain in cancer patients. Besides these clinical uses, MOR is one of common drugs of abuse. Several methods for determining MOR have been developed; liquid chromatography-mass spectrometry,[1] gas chromatography-mass spectrometry (GC-MS),[2] and high-performance liquid chromatography (HPLC) with fluorescence (FL),[3] chemiluminescence,[4] and electrochemical detections.[5] Although these methods are sensitive and can detect ng-μg/mL levels of MOR in biological materials, more sensitive method is required to analyze a small size of sample.

In this paper, we developed a highly sensitive HPLC-FL method for the determination of MOR using derivatization with 4-(4, 5-diphenyl-1H-imidazol-2-yl)benzoyl chloride (DIB-Cl). The method was applied to monitor the time-course of MOR concentrations in rat plasma samples after a single administration to rat.

METHODS

Chemicals

MOR-HCl was purchased from Takeda (Osaka, Japan). (+)-Cyclazocine used as the internal standard (IS) was purchased from Sigma (Tokyo, Japan). Ethyl acetate, acetonitrile of HPLC grade were purchased from Wako (Osaka, Japan). Water was deionized and passed through a water purification system (ADVANTEC GSR-500, Toyo, Tokyo, Japan). Solid-phase extraction (SPE) was carried out by a Bond Elute cartridge (50 mg C_{18}, 1 mL; Varian, USA).

HPLC system and chromatographic conditions

The separation of DIB derivatives of morphine and IS were performed using an HPLC system (Shimadzu, Kyoto, Japan) consisting of two pumps (LC-10AT$_{VP}$) with a system controller (PX-8010), a recorder (FBR-2), a FL detector (RF-550) set at λex=355 nm and λem=486 nm, and a Rheodyne 7125 injector (Cotati, CA, USA) with a 20-μL sample loop. In plasma analysis, the mobile phases used were a mixture of acetonitrile-0.1 M acetate buffer (pH5.4) (50:50, v/v, MP1) with a flow rate of 1.0 mL/min and acetonitrile (MP2). The separation program was set as follows: the flow rate of MP2 was set at 0 mL/min from 0 to 29 min, rapidly increased to 2 mL/min within 1 min (29-30 min), held for 10 min, and backed to the initial condition.

Plasma samples

Male Wistar rats (280-300 g) were used for *in vivo* experiments. Blood samples were centrifuged at 2000 g for 10 min at 20 °C and the resultant plasma samples were kept at −20 °C until use. After 10 μL of 0.1 (or 0.7) μM of (+)-cyclazocine solution in methanol were evaporated with N_2 gas, 100 μL of plasma was added to the residue, and extracted with 1.0 mL of ethyl acetate. After centrifugation for 10 min at 1000 g and 20 °C, 800 μL of the organic layer were transferred into a vial and evaporated to dryness. The residue was applied to derivatization with DIB-Cl.

Derivatization with DIB-Cl

The residues of the evaporated plasma samples were derivatized as follows: 25 μL of 0.4 M carbonate buffer (pH10) and 100 μL of 5 mM DIB-Cl suspension in acetonitrile were added to the residue, vortex mixed and then stand for 10 min. The reaction was stopped by adding 10 μL of aqueous ammonia (25%).

SPE for DIB derivatives

SPE cartridges were conditioned with 2 mL each of acetonitrile and water. Derivatized sample was applied to SPE cartridge, which was washed with 400 μL of acetonitrile-water (1:1, v/v) followed by 1600 μL of acetonitrile-sodium acetate solution (1:1, v/v). After elution with 400 μL of methanol-conc.HCl (99:1, v/v), the eluate was allowed to dry with N_2 gas. The residue was reconstituted with 100 μL of mobile phase A, and 20 μL of the resultant were injected into the HPLC system.

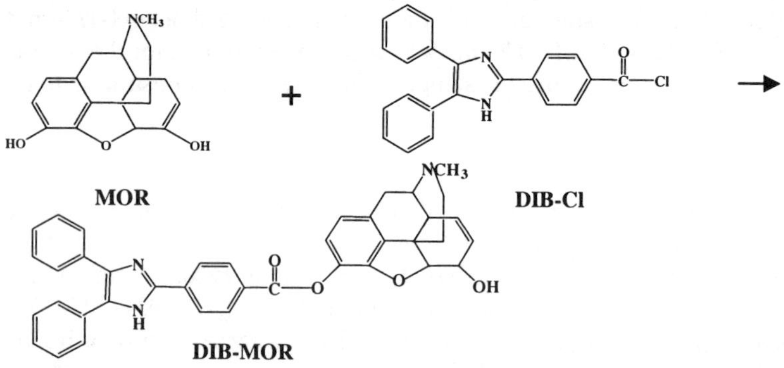

Figure 1. Reaction scheme for derivatization of MOR with DIB-Cl.

RESULTS

Derivatization conditions with DIB-Cl

The reaction scheme of MOR derivatization with DIB-Cl is shown in Fig.1. MOR showed maximum peak height at a DIB-Cl concentration of 0.5 mM and was constant to 7.5 mM while carbonate buffer showed maximum yield at 0.25 M with 0.5 mM DIB-Cl, and these conditions were used for the following experiments. The pH of carbonate buffer was adjusted to 8.5-11. The peak heights of DIB-MOR were similar in the studied pH range; hence pH 10 was chosen. The yield of DIB-MOR at 4 °C, room temperature, 60 °C were similar, and the reaction was essentially complete within 5 min; reaction time for 10 min at room temperature was selected.

Chromatogram of DIB-MOR

Typical chromatograms of plasma are shown in Fig. 2; normal plasma (A) and spiked plasma (B) with MOR at a concentration of 5.2 ng/ml, respectively. In plasma sample, DIB-Cl reactivity of cyclazocine was lower than MOR. As a result, 5 mM DIB-Cl was used for plasma sample. The retention times for the DIB-MOR and DIB-cyclazocine (as IS) were 11.0 and 24.0 min, respectively.

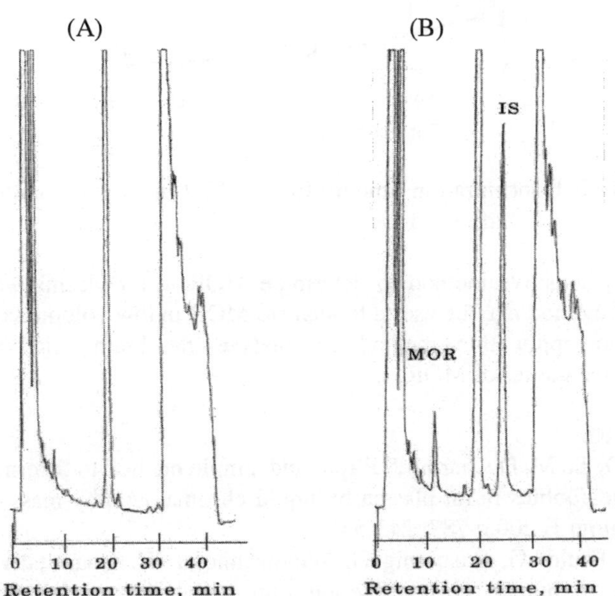

Figure 2. Chromatograms of normal plasma (A) and spiked plasma (B) with 5.2 ng/mL of MOR.

Method validation

The calibration curve of MOR in rat plasma was linear in the range of 0.5-540 ng/mL plasma (r =0.998) with the LOD was 0.09 ng/mL (5.0 fmol on column). The proposed method is more sensitive compared to other reported methods for MOR determination in plasma; 0.69 ng/mL (2.44 nM)[1] (LC-MS), 0.78 ng/mL (GC-MS)[2] and 1.0 ng/mL (HPLC-FL)[3]. The intra-day RSD ranged from 5.3 to 8.2% for spiked rat, while the inter-day RSD ranged from 7.9 to 9.4%. The recovery following the liquid-liquid extraction of spiked plasma was more than 92%.

MOR pharmacokinetics (PK)

The profile of MOR concentration time-course is shown in Fig. 3. C_{max} (ng/mL), T_{max} (min) and $T_{1/2}$ (min) were 610±212, 10±3 and 62.8±8, respectively. AUC_{inf} (ng/mL·min) was 52938±15548. C_{max} and $T_{1/2}$ are similar to those reported.[1]

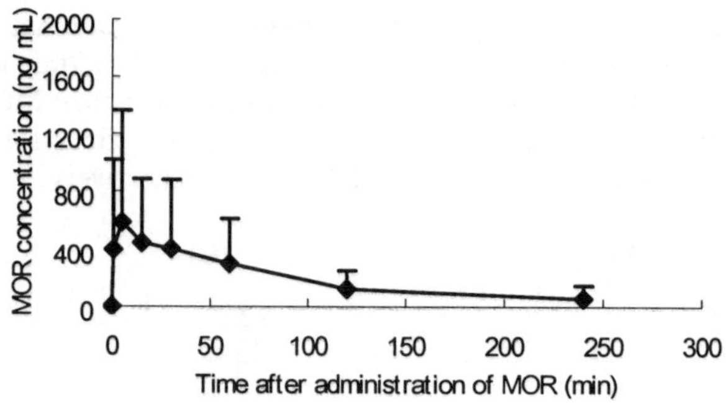

Figure 3. Concentration-time profile of MOR in rat plasma after single administration of MOR (2.5 mg/kg, i.p.)

A highly sensitive method to determine MOR in rat plasma was developed. The proposed method may be useful to analyze MOR in low volume samples. We intend to apply this method to analyze rat brain microdialysates for pharmacokinetic studies of MOR.

REFERENCES

1. Projean D, Tu M, Ducharme J. Rapid and simple method to determine morphine and its metabolites in rat plasma by liquid chromatography-mass spectrometry. J Chromatogr B 2003; 787: 243-53.
2. Leis HJ, Fauler G, Raspotnig G, Windischhofer W. Quantitative analysis of morphine in human plasma by gas chromatography-negative ion chemical ionization mass spectrometry. J Chromatogr B 2000; 744: 113-9.
3. Huwyler j, Rufer S, Kusters E, Drewe J. Rapid and automated determination of morphine and morphine glucuronides in plasma by on-line solid phase extraction and column liquid chromatography. J Chromatogr B 1995; 674: 57-63.
4. Abbott RW, Townshend A. Determination of morphine in body fluids by high-performance liquid chromatography with chemiluminescence detection. Analyst 1987; 112: 397-406.
5. Liaw WJ, Ho AT, Wang JJ, Hu OYP, Li JH: Determination of morphine by high-performance liquid chromatography with electrochemical detection application to human and rabbit pharmacokinetic studies. J Chromatogr B 1998; 714: 237-45.

LUMINESCENCE PROBES FOR SENSITIVE AND SPECIFIC OPTICAL IMAGING

A RODA[1], M GUARDIGLI[1], P PASINI[1], M MIRASOLI[1], E MICHELINI[1],
L CHARBONNIERE[2], R ZIESSEL[2]

[1]*Dept of Pharmaceutical Sciences, University of Bologna, Bologna 40126, Italy*
[2]*Laboratoire de Chimie Moléculaire, ECPM, ULP, Strasbourg 67087, France*
Email: aldo.roda@unibo.it

INTRODUCTION

The detection and localization at microscopic level of analytes in single cells and tissue samples is one of the most exciting challenges for bioanalytical chemistry.

This is usually performed by conventional prompt fluorescence microscopy, using immunohistochemical and *in situ* hybridization reactions with fluorescent labelled probes. However, fluorescence microscopy suffers from two main drawbacks that reduce its sensitivity, i.e. the autofluorescence of the sample and the light scattering in the apparatus. Several alternative detection principles have been proposed in order to increase the detectability of the labeled probes. For example, chemiluminescent (CL) enzyme labelled probes allow the sensitive localization of DNA sequences and antigens, due the high specificity and low background noise of the CL reaction. They also provide high spatial resolution and the possibility of quantifying the amount of analyte on a given surface area.[1]

Time-resolved fluorescence (TRF) microscopy represents another promising alternative to conventional fluorescence microscopy. This technique relies on the use of fluorescent labels with long (micro- or milliseconds) luminescence lifetimes. The sample is excited by a pulsed light source, and the emission is measured with a suitable delay after excitation, thus allowing for vanishing of the sample autofluorescence. Due to their peculiar photophysical properties, luminescent lanthanide chelates are the most suitable labels for this technique. Unfortunately, only a few, expensive labels are commercially available. We have thus developed a new lanthanide chelating ligand able to form stable and luminescent Eu^{3+} and Tb^{3+} complexes and suitable for binding to primary amino groups of biomolecules.[2] Using a conventional epifluorescence microscope, modified and equipped with suitable electronics in order to perform TRF imaging, we have compared the analytical performance of the novel TRF probe with those of fluorescent and CL enzyme labels.

METHODS

Imaging experiments were performed using an epifluorescence microscope (BX 60, Olympus Optical, Tokyo, Japan) and an ultrasensitive, cryogenically cooled CCD camera (LN/CCD, Princeton Instruments, Roper Scientific, Trenton, NJ). A standard instrument setup was used for prompt fluorescence measurements with fluorescein-labeled probes. For TRF measurements the microscope was equipped with a pulsed excitation source (L7684 Xenon flash lamp, Hamamatsu Photonics K.K., Shimokanzo, Japan), triggered by an optical chopper located in the light emission pathway, and a wide band UV filter cube. The flash lamp was operated at 100 Hz and the delay between excitation and luminescence measurement was set to 0.5 ms.

Horseradish peroxidase (HRP, Sigma-Aldrich Co., St. Louis, MO), fluorescein-labeled immunoglobulin (2.3:1 labeling ratio, Dako, Glostrup, Denmark) and Tb^{3+} chelate-labeled BSA, synthesized and characterized as previously described[2] (5.0:1 labeling ratio), were used as model samples. Samples were spotted on a nitrocellulose membrane (Hibond ECL, Amersham Biosciences, Little Chalfont, England) by means of a glass slide microarrayer (BioGene, Kimbolton, UK). The amount of sample in each spot (diameter ~ 600 μm) was evaluated from the volume of solution deposed (~3 nL). CL detection of HRP was performed using the ECL® substrate (Amersham Biosciences). In order to obtain good signal-to-noise ratios, acquisition times of 5 s, 30 s and 60 s were used for prompt fluorescence, TRF and CL images, respectively. Quantitative evaluation of the images was done using the Metamorph image analysis software (Universal Imaging Corporation, Downington, PA).

RESULTS
The new ligand (Fig. 1) is based on a tridentate metal-coordinating and luminescence-sensitizing unit, which takes advantage of both the light absorption and energy transfer ability of the 2,2'-bipyridine chromophore and the coordinating ability of the carboxylate anion. The introduction of two units in a glutamic acid skeleton allowed to obtain a ligand able to form stable lanthanide complexes (log K_{cond} = 16.5 at pH = 7.0), that also contains a N-hydroxysuccinimidyl ester group for linking to primary amino groups.

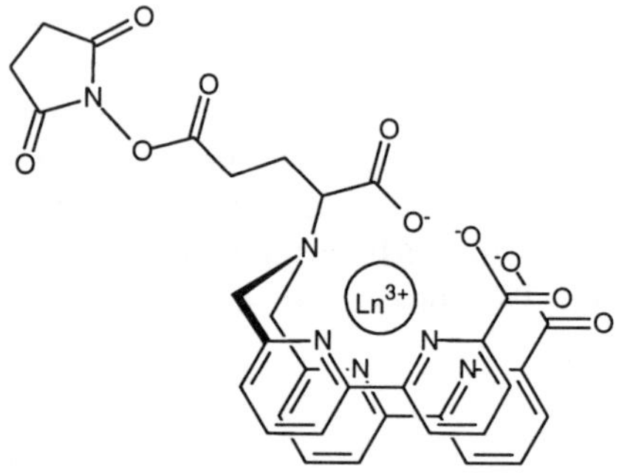

Figure 1. Structure of the lanthanide chelates

The Eu^{3+} and Tb^{3+} complexes are characterized by long luminescence lifetimes (0.6 and 1.5 ms) and high emission quantum yields (0.08 and 0.30). Their photophysical properties do not change upon binding to proteins, thus making these complexes potentially suitable for TRF imaging applications.

Comparisons of the performance of the different labels have been done by evaluating their limits of detection, defined as the amount of label (in molecules/μm^2) that gives a signal corresponding to the average background signal plus two standard deviations. In the case of the CL detection the kinetics of the luminescent signal was also taken into account, and the most intense signals (obtained immediately after the addition of the CL substrate) were used. Limits of detection of 8000, 1500, and 500 molecules/μm^2 were obtained for fluorescein, Tb^{3+} chelate and HRP labels, respectively (Fig. 2). CL detection clearly presents the lowest limit of detection, while the limit of detection for TRF is about three times that of CL. The low limit of detection obtained with CL is mainly due to its extremely low background signal rather than to a strong emission (despite the longest integration time, CL measurements gave the lowest signals). It should be also observed that the performance of TRF could approach that of CL in terms of detectable number of labeled biomolecules. In fact, CL enzymes require very low labelling ratios (e.g. 1:1), while TRF labels, thanks to their low molecular masses (of the order of 1000 Da), allow multiple labelling, thus increasing the signal to mass ratio of the probe.

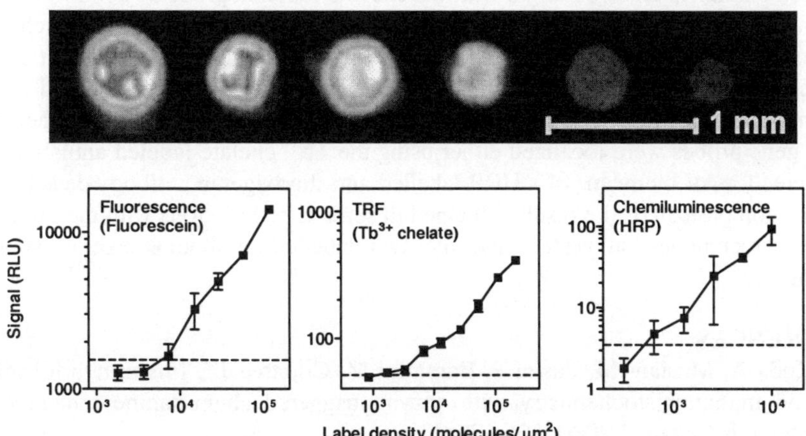

Figure 2. (Top) CL image of spots containing decreasing amounts of HRP (from 10^4 to 300 molecules/μm^2). (Bottom) Signals obtained from spots containing different amounts of label. The dashed lines represent the signal level corresponding to the limit of detection for each label

We have also investigated the possibility to perform the simultaneous detection of differently labelled biospecific probes. The measurement of sample spots using the different detection techniques (Fig. 3) indicated that, thanks to the different excitation/emission wavelengths and lifetimes of fluorescein and Tb^{3+} chelate and to the high specificity of the HRP-catalyzed CL reaction, each label can be separately detected and quantified. In addition, signals measured in spots containing two mixed labels indicated the absence of any interference between labels.

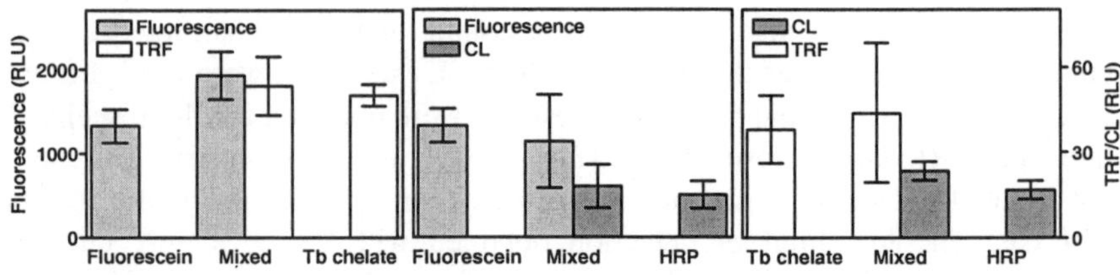

Figure 3. Signals obtained from spots containing fluorescent, TRF or CL probes, either mixed or alone

These experimental data indicated that the lanthanide complexes of the new ligand, particularly the Tb^{3+} one, are suitable for application as luminescent labels in TRF microscope imaging techniques, and that they could allow to achieve limits of detection similar to those obtained with CL enzyme-labelled probes.

We have thus synthesized and characterized a conjugate of the Tb^{3+} chelate with an anti-digoxigenin antibody and used this probe (conjugation ratio about 2.5:1) for the detection of human papillomavirus (HPV) nucleic acids in cells and tissue sections. Samples were hybridized with digoxigenin-labeled gene probes, then the bound gene probes were localized either using the Tb^{3+} chelate-labeled antibody and TRF detection, or by means of a HRP-labelled anti-digoxigenin antibody detected by CL. The comparison of the results obtained in serial tissue sections with the different detection techniques suggested that the two labelled antibodies exhibit similar detectability.

REFERENCES

1. Roda A, Musiani M, Pasini P, Baraldini M, Crabtree JE. In situ hybridization and immunohistochemistry with enzyme-triggered chemiluminescent probes. Methods Enzymol 2000; 305:577-90.
2. Weibel N, Charbonnière LJ, Guardigli M, Roda A, Ziessel R. Engineering of highly luminescent lanthanide tags suitable for protein labeling and time resolved luminescence imaging. J Am Chem Soc 2004; 126:4888-96.

MODELLING OF THE SIGNAL INTENSITY IN THE VARIOUS REACTION CHAMBERS OF THE NO-O$_3$ CHEMILUMINESCENCE NITROGEN OXIDES MONITOR TO OBTAIN HIGHER SENSITIVITY

H SAWADA, K OKITSU, N TAKENAKA, H BANDOW

Grdt. Schl. Eng., Osaka Pref. Univ., 1-1 Gakuen-cho, Sakai 599-8531, Japan
E-mail: hsawada@ams.osakafu-u.ac.jp

INTRODUCTION

Today, we often use the NO-O$_3$ chemiluminescence nitrogen oxides (NO$_x$) analyzer for determining NO$_x$ concentration in the atmosphere. In the atmosphere, especially at rural sites, NO$_x$ dominates the formation of O$_3$ known as a toxic substance.[1-3] It is important that we understand atmospheric chemistry including O$_3$ formation and the changes in atmospheric NO$_x$ concentration. High sensitivity and quick response are required on the NO-O$_3$ chemiluminescence NO$_x$ analyzer for that purpose. However, commercial NO$_x$ analyzers generally have insufficient sensitivity for the measurement of NO$_x$ concentration at the low levels, which are often observed in remote region such as rural sites.

In previous studies, optimizations of chemiluminescence NO$_x$ analyzers were carried out.[4,5] For example, the surface in the reaction chamber had been coated with gold to improve the optical collecting ratio. In this study, we examined the change in sensitivity (signal intensity) caused by altering the inlet of the reaction chamber and the form of the gas mixture flow in the reaction chamber. The model calculation of the flows has given that NO has not completely reacted with O$_3$ within the reaction chamber.

METHODS

Principles

The luminescence from electronically excited NO$_2$ (NO$_2$*) is measured in NO-O$_3$ chemiluminescence NO$_x$ analyzer, and NO$_x$ concentration is calculated from this luminescence intensity. The following reaction eqs. show a series of NO-O$_3$ chemiluminescence reactions.

$$NO \quad + \quad O_3 \quad \rightarrow \quad NO_2{}^* \quad + \quad O_2 \qquad (1)$$
$$NO \quad + \quad O_3 \quad \rightarrow \quad NO_2 \quad + \quad O_2 \qquad (2)$$
$$NO_2{}^* \quad \rightarrow \quad NO_2 \quad + \quad hv \qquad (3)$$
$$NO_2{}^* \quad + \quad M \quad \rightarrow \quad NO_2 \quad + \quad M \qquad (4)$$
$$NO_2{}^* \quad \rightarrow \quad NO_2 \qquad (5)$$

The chemiluminescence shown as 'hv' in reaction eq. (3) is detected by a photomultiplier tube (PMT). Eqs. (4) and (5) show NO$_2$* falls to the ground state without chemiluminescence.[6] When the pressure in the reaction chamber is constant, namely, the concentration of the third body 'M' is constant, the 'hv' is proportional to NO_2^* concentration from eq. (3). Furthermore, when O_3 concentration is larger than NO concentration, the NO_2^* concentration is proportional to O_3 concentration from eq. (1). Therefore, the 'hv' is in proportion to NO concentration, and detecting the chemiluminescence gives the atmospheric NO concentration. NO_2 which is the other major species of NO_x is reduced to NO by a reducing agent, and it is measured by the NO_x analyzer in the same manner as NO.

Experiments

In this study, we examined 4 types of reaction chambers, and compared the signal intensities among them. The structure of one of the reaction chambers(RC1)[4] is shown in Fig. 1. Sample and O_3/O_2 gases flow into the reaction chamber radially from the cylinder wall of the reaction chamber, and converge on the center part of the chamber in front of photoelectric surface of the PMT detector. Three other types of chambers were examined (not shown here), but briefly RC2 has 2 separate tubes for each gas facing head-on in front of the photoelectric surface, RC3 has also 2 separate tubes but each tube is set the same distance from the center of the chamber to make the gases swirl in the chamber, and in RC4 both gases are blown against the photoelectric surface from a concentric double-tube set on the center of the chamber. Flow rate of sample gas O_3/O_2 gas were 1900 cm³ min⁻¹ and 100 cm³ min⁻¹, respectively. O_3 in eqs. (1) and (2) was prepared from O_2 gas through an electric discharge O_3 generator. Pressure in the reaction chamber was kept constant at 40 Torr in all experiments.

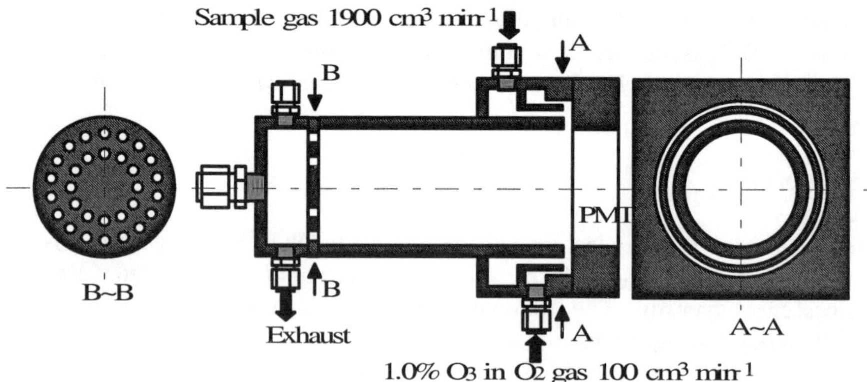

Figure 1. Schematic view of reaction chamber, RC1.[4]

RESULTS

The signal intensities of 4 type chambers are shown in Fig. 2. It shows the maximum attainable intensities for each type of chamber and the intensities are normalized to RC1 (100 %) for comparison. This result shows that signal intensities of RC2, 3 and 4 are significantly smaller than that of RC1. In RC2 and RC3, both gases are mixed in the detection zone, but in RC1, both gases are mixed before being introduced in the detection zone. Furthermore, total gas flow rate in RC2 and RC3 near the photoelectric surface was much faster than that of RC1 because of the difference of the opening areas of the inlet into the chamber. This result indicated that the mixing condition and gas flow rate may affect the sensitivity of a NO_x analyzer, because total gas flow rate in RC4 was faster than that in RC2 although both gases are mixed before being introduced into the detection zone.

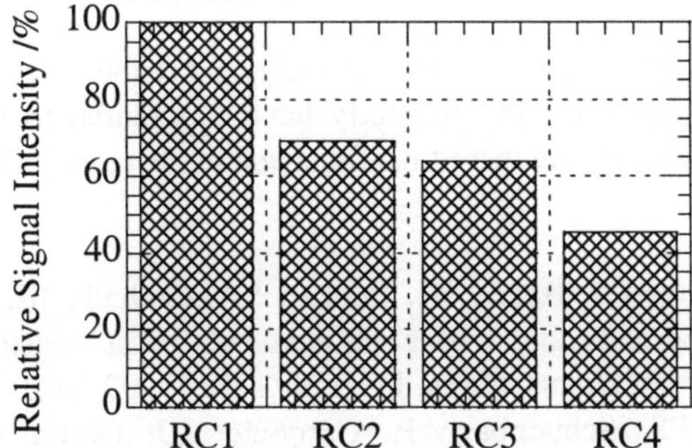

Figure 2. Relative signal intensties of 4 types of reaction chambers

For RC1, we modelled the gas flow of the mixed gas to calculate the signal intensity in the reaction chamber. In this model, the space of the reaction chamber is divided into several parts, and the concentration of NO_2* [7,8] and the intensity of illumination to the detector surface is calculated in each divided part. The sum of the calculated values for each part gives a theoretical signal intensity. The conformity of this model is verified experimentally. In the RC1, we set up the disk with 20 mm diameter facing parallel to the photoelectric surface of the PMT, and changed the distance between the disk and the photoelectric surface. The disk was blackened to remove the reflection of chemiluminescence on the surface. We calculated the signal intensity for the same conditions. The results of the measured and the calculated signal intensities are shown in Fig. 3. The horizontal axis shows the distance between the disk and the photoelectric surface. The ratio of both vertical axes indicating the signal intensity is adjusted to be directly compared. The model reproduces the general feature of the actual change in intensity, although both ratios at the maximum and minimum values are different. We suppose that this discrepancy between the model are mainly due neglecting the reflection of chemiluminescence on the wall of the reaction chamber in the model. When the distance is large, the effect of reflection

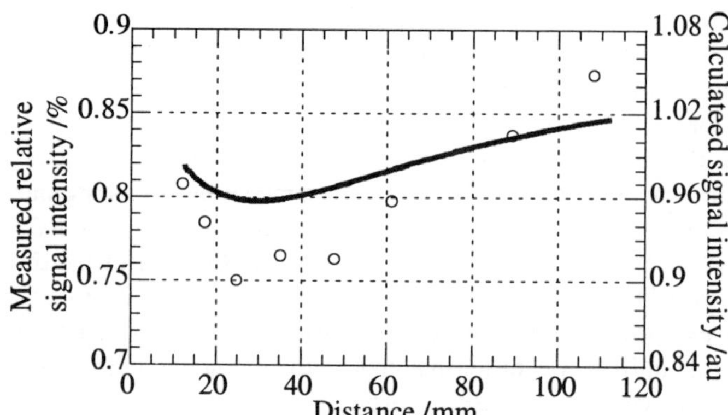

Figure 3. Signal intensitis of actual measurement and the model

on the wall may not be negligibly small. This model result demonstrates that even the best reaction chambers examined in this study have a potential to increase signal intensity by taking into account the rest of useful space in the reaction chamber.

REFERENCES

1. Carpenter LJ, Monks PS, Bandy BJ, Penkett SA, Galbally IE, Meyer CP. A study of peroxy radicals and ozone photochemistry at coastal sites in the Northern and Southern Hemispheres. J Geophys Res 1997; 102: 25417-27.

2. Zanis P, Monks PS, Schuepbach E, Carpenter LJ, Green TJ, Mills GP, Bauguitte S, Penkett SA. In-situ ozone production under free tropospheric conditions during FREETEX 98 in the Swiss Alps. J Geophys Res 2000; 105: 24223-34.

3. Salisbury G, Monks PS, Bauguitte S, Bandy BJ, Penkett SA. A seasonal comparison of the ozone photochemistry in clean and polluted air masses at Mace Head Ireland. J Atom Chem 2002; 41: 163-87.

4. Ridley BA, Grahek FE. A small, low flow, high sensitivity reaction vessel for NO/O_3 chemiluminescence detectors. NCAR 1990; 7.

5. Steffenson DM, Stedman DH. Optimization of the operating parameters of chemiluminescent nitric oxide detectors. Anal Chem 1974; 46: 1704-09.

6. Myers GH, Silber DM, Kaufman F. Quenching of NO_2 fluorescence. J Chem Phys 1966; 44: 718-23.

7. Bradburn GR, Lilenfeld HV. Absolute emission rate of the reaction between nitric oxide and atomic oxygen. J Phys Chem 1988; 92: 5266-70.

8. Clough PN, Thrush BA. Mechanism of chemiluminescent reaction between nitric oxide and ozone. J Chem Phys 1976: 914-25.

DEVELOPMENT OF HIGHLY SENSITIVE ANALYSIS OF GLYCATED PROTEIN IN HUMAN HAIR BY LUMINOL CHEMILUMINESCENCE

T YAJIMA[1], K ITO[1], R ITO[1], K INOUE[1], K MASUBUCHI[2], Y YOSHIMURA[1], S YAMADA[2], K ATSUDA[2,3], H KUBO[3], H NAKAZAWA[1]

[1]*Department of Analytical Chemistry, Hoshi University, Tokyo 142-8501, Japan*
[2]*Department of Pharmacy and Department of Internal Medicine,*
Kitasato Institute Hospital, Tokyo, 108-8641, Japan
[3]*Department of Analytical Chemistry, Kitasato University, Tokyo, 108-8641, Japan*

INTRODUCTION

Glycated proteins and glucose in the blood have been widely used as a glycemic control indicators in the medical field. The glycated protein exists not only in the blood but also in the skin, hair, etc. It has been reported that diabetic patients have an increase of the glycation.[1,2] Hair is useful as a biological sample in which non-invasive sampling and long-term preservation are possible in the medical field. It was reported that the amount of glycated proteins in the hair has correlated with that of blood, and measurement of the glycated protein in the hair is useful as a long term glycemic control indicator.[3] However, the analytical methods for glycated protein in hair have some disadvantages, such as the tedious procedure and necessity of large amounts of hair (5-10 mg).[3,4] In the present study, a highly sensitive and simple method was developed for diabetes screening by luminol chemiluminescence of hair using a microtiterplate assay. We also investigated correction of experimental errors produced by collection, weight of hair, and external factors such as permanent wave agent. Although cystine is used for the compensation of error,[4,5] it can be influenced by permanent wave agent or a decolorant. We therefore used arginine as a new and useful compensation factor in conjunction with a highly sensitive fluorescence method[6] for the diagnosis of diabetes.

METHODS

Subject

Scalp hairs and blood were sampled from healthy subjects (n=29) and diabetic patients (n=7). The blood samples were assayed for glycohemoglobin (HbA$_{1c}$).

Materials

Analytical grade sodium hydroxide, ethanol, potassium hexacyanoferrate (II) trihydrate, potassium hexacyanoferrate (III), hydrochloric acid, L-arginine and ninhydrin (triketohydrindene hydrate) obtained from Wako Pure Chemical Co. (Osaka, Japan). Analytical grade luminol (3-aminophthalhydrazide, sodium salt hemihydrate) obtained from Sigma-Aldrich Japan (Tokyo, Japan). Precimat$^{®}$ fructosamine (glycated polylysine ; GPL) by Roche Diagnostics K.K. (Tokyo, Japan) was used as standard of glycated protein. The white microtiterplate was purchased from Nalge Nunc International K.K. (Tokyo, Japan).

Preparation of sample solution One centimeter hair sample was rinsed with 1 mL of ethanol at 37 °C for 30 min. After removing ethanol using a capillary pipette, the sample was dissolved in 1 mL of 600 mmol/L sodium hydroxide at 95 °C for 10 min.

Measurement of glycated proteins
Sample solution is diluted 40 times by 600 mmol/L sodium hydroxide, the diluted sample solution 50 μL is poured into the plate. After addition of a luminescence solution (40 mmol/L potassium hexacyanoferrate (II) and 0.1 mmol/L potassium hexacyanoferrate (III) containing 5 μmol/L luminol solution), luminescence intensity was measured. Calibration curve was obtained by measurement of the chemiluminescence intensity using GPL as a standard, and the relative amount of the glycated protein in hair was computed.

Determination of arginine
A 700 mmol/L hydrochloric acid (250 μL) and distilled water (500 μL) are added to 250 μL sample solution. Furthermore, 2mL of 15 mmol/L ninhydrin solution and 1 mol/L sodium hydroxide solution are added to sample solution. After mixing of the solution, it is left for 7 min under dark, and fluorescence measurement (Ex. 302 nm, Em. 500 nm) was performed.

Utility of arginine as a compensation factor
Utility of arginine as a compensation factor for confounding factors was investigated by comparison with RSD values. The influence of a shampoo and a permanent wave used in daily life was also examined in respect with each hair, which was cut into pieces at intervals of 1 cm collected from six healthy subjects.

Measurement of glycated protein in hair sample
The glycated protein of hair collected from the healthy subjects (n=29) and the diabetic patients (n=7) was measured. And glycated proteins of diabetic patients were compared with that of healthy subjects.

The correlation of the value of glycated protein by compensation with arginine and the levels of HbA_{1c} in blood were examined. In addition, the glycated protein that shifts to the tip of hair due to growth, and past levels of HbA_{1c} were examined.

RESULTS

By compensation of the experimental error due to sampling, RSD values of glycated protein in hair for six subjects have been improved to less than 15% by use of arginine (Table 1). Moreover, when the influence of the damage in daily life was considered, all six persons' RSD values have been improved by compensation (Table 2). Therefore, it was considered that arginine is useful as a compensation factor. When the glycated protein of human hair was compared, the diabetic patients showed high levels that were significantly different from healthy subjects (P<0.003). The glycated protein in hair correlated well with the HbA_{1c} in blood (r=0.980) When

the glycated protein of human hair was compared, the diabetic patients showed high levels that were significantly different from healthy subjects (P<0.003).
The glycated protein in hair correlated well with the HbA_{1c} in blood (r=0.980)

Table 1. Effect of arginine on compensation of glycated protein in hair collected at different five places from six subjects.

Subjects		A	B	C	D	E	F
RSD Value (%)	Non-compensated	10.6	23.0	39.6	14.1	22.1	22.7
	Compensation by arginine	10.1	8.2	13.8	6.3	7.8	10.5

Table 2. Effect of arginine on compensation of glycated protein in hair pieces obtained at intervals of 1 cm.

Subjects		a	b	c	d	e	f
RSD Value (%)	Non-compensated	10.2	16.0	8.3	39.9	17.3	5.8
	Compensation by arginine	7.9	8.5	5.5	9.9	10.0	2.4

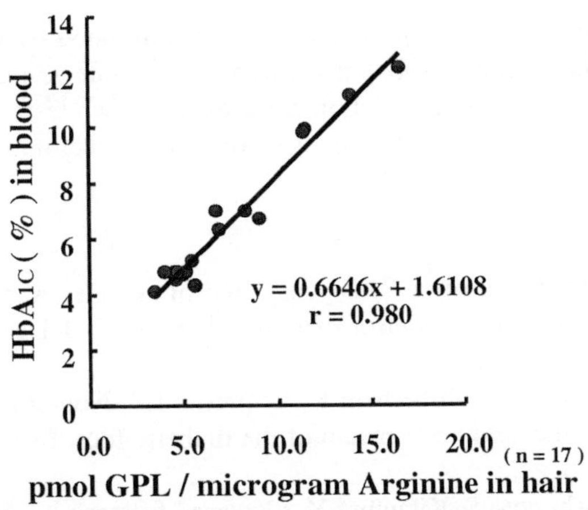

Figure 1. Correlation between glycated protein in hair and HbA_{1c} in bl

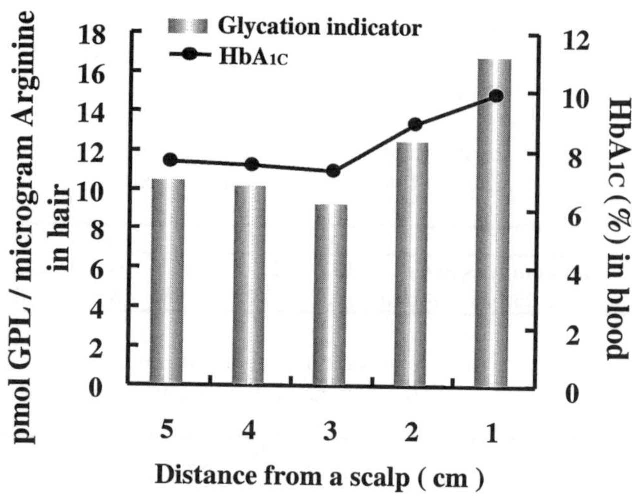

Figure 2. Past HbA1c in blood and glycated protein distribution in hair.

The distribution of glycated protein in hair and HbA_{1c} in blood over several months is shown in Fig 2. As the results show, it seems to be possible to evaluate past states of diabetes mellitus. The good correlation between these two tests suggests that the proposed method could be useful in monitoring of diabetes mellitus as an alternative to HbA_{1c} in blood.

REFERENCES

1. Sueki H, Nozaki S, Kuroiwa Y. Glycosylated of skin,nail and hair:application as an Index of long-term control of diabetes mellitus. J Dermatol 1989; 16: 103-10.

2. Delibridge L, Elles CS, Robertson K, Lequesne LP, Non-enzymatic glycation of keratin from the stratum corneum of the diabetic foot. Br J Dermatol 1985; 112: 547-54.

3. Oimomi M, Nishimoto S, Kitamura Y, Increased fructose-lysine of hair protein in diabetic patients. Klin Wochenschr 1985; 63: 728-30.

4. Kobayashi K, Igami H, Glycation index of hair for non-invasive estimation of diabetic control. Bio Pharm Bull 1996; 19: 487-90.

5. Kobayashi K, Yoshimoto K, Uchida K, Determination of glycated proteins in biolosical samples based on colorimetry of 2-keto-glucose released with hydrazine. Bio Pharm Bull 1994; 17: 365-9.

6. Kyoritsu shuppan Co., Ltd. Chemistry of proteins, 1976; 1: 288-90.

FLUORESCENCE SENSOR PEPTIDE FOR PROTEIN PHOSPHORYLATION

KENJI YOKOYAMA[1], KAZUYA ITODA[1,2]

[1] *Research Center of Advanced Bionics, National Institute of Advanced Industrial Science and Technology; Tsukuba Central 4, Tsukuba, Ibaraki 305-8562, Japan*
[2] *Japan Advanced Institute of Science and Technology; Tatsunokuchi, Ishikawa 923-1292, Japan*
e-mail: ke-yokoyama@aist.go.jp

INTRODUCTION

Living cells possess an intracellular signal transduction system that includies many biological reactions. The purpose of this system is to allow cells to respond to their outer environment. Protein phosphorylations are one of the most important and versatile information processing signals in the transduction system. We are currently focusing on a peptide as a novel biosensor material and have investigated several sensor peptides that can be applied to analyzing intracellular signal transduction. We have designed a sensor peptide based on fluorescence resonance energy transfer for analyzing protein kinase activity (Fig. 1). For instance, we have synthesized a sensor peptide modified with EDANS and dabcyl for monitoring MAPKK activity.

A cyclic AMP-dependent protein kinase (PKA) is one of the most important kinases. The kinase is activated with cyclic AMP which is synthesized intracellularly in response to various extracellular stimuli such as hormones, cytokines. The activated kinase plays an important direct role in activating and regulating metabolic enzymes and several functional proteins. It is reported that disorder of PKA signaling is strongly related to many diseases.[1,2] In this work, we evaluated fluorescence sensor peptide for monitoring PKA activity.

METHODS

We synthesized 10mer and 15mer peptides, which included the amino acid sequence of the phosphorylated site, ArgArgAlaSerLeu, of PKA. Ser is phosphorylated by PKA in the cell. EDANS-modified 10mer and 15mer peptides were synthesized on bead.[3] After cleavage and deprotection of peptide side chain, dabcyl group was attached at the N-terminal α-amino group. Ser-phosphorylated sensor peptides were also synthesized. The synthesized sensor peptides were purified using HPLC and identified by MALDI-TOF MS (Table 1).

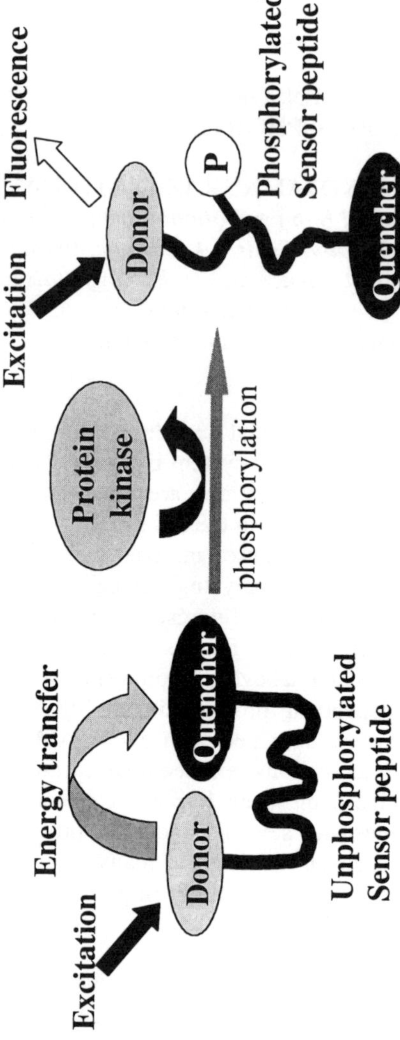

Figure 1. Principle of fluorescence sensor peptide for protein phosphorylation

RESULTS AND DISCUSSION

The fluorescence properties of Peptide 1 and 1p was investigated. Energy transfer from EDANS to dabcyl, quenching by dabcyl, was reduced by phosphorylation of Ser (Fig. 2). Peptide main chain was stretched by increase in hydrophilicity, and the distance between the fluorescence dye and quencher increased.

Table 1. Amino acid sequences of the fluorescence sensor peptides.

Peptide 1	dabcyl-GLRRASLGE(EDANS)G
Peptide 1p	dabcyl-GLRRA[pS]LGE(EDANS)G
Peptide 2	dabcyl-GDENLRRASLGEDE(EDANS)G
Peptide 2p	dabcyl-GDENLRRA[pS]LGEDE(EDANS)G
Peptide 3	dabcyl-GRRRLRRASLGRRE(EDANS)G
Peptide 3p	dabcyl-GRRRLRRA[pS]LGRRE(EDANS)G
Peptide 4	dabcyl-GRRRLRRASLGEEE(EDANS)G
Peptide 4p	dabcyl-GRRRLRRA[pS]LGEEE(EDANS)G

pS: phosphorylated serine

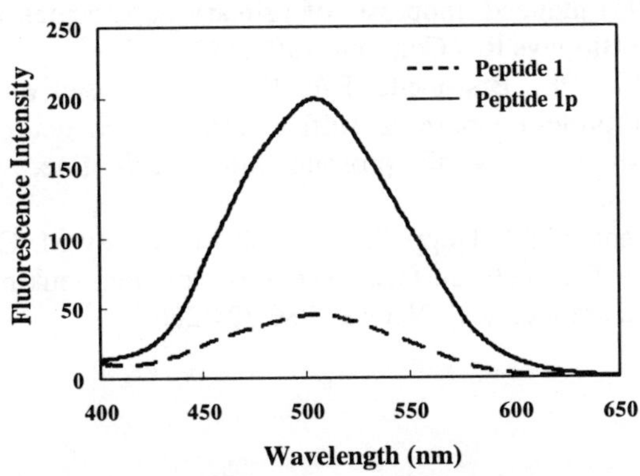

Figure 2. Fluorescence spectra of Peptides 1 and 1p

Fluorescence intensity change of sensor peptide as a result of enzymatic phosphorylation of PKA was investigated. Fluorescence intensity increased with increasing enzyme reaction time (Fig. 3). The relative fluorescence intensity after an enzyme reaction time of 120 min was 4.1 times greater than that of Peptide 1.

The ratio of phosphorylated Peptide 1 was investigated using HPLC. This showed that 96 % of Peptide 1 was phosphorylated by PKA in 120 min. These data clearly indicate that the phosphorylation of Peptide 1 by PKA can be monitored with fluorescence intensity changes.

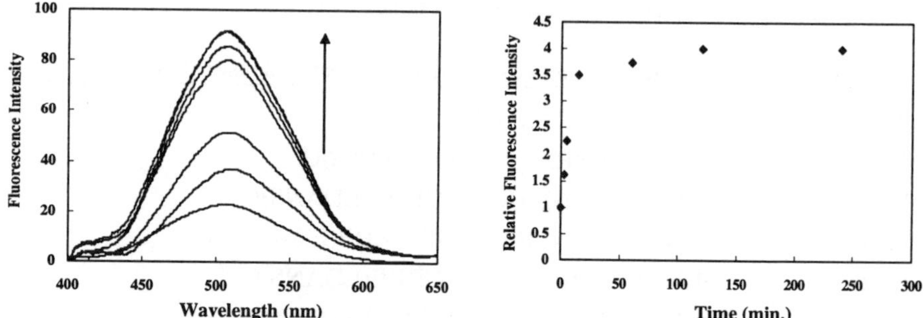

Figure 3.   Fluorescence spectrum change of Peptide 1 by enzymatic phosphorylation

REFERENCES

1.   Fladmark KE, Gjertsen BT, Doskeland SO, Vintermyr OK. Fas/APO-1(CD95)-induced apoptosis of primary hepatocytes is inhibited by cAMP. Biochem Biophys Res Commun 1997;232:20-5.
2.   Cox ME, Deeble PE, Bissonette EA, Parsons SJ. Activated 3',5'-cyclic AMP-dependent protein kinase is sufficient to induce neuroendocrine-like differentiation of the LNCaP prostate tumor cell line. J Biol Chem 2000;275:13812-8.
3.   Dower SK, Kronheim SR, Hopp TP, Contrell M, Deeley M, Gillis S, Henny CS, Urdal DL. The cell surface receptors for interleukin-1 alpha and interleukin-1 beta are identical. Nature 1986;324:266-9.

PART 9

ANTIOXIDANTS, REACTIVE OXYGEN SPECIES & PHAGOCYTOSIS

THE CHEMILUMINESCENT MEASUREMENT OF THE BLACK AND GREEN TEA ANTIOXIDANT CAPACITY AND THE COMPARISON WITH THEIR ANTIMICROBIAL ACTIVITY

M BANCÍŘOVÁ[1], I ŠNYRYCHOVÁ[2]

[1] *Dept of Physical Chemistry, Faculty of Science, Palacký University, Tř. Svobody 26, 771 46 Olomouc, Czech Republic*
[2] *Dept of Botany, Faculty of Science, Palacký University Šlechtitelů 11, 783 71 Olomouc, Czech Republic*
Email: bancir@aix.upol.cz

INTRODUCTION

The lifetime of reactive oxygen species (ROS) is extremely short, and if a physiological acceptor does not immediately neutralize them, ROS can damage biological systems. All aerobic organisms have developed more or less complex systems to neutralize them before their potentially harmful effect is activated. Nutritional elements are also extremely important. Foods that have potential or definite antioxidant capacities are mainly vegetables and fruits, as well as beverages like red wine, tea and beer.

Tea is a source of epigallocatechin gallate in green tea, and theaflavin and the associated thearubigins, in black tea. For example, cancer in the colon, breast, prostate and pancreas may be caused by a new class of carcinogens - the heterocyclic amines, formed during the broiling or frying of creatinine-containing foods, including fish and meats. Their formation and action can be inhibited by antioxidants such as those in soy and tea. Black tea is a powerful chemopreventor of reactive oxygen species and was found to be more efficient than green tea.

A sensitive and simple chemiluminescent (CL) method for measuring antioxidant activity was developed. The determination of TEAC (Trolox equivalent antioxidant capacity) is based on the inhibition of CL intensity of luminol by an antioxidant. Antimicrobial activity was tested as a minimal inhibitory concentration (MIC) by broth microdilution method on gram-positive (*Enterococcus faecalis, Staphylococcus aureus*) and gram-negative (*Pseudomonas aeruginosa, Escherichia coli*) bacterial strains.

MATERIAL AND METHODS

Reagents: luminol (5-amino-2,3-dihydro-1,4-phthalazinedione), horseradish peroxidase (HRP), E.C. 1.11.1.7., Grade II, hydrogen peroxide, Trolox (6-hydroxy-2,5,7,8-tetramethylchroman-2-carboxylic acid 98% pure, water soluble vitamin E analogue used as reference antioxidant to determine antioxidant capacity) were Sigma-Aldrich (Germany) compounds.

Preparation of tea solution: leaves were dissolved in hot water (under 100 °C) for 3 min (2 g/100 mL). The extracted solutions were filtered and allowed to cool at room temperature. The tea solutions were neither made fresh or frozen.

Chemiluminescent mixture: 100 µL of the 10 mM hydrogen peroxide solution was added to 200 µL of the luminol solution (2.0 mM). This mixture (10 mL) was protected against the light with aluminum foil. *Calibration curve:* from the stock solution 1 mM Trolox in 0.1 M phosphate buffer, pH 7.6, stored in a freezer until use, several working solutions were prepared in the range 0.1-1 mM.

Chemiluminescent assay: 20 µL of the peroxidase solution was added to 180 µL of the chemiluminescent mixture to trigger the luminol oxidation reaction. This mixture was the reference system representing 100% light emission in the absence of inhibition by the sample or standard antioxidant solutions. To determine the total antioxidant capacity of the samples, 20 µL of the sample or standard solutions were injected into the microplate. The light signal was immediately inhibited, reaching values close to zero, which were maintained for a certain time then gradually increased to return to values close to those observed prior to sample injection. The time between sample injection and the return to maximal emission was measured. The value obtained (in seconds) was a function of the antioxidant capacity of the sample examined. This antioxidant capacity was expressed, fitting the times obtained on the relative calibration curve, in concentration of Trolox (mM).

Antimicrobial activity was tested on standard strains of *Enterococcus faecalis* 4224, *Staphylococcus aureus* 3953 and 4223, *Pseudomonas aeruginosa* 3955 and *Escherichia coli* 3954, 3988 and 4225 (Czech Collection of Microorganisms, Brno). Microtiter wells containing serial twofold dilutions of samples are incolulated with a standard inoculum of the bacterium in question, the plates are incubated overnight, and the wells are then examined for the presence of bacterial growth. The lowest of each sample dilution series that prevents bacterial growth is considered to be the minimum inhibitory concentration (MIC) of the sample.

RESULTS

The Table 1 shows the MIC dilution of the different types of tea (they were obtained from the common tea-rooms). The tea solutions were made fresh

Table 1 MIC of the tea samples. The numbers show the efficient dilution of tea samples on bacterial strains

	Bacterial strain	Gram-positive			Gram-negative			
		3953	4223	4224	4225	3988	3954	3955
1	SHOU MEE	2					4	2
2	PU – ERH						2	
3	CHE XANH	2				2	2	4
4	SHUI HSIEN	2					2	2
5	CEYLON BOP HIGH FOREST	2	4				2	2
6	ZHU CHA	4				2	2	2
7	UVA	2						2
8	SENCHA KJOTO	2				2	4	4
9	TZA´- O- LONG	2				2	2	2
10	DARJEELING	4				2		2
11	GUN POWDER					4		2
12	CHINA OP YUNNAN						2	
13	ROOIBOS SUPERIOR	2	2				2	
14	MATE RANHCHO	2	2	2	2	4	4	2
15	NEPAL 1 FIKKAL ILLAM					2	2	4
16	TURKEY BOP		2					
17	VIETNAM		2			2		2
18	JAPAN HOJICHA			2	2	2	2	4
19	JAVA OP							
20	NILGIRI- SOUTH INDIA FOP					2		2
21	FORMOSA FINE OOLONG							2
22	GRUSIA OP							
23	EARL GREY							2
24	ASSAM OP KHONGEA	2	2					
25	ASSAM TG FOP-1CL							
26	LAPACHO							
27	LUNG CHING					2		
28	KOKEICHA					2		
29	JAVA OP TALOON					2		2
30	BANCHA							
31	GOLDEN TIPPY					2		
32	LAPSANG SOUCHONG-	2	2					
33	PAHARGOOMIAH				2	2		4
34	GOLDEN MARINYM				4	4		8

The graph (Fig. 1) shows the dependence of TEAC (Trolox Equivalent Antioxidant Capacity) on the different types of tea. The tea solutions were neither made fresh - the white column - nor frozen for one year - the grey column.

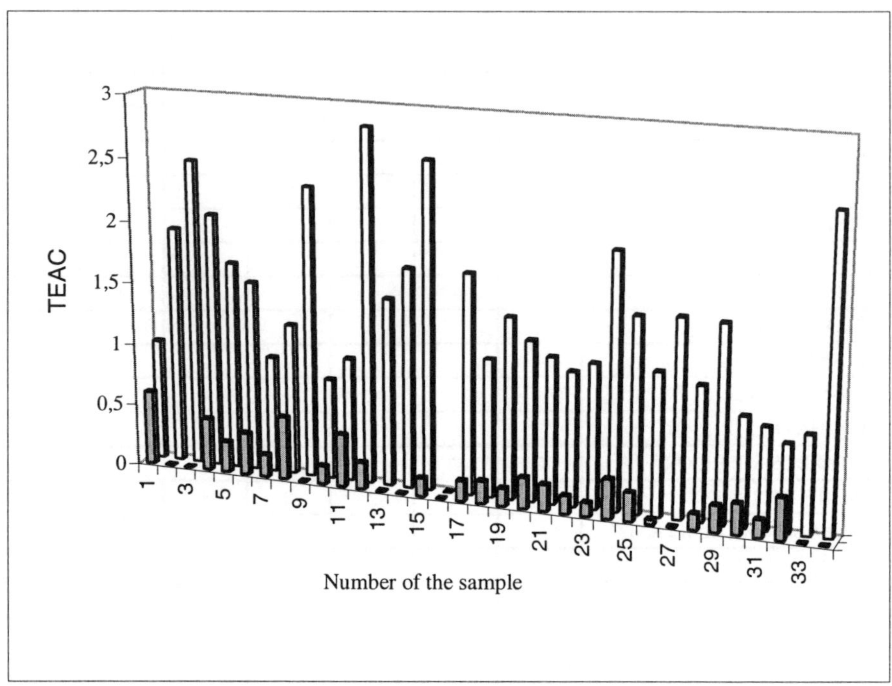

Figure 1. The dependence of TEAC on the type of tea (the number of sample). The white columns are the fresh tea solutions; the grey columns are the frozen tea solution (for one year)

CONCLUSIONS
Our results do not confirm relevant predominance of black tea over green tea or *vice versa* in their antioxidant capacity or antimicrobial activity

ACKNOWLEDGMENTS
This research was supported by the grant from Ministry of Education MSM 153100008.

UNIVERSAL CHEMILUMINESCENT ASSAY FOR OXIDATIVE AND ANTIOXIDATIVE PROCESSES IN CHEMICAL AND BIOLOGICAL MEDIA: FUNDAMENTALS AND APPLICATION ASPECTS

VA BELYAKOV, GF FEDOROVA, VV NAUMOV,
AV TROFIMOV, RF VASIL'EV

*Institute of Biochemical Physics, Russian Academy of Sciences,
119991 Moscow, Russia*

INTRODUCTION

It has been long known that free-radical chain oxidation of hydrocarbons is followed by chemiluminescence (CL).[1] In the oxidation of hydrocarbons with a single oxidizable group, excited-state generation comes about from the disproportionation of peroxy radicals (ROO$^\bullet$). This termination step of the oxidation chain, which proceeds through the tetroxide intermediate,[1] yields three products, namely, ketone (R_{-H}=O), alcohol (ROH) and molecular oxygen (Eq. 1).

$$\text{ROO}^\bullet + \text{ROO}^\bullet \rightarrow R_{-H}\text{=O} + \text{ROH} + O_2 + h\nu \tag{1}$$

If hydrocarbon does not possess chromophore groups with low-lying excited states, the CL emission derives from the $^3(n,\pi^*)$–excited carbonyl moiety of ketone formed in the reaction.

Herein, we discuss the pertinent details of the CL approach to monitor oxidation processes with the emphasis on oxidation in the presence of antioxidants. The reason for this emphasis resides in a paramount role of antioxidants in numerous areas of biology, material science, chemical and analytical technologies. The mere fact that antioxidants suppress oxidation reactions and thereby quench the light emission opens a direct opportunity for the use of CL in the antioxidant analysis. The experimental approach is based on the competition between disproportionation of peroxy radicals (Eq. 1) giving rise to light emission and scavenging the peroxy radicals by antioxidants (A) [Eq. 2] resulting in the CL quenching. The extent of

$$\text{ROO}^\bullet + \text{A} \rightarrow \textit{Inactive Products} \tag{2}$$

quenching and the kinetics of the CL recovery upon gradual consumption of the antioxidant depend on the antioxidant *reactivity* towards peroxy radicals (strength of the antioxidant) and its *concentration*.

METHODS

The CL measurements were performed with a self-made photometric device as reported before.[2] The probe CL solution consisted of the following components: (a) initiator, a peroxide or azo compound, which serves as a source of free radicals, (b) hydrocarbon (*e.g.*, Ph$_2$CH$_2$, PhCHMe$_2$, PhEt) being oxidized through ROO$^\bullet$ as a chain carrier, (c) solvent, *e.g.*, PhH, PhCl, MeCN, Bu^{t}OH, (d) dissolved oxygen, (e) fluorescer, which enhances the light intensity through energy transfer from poorly luminescent triplet carbonyl product; the best are 9,10- dibromoanthracene and europium chelates.[1]

RESULTS

The *versatility* of the presented approach is substantiated by the possibility of examining analytes dissolved in *different* phases. Since most of organic antioxidants are well-soluble in organic solvents, they easily diffuse into the probe solution even if the drop of analyte and the probe organic solution are not mixable and, thereby, do not form the same phase. The examples considered herein illustrate this contention.

Analyte dissolved in aqueous medium

Fig. 1 demonstrates a possibility to examine water-soluble analytes. Two different forms of the same phenolic antioxidant soluble *in chlorobenzene* (Fig. 1, curve 1) and *in water* (Fig. 1, curve 2) were injected in the same amount into the same probe organic solution. As it is seen from Fig. 1, in both cases results are very similar: the chemiluminescence time profiles are nearly the same. Thus, antioxidant easily diffuses from water into the organic phase of the probe solution and inhibits oxidation reaction there, which is manifested by the CL quenching.

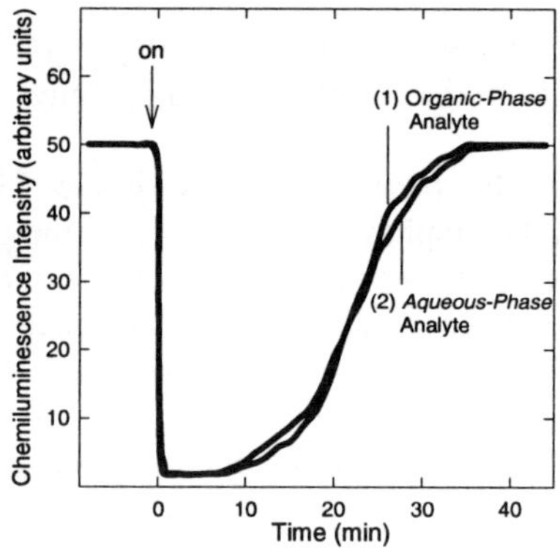

Figure 1. CL time profiles upon injection ("on") of two forms of the *same* antioxidant (7.0×10^{-6} M) into the probe solution [50 % cumene in PhCl in the presence of AIBN as an initiator; initiation rate was 1.0×10^{-8} Ms^{-1} at 333 K]. Curve 1 displays the response to the injection of γ-(4-hydroxy-3,5-di-*t*-butyl-phenol) propionic acid taken from its *organic-phase* (PhCl) stock solution, while curve 2 refers to the effect of the potassium salt of this acid taken from its *aqueous* solution.

Analyte in gaseous phase

Fig. 2 illustrates the possibility to examine volatile antioxidants present in gaseous phase, namely in ambient air, by means of air blowing through a probe organic-phase CL solution. Antioxidative species easily pass from the gaseous phase into the liquid

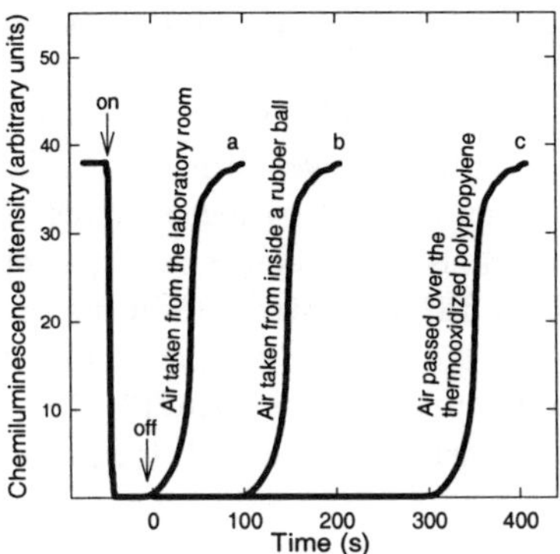

Figure 2. Effect of volatile antioxidant on CL generated in oxidation of Ph_2CH_2 (10% in PhCl) initiated by AIBN at 293 K (initiation rate was 1.8×10^{-11} Ms^{-1}). The antioxidant-containing air was introduced ("on") into the solution and then bubbling was ceased ("off").

organic solution, which is manifested by an *abrupt* drop of the CL intensity on blowing.[2] When blowing was switched off, the light intensity recovered after a "dark" (induction) induction period (Fig. 2), which manifests a consumption of the antioxidant. This behavior is *identical* to what is normally observed when an antioxidant analyte is initially dissolved in the liquid phase (*cf.* Fig. 1).

Analyte in solid phase

Fig. 3 provides an example that liquid organic-phase probe CL mixtures are suitable

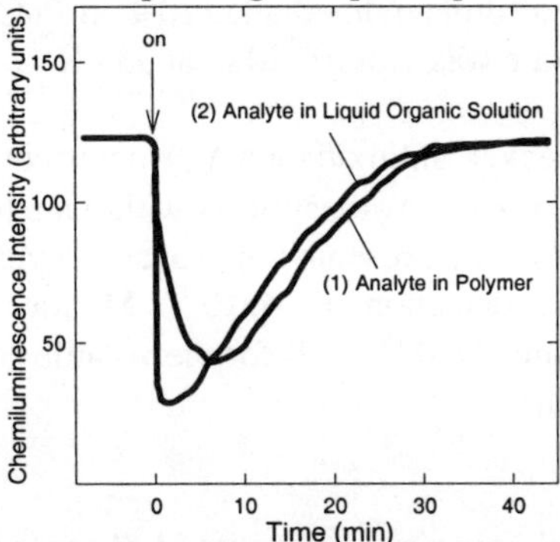

Figure 3. CL upon introduction ("on") of 19.8 mg of 250-μm polypropilene film containing 2,6-di-*t*-butyl-4-methyl-phenol [2.18×10^{-2} mol/kg] (curve 1); concentration of the antioxidant diffused into probe solution amounts to 8.65×10^{-5} M. Curve 2 refers to CL response to the injection of the equal amount of the same antioxidant taken from liquid (PhCl) solution. CL was generated in oxidation of PhEt (50% in PhCl) initiated by AIBN at 333 K (initiation rate was 2.52×10^{-7} Ms^{-1}).

    Belyakov VA et al.

to examine analytes present in solid phase. In this experiment, samples of the same phenolic antioxidant in polymeric film (Fig. 3, curve 1) and in the PhCl solution were introduced into the same probe CL mixture. Although the CL-quenching profiles do not coincide due to a slower diffusion of the antioxidant from polymer into liquid solution, the areas covered by these curves have been found to be about the same. Thus, similar amounts of light have been quenched in both cases.

Antioxidants of essentially different strength in the same analyte

Fig. 4 illustrates exemplary the case of a separate determination of a *strong* and a *weak* antioxidants in a red-pepper extract. With Ph_2CH_2 (fast self reaction of ROO˙), only strong antioxidant (A_S), namely, vitamin E is detected, while with cumene (slow

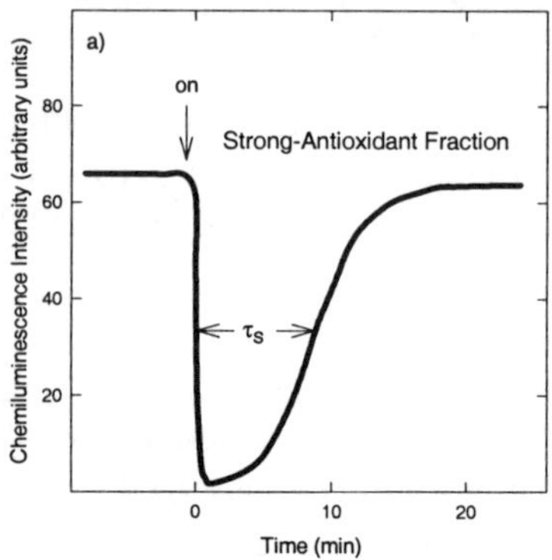

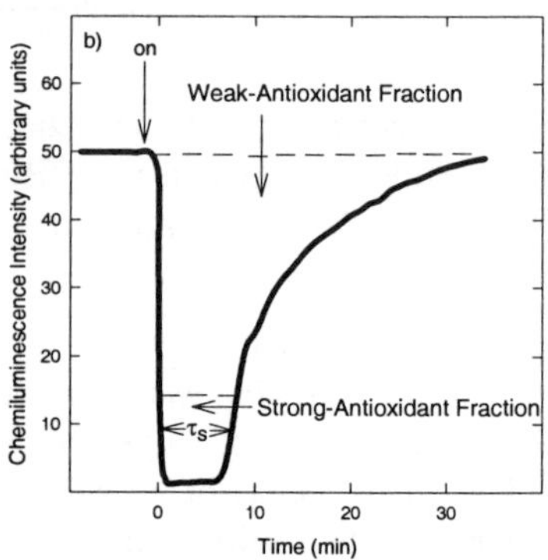

Figure 4. CL time profiles upon injection ("on") of a red-pepper extract into the probe solutions of (a) diphenylmethane (10% in PhCl) and (b) cumene (50 % in PhCl); initiation rate was 1.0×10^{-8} Ms^{-1} at 333 K.

ROO˙ self reaction) a weaker antioxidant (A_W) contributes to the quenching and becomes the only quencher when A_S disappears within the time τ_S.

The high sensitivity of the presented approach is evident from the fact that the limit of the antioxidant detection is 4×10^{-12} M for chromane C_I (synthetic α-tocopherol derivative) and 2×10^{-14} M (!) for the volatile "technogenic" antioxidants occurring in the ambient air.[2]

REFERENCES

1. Vasil'ev RF. Chemiluminescence in liquid-phase reactions. Progr Reaction Kinetics 1967; 4: 305-52.
2. Belyakov VA, Vasil'ev RF, Trofimov AV, Fedorova GF. Extremely efficient antioxidants evolved into environment from aging polymeric materials: chemiluminescence monitoring. In: Minisci F. ed. Free Radicals in Biology and Environment. Dordrecht: Kluwer Academic Publishers, 1997: 233-50.

EFFECT OF ANAESTHESIA WITH PROPOFOL AND REMIFENTANIL ON WHOLE BLOOD CHEMILUMINESCENCE: DISCRIMINANT ANALYSIS OF THE DATA

P DE SOLE, C ROSSI , R SCATENA

Institute of Biochemistry and Clinical biochemistry, Catholic University
Largo A. Gemelli, 8 – 00168 Roma "Italy"
Email: p.desole@rm.unicatt.it

INTRODUCTION

General anaesthesia is obtained by means of a variety of structurally unrelated compounds that are able to induce a global but reversible depression of central nervous system function, resulting in the loss of response to and perception of external stimuli.

Operationally, anaesthetics can be classified as inhalational or intravenous. The most used compounds of this last group are thiopental, propofol and ketamine; recently, new synthetic opioids related to phenylpiperidines (phentanyl and its congeners, sufentanil, alfentanil, remifentanil) are also used.

At the molecular level intravenous anaesthetics interact mainly with $GABA_A$ receptors (thiopental, propofol...) or with other ligand-gated ion channels (ketamine); synthetic opioids (phentanyl congeners) act in a more specific way as agonists of μ-opioid receptors. However, all the anaesthetics, in addition to their action at the level of central nervous system, have also a variety of effects at other levels; in particular, a temporary impairment of the immune system involving phagocytes can be induced.

Many papers describe dose-related in-vitro effects of propofol and remifentanil on different neutrophil (PMN) functions,[1-4] while in-vivo effects seem to be less clearly evident.[5-8]

In the present study the effect of a short-time intravenous anaesthesia with propofol and remifentanil on PMN chemiluminescence (CL) activity is measured in a group of young patients undergoing surgery for strabismus.

PATIENTS AND METHODS
Patients

Nine patients (age 7-12 years) were anaesthetised for strabismus surgery with 3 mg kg^{-1} propofol and 0.3 $\mu g\ kg^{-1}\ min^{-1}$ remifentanil for induction phase and 100-200 $\mu g\ kg^{-1}\ min^{-1}$ propofol and 0.3-0.5 $\mu g\ kg^{-1}\ min^{-1}$ remifentanil for maintenance phase. Peripheral blood was drawn before, at the end of anaesthesia, 3 and 24 h after surgery.

Neutrophil chemiluminescence

PMN CL was measured for 120 min in a modified whole blood system[9] containing in 1.0 mL final volume 0.5 μL of blood, 100 nmoles luminol or lucigenin (Sigma) in presence or absence of 0.5 mg opsonized zymosan or 150 nmoles PMA.

Statistical analysis

Statistical analysis was performed using a commercially available statistical program (Statgraphics Plus; Manugistics, Inc., Maryland, USA)

RESULTS

Table 1 shows the values of the CL parameters and of the PMN blood count (mean±SD) measured at different times from anaesthesia.

Table 1. Statistical analysis of CL parameters and of PMN blood count before and after different times from anaesthesia

	T_0	T_1	T_3	T_{24}	P
LucBI (*)	514±174	476±271	505±226	540±196	NS
LucBT (min)	40±21	40±21	33±21	38±20	NS
LucPI (*)	929±143	931±292	879±117	911±320	NS
LucPT (min)	47±17	46±15	36±9	52±20	NS
LumPI (*)	1105±276	1234±373	1016±374	1140±459	NS
LumPT (min)	41±18	41±18	35±21	39±18	NS
LumZI (*)	1454±	1624±	1648±	1849±	NS
LumZT (min)	41±5	46±6	35±5	41±6	0.005
PMN (μl^{-1})	3336±1210	3457±1544	4988±1828	4667±1679	NS

Luc = lucigenin-dependent CL   Lum = luminol-dependent CL
I = integral CL (counts/120min/PMN)   T = Tmax (time of maximum CL) (min)
B/P/Z = basal/PMA/Zymosan CL   (*) counts/120 min/PMN
T_0 T_1 T_3 T_{24} time from anaesthesia   P = probability of the K-W test

As clearly shown in the table no significant difference is obtained, except for the time of peak CL in presence of luminol and zymosan (LumZT). In order to better investigate any possible effect of intravenous anaesthesia on the PMNs, the CL data and PMN blood counts were analyzed by discriminant analysis. In Table 2 the P-values of the discriminating functions obtained with different associations of the parameters analyzed are reported.

The analysis of the discriminating functions clearly indicates that while the lucigenin-dependent CL parameters do not generate any statistically significant discriminating functions, the luminol-dependent parameters, mainly those in presence of zymosan, generate two statistically significant functions.

Table 2. Discriminant analysis of CL parameters and PMN blood count

Variables	Function 1	Function 2	Function 3
	P value	P value	P value
LucBI, LucBT, LucPI, LucPT, PMN	NS	NS	NS
LumPI, LumPT, LumZI, LumZT, PMN	0.0024	NS	NS
LumZI, LumZT, PMN	0.0004	0.0272	NS
LumPI, LumPT, PMN	NS	NS	NS

Labels as in table 1

Because the discriminant analysis seems to indicate that anaesthesia is able to induce a modification of luminol-dependent CL in presence of zymosan without any modification of lucigenin-dependent CL, we compared the luminol-dependent zymosan-stimulated CL (LumZI) with the lucigenin-dependent resting CL (LucBI) measured at different time intervals from anaesthesia (Figure 1a: time<2 h; Figure 1b: time>2 h). A clear change of the slope of the regression line is obtained.

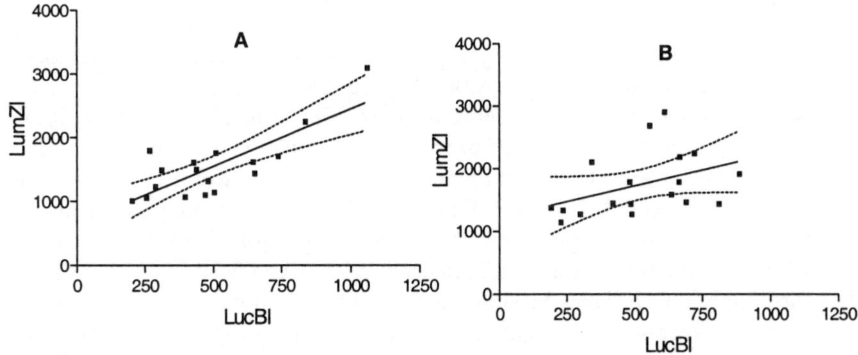

Figure1. Linear regression between zymosan-stimulated luminol-dependent CL (LumZI) and lucigenin-dependent basal CL (LucBI)
(A) CL values before and after 1 h of anaesthesia
(B) CL values after 3 and 24 h of surgery

DISCUSSION

Although intravenous anaesthetics can inhibit in-vitro the activity of PMNs,[1-4] their in-vivo effect on the phagocyte-dependent immune system is not clearly defined especially in cases of short term anaesthesia.[5-8] Our study indicates that PMN CL, if analyzed by an appropriate statistical test, seems to be modified also by short term intravenous anaesthesia. In fact, both discriminant analysis (Table 2) and regression analysis between luminol- and lucigenin-dependent CL (Figure 1) indicate that PMN CL is modified also after many hours of a short time intravenous anaesthesia. In particular the variation of regression slope between luminol-dependent zymosan stimulated CL and lucigenin-dependent basal CL could be indicative of a modification of the extra/internal production of reactive oxygen species associated to the respiratory burst. A possible mechanism of action for this long-term effect could be related to the hydrophobic interaction of propofol with cell membrane.[10]

Although the effects of intravenous anaesthesia on PMNs are temporary, their impact on subjects with immune deficit should not be underestimated. Moreover, the results of this study clearly show that an appropriate statistical test allows to get useful information also in presence of only few subjects.

REFERENCES

1. Murphy PG, Ogilvy AJ, Whiteley SM. The effect of propofol on the neutrophil respiratory burst. Eur. J Anesthesiol 1996; 13: 471-3.
2. Jaeger K, Scheinichen D, Heine J, Andre M, Bund M, Piepenbrock S, Leuwer M. Remifentanil, fentanyl, and alfentanil have no influence on the respiratory burst of human neutrophils in vitro. Acta Anaesthesiol Scand 1998; 42: 1110-3.
3. Mikawa K, Akamatsu H, Nishina K, Shiga M, Maekawa N, Obara H, Niwa Y. Propofol inhibits human neutrophil functions. Anest Analg 1998; 87: 695-700.
4. Nagata T, Kansha M, Irita K, Takahashi S. Propofol inhibits FMLP-stimulated phosphorilation of p42 mitogen-activated protein kinase and chemotaxix in human neutrophils. Brit J Anaesth 2001; 86: 853-8.
5. Sammartino M, Mignani V, Morelli-Sbarra G, Carducci P, Perotti V, Ranieri R, De Sole P, Fresu R. Effect of propofol on granulocyte function evaluated with chemiluminescence in surgical patients. Minerva Anestesiol 1991; 57: 594-5.
6. Erskine R, Janicki PK, Ellis P, James MF. Neutrophils from patients undergoing hip surgery exhibit enhanced movement under spinal anaesthesia compared with general anaesthesia. Can J Anaesth 1992; 39: 905-10.
7. Pirttikangas CO, Salo M, Peltola O. Propofol infusion anaesthesia and the immune response in elderly patiens undergoing ophthalmic surgery. Anaesthesia 1996; 51: 318-23.
8. Heine J, Jaeger K, Osthaus A, Weingaertner N, Munte S, Piepenbrock S, Leuwer M. Anaesthesia with propofol decreases FMLP-induced neutrophil respiratory burst but not phagocytosis compared with isoflurane. Brit J Anaesth 2000; 85: 424-30.
9. De Sole P, Lippa S, Littarru GP. Whole blood chemiluminescence: a new technical approach to assess oxygen dependent microbicidal activity of granulocytes, J Clin Lab Autom 1983; 3: 391-400.
10. Tsuchiya H. Structure-specific membrane-fluidizing effect of propofol. Clin Exp Pharmacol Physiol 2001; 28: 292-9.

CHEMILUMINESCENT MICROSPHERES FOR MEASURING REACTIVE OXYGEN SPECIES IN PHAGOCYTOSIS

S HOSAKA, Y HOSAKA, K ICHIMURA

Dept of Applied Chemistry, Tokyo Polytechnic University,
Atsugi 2430297 Japan

INTRODUCTION

Uchida et al covalently bound luminol to polymer microspheres and measured chemiluminescence (CL) in the phagocytosis of the luminol-bound microspheres by macrophages.[1,2] They concluded that the CL generated during the phagocytosis was caused by reactive oxygen species (ROS) reacting with microspheres undergoing phagocytosis, hence the intensity of the CL reflected the microbicidal activity of the cells.[3] Luminol-bound microspheres and an analogue, ABEI-bound microspheres, have been applied to the measurement of ROS produced in phagocytic cells in various cases.[4] It has not been clarified, however, which ROS were measured in this assay.

Recently, an analogue of *Cypridina* luciferin, CLA or methoxylated CLA (MCLA) has been preferred to luminol as a chemiluminescent probe by many researchers because of its high sensitivity to superoxide anion and singlet oxygen at the physiological pH. For example, Oosthuizen et al reported that CL response to superoxide anion or singlet oxygen was very low in either case when luminol was used as a probe because the pH optimum for this probe was above 9.0, and they preferred MCLA for physiological assessments.[5]

Following our previous work with luminol-bound polymer microspheres, we attempted to prepare new chemiluminescent microspheres containing CLA or its analogues. The properties of CLA immobilized polymer microspheres were previously reported.[6] In this paper, we report the immobilization of MCLA and FCLA (the conjugate of fluorescein and CLA) onto polymer microspheres.

MATERIALS AND METHODS

MCLA immobilized microspheres (MCLA-ms) were prepared as described below. As monomers, 5.60 g of methyl methacrylate, 7.67 g of 2-hydroxyethyl methacrylate, 1.02 g of tri(ethylene glycol) dimethacrylate were dissolved in 30 g of ethyl propionate, and then 30 mg of 2,2'-azobis(2,4-dimethyl-4-methoxyvaleronitril) was added to the solution as a polymerization initiator. The reaction solution was kept at 45 °C under argon atmosphere for an hour without agitation. The diameter of precipitated polymer microspheres ranged within 0.5~1.0 μm. In 10 mL of dimethyl sulfoxide, 10 mg of MCLA was dissolved and 200 mg of the polymer microspheres were suspended. The mixture was stirred at room temperature for 72 h. The microspheres were washed with water more than 10 times by repeated re-suspension and centrifugation. The amount of MCLA in the polymer microspheres was calculated to be 4.4 μmol/g from the nitrogen content that was 185 ppm.

FCLA immobilized microspheres (FCLA-ms) were prepared as described in the following. The microspheres of a copolymer of glycidyl methacrylate were prepared as described in the previous paper.[7] The diameter of the polymer microspheres was about 2 μm. Then, 11 g of sodium azide was dissolved and 1.1 g of the polymer microspheres was suspended in 110 mL of distilled water. The mixture was refluxed with stirring for 8 h; the polymer microspheres were washed with distilled water. The nitrogen content of thus azidated polymer microspheres was 17.0 %. In order to convert azide groups to amino groups in the polymer, 0.5 g of the polymer microspheres were suspended in 50 mL of distilled water and mixed with 5 mL of 17 % DL-dithiothreitol solution and stirred at 80 °C for 3 h. The nitrogen content was reduced to 5.66 % by this treatment. In 5 mL of 5 % $NaHCO_3$ solution, 5 mg of FCLA and 10 mg of 4-(4,6-dimethoxy-1,3,5-triazin-2-yl)-4-methylmorpholinium chloride (DMT-MM) were dissolved, then 50 mg of aminated polymer microspheres were suspended; the reaction mixture was stirred at room temperature for 4 h.

Chemiluminescence was measured in triplicate at 23 °C under moderate agitation by use of Luminescence Reader BLR-201 (Aloka Co. Ltd; Tokyo Japan). In a glass test tube, 100 μL of chemiluminescent probe suspension, 450 μL of distilled and deionized water, 100 μL of 10-fold concentrated PBS and 250 μL of 1 U/mL solution of xanthine oxidase from butter milk (XOD) were mixed; about 30 s after the start of CL measurement following the set of the test tube in the instrument, 100 μL of 1.67 mM hypoxanthine (HPX) solution was injected into the test tube through a rubber cap by a syringe. The initial concentration of MCLA or FCLA was 1 μM. The pH of the reaction mixture was 7.3 ± 0.1.

RESULTS AND DISCUSSION

As it was difficult to bind MCLA covalently to polymer microsphere, physical immobilization was studied. Figure 1 shows the typical time course of the CL intensity of MCLA-ms dispersion compared with that of the filtrate of the same dispersion when incubated with the same ROS generator. The accumulated values of the CL intensity in these measurements are shown in Table 1. The ratio of the accumulated value of the filtrate to that of the dispersion was 2.6%. The desorption of MCLA from the carrier polymer microsphere can be substantially neglected, although very weak CL was observed due to slight desorption.

Figure 2 shows the typical time course of the CL intensity of FCLA-ms dispersion compared with that of the filtrate of the same dispersion when incubated with the same ROS generator. The accumulated values of the CL intensity in these measurements are also shown in Table 1. FCLA was covalently bound to the polymer through amide group formed by the condensation of the carboxyl group of FCLA and the amino group of the polymer. As expected, FCLA was not desorbed from the carrier.

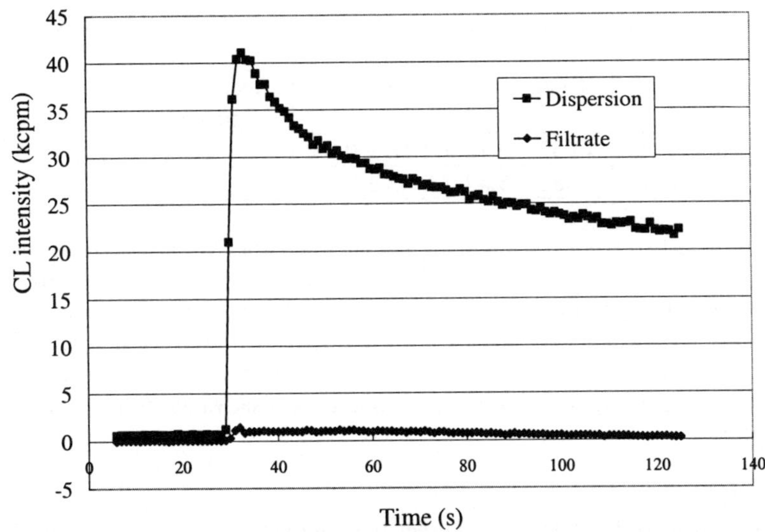

Figure 1. Time courses of the CL intensity of MCLA-ms dispersion and the filtrate, when ROS was generated by O_2/HPX/XOD.

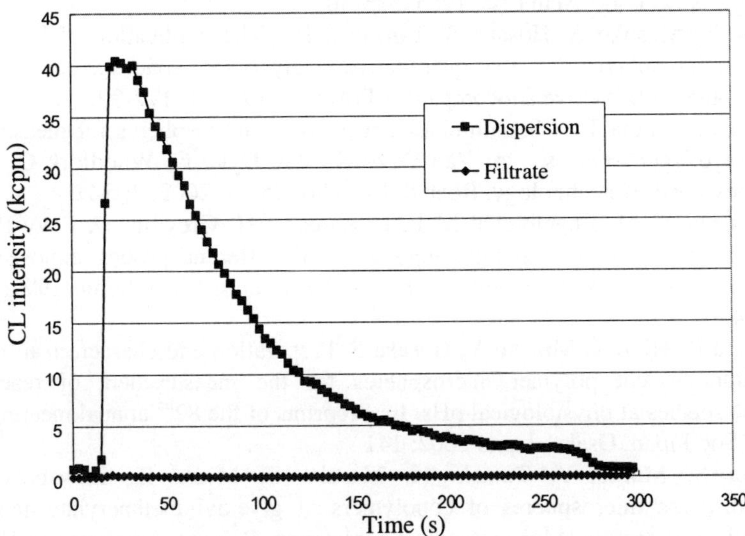

Figure 2. Time courses of the CL intensity of FCLA-ms dispersion and the filtrate, when ROS was generated by O_2/HPX/XOD.

Table 1. Accumulated values of chemiluminescence caused by O_2/HPX/XOD

	Dispersion	Filtrate	Accumulation time (s)
MCLA-ms	44.32 ± 2.16	1.16 ± 0.03	125
FCLA-ms	56.82 ± 5.42	0.42 ± 0.19	300

Either of these two kinds of chemiluminescent microsphere was poorly sensitive to singlet oxygen produced by endoperoxide of disodium (1,4-dinaphthylidine)-3,3'-dipropionate (NDPO$_2$), although both MCLA and FCLA themselves are highly sensitive to this ROS (data not shown). This property of these chemiluminescent microspheres is interesting from a point of view that it may be possible to measure O_2^- in distinction from $^1O_2$, while the same property is disadvantageous to measure all kinds of ROS.

REFERENCES

1. Uchida T, Kanno T, Hosaka, S. Direct measurement of phagosomal reactive oxygen by microsphere-bound luminol. J Immunol Methods 1985; 77: 55-61.
2. Uchida T, Hosaka, S. Direct measurement of phagosomal reactive oxygen by microsphere-bound luminol. In: Richards S, Kojima M. eds. Macrophage Biology. New York: Alan Liss, 1985: 545-50.
3. Uchida T, Masuko S, Hosaka S, Tanzawa, H. The application of luminol-bound microspheres for the quantitative analysis of toxic oxygen within phagosomes. J Bioactive Biocompatible Polymers 1986; 1: 172-80.
4. Hosaka S, Uchida T. Chemiluminescent polymer microspheres for measuring reactive oxygen species. In: Van Dyke K, Van Dyke C, Woodfork C. eds. Luminescence Biotechnology. Boca Raton: CRC Press, 2002: 305-20.
5. Oosthuizen M M, Engelbrecht M E, Lambrechts H, Greyling D, Levy R D. The effect of pH on chemiluminescence of different probes exposed to superoxide and singlet oxygen generators. J Biolumin Chemilumin 1997; 12: 277-84.
6. Ichimura K, Hirai T, Mizuno Y, Hosaka S. Preparation and characterization of chemiluminescent polymer microspheres for the measurement of reactive oxygen species at physiological pHs. In: Preprints of the 82nd annual meeting of Chem Soc Japan, Osaka, Japan 2002; 141
7. Hosaka S, Murao Y, Tamaki H, Masuko S, Miura K, Kawabata Y. Monodisperse microspheres of copolymers of glycidyl methacrylate and its derivatives as materials for biomedical application. Polymer International 1993; 30: 505-11.

EFFECTS OF CALORIC RESTRICTION AND AGING ON THE GENERATION OF REACTIVE OXYGEN SPECIES IN RAT LIVER MITOCHONDRIA AND PEROXISOMES

I IMADA, EF SATO, R KONAKA, M NISHIKAWA, Y KIRA,
A-M PARK, Q LI, M INOUE

Department of Biochemistry and Molecular Pathology, Osaka City University Medical School, 1-4-3 Asahimachi, Abenoku, Osaka, 545-8585 Japan
E-mail:imada@med.osaka-cu.ac.jp

INTRODUCTION

A significant fraction of molecular oxygen utilized in aerobic organisms is converted to reactive oxygen species (ROS) in and around mitochondria, endoplasmic reticulum, peroxisomes, and cytosol (Fig. 1).[1] Long lasting production of ROS results in the accumulation of DNA damage both in the nucleus and the mitochondria. Although caloric restriction has been shown to retard the aging process in various organisms,[2] the underlying mechanism remains unknown. The present work describes the effects of caloric restriction and aging on ROS generation by subcellular fractions of rat liver using a highly sensitive chemiluminescence (CHL) probe 8-amino-5-chloro-7-phenylpyrido[3,4-d]pyridazine-1,4-(2H,3H)dione, L012.[3]

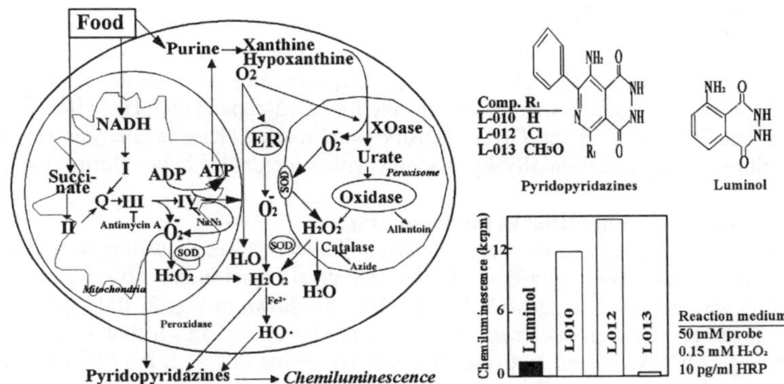

Figure 1. Cellular generation and metabolism of ROS in rat liver, and their determination by pyridopyridazines

MATERIALS AND METHODS

Luminol, lucigenin and MCLA were commercially obtained and L012 was from Takeda Chem. Ind. (Osaka, Japan). Male Wistar rats, 10-weeks-old were supplied from SLC (Shizuoka, Japan), housed under conventional conditions at 24±2 °C, and fed *ad libitum*. One group was subjected to starvation. Mitochondria were isolated from rat liver according to the method of Hogeboom using a medium containing 0.25 M sucrose, 10 mM Tris-HCl (pH 7.4) and 0.1 mM EDTA.[4] EDTA was omitted in the final wash and the mitochondrial samples were suspended in 0.25 M sucrose containing 10 mM Tris-HCl (pH 7.4) at 50-60 mg protein/mL.

Isolation of peroxisomes

Highly purified mitochondria and peroxisomes were prepared as described previously.[5] The liver tissue was homogenized in a solution containing 0.25 M sucrose, 1 mM EDTA, 0.1 % ethanol, 5 mM HEPES-KOH (pH 7.4), and 0.2 mM phenylmethylsulfonyl fluoride. The light mitochondrial fraction obtained by differential centrifugation of the homogenate was layered on top of a 12-mL Nycodenz linear gradient (density, 1.15 to 1.25 g/mL) with a 1-mL cushion of

Nycodenz (1.3g/mL) in an RP55VF tube (Hitachi, Tokyo). After centrifugation at 19,3000g for 90 min at 4 °C, fractions corresponding to mitochondria and peroxisomes were collected, and stored at -20 °C until use.

Assay for oxygen consumption and ROS generation

Oxygen consumption by isolated mitochondria was determined polarographically using a Clark type oxygen electrode fitted to a 2 mL water-jacketed closed chamber. Isolated mitochondria (1 mg protein/2 mL) were suspended in the reaction medium consisting of 0.2 M sucrose, 10 mM KCl, 1 mM $MgCl_2$, 2 mM sodium phosphate and 10 mM Tris-HCl (pH 7.4). Oxygen consumption was monitored in the presence of 5 mM succinate and 200 µM ADP. During the incubation, CHL intensity was recorded continuously for 10 to 15 min using a Luminescence Reader BLR-201 (Aloka, Tokyo, Japan). The mitochondrial suspensions (20-50 µL) were incubated at 25 °C for 1 min in 500 µL of reaction medium (pH 7.4) in the presence of either 200 µM L012, 1 mM luminol, or 4 µM MCLA. The reaction was initiated by adding 0.1 mM succinate or 0.1 mM glutamate, 0.2 mM ADP, 0.1mM azide and 1 mM urate.

RESULTS

Analysis of ROS generation by chemiluminescence probes

Pyridopyridazine compounds, L010, L012 and L013 were found to develop CHL in a hydrogen peroxide-horseradish peroxidase system.[6] We found that , among various probes used, L012 developed the strongest CHL under the present experimental conditions; its intensity was higher than that of luminol by about ten fold (Fig. 1). L012 is a luminol-like cyclic hydrazide derivative having 2,3-dihydro-1,4-pyridopyridazine dione in its structure. Accordingly, L012 is expected to develop CHL by stabilizing its oxidatively formed dianion. To elucidate the chemical nature of the reactive species responsible for L012-dependent CHL, effects of various scavengers and inhibitors on the CHL intensity were studied. Kinetic analysis revealed that L012 developed CHL preferentially by reacting with superoxide, hypochlorite and hydroxyl radical.[3]

Respiration and ROS generation in mitochondria

In the presence of substrate and inorganic phosphate, mitochondria showed typical state III and IV respiration after ADP addition. Under identical conditions, ROS generation by mitochondria was analyzed using MCLA and L012. As shown in Fig. 2, ROS generation was increased in the presence of succinate and glutamate by a mechanism that was inhibited by ADP. Although MCLA showed a strong CHL similar to that of L012, the blank value for the former was extremely high. Thus, L012 permits studies on the ROS generation by mitochondria under different respiratory states.

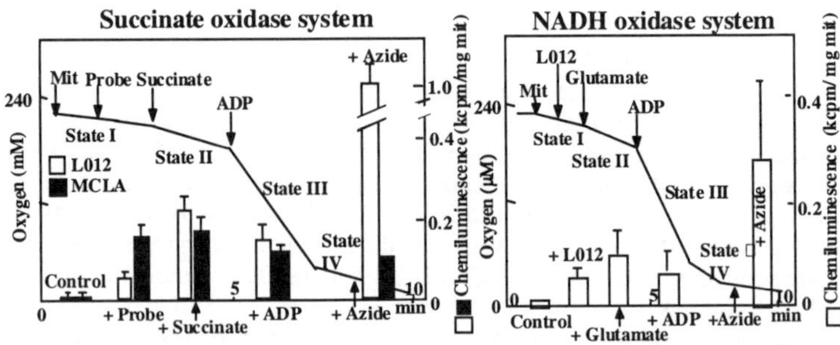

Figure 2. Mitochondrial respiration and ROS-dependent chemiluminescence with various probes

L012-CHL was increased by respiratory substrates via a mechanism that was inhibited by either SOD or deferoxamine, suggesting that superoxide and hydroxyl radicals were generated during state II respiration. Inhibition of mitochondrial complex IV by azide further increased the L012-CHL. ESR analysis revealed the generation of hydoxyl radical by a mechanisms that was inhibited by the presence of ascorbate. (data not shown)

ROS generation in peroxisomes

Peroxisomes are often cofractionated in the light mitochondrial fraction. Because of the presence of uricase, peroxisomes generate ROS in the presence of urate. Kinetic analysis using various CHL probes revealed that urate-dependent ROS generation could be analyzed by using urate oxidase (Fig. 3).

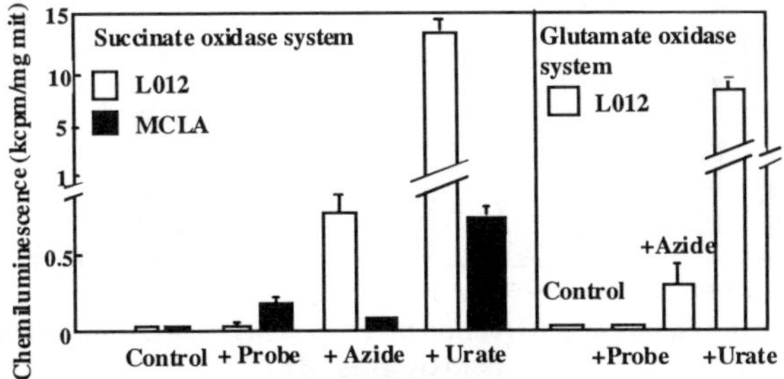

Figure 3. Determination of urate-dependent ROS generation in mitochondrial fractions with various probes

To evaluate the contribution of peroxisomes in the L012-CHL observed with mitochondrial fractions, we used highly purified peroxisomes and mitochondria.[5] As expected, in the presence of urate, highly purified peroxisomes also showed L012 CHL as strongly as that observed with the mitochondrial fraction. The peroxisome-free mitochondria did not develop L012-CHL even in the presence of azide and urate (Fig. 4A). Although uric acid is a potent scavenger for hydroxyl radical, mitochondrial fractions were generally contaminated with peroxisomes which contain urate oxidase and, hence, they formed hydrogen peroxide and hydroxyl radical in the presence of urate. As shown in Fig. 4B, biochemical analysis confirmed that peroxisomal uricase in the mitochondrial fraction was responsible for the enhanced ROS generation particularly in the presence of urate. The CHL intensity was suppressed by ascorbate but not by SOD. It has been reported that various hemoproteins play a role in the removal of hydrogen peroxide.

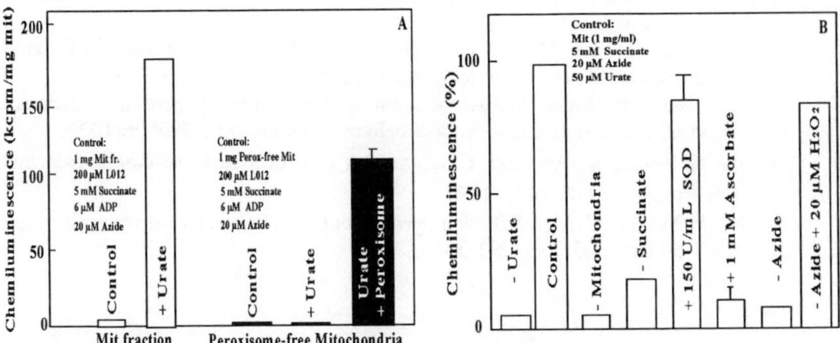

Figure 4. Urate dependent ROS generation by mitochondria and peroxisomes

Western blotting analysis also revealed the presence of urate oxidase in the crude mitochondrial fractions in our experiments (data not shown).

Effects of caloric restriction and aging on ROS generation in subcellular fractions
To clarify the effects of aging and caloric restriction on the relationship between energy
metabolism and ROS generation, properties of liver mitochondria from young (19-weeks-old)
and aged (two-years-old) rats were analyzed with and without starvation. Although
mitochondrial generation of ROS was higher with young than with aged rats, it decreased
significantly in both animal groups after starvation for 2 to 3 days (Fig. 5A).

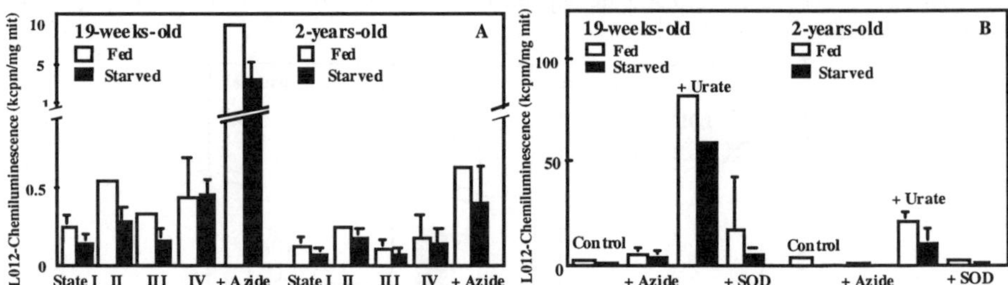

Figure 5. Effects of aging and starvation on ROS generation by mitochondria (A) and
peroxisomes (B)

ROS generation by peroxisomes was also higher with young rats than with aged animals. The
CHL   decreased in both animal groups after starvation for 2 to 3 days even in the presence of
urate (Fig. 5B).

These results suggest that ROS generation is greater in and around mitochondria and
peroxisomes in young rats than in aged animals and that caloric restriction suppresses ROS
generation in mitochondria but not in peroxisomes in both animal groups.

REFERENCES

1.   Boveris A, Oshino N, Chance B. The cellular production of hydrogen peroxide. Biochem
     J 1972; 128: 617-30.
2.   Lee C-K, Pugh TD, Klopp RG, Edwards J, Allison DB, Weindurch R, Prolla TA. The
     impact of α-lipoic acid, coenzyme Q_{10} and caloric restriction on life span and gene
     expression patterns in mice. Free Rad Biol Med 2004; 36:1043-57.
3.   Imada I, Sato EF, Miyamoto M, Ichimori Y, Minamiyama Y, Konaka R, Inoue M.
     Analysis of reactive oxygen species generated by neutrophils using a chemiluminescence
     probe L-012. Anal Biochem 1999; 271: 53-8
4.   Nishikawa M, Sato EF, Kuroki T, Inoue M. Role of glutathione and nitric oxide in the
     energy metabolism of rat liver mitochondria. FEBS Lett 1997; 415: 341-5.
5.   Kira Y, Sato EF, Inoue M. Association of Cu,Zn-type superoxide dismutase with
     mitochondrial and peroxisomes. Arch Biochem Biophys 2002; 399: 96-102.
6.   Masuya H, Kondo K, Aramaki Y, Ichimori Y. Pyridopyridazine compounds and their
     use. Eur Patent Appl 49,1477, 1992.
7.   Venditti P, Masullo P, Meo SD. Hemoproteins affect H_2O_2 removal from rat tissues. Int J
     Biochem Cell Biol 2001; 33: 293-301.

NONCOMPETITIVE INHIBITION OF LANTHANIDE-INDUCED OXIDATIVE BURST BY ZINC IN TOBACCO BY-2 CELLS: A CHEMILUMINESCENT ANALYSIS

T KAWANO, T KADONO, SC YANG, S MUTO

[1]*Graduate School of Environmental Engineering, The University of Kitakyushu, Kitakyushu 808-0135, Japan*
[2]*Nagoya University Bioscience Center, Nagoya Univ, Nagoya 464-8601*
Email: kawanotom@env.kitakyu-u.ac.jp

INTRODUCTION

Cations of Al and rare earth elements (REEs) are phyto-toxic.[1] The toxicity of such cations has been summarized as followings: (1) metal cations induce the production of superoxide (O_2^-) in plant cells; (2) the metal cations with higher valence induce greater oxidative burst (OXB); (3) with higher ion valence, the concentration required for maximal response is minimized; (4) the induced OXB is sensitive to inhibitors of NADPH oxidase.[2] To date, non-redox metal cations such as those of Al,[3] La,[1,4] Ce,[5] and Gd,[1,4] have been shown to induce OXB.

It has been shown that Zn^{2+} inhibits the REE-induced OXB.[4] Zn is normally present in plants at high level,[6] and its deficiency is one of the most widespread micronutrient deficiencies in plants, causing severe reductions in crop production.[7] Increasing studies indicate that oxidative damage to plants caused by reactive oxygen species (ROS) results from a deficiency of Zn.[7,8] A hypothetical model explains that the O_2^--generating activity of NADPH oxidase is sensitive to Zn^{2+}, thus the REE-dependent stimulation could be inhibited or retarded by Zn^{2+}.[4] In this study, we tested the effect of Zn^{2+} supplementation on REE-induced OXB in tobacco cell culture, measured with the O_2^--specific chemiluminescence (CL) of a *Cypridina* luciferin analog (CLA). CLA-CL specifically indicates the generation of O_2^- (and $^1O_2$ with a lesser extent) but not that of other ROS.[9]

Here, the mode of Zn action against the REE-induced OXB was assessed with Lineweaver-Burk kinetics and possible eco-physiological roles for Zn in interaction with other metal cations in plants are discussed.

MATERIALS AND METHODS

CLA (2-methyl-6-phenyl-3,7-dihydroimidazo[1,2-*a*]pyrazin-3-one) was purchased from Tokyo Kasei Kogyo Co. (Tokyo, Japan). All other reagents were from Sigma (St. Louis, MO, USA).

BY-2 tobacco culture was propagated as previously described.[3] Cells were harvested 1-12 d after sub-culturing, and used for experiments after addition of 10 μM CLA. LaCl$_3$ solution (various concentration) was added to cell suspension in glass tubes placed in a CHEM-GLOW Photometer (AMINCO., Silver Spring, MD, USA), and OXB was monitored by CLA-CL, and expressed as relative CL units (rcu) as previously described.[2] Cells were treated with ZnSO$_4$ (0.3, 1, 3 mM), 2 min prior to addition of $LaCl_3$.

For Lineweaver-Burk kinetic analysis, the reciprocals of CLA-CL yield (1/rcu) were plotted as function of the reciprocals of La^{3+} conc. $(1/[La^{3+}])$. Data in the absence and presence of Zn^{2+} were co-plotted and the mode of Zn^{2+} action was graphically analyzed.

RESULTS AND DISCUSSION

Effect of culture age on the sensitivity to La^{3+} was examined using the differently aged cultures (Fig. 1). In young cultures, OXB was induced by lower concentration (0.1-1 mM) of La^{3+}, but higher La^{3+} concentration was shown to be inhibitory for unknown reasons. Mature cultures (5-12 d-old) were shown to be less sensitive to low La^{3+} concentration but inhibition by higher La^{3+} conc. was no longer observed. In the range of concentration examined, La^{3+}-induced OXB was dose-dependent in the aged cultures.

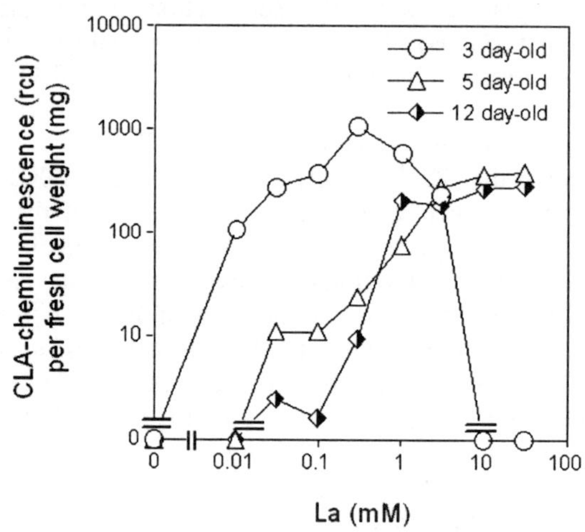

Figure 1. Effect of culture age on the La-induced OXB in tobacco culture

Making use of linear dose-dependency in the 5 d-old culture, the *in vivo* kinetic analysis was carried out by assuming La^{3+} as a ligand and Zn^{2+} as an inhibitor (Fig. 2). The reciprocals of CLA-CL yield were plotted against the reciprocals of La^{3+} concentration $(1/[La^{3+}])$. Here, V_{max} for the La^{3+}-induced response in the absence of Zn^{2+} was calculated to be 58.8 rcu. By 0.3, 1.0 and 3.0 mM of Zn^{2+}, V_{max} was lowered to 30.3, 15.6 and 10.2 rcu, respectively. In the competitive interactions, the V_{max} should not be altered.[10] Therefore the mode of Zn^{2+} action against La^{3+} does not correspond to a characteristic competitive inhibition. The apparent K_m values obtained for La^{3+} in the presence of 0, 0.3, 1.0 and 3.0 mM Zn^{2+} were 3.12, 3.42, 3.37 and 4.17 mM, respectively. Given that K_m is lowered only in the uncompetitive interactions, while in the noncompetitive interactions the apparent K_m should not be altered.[10] The K_m for La^{3+} was not lowered by any concentration of Zn^{2+}, thus strongly indicating that the mode of Zn^{2+} action against La^{3+}-induced OXB is not uncompetitive inhibition. Here we can conclude that the mode of La-Zn interaction observed here is most likely noncompetitive inhibition.

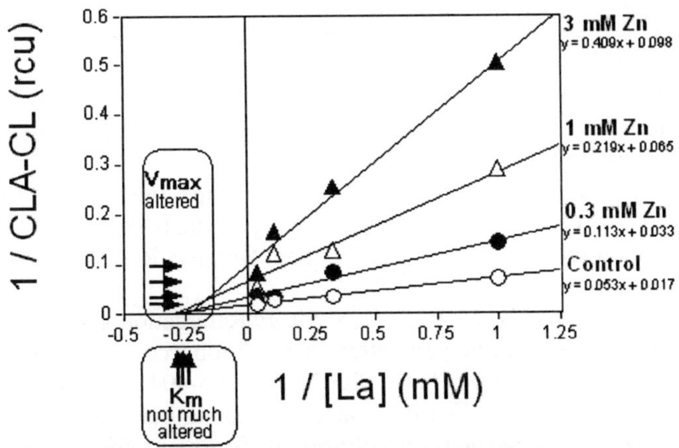

Figure 2. Noncompetitive inhibition of the La-induced OXB by Zn

There may be a possible receptor for REE on the plant cell surface (most likely NADPH oxidase),[2] and binding of REE to the receptor possibly leads to an immediate OXB. The noncompetitive mode of interaction between La and Zn implies that Zn may bind to both the La-bound form and La-free form of the putative receptor, thus expressed as; R + I ↔ RI, RL + I ↔ RLI; where R is the receptor, L is a ligand (La), I is an inhibitor (Zn). RI, RL and RLI are receptor complexes with other factors. This result suggests that the site of Zn binding on the target molecule is different from the REE binding site. In the future, this model must be examined in the purified and reconstituted O_2^--generating systems.

Effect of Zn^{2+} on protection of cells from oxidative damage has been reported in human systems. For an example, inhibitory action of Zn^{2+} on apoptotic cell death is tightly related to its role in protecting cell membranes and DNA from damaging attack by ROS.[11] In addition, Zn^{2+} exerts a strong inhibitory effect on the OXB by NADPH oxidase.[12,13] Interestingly, Zn action against cation-induced OXB in human has been reported. The O_2^--generating NADPH oxidase from phagocytic membrane was shown to be activated in the presence of divalent cations (Ca^{2+}, Mg^{2+}), and this activity was shown to be highly sensitive to Zn^{2+}.[14]

Several reports have suggested that plant NADPH oxidases are also possible targets of Zn-dependent inhibition.[8] Despite numerous studies examining the impact of Zn-deficiency, the impacts of Zn^{2+} supplementation on NADPH oxidase-dependent OXB in living plant cells or tissue have not been tested until recently. Our previous work provided the first evidence that Zn^{2+} inhibits and the metal cation (including REEs)-stimulated NADPH oxidase-dependent OXB in living plant cells.[4] Zn^{2+} may therefore protect the cells from the metal cation-dependent oxidative damages.

REFERENCES

1. Kawano T. Biological actions of rare earth element ions in plants: a review. ITE Lett 2003; 4: 44-8.
2. Kawano T, Kawano N, Muto S, Lapeyrie F. Cation-induced superoxide generation in tobacco cell suspension culture is dependent on ion valence. Plant Cell Environ 2001; 24: 1235-41.
3. Kawano T, Kadono T, Furuichi T, Muto S, Lapeyrie F. Aluminum-induced distortion in calcium signaling involving oxidative bursts and channel regulations in tobacco BY-2 cells. Biochem Biophys Res Commun 2003; 308: 35-42.
4. Kawano T, Kawano N, Muto S, Lapeyrie F. Retardation and inhibition of the cation-induced superoxide generation in BY-2 tobacco cell suspension culture by Zn^{2+} and Mn^{2+}. Physiol Plantar, 2002; 114: 395-404.
5. Yuan YJ, Li JC, Ge ZQ, Wu JC. Superoxide anion burst and taxol production induced by Ce^{4+} in suspension cultures of *Taxus cuspidate*. J Mol Catal B: Enzym 2002; 18: 251-60.
6. Santa Maria GE, Cogliatti DH. Bidirectional Zn-fluxes and compartmentation in wheat seedling roots. J Plant Physiol 1998; 132: 312-15.
7. Cakmak I. Possible roles of zinc in protecting plant cells from damage by reactive oxygen species. New Phytol 2000; 146: 185-205.
8. Cakmak I, Marschner H. Enhanced superoxide radical production in roots of zinc deficient plants. J Exper Bot 1988; 39: 1449-60.
9. Nakano M, Sugioka K, Ushijima Y, Goto T. Chemiluminescence probe with *Cypridina* luciferin analog, 2-methyl-6-phenyl-3,7-dihydroimidazo[1,2-a]pyrazin-3-one, for estimating the ability of human granulocytes to generate O_2^-. Anal Biochem 1986; 159: 363-9.
10. Dixon M, Webb, EC. Enzymes, 3rd edn. New York: Academic Press, 1979.
11. Parat MO, Richard MJ, Poller S, Hadjur C, Favier A, Beani JC. Zinc and DNA fragmentation in keratinocyte apoptosis: first inhibitory effect in UV B irradiated cells. J Photochem Photobiol B Biol 1997; 37: 101-6.
12. Hammermüller JD, Bray TM, Bettger WJ. Effect of zinc and copper deficiency on microsomal NADPH-dependent active oxygen generation in rat lung and liver. J Nutr 1987; 117: 894-901.
13. Bray TM, Bettger WJ. The physiological role of zinc as an antioxidant. Free Rad Biol Med 1990; 8: 281-91.
14. Suzuki H, Pabst MJ, Johnston RBJr. Enhancement by Ca^{2+} or Mg^{2+} of catalytic activity of the superoxide-producing NADPH oxidase in membrane fractions in human neutrophils and monocytes. J Biol Chem 1985; 260: 3635-9.

LOPHINE PEROXIDES AS AN EFFICIENT
ORGANIC SOURCE OF SINGLET OXYGEN

M KIMURA, G LU, H IGA, H NISHIKAWA

Department of Chemistry, Okayama University, 700-8530, Japan

kimuram@cc.okayama-u.ac.jp

INTRODUCTION

Singlet oxygen ($^1O_2$) is a molecular oxygen species in an excited state. It is widely recognized as a reactive species both in organic systems and in biological systems. As singlet oxygen sources are limited, any new efficient source is desirable. Lophine peroxide **1a** (2,4,5-triphenyl- 4-hydroperoxy-4H-isoimidazole) is derived from lophine **2a** by photo-oxygenation, and it has been only known as a chemiluminescent system for many years. However, recently lophine peroxides have been known as a singlet oxygen source. This paper will discuss the generation of $^1O_2$ from lophine peroxides and the substituents effect on formation of $^1O_2$. In order to examine the effect of 2-(p-)substituents on the formation of $^1O_2$, lophine peroxides **1a**, 2-(p-nitrophenyl)-4,5-diphenyl-4-Hydroperoxy- 4H-isoimidazole **1b** and 2-(p-dimethyl-aminophenyl)-4,5-diphenyl- 4-Hydroperoxy-4H-isoimidazole **1c** were studied. (Scheme 1)

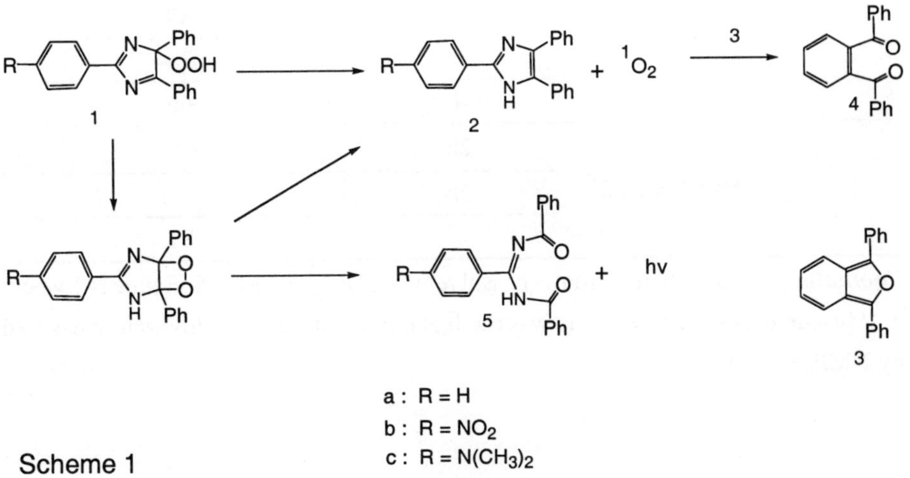

a : R = H
b : R = NO2
c : R = N(CH3)2

Scheme 1

303

METHODS

The sensitized photo oxygenation would be an effective method to prepare the desired hydroperoxides **1a-1c**.[1,2] A mixed solution of **2a** in dichloromethane and a catalytic amount of methylene blue in methanol was cooled to –78 ℃, and then irradiated with a sunlamp through UV-cutoff filter under an oxygen atmosphere for 1.5 h, which was followed by TLC. **2a** was exclusively transformed into **1a**. After the chromatographic purification to eliminate the sensitizer, **1a** was isolated as colorless granules in 93% yield.[1] The structure of **1a** was characterized by ^{1}H NMR, ^{13}C NMR, IR, and mass spectral analysis. The other imidazoles, **2b** and **2c**, were similarly oxygenated to give the corresponding hydroperoxides **1b** and **1c** in high yields.[2]

RESULTS

In order to quench $^1O_2$, 1,3-diphenylisofuran **3** was added in the reaction systems. o-Dibenzoylbenzene **4** was generated as the quenching product. The results are summarized in Table 1.

Table 1. The results of quenching with **3**[a,b]

Entry	Molar ratio	Products	Conversion (%)
1[c]	**1a : 3 = 1 : 1**	2a	56
		5a	34
		4	97
2[c]	**1a : 3 = 1 : 2**	2a	68
		5a	32
		4	66
3[d]	**1b : 3 = 1 : 2**	2b	84[e]
		5b	-
		4	81

Footnote:- a. Room temperature. b. Solvent is $CDCl_3$. c. Reaction time is 1 week. d. Measured by NMR. e. This was the isolated yield, other yields were measured by NMR.

A solution of **1** and **3** in $CDCl_3$ was transferred into an NMR tube. The solution was de-aerated for 1 min with supersonic wave and replaced with N_2 for 3 min. The NMR tube was sealed 1 week for **1a** or 2 weeks for **1b**.

Product **4** is formed by two successive reactions with **3**. A singlet oxygen molecule produces two molecules of **4** as shown in Scheme 2. The results in entry 1 and 2 confirmed such a process.

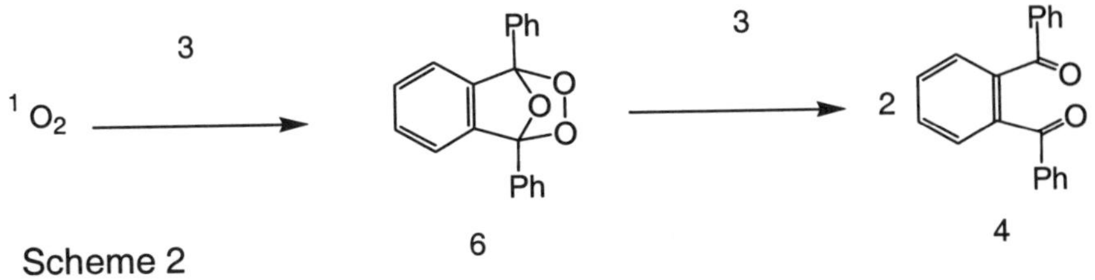

Scheme 2

By comparing the conversions of **1a** and **1b**, it was clear that **1b** generated more $^1O_2$ than **1a**. The NO_2 group attributed to the efficient formation of $^1O_2$. The reaction of **1b** was slower than that of **1a** due to the electron-attracting effect of nitro group. The effect of $N(CH_3)_2$ group had been studied in our lab.[3] **1c** was unfavorable to the formation of $^1O_2$ and more efficient chemiluminescent system than **1a**.[3,4]

CONCLUSIONS

1a and **1b** are good $^1O_2$ sources. The electron-attracting group (NO_2) contributes to the formation of $^1O_2$ in high yield, while the electron-attracting group [$N(CH_3)_2$] has a contrary effect.

REFERENCES

1.   Davidson D, Weiss M, Jelling M. The action of ammonia on benzil, J Org Chem 1937: 2: 319-27.

2.   Cook AH, Jones DG. Experiments in the triazine and the glyoxaline series. J Chem Soc 1941; 278-82.

3. Kimura M, Nishikawa N, Kura H, Lim H, White EH. Maximization of the chemiluminescence efficiency of 1,4,5-triarylhydroperoxy-4H-isoimidazoles, Chem Lett 1993: 505-8.

4. Boyatzis S, Nikokavouras J. Lophines in micellar environments: Spectroscopic behaviour and Chemiluminescence. J Photochem Photobiol A: Chem 1993; 74: 65-73.

EFFECTS OF FORCED EXERCISE STARTED FROM DIFFERENT AGES ON CHEMILUMINESCENT RESPONSE AND CYTOKINE EXCRETION OF ALVEOLAR MACROPHAGE

T KUMAE[1], H ARAKAWA[2]

[1]*Div of Health Promotion & Exercise, National Institute of Health & Nutrition, 1-23-1 Toyama, Shinjuku-ku, Tokyo 162-8636, Japan*
[2]*Dep of Education & Training Technology, National Institute of Public Health, 2-3-6 Minami, Wako, Saitama 351-0197, Japan*
Email: kumae@nih.go.jp

INTRODUCTION

The respiratory system is the principal organ in air-breathing animals. Lungs have a large surface area in order to perform the gas exchange and airborne antigens present in the air enter the lungs. The alveolar macrophage (AM) is situated at the air-tissue interface in the alveoli and alveolar ducts, and is the first cell type to encounter inhaled foreign materials. AMs are the most important phagocytic cells residing in the lungs, and can release cytokines in the lungs. Cytokines are potent intercellular molecules that regulate inflammation and immune responses. Cytokines act locally at extremely low concentrations in a paracrine or autocrine manner, and can also cause a priming of phagocytic cells such as neutrophils and AMs.

Numerous studies of physical exercise and immunity have been published in the past few years. Recent research trends indicate that moderate exercise may increase immune responsiveness, but high-level competition sport, especially if it involves extensive endurance training, may lead to a degree of immunosuppression. We previously reported the measurement technique for chemiluminescence using a parallel luminometer with a cooled charge-coupled device (CCD) camera together with a new software which decrease background noise.[1] The principal objective of this study is prevention or decreasing the risk of injury/disease in healthy aged persons engaging in endurance training for the purpose of increasing their health levels. In this study, to evaluate effects of training started from different ages on non-specific immunity, we measured reactive oxygen species (ROS) generation from rat AM by the chemiluminescent technique and also measured cytokine concentrations in supernatants of AM cultured medium.

MATERIALS AND METHODS

Male Wistar JcI strain (SPF, five weeks old) rats were obtained from Nippon Clea Breeding Laboratories. The rats were maintained in SPF condition on a 12-h light-dark cycle, and were allowed rodent food (F-2; Funabashi Farms) and water ad libitum during the experiment. We divided the rats into following three groups; first group of rats started training at 5 weeks old (Group A), second group of rats started at 11 weeks old (Group B), and third group started at 17 weeks old (Group C). Each group was divided into 2 sub-groups: 1) forced training group (Forced); exercised on a treadmill (35 m/min, 30 min/day, 5 days/week) and 2) control group (Control); housed under sedentary condition, same as usual breeding condition for rats. Training by a treadmill was continued for 12 weeks. After 6- or 12-week training, the rats were anesthetized with pentobarbital sodium (40 mg/kg, ip), and AMs were collected by the broncho-alveolar lavage method.[1] Cells in the broncho-alveolar lavage fluid (BALF) were counted differentially according to cell diameter, i.e., 6.0 to 8.0 µm, 8.0 to 10.0 µm, 10.0 to 12.0 µm, and 12.0 to 16.0 µm, by a Coulter Counter ZM (Coulter Electronic Inc.). AMs were suspended in the Dulbecco's MEM after adjustment of cell numbers to 6×10^5 cells/mL. AMs were placed in each well of a 96-well black microtiter plate (Greiner Japan Co.) and were cultured for 12 h. Supernatants of AMs were collected and stored for the cytokine measurements. Lucigenin-dependent chemiluminescence (LgCL) were measured by a parallel luminometer (Alpha-Basic 47, Tokken Inc.) using opsonized zymosan as stimuli. The chemiluminescence was indicated by peak height (PH) and peak time (PT). Concentrations of tumor necrosis factor-alpha (TNFa), interleukin-1-beta (IL-1b), and interferon-gamma (IFNg) were measured by ELISA kits (BioSource Inc.).

RESULTS AND DISCUSSION

The Forced groups in the all three groups were significantly smaller than the Control groups in body weights after 6- and 12-week training.

The Control groups showed significant changes in the cell populations in BALF. Total cell numbers in BALF increased significantly at 11 weeks old. The cell populations in smaller size cells, diameters 6.0 to 8.0 µm and 8.0 to 10.0 µm, were decreased significantly after 11 weeks old. To the contrary, the mean levels of larger size cells, diameter 10.0 to 12.0 µm, were increased significantly after 11 weeks old. In the Group A, after 6-week training, the level of total cell number in BALF of the Forced group was decreased significantly than that of the Control group. Significant differences in the cell populations between the Control and Forced groups were observed in the Groups A and B. These results suggest that the 11 weeks old is the turning point of maturation in the lung immunity, and the forced exercise affects the cell populations in younger animals.

In this study, LgCL was employed for the detection of superoxide generation as the indicator of AM activity. Changes of chemiluminescent response of AM are shown in Table 1. The mean levels of PT of LgCL were almost same levels. In the Control group, no significant difference in the mean levels of PT was observed in the entire experiment. The mean level of PT in the 6-week training group of Group C was delayed significantly (p < 0.05) than that of the Control group. Without this exception, no practical difference among the Control and Forced groups was observed in the mean levels of PT. To the contrary, the mean levels of PH of LgCL in the Control groups were significantly enhanced according to ageing. Except for Group B, there was no practical difference in PH between the Control and Forced groups. The mean level of PH in the 12-week training group of Group B was suppressed significantly.

Table 1. Changes of chemiluminescent response

Group	Control		Forced (A)		Forced (B)		Forced (C)	
Age	PT	PH	PT	PH	PT	PH	PT	PH
5 weeks	32.7	491	32.7	491	--	--	--	--
11 weeks	30.4	493	28.1	526	30.4	493	--	--
17 weeks	30.8	655	29.1	676	29.2	732	30.8	655
23 weeks	34.0	797[##]	--	--	32.8	499**	37.1*	888
29 weeks	33.0	1015[###]	--	--	--	--	38.7	1142

Numbers represent mean values.

PT: min, PH: counts/sec

Statistical significances between the Control and Forced groups are represented as follows: *; p<0.05, **; p<0.01

Statistical significances between the 5 weeks and other weeks old in the Control group are represented as follows: [##]; p<0.01, [###]; p<0.001

Changes of cytokine excretions of AM are summarised in Table 2. In the Control group, the mean level of TNFa of the 17 weeks old was increased significantly (p < 0.05) than that of the 5 weeks old. In the Group B, the mean level of TNFa of the Forced group was decreased significantly (p < 0.05) after 6-week training but increased significantly (p < 0.01) after 12-week training than that of corresponding Control group. It has been reported that AMs are capable of an enhanced production of TNFa contrary to blood monocytes (MOs). TNFa released from AMs was 2- to 3-fold more abundant than from MOs. On the other hand, AMs produce considerably less IL-1b than their precursors, MOs. IL-1b secretion is down-regulated when MOs mature into AMs. However, in the Control group, no significant difference among the different age groups was observed in the mean levels of IL-1b. In the Group B, the mean level of IL-1b of the Forced group was decreased significantly (p < 0.05) after 12-week training than that of the Control group. The mean levels of IFNg were practically same levels in the Control group. The 6-week training group in the Group A and the 12-week training group in the Group B showed significantly increased levels of IFNg than the corresponding Control group, p < 0.01 and p < 0.01, respectively. It has been studied that AMs function to a cytocidal or inflammatory state is regulated coordinatedly by production of endogenous cytokines. These primed cells can become activated to release ROS. AM activation, however, is complicated and specific. The significant suppression of the mean level of PH observed in the 12-week training group of Group B may be reflected these changes of cytokine excretions.

Table 2. Changes of cytokine excretions

Group	Cytokine (pg/mL)	Age (weeks)				
		5	11	17	23	29
Control	TNFa	1398	2320	3428[#]	2544	2330
	IL-1b	110.4	80.4	83.4	101.6	67.5
	IFNg	8.9	7.6	8.1	10.8	5.2
Forced (A)	TNFa	1398	3847	3949	--	--
	IL-1b	110.4	88.6	87.4	--	--
	IFNg	8.9	9.8*	11.1	--	--
Forced (B)	TNFa	--	2320	2451*	4780**	--
	IL-1b	--	80.4	76.6	64.2*	--
	IFNg	--	7.6	9.2	13.4**	--
Forced (C)	TNFa	--	--	3428	3836	1982
	IL-1b	--	--	83.4	80.2	62.5
	IFNg	--	--	8.1	10.4	5.3

Numbers represent mean values.
Statistical significances between the Control and Forced groups are represented as follows: *; p<0.05, **; p<0.01
Statistical significance between the 5 weeks and other weeks old in the Control group is represented as follows: [#]; p<0.05

Largest number of cells and changes of cell populations in BALF were observed at 11 weeks old. Furthermore, significant increase of TNFa, decrease of IL-1b, and increase of IFNg between the Control and the 12-week training groups of Group B were observed. These results suggest that forced training started from the 11 weeks old, is thought to be the turning point of lung immunity, affect AM activity, and the chemiluminescent technique is useful to evaluate the changes of AM activity.

REFERENCES

1.   Kumae T, Arakawa H. Assessment of training effects on activity levels of alveolar macrophage in matured rats using chemiluminescent technique. Luminescence 2003; 18: 61-6.

VISUALIZATION OF SUPEROXIDE GENERATED FROM COLONIES OF *CANDIDA ALBICANS*

S MASUI[1], T MAJIMA[1], S ITO-KUWA[2], K NAKAMURA[2], S AOKI[2]
*[1]POLA Chemical Industries Inc, 560 Kashio-cho, Totsuka-ku,
Yokohama, 244-0812, Japan*
*[2]Advanced Research Center, Nippon Dental University, 1-8 Hamaura-cho
Niigata, 951-8580, Japan*
E-mail: smasui@pola.co.jp

INTRODUCTION

Studies reporting reactive oxygen species (ROS) production are much less common for mycetes than phagocyte cells. Aoki, *et al.* reported that paraquat (methyl viologen) induced respiration-dependent ROS production in *C. albicans*.[1] This work undertakes the visualization of ROS generation in *C. albicans,* using a photon-imaging instrument with a luciferin analogue.

METHODS

The strain, *C. albicans* K was isolated from a patient with oral candidiasis, and maintained on PYG (2% polypepton-1% yeast extract-2% glucose) agar.

Chemiluminescence images of the colony

The strain was precultured overnight in liquid PYG medium, at 37 °C with shaking. Cultures were then diluted, and 0.1 mL of the diluent, corresponding to roughly 50 cells was spread on PYG agar plates. Following incubation at 37 °C for 1 to 5 d, plates were observed under an ultra-low light image analyzer (ARGUS-50, Hamamatsu Photonics, Hamamatsu, Japan) equipped with a photon-counting CCD camera (C2400-30H). After photographing illuminated colonies under light, a mixture of 0.1 M PQ (paraquat, Nacalai Tesque, Japan) and 0.05 mM MCLA (chemiluminescence probe methyl-*Cypridina*-luciferin analogue (Tokyo Kasei, Japan)) (1:1) was gently layered onto the colonies. To examine the effects of SOD (super oxide dismutase from bovine erythrocytes, Sigma, USA), the enzyme was added to the PQ-MCLA mixture at 40 units/mL. The MCLA-dependent chemiluminescence due to ROS generated by the colonies was recorded for 5 min in a light-tight box. Measured chemiluminescence intensities were processed by the ARGUS software and displayed using pseudo-color images.

Chemiluminescence measurement

Photon counting was carried out using a previously described method.[1] In brief, cells were cultured and collected at the log phase and again at the stationary phase. The cells in a glass tube were placed in the ALOKA chemiluminescence reader, and MCLA and PQ were added sequentially. Observed chemiluminescence was expressed as count in counts per min (CPM).

Chemiluminescence images of 96-microtiter plate
Cells cultured in PYG liquid medium were divided into two tubes and collected via centrifugation. One tube was resuspended in a proline medium[2] to shift the morphology, while the other was resuspended in PYG liquid medium. Both tubes were then incubated at 37 °C, for 1 h, with shaking. 100 μL of the cells were pipetted into a 96-microtiter plate at $2x10^7$cells per well and added with 100 μL of probe solution (MCLA or MCLA+PQ). Observation was performed in the same manner as with colonies.

RESULTS AND DISCUSSION
Photon emission of MCLA-dependent chemiluminescence generated by colonies on PYG agar plates was observed, in parallel with the increase in the colony size after incubation for 1,3 and 5 d (Fig.1A-1C). Especially in the marginal regions (Fig. 1D-1F). However, weak photon emission was also observed in colonies treated with MCLA alone (Fig. 1 arrow head). These results indicate that *Candida* colonies

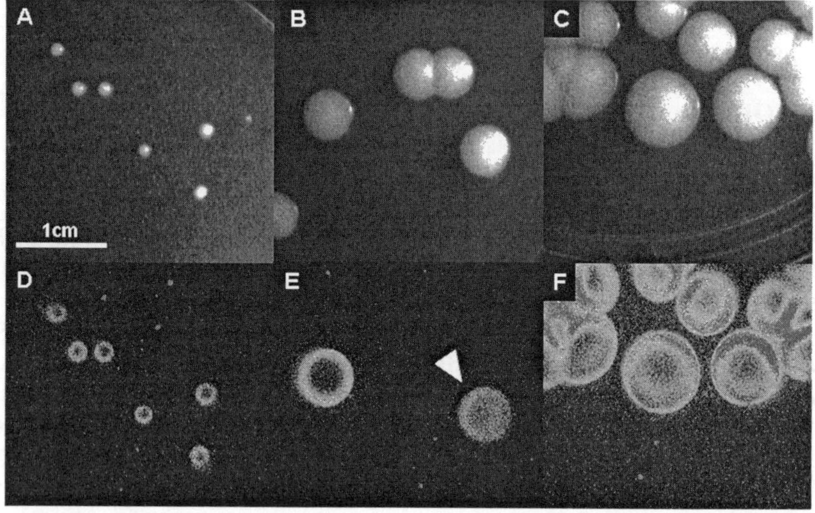

Figure 1. Colonies of *C. albicans* strain K and their chemiluminescence image

expand by division of metabolically active cells in the marginal regions, leaving aged cells in the central regions. The photon emission from colonies vanished after the addition of O_2^- scavenger, SOD (Fig. 2B). This result confirms that the observed light emission was due to superoxide anion. Application of the antioxidant, L-cysteine, at a concentration of 100 mM also extinguished colony photon emissions (Fig. 2D).

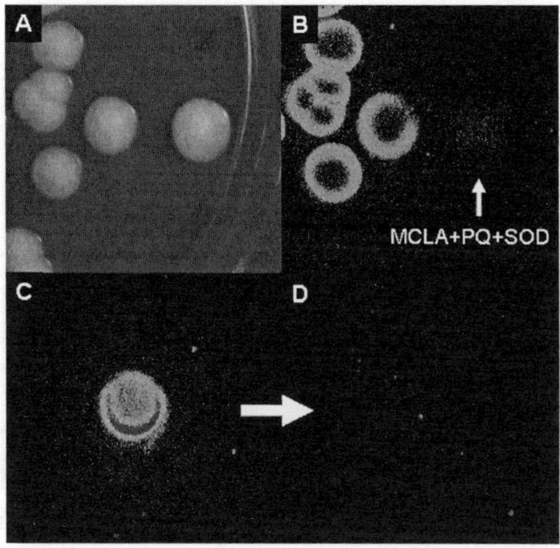

Figure 2. Effect of SOD and L-cysteine on photon emission by colonies

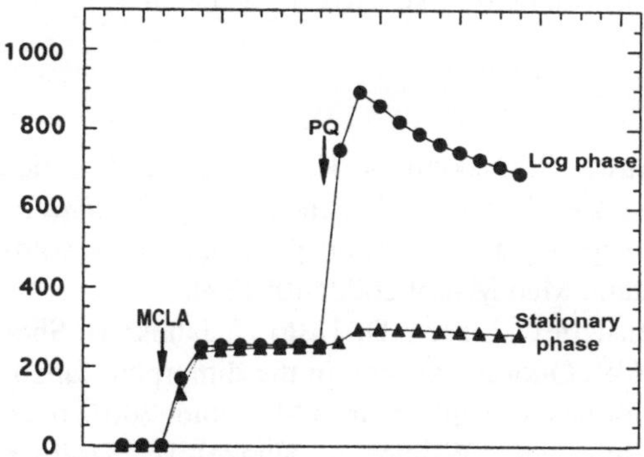

Figure 3. Chemiluminescence in log phase and stationary phase

As measured by the luminometer, the chemiluminescence intensity observed in log phase cells was greater than the amount in the stationary phase (Fig. 3).

Fig. 4 shows emission of MCLA-dependent chemiluminescence generated by *Candida* cells in the 96-microtiter plate. Strong photon emission was observed in both hyphal form and yeast form cells, treated with a mixture of MCLA and PQ. Strong photon emission observed in hyphal form cells treated with MCLA alone indicated endogeneous superoxide generation without stimulation by PQ. Schröter *et.al.* reported that the formation of hyphae, usually known as a major factor in *Candida* pathogenicity, was associated with markedly increased ROS formation.[3]

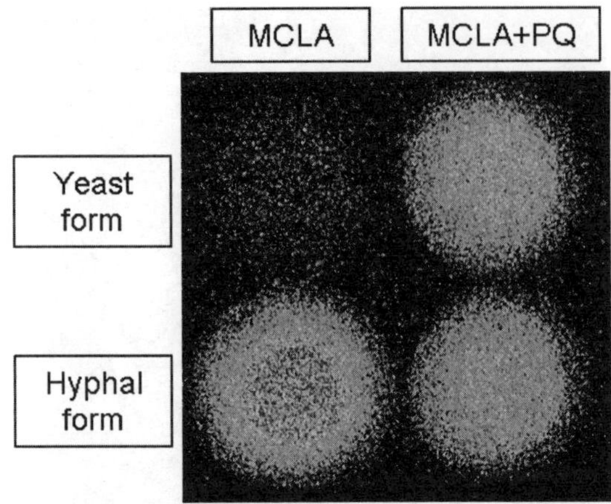

Figure 4. Chemiluminescence image by cells in 96-microtiter plate

To our knowledge, the present report is the first chemiluminescent visualization of ROS including superoxide generated by the hyphal form of *C. albicans*.

REFERENCES

1. Aoki S, Ito-kuwa S, Nakamura K, Nakamura Y, Vidotto V, Takeo K. Chemiluminescence of superoxide generated by *Candida albicans*: differential effect of the superoxide generator paraquat on a wild-type strain and a respiratory mutant. Med Mycol 2002; 40: 13-9.
2. Hornby JM, Jensen EC, Lisec AD, Tasto JJ, Jahnke B, Shoemaker R, Dussault P, Nickerson KW. Quorum sensing in the dimorphic fungus *Candida albicans* is mediated by farnesol. Appl Environ Microbiol 2001; 67:2982-92.
3. Schröter C, Hipler UC, Wilmer A, Künkel W, Wollina U, Generation of reactive oxygen species by *Candida albicans* in relation to morphogenesis. Arch Dermatol Res 2000; 292: 260-4.

EFFECTS OF VARIOUS ANTIFUNGAL AGENTS ON REACTIVE OXYGEN SPECIES GENERATION BY *CANDIDA ALBICANS*

S MASUI, T MAJIMA

POLA Chemical Industries Inc, 560 Kashio-cho, Totsuka-ku,
Yokohama, 244-0812, Japan
E-mail: smasui@pola.co.jp

INTRODUCTION

It is known that *Candida albicans* can generate reactive oxygen species (ROS), and that some antifungal agents affect the ROS generation. In this study, the effects of seven antifungal agents (terbinafine (allylamine), amorolfine (morfolin), ciclopirox olamine (hydroxypyridone), and four imidazole antimycotics, including miconazole, bifonazole, sulconazole and clotrimazole) on the ROS generation by *C. albicans* were examined on a 96-microtiter plate, using a fluorescence analyser.

MATERIALS AND METHODS

The strain *C. albicans* K was isolated from a patient with oral candidiasis. *C. albicans* ATCC 10231 was purchased from American Type Culture Collection. Cells ware maintained on PYG agar, comprised of 1% polypepton, 0.5% yeast extract, 1% glucose, and 1.5% agar.

Terbinafine (TBF) as an allylamine was purchased from Wako pure chemical (Osaka, Japan). Ciclopirox olamine (CPO), and four imidazole antimycotics, including miconazole (MCZ), bifonazole (BFZ), sulconazole (SCZ) and clotrimazole (CTZ) were purchased from Sigma (St. Louis, MO, USA). Amorolfin (AMF) was synthesized in our laboratories and identified by nuclear magnetic resonance. These agents were dissolved in dimethyl sulfoxide (DMSO) at various concentrations and added at a final concentration less than 1%.

ROS was measured via fluorometric assay with 2',7'-dichlorofluorescin diacetate (DCFH-DA; Sigma). *C. albicans* cells (hereafter referred to as 'cells') were cultured overnight in PYG liquid medium at 37 °C with shaking overnight and collected by centrifugation (1500 rpm, 10 min). The cell pellet was washed in Dulbecco's Phosphate-Buffered Saline without Mg^{2+}, Ca^{2+} (PBS), centrifuged, and resuspended with PBS. The cell suspension was adjusted to $2x10^7$ cells/mL and DCFH-DA was added at 10 µM. After incubating 15 ml of the cell suspension for 1 h at 37 °C in a 50 mL conical centrifuge tube, 100 µL volumes were pipetted into a 96-microtiter plate, with 100 µL volumes of PBS containing antifungal agent. Plate was then incubated for 4 h at 37 °C.

The fluorescence intensity of these wells was measured with a fluorometer (excitation: 485 nm; emission: 538nm; Walloc ARVO SX 1420, Perkin Elmer), and the mean value of duplicate samples was calculated.

RESULTS AND DISCUSSION

In order to investigate the profile of ROS generation, miconazole was added and fluorescence was measured. *C. albicans* strain K cells were incubated with various concentrations of miconazole. The resulting fluorescence intensity increased throughout the incubation period. The measurement point was thus set at 4 h after the start of the incubation (Fig.1). Fig. 2 shows representative experimental

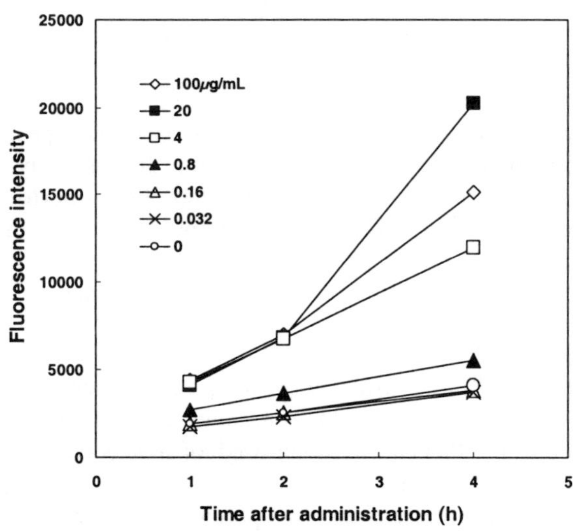

Figure 1. Time course of ROS generation by miconazole-treated
C. albicans K

measurements of the effect of antifungal agents on ROS generation by *C. albicans* strain K. These results demonstrate that: terbinafine effected suppression, amorolfine had little effect, miconazole and other imidazoles effected enhancement, and ciclopirox olamine effected extreme enhancement. This effect of terbinafine and miconazole has previously been reported.[1,2] Note that similar results on ROS generation were observed in *C. albicans* ATCC 10231 cells (data not shown). To examine the participation of glucose metabolism in ROS generation, glucose (10mM) was added to a separately prepared 96-microtiter plate. Although for 6 of the antimycotics little affect was observed on glucose additive ROS generation as compared to the basal levels, with CPO it effected suppression in strain K. An investigation of the effect of glucose metabolism on CPO induced ROS in *C. albicans* ATCC 10231 was also conducted. Fig. 3 shows the adverse results of the glucose addition on ciclopirox olamine (100 µg/mL)-induced ROS in strain K and ATCC 10231.

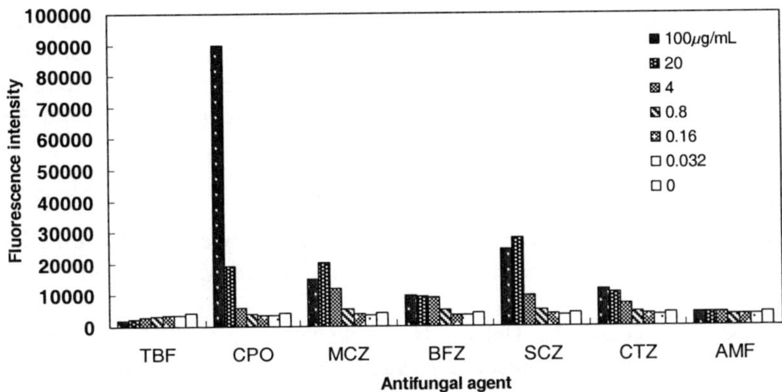

Figure 2. Difference in the antifungal agents induced ROS generation by
C. albicans K

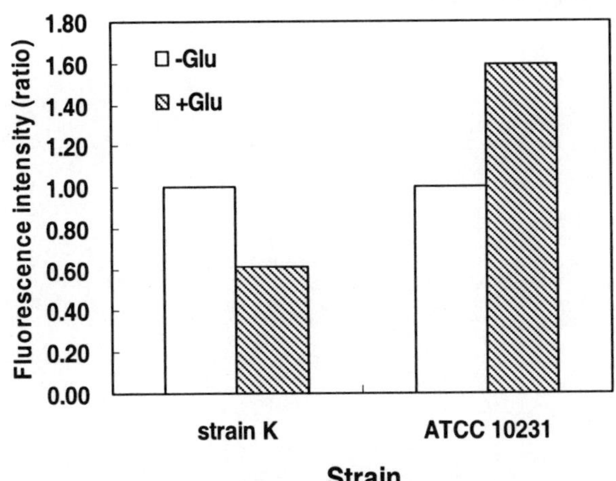

Figure 3. Effects of glucose on a CPO-induced ROS generation
in *C. albicans* K and ATCC 10231

CPO may act via chelation of trivalent metal cations (e.g., Fe^{3+}), for which it has a high affinity, or may act to inhibit metal-dependent enzymes (e.g., catalase/peroxidase), which participate in the intracellular degradation of peroxides[3] resulting in an increase in ROS generation.

Schröter *et al.*[4] reported that the ROS generation by *C. albicans* was affected by medium components. In particular, ROS generation was suppressed in YEPG

318

medium. Although the presence of antioxidative substitutes in the medium was not excluded, glucose included in the YEPG medium might act as an inhibitor for ROS generation, as observed with strain K in this study. Further studies are required to determine the cause of different CPO responses of *C. albicans* strain K and ATCC 10231.

REFERENCES

1. Sander CS, Hipler UC, Wollina U, Elsner P, Inhibitory effect of terbinafine on oxygen species (ROS) generation by *Candida albicans*. Mycoses 2002; 45: 152-5.
2. Kobayashi D, Kondo K, Uehara N, Otokozawa S, Tsuji N, Yagihashi A, Watanabe N. Endogenous reactive oxygen species is an important mediator of miconazole antifungal effect. Antimicrob Agents Chemother 2002; 46: 3113-7.
3. Gupta AK, Ciclopirox: an overview. Int J Dermatol 2001, 40: 305-10.
4. Schröter C, Hipler UC, Wilmer A, Künkel W, Wollina U, Generation of reactive oxygen species by *Candida albicans* in relation to morphogenesis. Arch Dermatol Res 2000; 292: 260-4.

CHEMILUMINESCENT ANALYSIS OF HYDROGEN PEROXIDE GENERATION FROM NATURAL ANTIMICROBIAL MATERIALS

N MATSUO[1], K SOMEYA[1], Y UEDA[1], H ARAKAWA[2], M MAEDA[2]

[1]*Lion Corporation, 100 Tajima, Odawara 256-0811, Japan*
[2]*Pharmaceutical Sciences, Showa University, Shinagawa-ku 142-8555, Japan*
Email: matsu705@lion.co.jp

INTRODUCTION

There are many kinds of natural antibacterial materials, which are commonly known and used in our daily lives. However, their antibacterial mechanisms are not clearly understood. Recently, it was reported that the catechin contained in green tea, a natural antimicrobial material, generates hydrogen peroxide (H_2O_2) continuously, in solutions using the peroxyoxalate chemiluminescent reaction.[1] This was established as a highly sensitive analytical method for H_2O_2 generation. In addition, the antimicrobial activities of tea catechin were higher than those of H_2O_2, in the same molar concentration.[1] We assumed that the generation of H_2O_2 from the other natural materials was one of the more important mechanisms of their antibacterial activity, in addition to tea catechin. To confirm this hypothesis, we investigated H_2O_2 generation from natural plant extracts that exhibited bacteriocidal activities by employing the chemiluminescent reaction technique.

MATERIALS AND METHODS
Chemicals

Alpha-terpinene, was from Extrasynthese (Genay, France). Epigallo-catechin-3-gallate (EGCg), was from Kurita Water Industries (Tokyo, Japan). Bis (2,4,6-trichlorophenyl) oxalate (TCPO), was from Tokyo Chemical Industries, Inc (Tokyo, Japan). Bovine serum albmin (BSA), was from Sigma Chemical Co. (St.Louis, USA). Tea tree oil, was a gift from Toyotama Koryo Co. (Tokyo, Japan). Other chemicals, were purchased from Wako Pure Chemical Industries (Osaka, Japan).

Preparation of standard solution for EGCg and H_2O_2

Standard solutions of EGCg (1 mmol/L) in water were freshly prepared and serially diluted with water. A stock solution of H_2O_2 (0.1 mol/L) in water was prepared and stored at 4 °C until used. The working solution was diluted with a 50 mmol/L phosphate buffer at pH 7.8 (PB) or a 100 mmol/L carbonate buffer at pH 10.0 (CB), immediately prior to use.

Preparation of sample solution for detection of H_2O_2 in plants extract

The dried plant leaves or roots (10 g) were immersed in a 100 mL solution, containing 30% or 100% ethanol, for 5 days at room temperature and were shaken intermittently. The supernatant was separated from the mixture by filtration and concentrated *in vacuo*. An adequate volume of ethanol was added to the concentration to prepare a 5% plant extract. The extract was diluted to an arbitrary concentration with PB or CB, and then incubated for 30 min at room temperature. Subsequently, the dilution was measured with the peroxalate chemiluminescence detection system. The concentration of H_2O_2 in the plant extract was determined by comparison to the standard curve obtained with the diluted H_2O_2 solution.

Measurement of antibacterial activity

Standard strains of *Staphylococcus aureus* (IFO12732) and *Escherichia coli* (IFO3972), cultured in SCD agar, and a 10^9 cfu/mL bacterial solution was prepared using 10% glycerol solution. The bacterial solution was diluted 10 times with a 100 mmol/L phosphate buffer (pH 7 or pH 10). Each plant extract was added to the solution, and the final concentration was 0.25%. These mixtures were incubated at 37 °C for 24 h. After the incubation period, the reacted mixtures were again diluted with fresh SCD broth on a 96-well plate. These plates were then incubated at 37 °C for 2 days. The turbidity (595 nm) of the broth in the plate was measured with a plate reader (Spectra image, Tecan, Austria), and the number of sterilized cells were estimated. We were able to define the antimicrobial materials that were able to sterilize over 10^4 cfu/mL of the bacteria used in this procedure.

Peroxalate chemiluminescence detection methods for H_2O_2

A sample solution (0.1 mL) was mixed with 0.1 mL of ANS solution (pH 9.0), which was prepared from 0.02% ANS and 0.1% BSA, 0.2 mol/L barbital, and 0.2 mL of 5 mmol/L TCPO in an ethyl acetate solution. The intensity of the luminescence was measured for 1 min after a period of 30 s, following mixing of the sample solution with a chemiluminescence analyzer CLA-FS1 (Tohoku Electronic Industrial Co., Sendai, Japan).

Identification of the H_2O_2 generated from natural plant extract.

A sample solution (0.1 mL) was mixed with a 0.1 mL ANS solution and a 0.2 mL TCPO in an ethyl acetate solution. Twenty seconds after the luminescence measurement was taken, 0.6 units of catalase were microinjected into the solution. We monitored the changes in the luminescence intensity and determined the H_2O_2 generation based on the decrease in the intensity.

RESULTS AND DISCUSSION

Antibacterial effect of natural plant extract

We measured the antibacterial properties of natural plant extracts against *E.coli* or *S.aureus*. Many of the plants showed antimicrobial activity (Table 1). The number of plant extracts that showed an antibacterial effect against gram-positive bacteria, was higher than that against gram-negative bacteria, at both pH 7 and pH 10. On the other hand, there were a larger number of plant extracts at pH 7 that had antimicrobial activity than at pH 10.

Table 1. Antibacterial activity and H_2O_2 generation of natural plant extracts

Samples	pH7			pH10		
	Antibacterial activity		H_2O_2 generation ratio (EGCg=1)	Antibacterial activity		H_2O_2 generation ratio (EGCg=1)
	Gram+	Gram-		Gram+	Gram-	
Melaleuca alternifolia	+	+	1.34	+	+	5.32
Rosa spp.	+	+		+	+	0.25
Coptis japonica	+	+		+	+	0.36
Magnolia officinalis	+	+	1.03	+		1.23
Sophora flavescens	+	+		+		1.17
Phellodendron amurense	+	+	0.72		+	0.09
Origanum vulgare	+	+	2.17			
Pimenta officinalis	+	+	1.13			
Mosla chinensis	+	+	1.53			
Laurus nobilis	+	+	0.58			
Amomum xanthioides	+	+				
Cnidium monnieri	+		0.97	+		0.40
Thymus vulgaris	+		0.76	+		1.30
Rosmarinus officinalis	+		0.96	+		1.18
Mahonia aqnifolium		+			+	0.23
Salvia officinalis	+		0.75			
Glycyrrhiza glabra	+		0.64			
Houttuynia cordata	+		0.35			
Angelica dahurica	+					
Atractylodis lancea	+					
Cayratia japonica	+					
Coleus forskohlii	+					
Psoralea corylifolia	+					
Solidago altissima	+					
Sophora japonica	+					
Zingiber officinale	+					
Caryophyllus aromaticus		+	0.76			

*Symbol "+" represent antibacterial activity which sterilized over 10000 cfu/mL.

Generation of H_2O_2 from natural plant extract

We investigated the H_2O_2 generation from natural plant extract, which showed antibacterial activities, using the peroxyoxalate chemiluminescent reaction.[1] As shown in Table 1, a number of plants showed H_2O_2 generation and several plants exhibited a higher H_2O_2 generation than that of EGCg (e.g. *Origanum vulgare*, *Mosla chinensis*, *Melaleuca alternifolia*, *Pimenta officinalis*, *Magnolia officinalis*). A previous study[1] reported that tea catechin could generate H_2O_2. Since these plants do not include significant quantities of catechins, the mechanism of H_2O_2 generation in these plants may differ from that in tea catechin. On the other hand, the luminescent intensity was inhibited by the addition of catalase (Fig. 1). The results prove that the luminescent intensity originated from the H_2O_2.

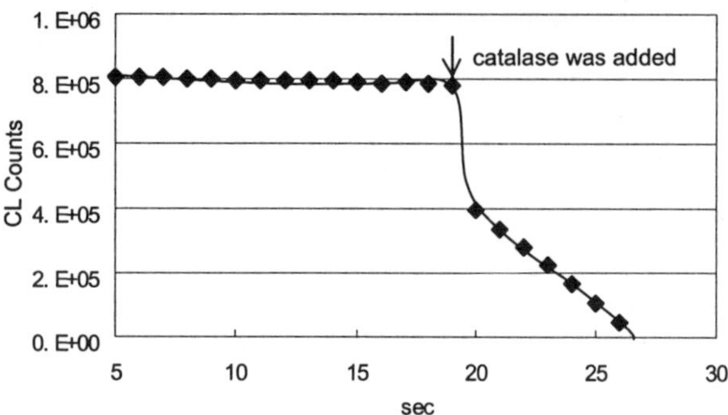

Figure 1. Reduction of chemiluminescent count from natural
plant extract by the addition of catalase.

Correlation between the antibacterial spectrum and H_2O_2 generation

In addition, we investigated the correlation between the antibacterial spectrum and H_2O_2 generation in these plant extracts. Consequently, the materials that generated a large amount of H_2O_2, showed antimicrobial effects against both gram-positive bacteria and gram-negative bacteria (Table 1). We conclude that several natural antimicrobial materials were able to generate H_2O_2, and this generation was considered to be one of the more important mechanisms that provide an antibacterial effect.

Effect of the components in *Melaleuca alternifolia* on H_2O_2 generation

We then measured H_2O_2 generation from terpinen-4-ol, gamma-terpinene, Alpha-terpinene, 1,8-cineole, and myrcene, which were ingredients in *M. alternifolia*.[2] However the entire extract showed a high H_2O_2 generation at both pH 7 and pH 10, and we could not confirm a distinct generation of H_2O_2 by the components. We must consider the possibility that the other components in *M. alternifolia* could contribute the generation of H_2O_2.

REFERENCES

1.   Arakawa H, Maeda M, Okubo S, Shimamura T. Role of hydrogen peroxide in bactericidal action of catechin. Biol Pharm Bull 2004, 27:277-81.
2.   Inoue Y, Shiraishi A, Hada T, Hamashima H, Shimada J, The antibacterial effects of myrcene on *Staphylococcus aureus* and its role in the essential oil of the tea tree (*Melaleuca alternifolia*). Natural Medicines 2004, 58:10-4.

INDUCTION OF DIFFERENTIATION IN HL-60 HUMAN LEUKEMIA CELLS BY HEMA, TEGDMA AND BIS-GMA

G NOCCA[1], P DE SOLE [1], G GAMBARINI[2], C CHIMENTI[3],
F DE PALMA[1], B GIARDINA[1,4], A LUPI[4]

[1]*Biochemistry and Cl Biochemistry Inst, School of Medicine,
Catholic University, Rome, Ialy*
[2]*School of Dentistry, La Sapienza Uniersityv, Rome, Italy–*
[3]*Chirurgical Sciences Dept, School of Medicine, L'Aquila University, Italy*
[4]*Ist. di Chimica del Riconoscimento Molecolare, C.N.R., Rome,-Italy*

INTRODUCTION

A number of chemically distinct compounds induce differentiation of the HL-60 human promyelocytic cell line[1] as, dimethyl-sulfoxide (DMSO),[2] 1,25-dihydroxy-vitamin D_3,[3] and bezafibrate.[4] During differentiation, promyelocytic HL-60 cells stop growing, become much smaller in size[5] and acquire the ability to form reactive oxygen species (R.O.S.) as shown by the appearance of the phorbol-12-myristate-13-acetate (PMA)-stimulated respiratory burst, detect by the chemiluminescence technique and other methods.[6]

In the last generation endodontic and restorative materials molecules such as 2-hydroxyethylmethacrylate (HEMA), triethyleneglycol dimethacrylate (TEGDMA) and 2,2-bis[4-(2-hydroxy-3-methacryloxy)-phenyl]propane (Bis-GMA) are present. In the clinical use these compounds are chemically or photochemically polymerised and the degree of polymerization - which is never complete[7]- determines the release of some quantity of uncured monomers in the pulpar cavity. Thus, the pulpal tissues can be exposed - through the dentinal diffusion - to these compounds that may cause inflammatory reactions and cellular damage, as confirmed by several reports.[8]

In this study we show that HEMA, TEGDMA and Bis-GMA act as differianting agents on HL-60 cell line.

MATERIALS AND METHODS

Unless indicated all chemicals and reagents (cell culture grade) were obtained from Sigma Chemical Co., Milan, Italy.

Cells culture

Human leukemic HL-60 cell line (Ist. Zooprofilattico, Brescia, Italy) was maintained at 37 °C under a humidified atmosphere of 5% CO_2 in RPMI 1640 with 10% (v/v) heat inactivated Fetal Calf Serum, 100 units mL^{-1} penicillin, 100 μg mL^{-1} streptomycin, and 2 mmol L^{-1} glutamine.

Cell viability

Cell viability was determined using trypan blue dye exclusion test.

All-trans retinoic acid and monomers treatment

The stock solutions were prepared immediately before use. TEGDMA, Bis-GMA, ATRA were dissolved in DMSO at various concentrations. HEMA was added at various concentrations ranging from 0.11 mM to 1.10 mM. The HL-60 exponentially growing cells (2×10^5 cells/mL) were set at day 0 in RPMI 1640 containing 1×10^{-3} mmol/L ATRA or different concentrations of each monomer for five days. The final concentration of the monomers were as follows: TEGDMA: 3.1×10^{-5}, 0.1×10^{-3}, 0.2×10^{-3}, 0.4×10^{-3}, 0.6×10^{-3}, 0.775×10^{-3}, 1.55×10^{-3} and 3.1×10^{-3} mol/L; Bis-GMA: 8.0×10^{-6} and 16×10^{-6} mol/L; HEMA: 0.11×10^{-3} and 1.10×10^{-3} mol/L; ATRA : 1.0×10^{-6} mol/L. The final DMSO concentration was the same in all samples (with the exception of HEMA treated cells) during the experiments (0.1% v/v).

Differentiation assay

The reactive oxygen species production, stimulated by PMA, essentially due to the activity of NADPH oxidase system, was adopted as cellular functional character in human HL-60 cell line. R.O.S. metabolism was studied by CL assay as described.[9] Assays were performed in triplicates at 25 °C. CL system contained in 1.00 mL final volume with modified KRP solution: 100 nmoles luminol, 1×10^5 cells without treatment (control) or treated with ATRA or monomers, in presence or absence of 1.50 nmoles PMA. The CL parameter considered for analysis was treated index: [Photons signal produced by treated cells]/[Photons signal produced by untreated cells] %.

Statistical analysis

All results are expressed as mean $\pm$ SD or $\pm$ SEM. The group means were compared by analysis of variance (ANOVA) followed, when appropriate, by a comparison of means by Student-Newman-Keuls test. $p < 0.05$ was considered significant.

RESULTS

Cell viability and inhibition of cell growth

Figure 1 shows the effect of different concentrations of xenobiotic agents on growth of HL-60 cells cultured for five days.

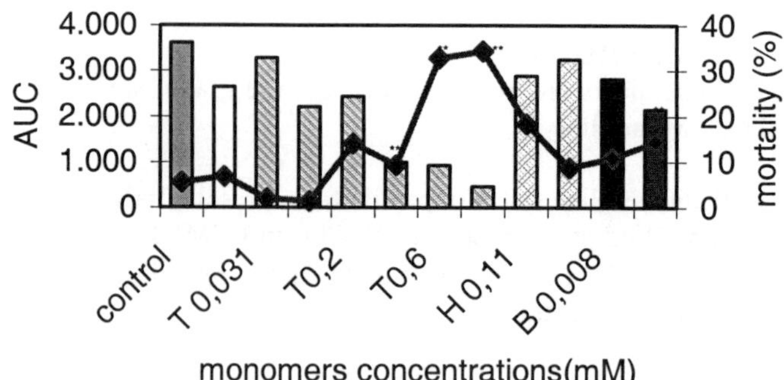

Figure 1. AUC and cell mortality of treated and untreated cells.**p<0.01

TEGDMA (at concentrations higher than 0.2 mM) and HEMA at concentration of 1.1 mM showed a significant inhibitory effect on growth curve of HL-60 cell line (p<0.01)In the same experimental conditions, ATRA 1 μmol/L, used a positive control, reduced cell growth of HL-60 about 23% (p<0.05). The cytotoxic activity of different monomers is calculated as percentage of dead cells during 96 hours of incubation.

The obtained results (Figure 1) elicited that the inhibitory effect of TEGDMA on cell growth depend on a cytotoxic action only at concentrations of 0.6 mM and 0.775 mM (p<0.01) while Bis-GMA action seems to be independent on a direct cytotoxic action of the monomer. In Figure 2 the CL response of HL-60 cell line is shown, results clearly indicate that HL-60 cells recover the oxidative burst in a dose-dependent manner, after stimulation with PMA.

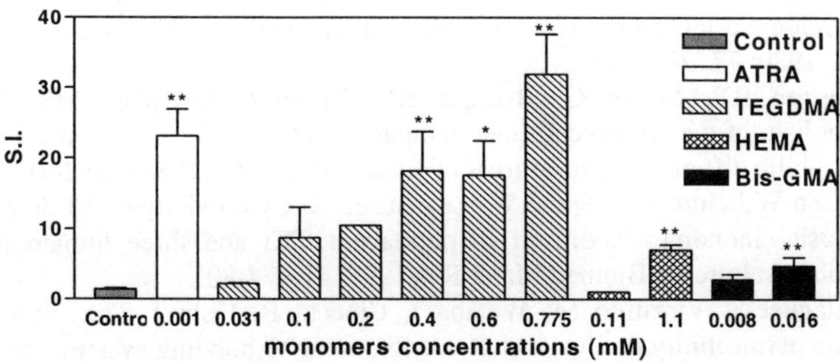

Figure 2. Stimulation index of treated and untreated cells. *=p<0.05**p<0.01

DISCUSSION

In this study we investigated the activity of three acrylate monomers on HL-60 cells by examining of some functional parameters. Ours results clearly indicate that TEGDMA and Bis-GMA, in a dose-dependent manner, reduces cellular proliferation in HL-60 leukemia cell line. This antiproliferative effect of these monomers seems to depend on a cytostatic activity with a cytotoxic effect only at 0.6 and 0.775 mM concentrations. Moreover, the effect on cell proliferation appears higher than that observed with ATRA. HEMA, on the contrary, do not shows any effect on the cellular proliferation. All monomers induce the differentiation process of HL-60 cell line as indicated by the presence of CL activity which is strictly dependent on the appearance of respiratory burst in human myeloid tumour cell lines.

In conclusion these data confirm that CL may be considered a rapid and useful method to evaluate differentiation in leukaemia cell lines. Furthermore, the results of the present study indicate further work is needed to investigate the mechanism of action of this monomers in cellular differentiation.

REFERENCES

1. Collins SJ, Gallo RC, Gallagher RE. Continuous growth and differentiation of human myeloid leukemic cells in suspension culture. Nature 1977; 270: 347-9.
2. Blair OC, Carbone R, Sartorelli AC. Differentiation of HL-60 promyelocytic leukemia cells: Simultaneous determination of phagocytic activity and cell cycle distribution by flow cytometry. Cytometry 1986; 7: 171-7.
3. Tanaka H, Abe E, Miyaura C, Kuribayashi T, Nishii Y, Suda T. 1,25-dihydroxycholecalciferol and a human myeloid leukaemia line (HL-60); the presence of a cytosol receptor and induction of differentiation. Biochem J 1982; 204: 713-9.
4. Scatena R, Nocca G, De Sole P, Rumi C, Puggioni P, Remiddi F, Bottoni P, Ficarra S, Giardina B. Bezafibrate as differentiating factor of human myeloid leukemia cells. Cell Death Differ 1999; 6: 781-7.
5. Breitman TR, Selonick SE, Collins SJ. Induction of differentiation of the human promyelocytic leukaemia cell line (HL-60) by retinoic acid. Proc Nation Acac Scien 1980; 77: 2936-40.
6. Thompson BY, Sivam G, Britigan BE, Rosen GM, Cohen MS. Oxygen metabolism of the HL-60 cell line: comparison of the effects of monocytoid and neutrophilic differentiation. Journal of leukocyte biology 1988; 43: 140-7.
7. Geurtsen W,Lehmann F Spahl W, Leyhausen G. Cytotoxicity of 35 dental resin composite monomers/additives in permanent 3T3 and three human primary fibroblast cultures. J Biomed Mater Res 1998. 41:474-80.
8. Bouillaguet S, Virgillito M, Wataha J, Ciucchi B, Holz J. The influence of dentine permeability on cytotoxicity of four dentine bonding systems, in vitro. J Oral Rehabil 1998; 25(1):45-51.
9. De Baeselier P, Schram E. Luminescent bioassay based on macrophage cell lines. Methods Enzymol 1986;133:507-30.

COMPARATIVE STUDY OF ROS SCAVENGERS BASED ON QUENCHING OF MCLA-DEPENDENT CHEMILUMINESCENCE

M OBUKI, J NAKAJIMA, M SUZUKI, K MIYAHARA, S HOSAKA

Dept of Applied Chemistry, Tokyo Polytechnic University, Atsugi
2430297 Japan

INTRODUCTION

Superoxide anion O_2^-, singlet oxygen $^1O_2$, hydrogen peroxide H_2O_2, hypochlorite anion ClO^-, and hydroxyl radical $HO^·$ are well known as important reactive oxygen species (ROS) *in vivo*. Various enzymes and antioxidants are reported to be effective as quenchers or scavengers for these ROS. Specificities of enzymatic scavengers are well established: for example, superoxide dismutase (SOD) and catalase are specifically effective against O_2^- and H_2O_2 respectively. On the contrary, specificities of non-enzymatic antioxidants as ROS scavengers are not satisfactorily clarified. One of the reasons for the complexity of evaluating the effectiveness of ROS scavengers is the disunity of the methods of measuring the effect. From this view point, we have undertaken the comparative study of ROS scavenger candidates by the measurement of quenching effect on MCLA-dependent chemiluminescence caused by ROS. CLA (*Cypridina* luciferin analogue) and MCLA (methoxylated CLA) have been confirmed to be highly sensitive to O_2^- and $^1O_2$ at physiological pH.[1, 2]

EXPERIMENTAL

Chemiluminescence (CL) was measured in triplicate at 23 °C under moderate agitation by use of Luminescence Reader BLR-201 (Aloka Co. Ltd; Tokyo Japan). When O_2/HPX/XOD was used as O_2^- generator, 500 μL of DIW, 100 μL of 10-fold concentrated PBS, 100 μL of 1.67 mM HPX solution, 100 μL of 1 μM MCLA solution, 100 μL of antioxidant solution were mixed in a glass test tube, and then the test tube was set in the Luminescence Reader. About 30 s after the start of CL measurement, 100 μL of 25 U/mL XOD solution was injected into the test tube through a rubber cap by a syringe. Here, HPX, XOD, DIW, and PBS denote hypoxanthine, xanthine oxidase, distilled and deionized water, and phosphate buffered saline, respectively.

KO_2 also was used to release O_2^-: 698 μL of DIW, 100 μL of 10-fold concentrated PBS, 100 μL of 0.5 μM MCLA solution, 100 μL of antioxidant solution were mixed in a glass test tube, and 2 μL of 20 mM KO_2 solution was injected into the test tube in the same way as above described. The KO_2 solution was prepared in the following way: 0.0071 g of KO_2 was dissolved in 5 ml of DMSO solution containing 0.0711 g of 18-crown-6.

When $NDPO_2$ was used as $^1O_2$ generator, 350 μL of DIW, 100 μL of 10-fold concentrated PBS, 100 μL of 0.5 μM MCLA solution, and 100 μL of antioxidant solution were mixed in a glass test tube; and about 30 s after the start of CL measurement, 50 μL of 200 mM $NDPO_2$ solution was injected into the test tube in the same way as above described. $NDPO_2$, the endoperoxide of disodium (1,4-dinaphthylidine)-3,3'-dipropionate[3] was used as the generator of $^1O_2$.

Hydroxyl radical HO· was generated by Fenton Reaction. In this case, 500 μL of DIW, 100 μL of 10-fold concentrated PBS, 100 μL of 0.5 μM MCLA solution, 100 μL of antioxidant solution, 100 μL of 50 μM H_2O_2 solution were mixed in a glass test tube; and 100 μL of 5 μM $Fe(NH_4)_2(SO_4)_2$ solution was injected into the test tube about 30 s after the start of CL measurement in the same way as above described.

In the experiments above, water soluble antioxidants such as ascorbic acid and glutathione were added to the reaction solution as an aqueous solution, while β-carotene and α-tocopherol were added as a chloroform solution emulsified with sodium dodecyl sulphate. Electrochemically reduced water was prepared by use of TRIMION TI-8000 (Nihon Trim Inc, Osaka, Japan) from water purified by deionization and reverse osmosis, containing 0.02 % NaCl. In the CL measurements in this case, 600 μL of electrochemically reduced water and same volumes of other reagents as above were used, and then total volume of the reaction mixture was adjusted to 1 mL by changing the volume of DIW.

RESULTS AND DISCUSSION

Fig. 1 shows the effect of water soluble scavengers on O_2^- dependent CL from MCLA by O_2/HPX/XOD. Ascorbic acid almost completely suppressed CL at the initial stage under these conditions, while the effect of glutathione at the same concentration was limited. On the other hand, the effect of ascorbic acid was inferior to that of glutathione in the case where ROS was $^1O_2$ produced by $NDPO_2$. β-Carotene and α-tocopherol significantly reduced CL when ROS was HO· generated by H_2O_2/Fe^{2+}. A typical time course of CL suppressed by α-tocopherol was shown in Fig. 2.

In Table 1, suppressive effects of antioxidants are compared based on the rate of decrease in accumulated CL. Ascorbic acid was most effective for O_2^- but less than glutathione for $^1O_2$. Evaluation of ascorbic acid in an HO· generating system was not feasible, because H_2O_2 reacted with ascorbic acid very rapidly.

β-Carotene and α-tocopherol were remarkably effective for HO· but much less effective for O_2^- and $^1O_2$. This result is understandable in consideration of the fact that both compounds are free radical scavengers rather than reducing agents. These characteristics were more clearly manifested in the experiments where the measurements were made in organic solvents. Details of the experiments in organic solvents will be reported elsewhere.

Shirahata et al reported that electrochemically reduced water quenched CL elicited from CLA by O_2^-.[4] The scavenging of ROS by electrochemically reduced water was ascribed to the presence of colloidal platinum.[5] In our study also, electrochemically reduced water almost completely abolished CL in any case of O_2^-, $^1O_2$ and $HO^.$ as shown in Table 1. We found, however, that electrochemically reduced water enhanced CL due to O_2/myeroperoxidase/Cl$^-$ (data not shown). The details of our study will be reported elsewhere.

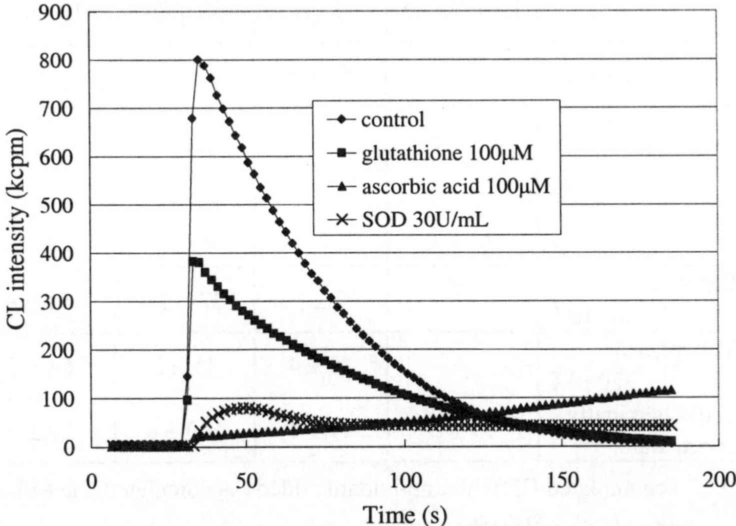

Figure 1. Effects of water soluble scavengers on CL from MCLA by O_2/HPX/XOD.

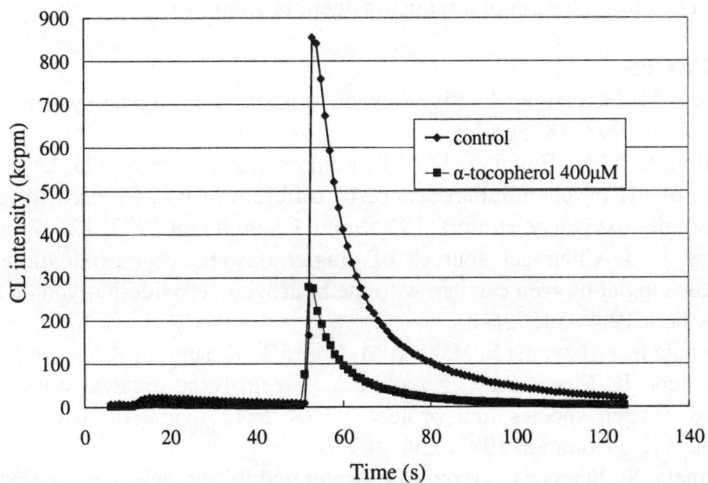

Figure 2. Effect of α-tocopherol on CL from MCLA by H_2O_2/Fe^{2+}

Table 1. Relative suppressive effect of antioxidants on MCLA-dependent CL*

Antioxidant		ROS generator			
		HPX/XOD	KO_2	$NDPO_2$	H_2O_2/Fe^{2+}
Ascorbic acid	10 μM		58 ± 8		
	100 μM	74 ± 3		35 ± 4	
Glutathione 100 μM		44 ± 1		62 ± 2**	
β-Carotene 400 μM			16 ± 15	27 ± 12	52 ± 8
α-Tocopherol 320 μM			15 ± 4	45 ± 2	64 ± 3
Electrochemically reduced water***		99 ± 0	99 ± 1	100 ± 0	99 ± 0

* (1 − accumulated CL with antioxidant added / accumulated CL without antioxidant) × 100 (%)

** MCLA concentration was 0.2 μM in these cases, while 0.05 μM in all other cases.

***The concentration of a reducing agent is unknown.

REFERENCES

1. Nakano M. Detection of active oxygen species in biological systems. Cell Mol Neurobiol 1998; 18: 565-79
2. Oosthuizen M M, Engelbrecht M E, Lambrechts H, Greyling D, Levy R D. The effect of pH on chemiluminescence of different probes exposed to superoxide and singlet oxygen generators. J Biolumin Chemilumin 1997; 12: 277-84.
3. Aubry J M. Chemical sources of singlet oxygen. 3. Peroxidation of water soluble singlet oxygen carriers with the hydrogen peroxide-molybdate system. J Org Chem 1989; 54: 726-8
4. Shirahata S, Kabayama S, Nakano M, Miura T, Kusumoto K, Gotoh M, Hayashi H, Otsubo K, Morisawa S, Katakura Y. Electrolyzed-reduced water scavenges active oxygen species and protects DNA from oxidative damage. Biochem Biophys Res Commun 1997; 234: 269-74.
5. Shirahata S. Sciences on reduced water exhibiting reactive oxygen species-scavenging effect and its medical application. Center News, Center of Advanced Instrumental Analysis, Kyushu University 2002; 19: 7-21

SPECIFIC DETECTION OF SINGLET OXYGEN USING VINYLPYRENE DERIVATIVES AS CHEMILUMINESCENT PROBE

K OHNO[1], Y HARYU[1], K NAKANO[1], J-M LIN[2], M YAMADA[1]

[1]*Dept. of Applied Chemistry, Graduate School of Engineering,*
Tokyo Metropolitan University, 1-1 Minami-Ohsawa, Hachioji, Tokyo 192-0397, Japan
[2]*Research Center for Eco-Environmental Sciences, Chinese Academy of Sciences,*
P.O. Box 2871, Beijing 100085, PR China
Email: ohnok@comp.metro-u.ac.jp

INTRODUCTION

Reactive oxygen species (ROS) such as hydrogen peroxide (H_2O_2), hydroxyl radical ($\cdot OH$), singlet oxygen ($^1O_2$) and superoxide ($\cdot O_2^-$) have been studied for their adverse effects in a biological system by electron spin resonance (ESR) and chemiluminescence (CL) methods. CL detection systems are of particular interest because of their advantages of high sensitivity, rapid and wide response with simple instruments. Some CL reagents, luminol, lucigenin and oxalate diesters, are often used for the detection of ROS,[1,2] but their selectivity to detect each ROS is insufficient. Up to now, a CL probe, *trans*-1-(2-methoxyvinyl)pyrene (*t*-MVP), for $^1O_2$ was reported,[3] and its CL properties (quantum yield, spectrum and reaction mechanism) were investigated. It is considered that by reacting with electrophilic $^1O_2$, *t*-MVP transforms into the corresponding dioxetane intermediate, and then its cleavage induces the excited state of 1-pyrenecarboxyaldehyde emitting light by the intramolecular CIEEL (chemically initiated electron exchange luminescence) mechanism. We also synthesized and characterized *t*-MVP for analytical use. As a result, by reacting it with $^1O_2$ generated from the NaOCl–H_2O_2 reaction[4] *t*-MVP emitted the weak and long-term CL for 100 sec, suggesting that the dioxetane intermediate is highly thermostable.

In order to improve the sensitivity and to perform real-time measurements of $^1O_2$, we developed a novel CL probe, *trans*-1-(2-methylsulfanylvinyl)pyrene (*t*-MSVP) and characterized its CL and analytical properties.

MATERIALS AND METHODS

<u>Instruments</u>: Batch experiments for obtaining CL profiles were performed using a Microtec NITI-ON Lumicounter 2500 (Chiba, Japan). Proton nuclear magnetic resonance (^1H-NMR) spectra were obtained on a JEOL JNM-EX270 spectrometer (Tokyo, Japan) with tetramethylsilane as an internal standard. Mass spectra (FAB-MS) were measured on a JEOL JMS-LX1000 (Tokyo, Japan) with *m*-nitrobenzyl alcohol as a matrix.

<u>Reagents</u>: Hydrogen peroxide (31%) was purchased from Kanto Chemical (Tokyo, Japan). Sodium hypochlorite (NaOCl) solution (8.5-13.5 w/v%) was from Nacalai tesque (Kyoto, Japan). (Methoxymethyl)tripheny phosphonium chloride was from

Aldrich Chemical. (Milwaukee, WI). 1-Pyrenecarboxyaldehyde was from Lancaster Synthesis (Lancashire, England). All other reagents were of analytical or guaranteed reagent grade and used without further purification. Water was purified using a Milli-Q reagent system (Millipore, Bedford, MA).

<u>Synthesis</u>: T*rans*-1-(2-Methoxyvinyl)pyrene (*t*-MVP) was synthesized by Wittig reaction according to the previous paper[3] with minor modification. In brief, 1-pyrenecarboxaldehyde and (methoxymethyl)triphenyl phosphonium chloride were dissolved in tetrahydrofuran (THF). To the mixture, potassium *tert*-butoxide was slowly added and the solution was stirred for 15 h at room temperature. The crude product was thrice purified by column chromatography on silica gel to afford *t*-MVP.

Trans-1-(2-Methylsulfanylvinyl)pyrene (*t*-MSVP) was synthesized as follows. After sodium hydroxide was suspended in dimethyl sulfoxide (DMSO), 1-pyrenecarboxaldehyde was added and the solution was stirred for 2 h at 90°C. The crude product was purified by column chromatography on silica gel to afford *t*-MSVP.

<u>CL measurement</u>: The NaOCl–H_2O_2 reaction was used as the source of $^1O_2$. NaOCl, H_2O_2 and SDS solutions were prepared with Britton-Robinson buffer (pH 7.4), and the CL probes were dissolved in 0.1 M SDS solution. Continuously monitoring the CL signals, to 100 μL of 0.1 mM the CL probes placed in a plastic cell, 100 μL of 20 mM H_2O_2 and 100 μL of each concentration of NaOCl were sequentially added with a dispenser attached on Lumicounter 2500. The generated CL signals were measured by counting photon for 100 sec (gate time: 0.1 sec) in triplicate.

RESULTS AND DISCUSSION

<u>Design and synthesis of a novel CL probe for $^1O_2$</u>: Reacting $^1O_2$ with *t*-MVP provided long-term CL, since a dioxetane with higher electron density is more thermostable.[5] We furthermore demonstrated that during the *t*-MVP-$^1O_2$ reaction, the CL signals were increased by heating the solution (data not shown). Therefore, a dioxetane with lower electron density may provide short-term and strong CL. The Hammett substituent constants[6] as an index controlling the electron density were employed to determine the substituent of a novel CL probe for $^1O_2$. *t*-MSVP was designed (σ_p = –0.27 for –OMe group of *t*-MVP, σ_p = 0.00 for –SMe group of *t*-MSVP), and then was readily and efficiently synthesized from 1-pyrenecarboxaldehyde and DMSO as the starting materials.

<u>Characterization of CL probes</u>: Both CL probes were subjected to reaction with $^1O_2$ and the CL was observed. As shown in Fig. 1, *t*-MSVP emitted short-term CL for 2-3 sec and the peak height was 50 times higher than that of *t*-MVP.

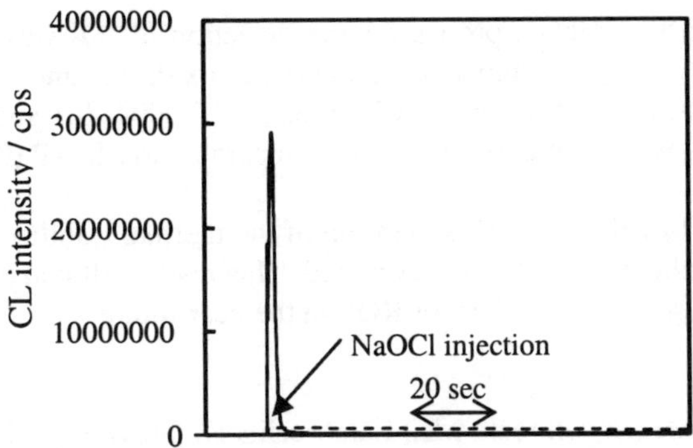

Figure 1. CL profiles induced by the reaction of *t*-MVP (broken line) and *t*-MSVP (solid line) with $^1O_2$ (NaOCl–H_2O_2).

The CL properties were investigated in detail to assess possible analytical uses. To obtain the calibration curves for $^1O_2$, water-soluble naphthalene endoperoxide (NEP)[7,8] which thermally releases $^1O_2$ at a constant concentration level was synthesized and used instead of the NaOCl–H_2O_2 reaction. As shown in Fig. 2, the calibration curves for $^1O_2$ were obtained as the function of the NEP concentration at 25°C. In the case of *t*-MSVP, linearity was shown over the concentration range of 2×10^{-6}–1×10^{-2} M and the lower limit of detection was 5×10^{-7} M which were superior to those of *t*-MVP.

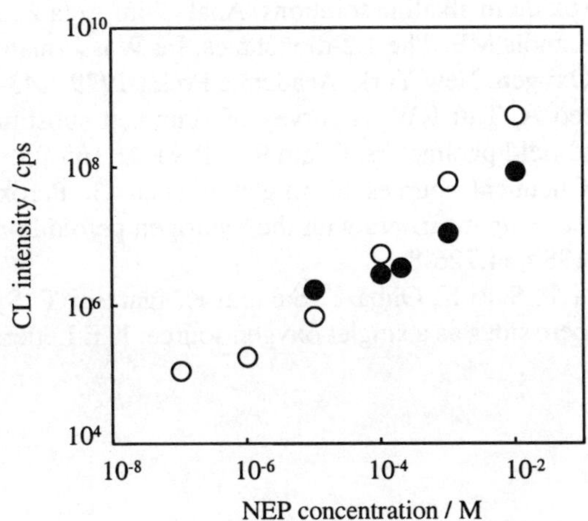

Figure 2. Calibration curves for $^1O_2$ using *t*-MVP (closed circle) and *t*-MSVP (open circle).

Next, the specificity of the probes for the detection of $^1O_2$ was examined by reaction with H_2O_2, $\bullet O_2^-$ (hypoxanthine–xanthine oxidase) and $\bullet OH$ (Fenton reaction). As a result, t-MSVP provided CL signals of 0.0004, 0.1 and 10% of that for $^1O_2$, respectively, which were excellent and comparable to t-MVP (0.008, 0.3 and 5%, respectively).

In conclusion, based on our consideration of the thermal stability of dioxetane, an alternative CL probe, t-MSVP, was developed. This result will make it possible to develop other and more useful probes for ROS in the near future.

REFERENCES

1. Oosthuizen MMJ, Ebgelbrecht ME, Lambrechts H, Greyling D, Levy RD. The effect of pH on chemiluminescence different probes exposed to superoxide and singlet oxygen generators. J Biolumin Chemilumin 1997;12:277-84.
2. Hosaka S, Itagaki T, Kuramitsu Y. Selectivity and sensitivity in the measurement of reactive oxygen species (ROS) using chemiluminescent microspheres prepared by the binding of acridium ester or ABEI to polymer microspheres. Luminescence 1999;14:349-54.
3. Thompson A, Canella KA, Lever JR, Miura K, Posner GH, Seliger HH. Chemiluminescence mechanism and quantum yield of synthetic vinylpyrene analogs of benzo[a]pyrene-7,8-dihydrodiol. J Am Chem Soc 1986;108:4498-504.
4. Almeida EA, Miyamoto S, Martinez GR, Medeiros MHG, Mascio PD. Direct evidence of singlet oxygen production in the reaction of acetonitrile with hydrogen peroxide in alkaline solutions. Anal Chim Acta 2003;482:99-104.
5. Bartlett PD, Landis ME. The 1,2-dioxetanes. In: Wasserman HH, Murray RW. eds. Singlet Oxygen. New York: Academic Press, 1979: 243-286.
6. Hansch C, Leo A, Taft RW. A survey of Hammett substituent constants and resonance and field parameters. Chem Rev 1991;91:165-95.
7. Aubry JM. Chemical sources of singlet oxygen. 3. Peroxidation of water-soluble singlet oxygen carriers with the hydrogen peroxide-molybdate system. J Org Chem 1989;54:726-8.
8. Liu W, Ogata T, Sato K, Ohba Y, Sakurai K, Igarashi T. Syntheses of water-soluble endoperoxides as a singlet oxygen source. ITE Letters 2001;2:98-101.

6,8-DIARYLIMIDAZO[1,2-*a*]PYRAZIN-3(7*H*)-ONES AS POTENTIAL CHEMILUMINESCENT pH/SUPEROXIDE DOUBLE SENSORS

R SAITO[1], N SUGA[1], A KATOH[1], S MAKI[2], T HIRANO[2], H NIWA[2]
[1]*Dept of Applied Chemistry, Seikei University, Musashino 180-8633, Japan*
[2]*Dept of Applied Physics and Chemistry, The University of Electro-Communications, Chofu 182-8585, Japan*
Email: saito@ch.seikei.ac.jp

INTRODUCTION

Due to its high reactivity, superoxide is blamed for many harmful events in living bodies, and recently has been found to be involved in apoptosis.[1] Such disease states usually result from homeostatic disorders such as alteration of cellular pH, so that development of sensitive sensors capable of detecting superoxide along with the local pH can provide helpful tools for mechanistic studies and diagnoses of related diseases. In recent years, 2-methyl-6-phenylimidazo[1,2-*a*]pyrazin-3(7*H*)-one, *Cypridina* luciferin analogue (CLA), and its derivatives have been developed as sensitive probe for superoxide in biological systems.[2-4] These probes can exhibit bimodal luminescence depending on the medium pH as shown in Scheme 1.[5] However, very few attempts have been made at developing such a double sensor probably because the two spectra largely overlap each other and this prevent us from estimating medium pH precisely from the luminescence intensity without any data processing. In this context, we have recently demonstrated the first example of potential pH/superoxide double sensors.[6] In our ongoing program on the design of chemiluminescent double sensors for superoxide and pH, we wish to report here synthesis of imidazopyrazinones (**1a-c**) and their chemiluminescence in phosphate buffer solutions triggered by superoxide generated from hypoxanthine-xanthine oxidase system under various pH conditions.

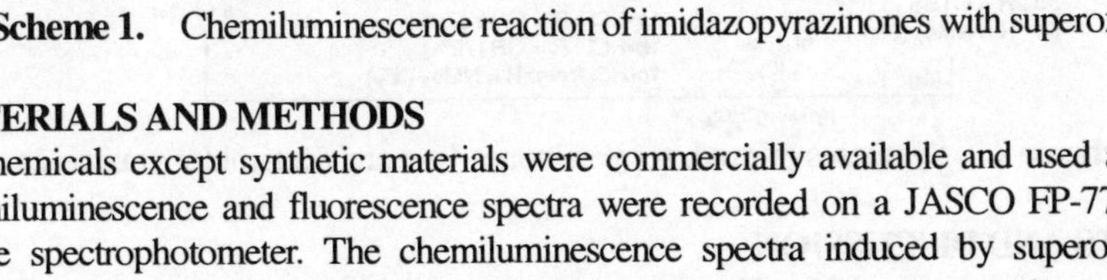

CLA: R_1 = H, R_2 = H	^{1}(**OCLA**$^-$)*: R_1 = H, R_2 = H	^{1}(**OCLA**)*: R_1 = H, R_2 = H
1a: R_1 = OMe, R_2 = 4-CF_3-C_6H_4	^{1}(**2a**$^-$)*: R_1 = OMe, R_2 = 4-CF_3-C_6H_4	^{1}(**2a**)*: R_1 = OMe, R_2 = 4-CF_3-C_6H_4
1b: R_1 = OH, R_2 = 4-CF_3-C_6H_4	^{1}(**2b**$^-$)*: R_1 = OH, R_2 = 4-CF_3-C_6H_4	^{1}(**2b**)*: R_1 = OH, R_2 = 4-CF_3-C_6H_4
1c: R_1 = NMe_2, R_2 = 4-CF_3-C_6H_4	^{1}(**2c**$^-$)*: R_1 = NMe_2, R_2 = 4-CF_3-C_6H_4	^{1}(**2c**)*: R_1 = NMe_2, R_2 = 4-CF_3-C_6H_4

Scheme 1. Chemiluminescence reaction of imidazopyrazinones with superoxide

MATERIALS AND METHODS

All chemicals except synthetic materials were commercially available and used as it was. Chemiluminescence and fluorescence spectra were recorded on a JASCO FP-777 fluorescence spectrophotometer. The chemiluminescence spectra induced by superoxide was measured in a mixture of 20-μL methanolic solution (1 mM) of an imidazopyrazinone derivative and 1 mL of 0.2 M phosphate buffer with various pH containing 100 mM KCl, 0.05 mM EDTA, and 0.15 mM hypoxanthine; light emission was started by adding xanthine oxidase (1.33 units/mL, 20 μL). Chemiluminescence intensity of **1a**, CLA, and luminol triggered by superoxide was measured in 10 mM Mops buffer, pH 7.3 (2.0 mL).[3] The other contents in the Mops buffer are the same as the phosphate buffer mentioned above. Imidazo-pyrazinones (**1a-c**) were synthesized from 2-amino-3,5-dibromopyrazine (**3**) *via* successive Suzuki-Miyaura coupling and obtained as hydrochloride salts as shown in Scheme 2. 2-Acetamido-pyrazine (**2a**), which is the expected product of the chemiluminescent reaction of **1a**, was prepared by acetylation of **5a**.[7]

2-Methyl-6-(4-methoxyphenyl)-8-(4-trifluoromethylphenyl)imidazo[1,2-*a*]pyrazin-3(7*H*)-one (1a) hydrochloride: mp 256 °C (decomp); ^{1}H-NMR (400 MHz, methanol-d_4) δ ppm 2.54 (s, 3H), 3.88 (s, 3H), 7.09 (d, *J* 8.8 Hz, 2H), 7.99 (d, *J* 8.3 Hz, 2H), 8.08 (d, *J* 8.3 Hz, 2H), 8.23 (d, *J* 8.8 Hz, 2H), and 8.66 (s, 1H); IR (KBr) ν_{max}/cm^{-1} 3423, 3053, 2933, 1663, 1609, 1461, 1329, 1258, 888, 836, and 708. Anal Calcd for $C_{21}H_{16}F_3N_3O_2$•HCl•0.2MeOH: C, 57.58; H, 4.06; N, 9.50. Found: C, 57.40; H, 3.78; N, 9.21. **2-Methyl-6-(4-hydroxyphenyl)-8-(4-trifluoromethylphenyl)imidazo[1,2-*a*]pyrazin-3(7*H*)-one (1b) hydrochloride**: mp 276 °C (decomp); ^{1}H-NMR (400 MHz, methanol-d_4) δ ppm 2.52 (s, 3H), 6.93 (d, *J* 8.8 Hz, 2H), 7.85-7.98 (m, 4H), 8.26 (d, *J* 8.0 Hz, 2H), and 8.62 (s, 1H); IR (KBr) ν_{max}/cm^{-1} 3420, 2926, 1613, 1568, 1515, 1385, 1324, and 843; HRMS Calcd for $C_{20}H_{14}F_3N_3O_2$: 385.1036. Found: 385.1018. **2-Methyl-6-(4-dimethylaminophenyl)-8-(4-trifluoromethylphenyl)imidazo[1,2-*a*]pyrazin-3(7*H*)-one (1c)**: ^{1}H-NMR (400 MHz, methanol-d_4) δ ppm 2.45 (s, 3H), 3.02 (s, 6H), 6.83 (d, *J* 9.0 Hz, 2H), 7.79-7.85 (m, 4H), 8.04 (d, *J* 8.1 Hz, 2H), and 8.42 (s, 1H); IR (KBr) ν_{max}/cm^{-1} 3420, 2910, 1612, 1518, 1367, 1323, and 846; HRMS Calcd for $C_{22}H_{19}F_3N_4O$: 412.1511. Found: 412.1492. **2-Acetamido-5-(4-methoxy-phenyl)-3-(4-trifluoromethylphenyl)pyrazine (2a)**: mp 213-214 °C; ^{1}H-NMR (400 MHz, CDCl$_3$) δ ppm 2.25 (s, 3H), 3.88 (s, 3H), 7.03 (d, *J* 8.3 Hz, 2H), 7.72 (s, 1H), 7.77 (d, *J* 8.8 Hz, 2H), 7.92 (d, *J* 8.3 Hz, 2H), 8.01 (d, *J* 8.8 Hz, 2H), and 8.75 (s, 1H); IR (KBr) ν_{max}/cm^{-1} 3443, 3204, 2840, 1677, 1607, 1442, 1330, 1251, 1071, and 843. Anal Calcd for $C_{20}H_{16}F_3N_3O_2$: C, 62.01; H, 4.16; N, 10.85. Found: C, 61.74; H, 3.92; N, 10.65.

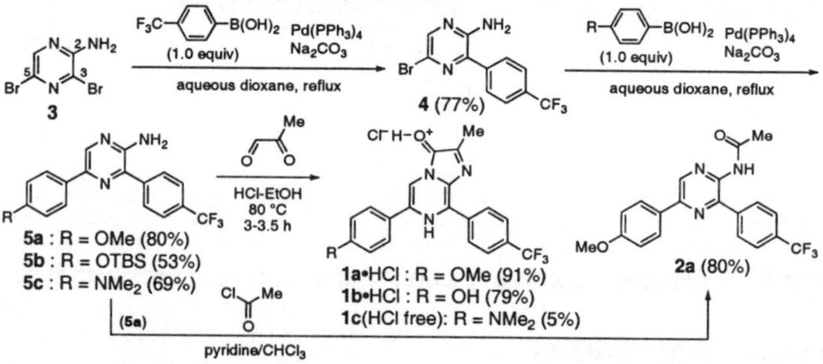

Scheme 2. Syntheses of imidazopyrazinones **1a-c** and acetamidopyrazine **2a**

RESULTS AND DISCUSSION

The reaction of compound **1a** with superoxide generated from hypoxanthine-xanthine oxidase system in phosphate buffer at pH 7.0 gave the light emission at 460 nm (Fig. 1A). As the pH rose, the emission at 540 nm intensified along with decrement of the neutral emission at 460 nm. The emission at 460 nm arose from the singlet-excited state of the corresponding acetamidopyrazine, i.e. ^{1}(**2a**)*, and the emission at 530 nm from its conjugate base, ^{1}(**2a**$^-$)*.

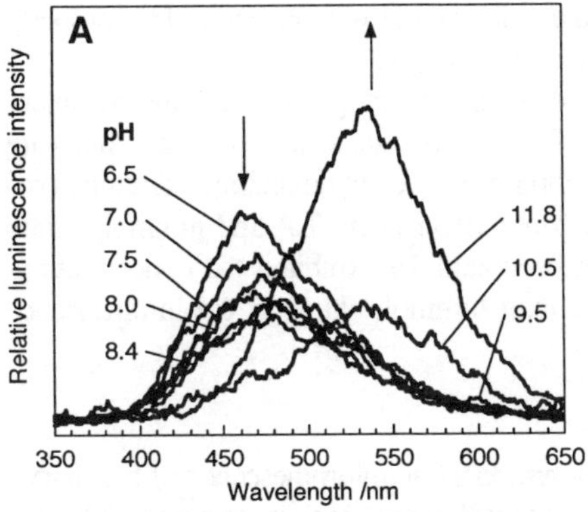
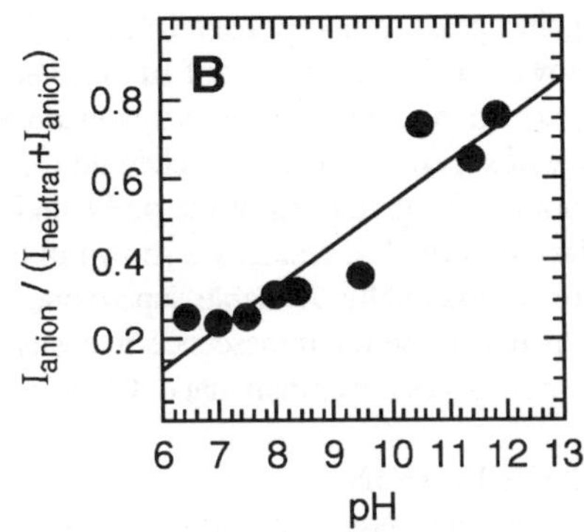

Figure 1. (A) Chemiluminescence spectra of **1a** induced by superoxide in phosphate buffer at 25 °C with various probe concentrations, and (B) plots of the ratio of the superoxide-triggered luminescence intensity, $I_{anion}/(I_{neutral}+I_{anion})$, for **1a** in Mops buffer at 25 °C.

This was confirmed by fluorescence of **2a** in DMSO in the presence and absence of NaOMe as a base. The neutral **2a** in DMSO without the additive exhibited fluorescence maximum at 430 nm, while the conjugate base of **2a**, i.e. **2a⁻**, emitted light with maximum at 557 nm in DMSO in the presence of NaOMe. Slight difference in the emission maxima between chemiluminescence and fluorescence is attributable to the variation of the solvents. Both **1b** and **1c** did not exhibit detectable luminescence in the reaction under the same conditions probably because the strong hydrophilic hydroxy and dimethyamino groups could interact with solvating water through the hydrogen bonding that deactivates the singlet-excited light emitters produced in the course of the chemiluminescent reaction. The largely bathochromic shift in the anionic luminescence of **1a** is supposed to be caused by the large energy stabilization in $^1(2a^-)^*$ with extended π-conjugation at the 3-position. The chemiluminescence of **1a** was quenched by addition of superoxide dismutase, confirming this luminescent reaction was surely caused by superoxide. Fig. 1B showed plots of the luminescence intensity ratio of the anion luminescence *versus* the sum of the bimodal luminescence, $I_{anion}/(I_{neutral}+I_{anion})$, for **1a**

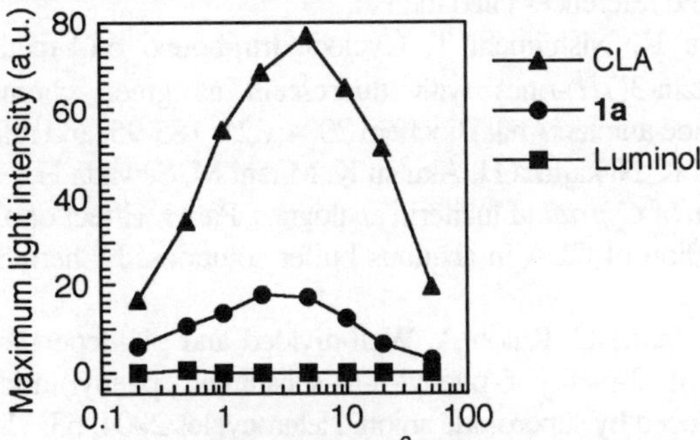

Figure 2. Effect of probe concentration on the intensity of the superoxide-induced chemiluminescence in Mops buffer (pH 7.3) at 25 °C.

against pH. A proportional relationship between the ratio and pH was observed. This property provides us the usefulness of **1a** as pH/superoxide double sensor.

To evaluate the sensitivity toward superoxide under physiological conditions, the intensity of superoxide-triggered chemiluminescence of **1a** was measured in a neutral buffer with various concentrations of the probe and a fixed composition of hypoxanthine-xanthine oxidase system. The results are shown in Fig. 2, in comparison with CLA and luminol, which are commercially available superoxide probes. Compound **1a** exhibited the concentration-dependent chemiluminescence intensity stronger than luminol, although the luminescence response was lower than that of CLA.

CONCLUSION

The present study demonstrates the pH-dependent bimodal chemiluminescence of 6,8-diaryl-imidazopyrazinone **1a** induced by superoxide. Despite its lower sensitivity to superoxide than CLA, **1a** showed enough sensitivity to superoxide in comparison with luminol and appears to be capable of superoxide detection. What the advantage of **1a** over CLA is that it exhibits the pH-dependent bimodal chemiluminescence, and, therefore, the direct estimation of the local pH in a cell or a droplet is feasible based on the luminescence ratio, $I_{anion}/(I_{neutral}+I_{anion})$. This property of **1a** allows us to apply **1a** for simultaneous detection of the two concurrent events, superoxide evolution and pH alteration, in biological systems.

REFERENCES

1. Kogure K, Morita M, Nakashima S, Hama S, Tokumura A, Fukuzawa K. Superoxide is responsible for apoptosis in rat vascular smooth muscle cells induced by alpha-tocophenyl hemisuccinate. Biochim Biophys Acta 2001; 1528: 25-30.
2. Goto T, Takagi T. Chemilluminescence of a *Cypridina* luciferin analogue, 2-methyl-6-phenyl-3,7-dihydroimidazo[1,2-*a*]pyrazin-3-one, in the presence of the xanthine-xanthine oxidase system. Bull Chem Soc Jpn 1980; 53: 833-4.
3. Shimomura O, Wu C, Murai A, Nakamura H. Evaluation of five imidazopyrazinone-Type chemiluminescent superoxide probes and their application to the measurement of superoxide anion generated by *Listeria* monocytogenes. Anal Biochem 1998; 258: 230-5, and references cited therein.
4. Teranishi K. Nishiguchi T. Cyclodextrin-bound 6-(4-methoxyphenyl)imidazo-[1,2-*a*±]pyrazin-3(7*H*)-ones with fluorescein as green chemiluminescent probes for superoxide anions. Anal Biochem 2004; 325: 185-95, and references cited therein.
5. Fujimori K, Nakajima H, Akutsu K, Mitani M, Sawada H, Nakayama M. Chemiluminescence of *Cypridina* luciferin analogues. Part 1. Effect of pH on rates of spontaneous autoxidation of CLA in aequous buffer solutions. J Chem Soc Perkin Trans 1 1993: 2405-9.
6. Saito R, Inoue C, Katoh A. Well-divided and pH-dependent bimodal chemilumine-scence of 2-methyl-6-phenyl-8-(4-substituted phenyl)imidazo[1,2-*a*]pyrazin-3(7*H*)-ones induced by superoxide anion. Heterocycles 2004; 63: 759-64.
7. Saito R, Hirano T, Niwa H, Ohashi M. Solvent and substituent effect of the fluorescent properties of coelenteramide analogues. J Chem Soc Perkin Trans 2 1997; 1711-6.

SIMULTANEOUS MEASUREMENT OF FLUORESCENCE AND CHEMILUMINESCENCE USING NEUTROPHIL-LIKE CULTURE CELLS

H SATOZONO, K KAZUMURA, S OKAZAKI, M HIRAMATSU
Hamamatsu Photonics K.K. 5000 Hirakuchi, Hamakita-city, Japan
E-mail: satozono@crl.hpk.co.jp

INTRODUCTION

It is well known that the superoxide anion from neutrophils kills bacteria and virus and plays an important role in biological defence. It is also well known that calcium ion is a intracellular mediator. But details about the relationship between superoxide anion and calcium ion are not clear. To reveal their relationship, it is important to measure the time-course of calcium ion concentration and superoxide anion genaration from neutrophil-like cells in real time. We have developed a novel method to measure fluorescence and chemiluminescence simultaneously in real time to investigate function of the cells.

EXPERIMENTS

Reagents

CLA(2-Methyl-6-phenyl-3,7–dihydroimidazo(1,2-)pyrazin-3-one, Tokyo Kasei Kogyo Co., Ltd.) was used as a chemiluminescent reagent for superoxide anion.[1,2] Flur-3-AM(1-[2-Amino-5-(2,7-dichloro-6-hydroxy-3-oxo-9-xanthenyl)phenoxy]-2-(2-amino-5-methylphenoxy)ethane-N,N,N',N'-tetraacetic acid, Dojindo Molecular Technologies Inc.) was used as a fluorescent calcium indicator. f-MLP(N-formyl-methionyl-L-leucyl-L-phenylalanine, Sigma) was used as agonist.

Instrument and measurement principle

Fig.1 shows the block diagram of the instrument for simultaneous measurement of fluorescence and chemiluminescence. A high-intensity blue LED was flashed according to the pulse generator. Two optical bandpass filters at 485nm are used, because the spectrum of LED is very broad. The sample emits chemiluminescence from CLA (385nm) and fluorescence from Fluo 3 (530nm). The emission is guided to PMT (R1635, Hamamatsu Photonics K.K.) through a band rejection filter. This filter rejects the scattering of excitation. Signals from the PMT were amplified and discriminated by Photon Counting Unit (C6465, Hamamatsu Photonics K.K.). The photon pulses from C6465 were entered to a multi-channel universal counter (PCI32-8M, CONTEC Co.,Ltd.) via hand-made prescaler. The universal counter counts the photon pulses on two memories. It was inserted into PCI slots of a PC and controlled by the measurement software.

Fig.2 shows the timing chart of simultaneous measurement of fluorescence and chemiluminescence. The pulse generator generates the timing pulse.

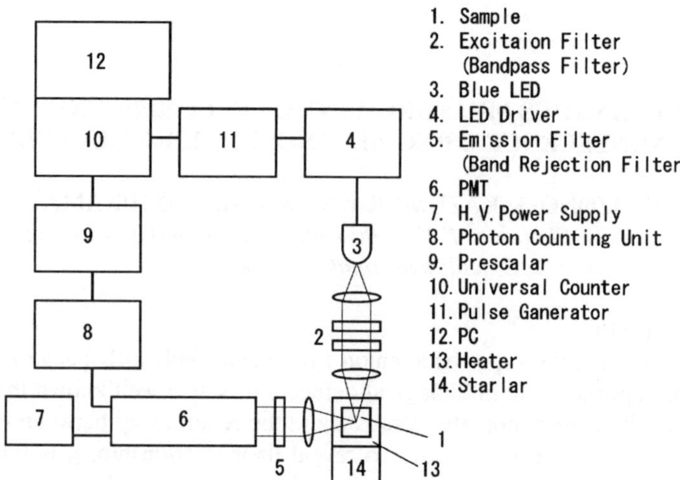

Figure 1. Block diagram of the simultaneous measurement of fluorescence and chemiluminescence.

When the timing pulse is low level, the excitation light does not irradiate the sample, thus the sample emits only chemiluminescence. The photons from the sample are counted on the memory channel 1.

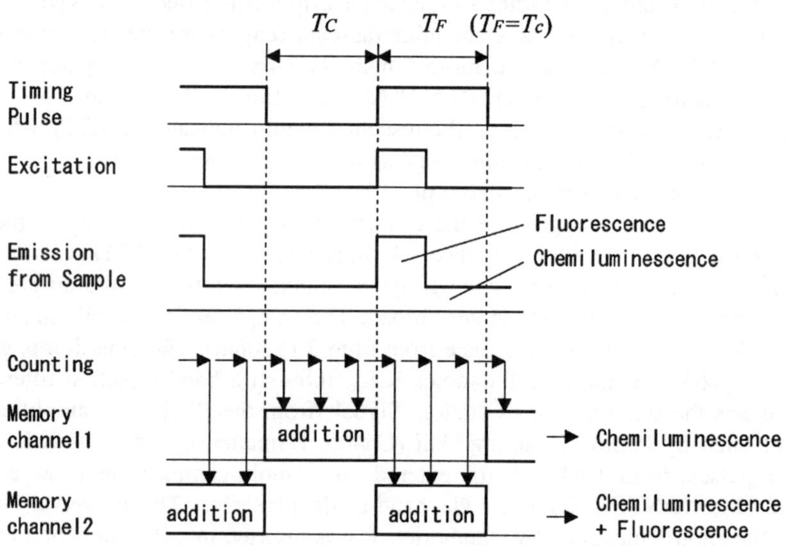

Figure 2. Timing Chart of the simultaneous measurement

When the timing pulse is raised, the excitation light is flashed and the sample emits both fluorescence and chemiluminescence. The photons from the sample are counted on the memory channel 2.As a result, the chemiluminescence intensity is obtained on the memory channel 1 and both chemiluminescence and fluorescence are obtained on the memory channel 2. The fluorescence intensity is obtained by subtracting the channel 1 from channel 2, because the width of both pulses is same.

The timing pulse frequency is 1kHz and it is faster than cell response. Thus the measurement of fluorescence and chemiluminescence are practically simultaneous. In addition the excitation width is 0.2 ms and it is shorter than the timing pulse width because fluorescence intensity is enough strong to measure in a short excitation time.

Sample preparation

The neutrophil-like cells were prepared by incubating THP-1 (human acute monocytic leukaemia cell line) (Dainippon Pharmaceutical Co., Ltd.) for 3 days at 37 °C in RPMI1640 medium which contains 10% FBS (Fetal Bovin Serum) and 0.5 mM dibutylyl cyclic AMP, which cell density was 3.0 x10^{5}cells/mL. Before measurement, the THP-1 cells were washed twice with RH (Ringer-Hepes) buffer, and suspended medium, treated with 3µM Flor-3-AM at 37 °C for 30 min in 5% CO_2 atmosphere and washed twice and suspended RH buffer again.

Assays

The sample contained 8.0x10^{5}cells/mL of the THP-1 cells, 1 mM $CaCl_2$ and 0.5µM CLA in RH buffer. The sample was stirred and incubated at 37°C during the measurement. After 5 min, the THP-1 cells were stimulated by injection of 1µM f-MLP.

RESULTS AND DISCUSSION

Fig. 3 shows the result of chemiluminescence and fluorescence time courses from the THP-1 cells in real time measurement. Chemiluminescence and fluorescence from the sample were clearly separated.

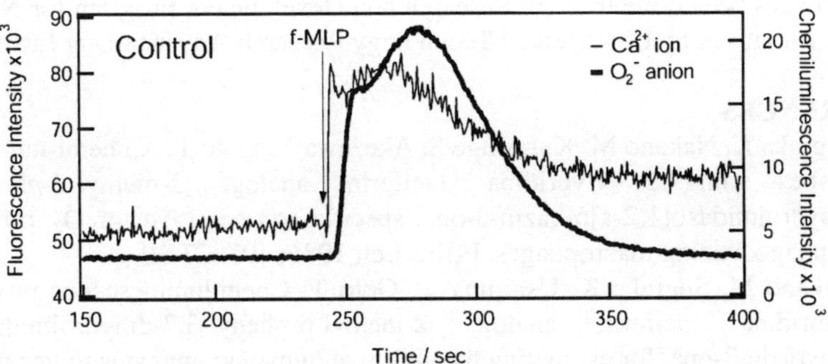

Figure 3. Time course of chemiluminescence and fluorescence from THP-1 cells

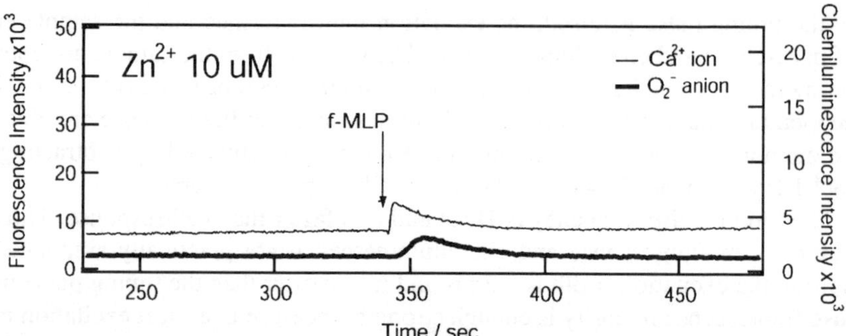

Figure 4. Time courses chemiluminescence and fluorescence from THP-1 cells with Zn^{2+} ion. Zn^{2+} ion concentration was 10 μM

The chmiluminescence curve was not affected by the excitation light and fluorescence, even though chemiluminescence was very weak.

Fig. 3 also shows the chemiluminescence rise is delayed for about 10 sec from the fluorescence rise. This suggests the generation of superoxide anion from the THP-1 cells is originated from the rise of intracellular Ca$_2^+$ ion concentration.

Fig. 4 shows the time courses from same cell with Zn^{2+} ion. Both of chemiluminescence and fluorescence are remarkably decreased.

This suggests Zn^{2+} ion inhibits both the rise of intracellular Ca^{2+} ion concentration and the generation of superoxide. As a result, we could measure the generarion of superoxide anion and the intracellular Ca^{2+} ion concentration of the neutrophil-like cells simultaneously by using our method and instrument. We found that the presence of Zn^{2+} ion inhibited both the generation of superoxide anion and Ca^{2+} ion influx.

ACKNOWLEDGMENTS

Our study has been done through Research and Development Program for New Bio-industry Initiatives by Bio-oriented Technology Research Advancement Institution.

REFERENCES

1. Sugioka K, Nakano M, Kurashige S, Akuzawa Y, Goto T. A chemi-luminescent probe with a Cypridina luciferin analog, 2-methyl-6-phenyl-3,7-dihydroimidazo[1,2-a]pyrazin-3-one, specific and sensitive for O$_2^-$ production in phagocytizing macrophages. FEBS Lett 1986; 197: 27-30
2. Nakano M, Sugioka K, Ushijima Y, Goto T. Chemiluminescence probe with Cypridina luciferin analog, 2-methyl-6-phenyl-3,7-dihydroimidazo[1,2-a]pyrazin-3-one, for estimating the ability of human granulcytes to generate O$_2^-$. Anal Biochem 1986;159:363-96

PURIFICATION OF ENVIRONMENT BY SINGLET OXYGEN

NOBUTAKA SUZUKI

Graduate School of Biosphere Science, Hiroshima University,
Higashi-Hiroshima 739-8528, Japan
E-mail: suzukin@hiroshima-u.ac.jp

INTRODUCTION

Singlet oxygen has a very short half-life time (10^{-6} sec in water and 10^{-3} sec in air). It travels only a few centimeters in water, then changes to ordinary molecular oxygen and leaves almost no residual toxicity in the environment. This is a demerit in an ordinary sense but a great merit at the same time as a disinfectant or a microbicide to kill microorganisms attaching to important cells such as human cells or cultured plant/animal cells.[1-3] Singlet oxygen would be a useful tool for purifying the environment without polluting or injuring important cells. As an example for the latter we showed that the penaeid white spot syndrome virus (WSSV) attaching on the eggs of Kuruma shrimp *(Penaeus japonicus)* can be eliminated by singlet oxygen without injuring the eggs.

There are many people having no good water to drink in the world. Over a million children die every year by drinking unsanitary water. We would like to produce a costless sanitary device to make good drinking water. We describe here some experimental works such as eliminating *Escherichia coli* in drinking water and on plant seeds; and eliminating "Aoko", a water-polluting weeds, *Microcystis aeruginosa* in the Lake Biwa that is supplying tap water for the Cities of Osaka and Kyoto, Japan by singlet oxygen.

GENERATION AND SOME PROPERTIES OF SINGLET OXYGEN

The energy level of singlet oxygen ($^1\Delta$g) lies 22.5 kcal/mol (= 1270 nm in wavelength) above its ground state,[4,5] and therefore, it gives near-infrared emission from a single molecule (eq. 1) and also gives red light from two molecules (eq. 2).

$$^1\Delta_g \longrightarrow {}^3\Sigma_g - 1270 \text{ nm} \qquad (1)$$
$$2\,^1\Delta_g \longrightarrow 2\,^3\Sigma_g - 633 \text{ nm} \qquad (2)$$

Half-life ($\tau_{1/2}$) is about 2.0-3.3 x 10^{-6} sec in water; and about 10^{-3} sec in air.[6] Its diffusion distance (Effective range) would be less than a few cm. These features are of great utility for sterilizing harmful microbes. It could not reach long distance from its generated points and does not give any harmful residues like many other disinfecting drugs after its going back to harmless molecular oxygen.

GENERATION METHODS OF SINGLET OXYGEN

Photosensitization using a dye is convenient and we employed this method for our work in environmental purification. Quantum efficiencies of generating singlet oxygen (Φ_{1O2}) using Rose Bengal, Methylene Blue, and Eosin, for examples, are known to be 0.80; 0.50; and 0.42, respectively. Two mechanisms have been suggested (Type I and Type II).[7, 8] In order to eliminate the direct reaction of substrate with photo-excited sensitizer, photoirradiation of an immobilized dye on a glass surface is used to give pure $^1\Delta_g$.[9]

Microwave discharge in oxygen gas and several chemical reactions also give singlet oxygen. Both reaction of H_2O_2 with NaOCl and thermolysis of the endoperoxide derivative of anthracene are representative chemical generation methods.[5] Matsuura and his coworkers developed a water-soluble naphthalene endoperoxide as a versatile generator under a mild conditions (30 °C, pH 7.8).[10]

SENSITIZING WAVELENGTH FOR PHOTOSENSITIZING METHODS (RB, MB, TiO₂)

The singlet-excited energy for singlet oxygen is 22.5 kcal/mol (1270 nm in wavelength). So light of shorter wavelength than 1270 nm is required to excite it and make singlet oxygen. Excitation at shorter wavelength as UV could result in very "active" singlet oxygen or more energetic active oxygen species like OH radical or superoxide anion radical. In order to obtain "neat singlet oxygen," it is very important that the triplet excited state of the dye has a slightly upper energy than 22.5 kcal/mol (1270 nm).

EFFECTS OF IRRADIATION ON LIVING THINGS AND THE ENVIRONMENT

Irradiating wavelength (especially shorter wavelength like UV light), when applied to the non-living things like walls, road surfaces, tiles, flushers, etc., is not harmful, if there is no human or valuable things in the environment. Visible light or near-infrared light is preferable, if singlet oxygen is applied in the presence of any valuable living things. That is the point for choosing Rose Bengal, phthalocyanine, Methylene Blue, or Eosin as the sensitizing dyes.

DYE-SOLUTION OR IMMOBILIZED DYE?

Singlet oxygen survives only a few cm or less. If you want to limit the reactive species in the possible active oxygen species to singlet oxygen as strictly as possible, because of its very short half-life time, it is important to use immobilized dyes. This can avoid pollution of the environment with the dye. If not, sometimes a dye solution would be useful, since the solution can penetrate into the organs or into the cells where the target microbes are latent, and singlet oxygen is generated very close to the target.

APPLICATIONS OF SINGLET OXYGEN TO STERILIZATION/ PURIFICATION

Many studies have been devoted to this purposes; i.e., (1) Sterilization of environments (virus, bacteria, and the other microbes); (2) Sterilization of them in the presence of human cells, bodies, or the other living things that are cultivated like plants, vegetables, fishes, shells, and shrimps or purifications of HIV or HIC in blood; (3) Cancer treatment.

Sterilization of environments (in vitro works)

There is a large body of literature devoted to this subject.[8, 11] Many microbes are known to be killed by numerous dyes on photoirradiation containing HIV, *E. coli, Helicobacter pylori.* We have tried and found that E. coli and *Microcystis aeruginosa* (a main polluting origin, "Aoko" in Lake Biwa and many water reservoirs) can be killed by singlet oxygen generated from photoirradiation of Rose Bengal or phthalocyanine (both by immobilized dyes and dye-solution) for the purpose of purifying drinking water.

Sterilization of microbes in the presence of human cells, bodies, or the other valuable living things (in vivo work)

Most of this of work has been devoted to make blood products free from viral infection like human immunodeficiency virus (HIV) or the hepatitis viruses B and C.[12, 13] For applying these methods to the agriculture and aquaculture regions, we tried and found that Penaeid White Spot Syndrome Disease (WSS) that has been present in most of aquacultural farms all over the world can be prevent completely by the singlet oxygen treatment using the immobilized RB dye and visible light, if the treatment was applied at the egg–stage.[1-3] We also found that this method can cure the white spot disease (caused by *Ichtyophthirius multifilis*) of goldfish (*Carassius auratus*). Many applications are under investigation in the large field of agriculture and aquaculture.

Cancer treatment (Photodynamic Treatment (PDT))

Many reviews have been written following the first report on PDT by Bellnier and Dougherty.[14, 15] Photodynamic therapy is based on the dye-sensitized photooxidation of biological matter in the target tissue (Foote, 1990). This requires the presence of a dye (sensitizer) in the tissue to be treated. Although such sensitizers can be naturally occurring constituents of cells and tissues, in the case of PDT, they are introduced into the organism as the first step of treatment. In the second step, the tissue-localized sensitizer is exposed to light of wavelength appropriate for absorption by the sensitizer. Through various photophysical pathways, also involving molecular oxygen, oxygenated products harmful to cell function arise and eventual tissue destruction results. These are to be regarded as one special case in the above (2).

ACKNOWLEDGEMENTS

The author cordially acknowledges Professors T. Itami (Miyazaki Univ., Japan), Y. Takahashi (Shimonoseki Univ. Fisheries, Japan), and T. Nagai (Tokyo Univ. Agriculture, Japan) for their great help to the Penaeid works. He also thanks his many students involved in the environmental work.

REFERENCES

1. Takahashi Y, Itami T, Maeda M, et al. Polymerase chain reaction (PCR) amplification of bacilliform virus (RV-PJ) DNA in *Penaeus japonicus* Bate and systemic ectodermal and mesodermal Baculovirus (SEMBV) DNA in *Penaeus monodon* Fabricius. J Fish Diseases 1996; 19: 399-403.

2. Itami T, Maeda M, Suzuki N, et al. Possible prevention of white spot syndrome (WSS) in Kuruma shrimp, *Penaeus japonicus*, in Japan. In: Flegel TW. eds. Advances in Shrimp Biotechnology. Bangkok, Thailand: BIOTEC, National Center for Genetic Engineering & Biotechnology, 1998: 291-5.

3. Suzuki N, Mizumoto I, Itami et al. Dye-sensitized inactivation of white spot syndrome virus attached to eggs of Crustaceans. In: Roda A, Pazzagli M, Kricka LJ, Stanley PE. eds. Bioluminescence and Chemiluminescence: Perspectives for the 21st Century. Chichester: J. Wiley & Sons, 1999: 559-62.

4. Kearns DR. Physical and chemical properties of singlet molecular oxygen. Chem Rev 1971; 71: 395-429.

5. Wasserman HH, Murray RW. eds. Singlet Oxygen. New York: Academic Press, 1979.

6. Bellus D. Physical quenchers of singlet molecular oxygen. Adv Photochem 1979; 11: 105-202.

7. Matsuura T. Oxygenation Reactions. Tokyo: Maruzen, 1977.

8. Foote CS. Photosensitized oxidation and singlet oxygen: Consequences in biological systems. In: Pryor W. ed. Free Radicals in Biology. New York: Academic Press, 1976.

9. Midden WR, Wang SY. Singlet oxygen generation for solution kinetics: Clean and Simple. J Am Chem Soc 1983; 105: 4129-35.

10. Saito I, Matuura T, Inoue K. Formation of superoxide ion via one-electron transfer from electron donors to singlet oxygen. J Am Chem Soc 1983; 105: 3200-6.

11. For example: Dahl TA, Midden WR, Hartman PE. Comparison of killing of gram-negative and gram-positive bacteria by pure singlet oxygen. J Bacteriol 1989; 171: 2188-94.

12. Dodd RY. The risk of transfusion-transmitted infection. N Engl J Med 1992; 327: 419-21.

13. Sloand EM, Pitt E, Klein HG. Safety of the blood supply. J Am Med Assoc 1995; 274: 1368-73.

14. Bellnier DA, Dougherty TJ. Membrane lysis in Chinese hamster ovary cells treated with hematoporphyrin derivative plus light. Photochem Photobiol 1982; 36: 43-7.

15. Bellnier DA, Dougherty. Protection of murine skin and transplantable tumor against PhotofrinII mediated photodynamic sensitization with WR-2721. J Photochem Photobiol 1989; 49: 369-72.

STUDY ON CHEMILUMINESCENT PROBES FOR SUPEROXIDE ANIONS : CONTROL OF CHEMILUMINESCENCE RESONANCE ENERGY TRANSFER BY CYCLOMALTOOLIGOSACCHARIDE (CYCLODEXTRIN)

K TERANISHI, T NISHIGUCHI

Faculty of Bioresources, Mie University, Kamihama, Tsu, Mie 514-8507, Japan
Email: teranisi@bio.mie-u.ac.jp

INTRODUCTION

Our studies have focused on achieving goals to obtain improved green-chemiluminescent probes, in comparison with 6-[4-[2-[N'-(5-fluoresceinyl)thioureido]-ethoxy]phenyl]-2-methylimidazo[1,2-*a*]pyrazin-3(7*H*)-one (FCLA),[1] for measuring superoxide anions. Herein, we describe the synthesis and luminescence properties of novel green-luminescent probes that feature a hypoxantine-xanthine oxidase system as a source of superoxide anions.

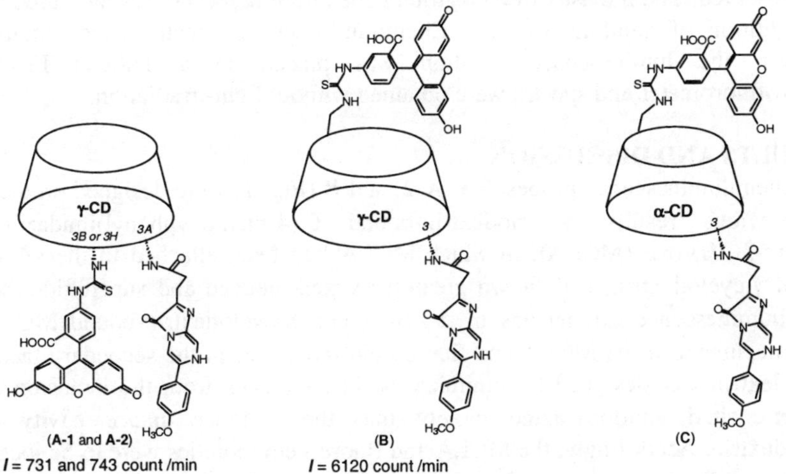

Figure 1. Probes **A-1**, **A-2**, **B**, and **C**, and superoxide-induced chemiluminescence intensity (*I*) at 1.0 μM probe concentration

MATERIALS AND METHODS

General procedure for synthesis and analysis of probes

Analytical and preparative HPLC were done using a JASCO Gulliver HPLC system with a MD-910 detector. A Cosmosil 5C18-MS column (4.6 mm x 150 mm) was used for the analytical HPLC. HPLC preparative chromatography was carried out with a Cosmosil 5C18-MS column (20 mm x 250 mm). Preparative open chromatography was conducted with a Fuji Silysia Chromatorex DM1020T ODS gel. Analytical TLC was performed on Merck Kieselgel 60 F254 precoated, glass-packed plates of 0.25 mm layer thickness and spots of compounds were visualized under a UV lamp or with a *p*-anisaldehyde-H_2SO_4-EtOH solution. ^{1}H NMR spectra were measured with a JEOL JNM-A500 spectrometer operating at 500 MHz. ^{13}C NMR spectra were measured with a JEOL JNM-A500 spectrometer operating at 125.65 MHz. Matrix-assisted laser desorption ionization-time of flight (MALDI-TOF) mass spectra (positive) were recorded on a Kratos Analytical Ltd. Kompact Discovery instrument using 2,5-dihydroxybenzoic acid as a matrix and an average of 50 laser shots per sample.

Measurement of superoxide-induced chemiluminescence intensities and spectra
Chemiluminescence intensities were obtained as follows: xanthine oxidase (0.37 units/mL, 40 μL) was added to the mixture consisting of 20 mM Mops/0.2 M KCl (pH 7.2, 0.5 mL), 0.3 mM hypoxanthine (0.5 mL), and 25 mM probe in water at 25 °C, then the reaction mixture was placed in an Aloka Luminescence Reader BLR-301 and chemiluminescent intensity time curves were obtained at 25 °C. Immediately after xanthine oxidase was added, the chemiluminescence with maximum intensity was observed. The intensity of background chemiluminescence was measured before the addition of xanthine oxidase. Chemiluminescence spectra were obtained as follows: the luminescence solution was placed in a JASCO FP-750DS spectrofluorometer and spectra were obtained without light-irradiation.

RESULTS AND DISCUSSION
The chemiluminescence probes **A-1**, **A-2**, and **B** (Fig. 1) were designed on the basis of previous results: γ-cyclodextrin-bound 6-(4-methoxyphenyl)imidazo[1,2-*a*] pyrazin-3(*7H*)-one (MCLA), in which MCLA had been attached to the secondary site of γ-cyclodextrin, had shown greatest oxygen-induced and superoxide-induced chemiluminescence efficiencies in α-, β-, γ-and δ-cyclodextrin-bound MCLAs. [2-4] The attachment of the MCLA and fluorescein molecules at the secondary face of γ-cyclodextrin was designed to minimize the interferences from the inclusion of the singlet-excited amidopyrazine moiety into the entrance and/or cavity of γ-cyclodextrin. Accordingly, the MCLA and fluorescein moieties were to be located at the A and B glucose units of the γ-cyclodextrin molecule. In accordance to this design, probes **A-1** and **A-2** were synthesized as shown in Scheme 1. Moerover, synthesis of probe **B**, which has the MCLA moiety at the secondary face and the fluorescein moiety at the primary face of γ-cyclodextrin, was achieved as shown Scheme 2.

Spectra of the superoxide-induced chemiluminescence of probes **A-1**, **A-2**, and **B** exhibited their luminescence maximum only at around 515 - 527 nm, which was due to luminescence from the fluorescein moiety, along with the absence of blue luminescence due to the MCLA moiety. These results clearly indicated that the superoxide-induced chemiluminescence of **A-1**, **A-2**, and **B** were generated from the

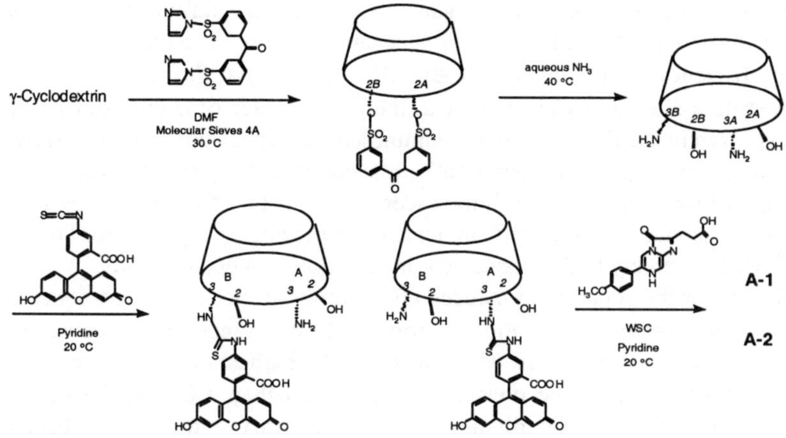

Scheme 1. Synthesis of probes **A-1** and **A-2**

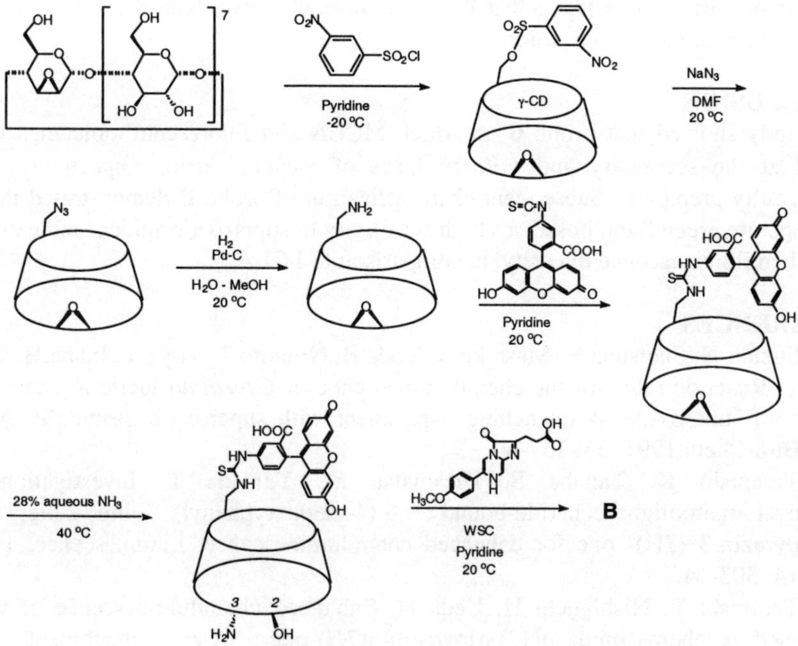

Scheme 2. Synthesis of probe **B**

fluorescein moiety, and that the singlet-excitation energy generated from the MCLA moiety was efficiently transferred to the fluorescein moiety, even the presence of cyclodextrin molecule. In order to gain insight into the influence of the distance between the singlet-excited amidopyrazine and fluorescein moieties to the energy transfer efficiency, a compound with the MCLA moiety at the secondary face of α-cyclodextrin and a fluorescein moiety at the primary face of α-cyclodextrin (Fig. 1, **C**) was synthesized. The chemiluminescence spectrum of **C** showed a chemiluminescence peak at around 460 nm accompanied by smaller peaks at around 520 nm, which was due to luminescence from the fluorescein moiety. The chemiluminescence spectrum of **C** indicated that the energy transfer from the singlet-excited MCLA moiety to the fluorescein moiety does not occur readily.

Results of the superoxide-induced chemiluminescence at a probe concentration of 1.0 μM, are summarized in Fig. 1. Probe **B** showed green-luminescence intensity that was 26 times that of FCLA, which was also the highest luminescence intensity in this present study. At probe concentrations of less than 1.0 μM, the ratio of the superoxide-dependent chemiluminescence intensity to the background chemiluminescence intensity for **B** was higher than that of FCLA. These high superoxide-induced chemiluminescence intensity and superoxide-specificity in low probe concentrations indicates that **B** can be more effective than FCLA towards the measurement of superoxide anions.

CONCLUSION

This study showed that probe **B**, in which MCLA and fluorescein molecules were bound at the secondary and primary faces of γ-cyclodextrin, respectively, was successfully prepared. Subsequent characterization of probe **B** demonstrated that it can generate green light, possesses high sensitivity to superoxide anions, and exhibits high chemiluminescence intensity, in comparison to FCLA.

REFERENCES

1.   Suzuki N, Suetsuna K, Mashiko S, Yoda B, Nomoto T, Toya Y, Inaba H, Goto T. Reaction rates for the chemiluminescence of *Cypridina* luciferin analogues with superoxide: A quenching experiment with superoxide dismutase. Agric Biol Chem 1991; 55: 157-60.
2.   Teranishi K, Tanabe S, Hisamatsu M, Yamada T. Investigation of cyclomaltooligosaccharide-bound 6-(4-methoxyphenyl) imidazole[1,2-a] pyrazin-3 (7H)- one for enhanced chemiluminescence. Luminescence, 1999; 14: 303-14.
3.   Teranishi T, Nishiguchi H, Ueda H. Enhanced chemiluminescence of 6-(4-methoxyphenyl)imidazo[1,2-*a*]pyrazin-3(7*H*)-one by attachment of cyclomaltooligosaccharide (cyclodextrin). Attachment of cyclomaltononaose (δ-cyclodextrin). Carbohydr Res 2003; 228: 987-93.
4.   Teranishi K. Cyclodextrin-bound 6-(4-methoxyphenyl)imidazo[1,2-*a*]pyrazin-3(7*H*)-one as chemiluminescent probe for superoxide anions. ITE Letts Batteries New Technol Med 2003; 4: 16-20.

LUMINOL-DEPENDENT CHEMILUMINESCENCE OF PERIPHERAL NEUTROPHILS FROM WORKERS EXPOSED TO LOW FREQUENCY ELECTROMAGNETIC FIELDS

ML VUOTTO[1], N SANNOLO[2], R MIRANDA[3], F LIOTTI[2], C DE SETA[2], D SPATUZZI[1], G RUGGIERO[1], M DI GRAZIA[2], P DE SOLE[4]

[1]*Dipartimento di Patologia Generale,* [2]*Dipartimento di Medicina Sperimentale, Seconda Università degli Studi di Napoli,* [3] *Servizio Sanitario della Polizia di Stato, Italy,* [4]*Istituto di Biochimica e Biochimica Clinica, Università Cattolica, Roma, Italy*
Via L. De Crecchio, 7 80138 Napoli, Italy
E-mail: marialuisa.vuotto@unina2.it

INTRODUCTION

Electromagnetic fields (EMF) can exert biological effects. Epidemiological studies hypothesized that EMF exposure may be linked to an increased risk of leukemia and cancer.[1] The production of reactive oxygen species (ROS) is considered a possible route for cellular damage in presence of environmental electromagnetic fields, through their effects on the spin procession rates of unpaired electrons with consequent effects on the radical lifetime.[1]

Phagocytes are, *in vivo*, the main source of free radicals and other ROS that are generated in defense against bacteria and in response to various stimuli.[2] For this reason, polymorphonuclear neutrophils (PMNs) are a useful model to study cell activation and the interference of EMF with signalling pathways.

Some authors reported the effects of *in vitro* 60 Hz 0.1 mT magnetic fields exposure on the phorbol 12-myristate-13-acetate (PMA)-induced oxidative burst in peritoneal elicited rat PMNs. The exposed cells showed more than 10% increase of fluorescence than unexposed ones. This work was an early observation that extremely low frequency electromagnetic fields (ELF) influence cellular events by free radical production.[3]

Studies on phagocyte activity after *in vitro* exposure to ELF showed that they can affect monocyte NO production[4] or induce PMN morphologic changes[5] and A_{2A} receptor expression.[6] Few data are available about the effects of ELF on PMN activity in exposed workers. In this study we evaluated the consequences of chronic exposition to ELF on PMNs, measuring their resting and stimulated chemiluminescence (CL) activity.

PATIENTS AND METHODS
Subjects, blood collection and PMN isolation

20 locomotive conductors, age 46.5 ± 4.5 years (mean ± SD), with professional exposure to ELF for more than 10 years, were recruited for this study. Exclusion criteria were:

body mass index >30, presence of chronic or acute diseases at recruitment, drug use (also herbal medicine), professional or residential exposure to ELF or other physical or chemical agents. All subjects were male. Seven subjects were smokers at the time of recruitment and six were formerly smokers. 20 controls were chosen from office workers with the same exclusion criteria established for ELF exposed workers. All controls were male and their age was (mean ± SD) 47.2 ± 4.9 years. 7 subjects were smokers at the time of recruitment and 7 smokers in the past times.

Samples were collected, before eating, between 08.00 and 09.00 a.m., to minimize day-time variability of phagocyte respiratory burst. PMNs were isolated using a discontinuous gradient of isotonic Percoll (Pharmacia)..

Electromagnetic field measurements

EMF have been recorded by the probe Radianse Innova BMM-3. Measurements were performed at the work station of the locomotive, at 13 cm and 55 cm from ground level. Values (μT) were the mean of 10 repeated determinations (lasting 5 minutes).

Neutrophil chemiluminescence measurements

CL assays were performed following De Sole protocol.[7] The reaction mixtures contained, in 1.0 mL final volume, 100 μL of isolated PMN suspension at 0.5x10[6] cell/mL^{-1} and 100 nmoles luminol (Sigma), in presence or absence of 0.5 mg opsonized zymosan (OZA) or 150 nmoles phorbol myristate acetate (PMA). The CL responses were evaluated as total counts x 90 min. The quantification of extra- and intracellular CL was performed according to Mundi protocol.[8] The extracellular CL emission was measured by adding 1 mM azide and 4U horseradish peroxidase (HRP) to the luminol-dependent CL assay reagents, whereas the evaluation of intracellular CL was performed by adding superoxide dismutase (200 U) and catalase (2000 U) to the same reagents.

Statistical analysis

The results of all experiments were expressed as mean ± SD. Levels of significance were determined using analysis of variance. Values of $p < 0.05$ were regarded as significant.

RESULTS

In the work areas, mean ELF electromagnetic field intensity was 0.63 ± 0.08 μT (mean ± SD) at 13 cm and 1.07 ± 0.12 μT at 55 cm from ground level.

CL test results are shown in Table 1. Total CL emission of resting PMNs from ELF-exposed subjects was higher than unexposed controls ($p < 0.05$). PMA and OZA elicited a CL emission lower in ELF-exposed subjects than in controls ($p < 0.05$).

Extracellular CL emission from resting PMNs of ELF-exposed subjects showed no differences versus controls. When stimulated, cells showed lower CL emission in ELF-exposed subjects than in controls ($p < 0.05$). On the contrary, intracellular CL emission in ELF-exposed subjects was higher than in controls ($p < 0.05$). When stimulated by PMA or OZA, PMNs from ELF-exposed workers showed no differences in intracellular emission compared to the controls.

Table 1. CL emission of isolated PMNs (^)

		PMNs		
		Resting	**PMA-activated**	**OZA-activated**
TotalCL	**ELF exposed**	1.93×10^3 $\pm 1.38 \times 10^2$ (*)	1.10×10^4 $\pm 1.86 \times 10^3$ (*)	8.25×10^4 $\pm 2.06 \times 10^3$ (*)
	controls	1.16×10^3 $\pm 1.12 \times 10^2$	1.75×10^4 $\pm 1.49 \times 10^3$	1.15×10^5 $\pm 2.11 \times 10^3$
ExtraCL	**ELF exposed**	2.94×10^4 $\pm 2.18 \times 10^3$	1.14×10^5 $\pm 2.55 \times 10^4$ (*)	6.81×10^5 $\pm 1.69 \times 10^4$ (*)
	controls	2.80×10^4 $\pm 1.11 \times 10^3$	2.46×10^5 $\pm 1.87 \times 10^4$	1.77×10^6 $\pm 9.45 \times 10^4$
IntraCL	**ELF exposed**	7.76×10^3 ± 532 (*)	7.11×10^5 $\pm 3.03 \times 10^4$	5.43×10^6 $\pm 7.41 \times 10^4$
	controls	5.90×10^3 ± 421	7.48×10^5 $\pm 1.24 \times 10^4$	6.22×10^6 $\pm 4.12 \times 10^4$

(^) counts/90 min/cell (mean±SD)   (*) p<0.05

DISCUSSION AND CONCLUSIONS

CL activity of resting PMNs was higher in exposed than in control subjects. This finding indicates that subjects chronically exposed to ELF have a basal production of ROS higher than unexposed controls. Compartmental studies showed that the basal intracellular chemiluminescence emission increased in ELF exposed subjects, whereas the extracellular one did not change when compared with controls.

Some authors found that weak ELF can increase the amplitude of oscillation of NADPH concentration in resting PMNs. This increase directly paralleled cell activation and shape modifications.[6] On the basis of these findings, we could hypothesize that the increased ROS production of resting PMNs from locomotive conductors could be of metabolic origin.

PMNs from ELF-exposed subjects emitted total CL significantly lower than controls when PMNs were activated by PMA or OZA. However, PMA- and OZA-stimulated intracellular CL did not show differences between the two groups of subjects while, on the contrary, extracellular emission was lower in ELF-exposed subjects.

PMA is a direct activator of PKC, a key enzyme implied in the cascade of events following the OZA-induced activation.[2] PKC activity was found altered in HL60 cells exposed to 60 Hz AC electric fields.[9] In this experiment, ELF, alone or combined with PMA, promoted a down-regulation of cytosolic PKC activity. Therefore, PKC down-regulation could be responsible for reduced activation in our experiments. We can also hypothesize that stimulated PMNs from exposed subjects respond less than unexposed ones because the cells have a basic, chronic, low grade of activation, as found in resting PMNs.

REFERENCES

1. Lacy-Hulbert A, Metcalfe JC, Hesketh R. Biological responses to electromagnetic field. FASEB J 1998; 12: 395-420.
2. Seymour JK. Oxygen metabolites from phagocytes. In: Gallin JI, Snyderman R. eds: Inflammation. Basic principles and clinical correlates. Philadelphia, Lippincot Williams & Wilkins, 1999; pp.721-68.
3. Roy S, Noda Y, Eckert V, Traber MG, Mori A, Liburdy R, Packer L. The phorbol 12-myristate 13-acetate (PMA)-induced oxidative burst in rat peritoneal neutrophils is increased by a 0.1 mT (60 Hz) magnetic field. FEBS Lett 1995; 376: 164-6.
4. Yoshikawa T, Tanigawa M, Tanigawa T, Imai A, Hongo H, Kondo M. Enhancement of nitric oxide generation by low frequency electromagnetic field. Pathophysiology 2000; 7: 131-5.
5. Varani K, Gessi S, Merighi S, Iannotta V, Cattabriga E, Spisani S, Cadossi R, Borea PA. Effect of low frequency electromagnetic fields on A2A adenosine receptors in human neutrophils. Br J Pharmacol 2002; 136: 57-66.
6. Kindzelskii AL, Petty HR. Extremely low frequency pulsed DC electric fields promote neutrophil extension, metabolic resonance and DNA damage when phase-matched with metabolic oscillators. Biochim Biophys Acta 2000; 1495: 90-111.
7. De Sole P, Fresu R, Frigieri L, Pagliari G, De Simone C, Guerriero C. Effect of adherence to plastic on pheripheral blood monocyte and alveolar macrophage chemiluminescence. J Biolumin Chemilumin 1993; 8: 153-8.
8. Mundi H, Bjorksten B, Svanborg C, Ohman L, Dahlgren C. Extracellular release of reactive oxygen species from human neutrophils upon interaction with *Escherichia coli* strains causing renal scarring. Infect Immun 1991; 59: 4168-72.
9. Holian O, Astumian RD, Lee RC, Reyes HM, Attar BM, Walter RJ. 1996. Protein kinase C activity is altered in HL60 cells exposed to 60 Hz AC electric fields. Bioelectromagnetics 1996; 17: 504-9.

DEVELOPMENT OF FIA-CHEMILUMINESCENCE METHODS TO EVALUATE QUENCHING EFFECTS AGAINST REACTIVE OXYGEN SPECIES

M WADA[1], M KATOH[1], H KIDO[2], MN NAKASHIMA[1],
N KURODA[1], K NAKASHIMA[1]

*[1]Graduate School of Biomedical Sciences, Nagasaki University, 1-14 Bunkyo-machi,
Nagasaki 852-8521, Japan*
*[2]Mitsubishi Chemical Corporation, Specialty Chemicals Company,
1000 Kamoshida-cho, Aoba-ku, Yokohama 227-8502, Japan*
E-mail: naka-ken@net.nagasaki-.ac.jp

INTRODUCTION

Oxidative stress induced by reactive oxygen species (ROS) is believed to be a primary factor in various diseases. Recently, a ROS quencher has become important for human health in proportion to understanding its quenching mechanism. Thus, food additives and functional foods with quenching effects have been widely used as health food supplements. These are considered to have potential healthy benefits. In this view, simple and rapid evaluation methods for a quenching effect against ROS are required to control properties of health foods. In our previous reports, the quenching effects of fluvastain, its metabolites[1] and rosemary extracts[2] were evaluated by batch-CL methods based on the chemiluminescence reaction between luminol and ROS. However, they consumed relatively large volumes of reagents, were tedious and showed low reproducibility.

In this study, therefore, flow injection analysis-chemiluminescence (FIA-CL) methods to evaluate quenching effects of functional foods against ROS such as singlet oxygen ($^1O_2$) and hydroxyl radicals ($\cdot$OH) were developed. The applicabilities of these methods were confirmed by evaluating quenching effects of grape seed extracts and their isolated polyphenols.

METHODS

Grape seed extracts used were commercially available. Polyphenols in grape seed extracts selected were chalcone, pelargonidin, cyanidin, delphinidin and resveratrol. All samples were dissolved in DMSO or DMF to prepare 0.5-2.0 mg/mL for $^1O_2$ and 0.25-2.0 mg/mL for $\cdot$OH. The quenching activities of the samples for ROS were measured by the FIA-CL methods. The proposed methods are based on the CL reaction between luminol and ROS. A Fenton-reaction and a H_2O_2-NaBr-lactoperoxidase (LPO) system were utilized to prepare $\cdot$OH and $^1O_2$, respectively.

Sample solution for FIA: [$^1O_2$] To 6 μL of sample in DMSO in a test tube, 300 μL each of 0.1% H_2O_2, 80 mM NaBr and 0.08 mM luminol in 100 mM acetate buffer (pH 4.5) solution were added. The mixture was incubated at 37 °C for 10 min and then injected into the FIA system. [$\cdot$OH] To 6 μL of sample in DMF, 300 μL of

0.05% H_2O_2 in Hepes buffer (pH 7.4), and 600 μL of 1.2 mM luminol in Hepes buffer were added. After incubation at 37 °C for 10 min the mixture was injected into FIA.

The FIA system consisted of two CCPD chromatographic pumps (Tosoh, Tokyo, Japan), an 825-CL detector (Jasco, Tokyo), a 7125 injector with a 100-μL of sample loop (Rheodyne, Cotati, CA,USA) and an R-111 recorder (Shimadzu, Kyoto, Japan).

FIA conditions: Carrier solution, 0.1 M acetate buffer (pH 4.5) for $^1O_2$ and 0.1 M Hepes buffer (pH 7.4) for ·OH; CL reagent, 10 μg/mL LPO in acetate buffer for $^1O_2$ and 8 mM diethylenetriaminepentaacetic acid and 200 μM $FeCl_2$ in Hepes buffer for ·OH. The flow rates of carrier solution and CL reagent for both ROS were set at 0.5 and 0.1 mL/min, respectively. The lengths of mixing coil (0.25 mm, i.d.) used were 12 cm for ·OH and 23 cm for $^1O_2$.

Quenching effect was calculated by the following equation: quenching effect %=100-{(CL intensity of sample/CL intensity of blank)}x100. The increase in the value means an increase in quenching effect.

RESULTS

FIA conditions such as concentration of luminol, flow rates of carrier solution and CL reagent were optimized for $^1O_2$ and ·OH by measuring blank CL intensity, respectively. In both ROS, 0.5 mL/min of carrier solution and 0.1 mL/min of CL reagent gave maximum and constant CL intensity. Luminol concentrations for $^1O_2$ and ·OH were selected 0.08 mM and 1.2 mM as optimum, respectively. These conditions gave almost maximum and constant blank CL intensity, and the precision of 5 replicate measurements for ·OH and $^1O_2$ as the relative standard deviation were less than 1.3%.

Applicabilities of the proposed methods were confirmed by measuring quenching effects of grape seed extracts, chalcone, pelargonidin, cyanidin, delphinidin and resveratrol. In Fig. 1, polyphenols (except chalcone) showed higher quenching effects on ·OH than grape seed extracts. Quenching effect of resveratrol was the highest in all polyphenols tested. A typical recorder response of pelargonidin on ·OH was shown in Fig. 2.

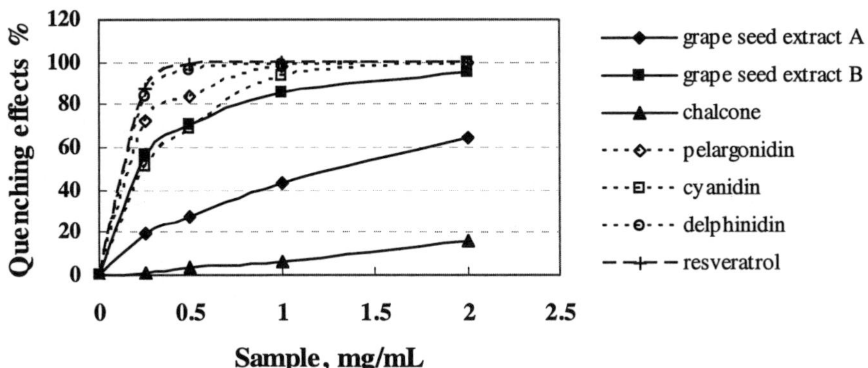

Figure 1. Quenching effects of grape seed extracts and polyphenols against ·OH

Quenching effects against $^1O_2$ were ranging from 3.1 to 100%. Quenching order of polyphenols corresponded well to that on ·OH. Chalcone had no quenching effect on both ·OH and $^1O_2$. The quenching effects of samples were summarized in Table 1.

Table 1. Quenching effect of grape seed extracts and polyphenols

Sample	Quenching effect %, mean ± S.D.*	
	·OH**	$^1O_2$**
Grape seed extract A	19.2±7.3	14.9±2.1
B	56.6±8.6	52.0±4.1
chalcone	1.1±1.7	3.3±0.6
pelargonidin	72.2±1.4	78.5±5.1
cyanidin	51.3±4.9	73.3±8.8
delphinidin	80.2±7.3	90.2±6.8
resveratrol	87.7±2.4	90.1±1.5

*n=3; ** Sample concentration=0.25 mg/mL for ·OH and 0.5 mg/mL for $^1O_2$.

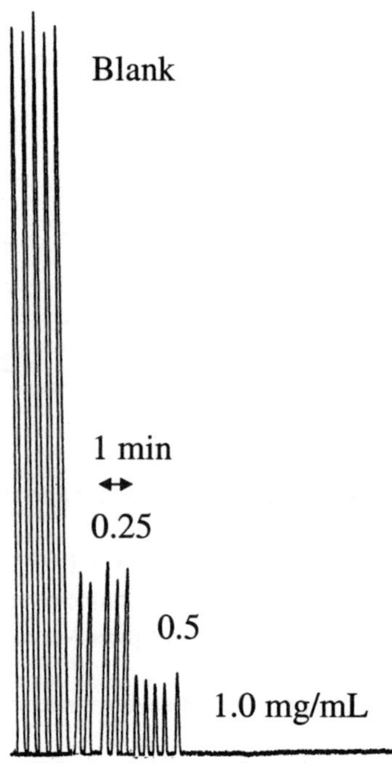

Figure. 2. Recorder responces of pelargonidin

The FIA-CL methods were developed for measurement of quenching effects of grape seed extracts and polyphenols to ROS. Rapid and precise measurement for ·OH and $^1O_2$ could be achieved by the proposed methods (2 or 3 injections/min).

REFERENCES

1. Nakashima A, Ohtawa M, Iwasaki K, Wada M, Kuroda N, Nakashima K. Inhibitory effects of fluvastain and its metabolites on the formation of several reactive oxygen species. Life Sci. 2001; 69: 1381-89.
2. Wada M, Kido H, Ohyama K, Kishikawa N, Ohba N, Kuroda N, Nakashima K. Evaluation of quenching effects of non-water soluble and water-soluble rosemary extracts against active oxygen species by chemiluminescent assay. Food Chem 2004; 87: 261-67.

SYNTHESES AND PROPERTIES OF CELL-MEMBRANE PERMEABLE LUCIGENIN DERIVATIVES FOR THE ASSAY OF INTRACELLULAR SUPEROXIDE

S YAMADA, N KOHSAKA, M IWAMURA

Dept of Biomolecular Science, Toho University, Chiba 274-8510, Japan
Email: sachiko@biomol.sci.toho-u.ac.jp

INTRODUCTION

Several chemiluminescent compounds have been used for the analyses of reactive oxygen species (ROS) such as superoxide and hydrogen peroxide, generated in biosystems. Among them lucigenin (10,10'-dimethyl-9,9'-biacridinium dinitrate, LUC) is one of the most useful chemiluminescent probes, because of its specific reactivity for superoxide, which is firstly formed from molecular oxygen in living cells and converted into the other ROS.[1, 2] However, LUC cannot be used for the assay of intracellular superoxide due to its cell-membrane impermeability.[3, 4] In addition, it has been known that LUC forms an intramolecular charge-transfer complex between biacridinium dication and counter anions and that the CT characters depend on the electron donor abilities of the anions.[5, 6] We assumed that this impermeability is mainly due to hydrophilic property of LUC and electrostatic interactions between nitrate anions of LUC and negative charges of membrane surfaces. That is, LUC derivatives consisting of appropriate anions might be amphiphilic and less ionic to become cell-membrane permeable. Therefore in order to find suitable anions, we synthesized LUC derivatives having different kinds of anions from the nitrate anions. In this paper we will report the syntheses of LUC derivatives and their cell-membrane permeabilities and further reveal their properties related to cell-membrane permeabilites.

X; ionic ⟹ less ionic (CT complex)

Cell-membrane impermeable ⟹ permeable

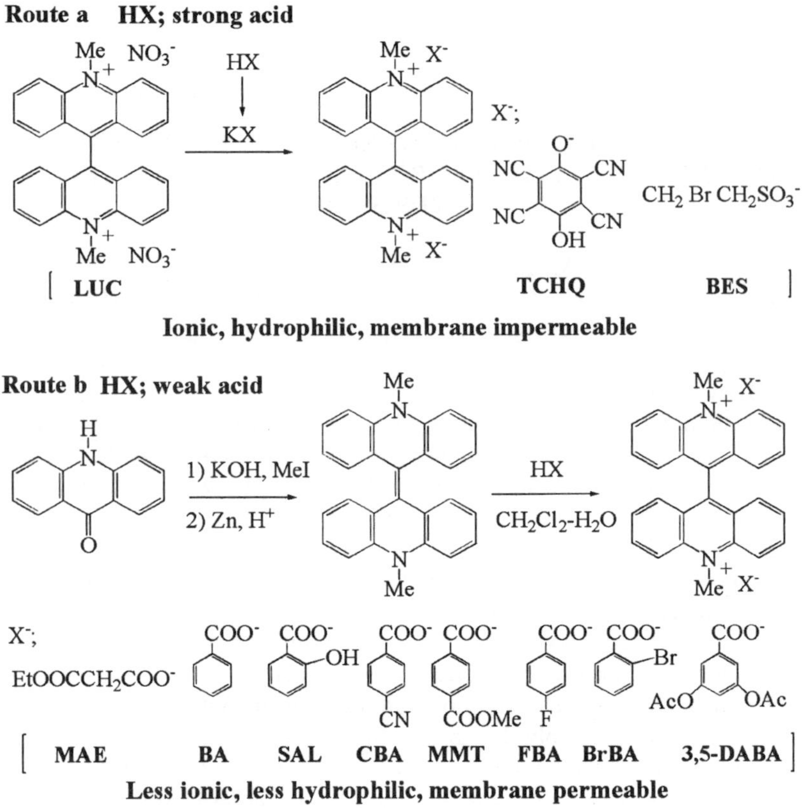

Route a   HX; strong acid

Ionic, hydrophilic, membrane impermeable

Route b HX; weak acid

Less ionic, less hydrophilic, membrane permeable

Scheme 1. Syntheses of lucigenin derivatives and their cell- mebrane permeabilities.

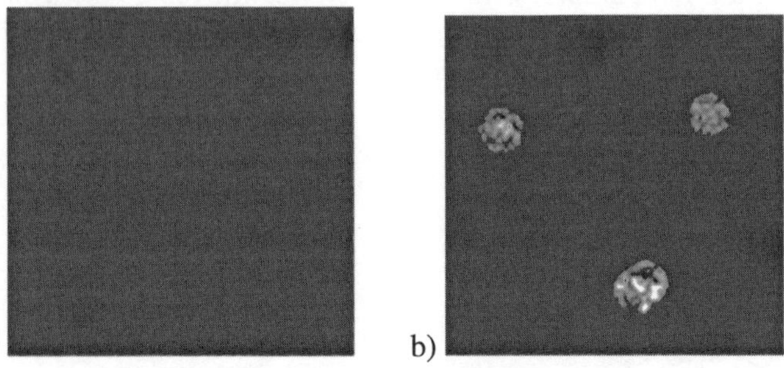

Figure 1. The dark-field images of mouse neutrophils (5×10^4 cells in RPMI 500 μL) incubated with a) LUC and b) MMT (final concentration 25 μM).

MATERIALS AND METHODS

LUC derivatives were synthesized by either **route a** or **b** as shown in Scheme 1. In **route a,** the nitrate ions of LUC was directly exchanged into other weaker electron donating anions to afford LUC derivatives. In **route b**, starting material, 9(10)-acridone was methylated followed by reductive coupling to afford 10,10'-dimethyl-9,9'-biacridinylidene, which was then oxidized automatically to LUC derivative in the presence of the corresponding acid in a biphasic CH_2Cl_2-H_2O solvent system. The aqueous layer was evaporated to afford LUC derivative in pure solids. The structures of the synthesized derivatives were determined by NMR, MS spectra and elemental analyses. The UV-VIS absorption spectra were measured by a Hitachi spectrophotometer (U-3010) and the fluorescence spectra were observed on a Fluoro-max spectrofluorometer. The chemiluminescence spectra were carried out with Otsuka multichannel photodetector (MCPD 7000). The cell-membrane permeability was determined on mouse neutrophils with a confocal laser-scanning micrometer (OLYMPUS IX 71).

RESULTS AND DISCUSSION

In order to find amphiphilic, less hydrophilic derivatives we used various kinds of organic acids as the source of counter anions of LUC derivatives. Strong acids such as 2-bromoethanesulfonic acid and tetracyanohydroquinone readily formed LUC derivatives, BES and TCHQ in route **a**. On the other hand, LUC derivatives from rather weak acids such as malonic acid monoethyl ester and benzoic acid could not be purified by conventional methods. Therefore we developed a new synthetic method, route **b** to afford pure LUC derivatives, MA, MAE, BA, SAL, BrBA, CBA, MMT, FBA, 3,5-DABA. All of them showed characteristic absorption, fluorescence and chemiluminescence spectra of bicridinium di-cation as LUC itself. These features indicate that these derivatives are as useful chemiluminescent probes as LUC. The specific reactivity of MMT toward superoxide among ROS (O_2^-, H_2O_2, HClO, OH, $^1O_2$) was examined by using chemically produced ROS.[7] It is confirmed that MMT has the specific reactivity as well as LUC.

The cell-membrane permeability of these compounds (Scheme 1) was examined using mouse neutrophils with comparison of bright and dark images observed by a confocal laser-scanning microscope. Figure 1 shows the dark field images of the neutrophils incubated with LUC and MMT, respectively. Although the image of LUC (a) has no spots, the image of MMT (b) has fluorescent spots at several positions, where the cells are localized in the bright field. It confirms that MMT is incorporated into the cells, but LUC is not, and shows that MMT is cell-membrane permeable.

Next, we studied the correlation between the cell-membrane permeability and chemical and photophysical properties of these compounds. These compounds showed stronger CT absorptions in 500 − 700 nm than LUC. And further, the

                                                    Yamada S et al.

fluorescence quantum yields and redox potentials were lower than those of LUC. It apparently indicates that these compounds are strong CT complexes.

It is noted that the conjugated acids of the anions are weakly acidic (pKa 2.8 – 5.0) and less hydrophilic (log P 1.5 – 5.0) than nitric acid (pKa - 1.8 and log P -0.13). The results on the bioassays using these derivatives will be reported in the near future.

ACKNOWLEDGEMENT
We wish to thank Prof. Yoshiro Kobayashi and Mr. Soichiro Sasaki for the bioassays.

REFERENCES
1. Mckinney K A, Lewis S E M, Thompson W. Reactive oxygen species generation in human sperm; Luminol and lucigenin chemiluminescence probes, Arch. Andrology, 1996; 36:119-25.
2. Műnzel T, Afanas'ev I G, Kleschyov A L, Harrison D G. Detection of superoxide in vascular tissue. Arterioscher Thromb Vasc Biol 2002; 22: 1761-8.
3. Parij N Nagy A-M, Fondu P, Neve J, Effects of non-steroidal anti-inflamamatory drugs on the luminal and lucigenin amplified chemiluminescence of human nuetrophils. Euro J Pharm 1998; 352: 299-305.
4. Dyke K V, Allender P, Wu L, Gutierrez J, Garcia J, Aredekani A, Karo W. Luminol- or lucigenin-coated micropolystyrene beads, a single reagent to study opsonin-independent phagocytosis, by cellular chemiluminescence; Reaction with human neutrophils, monocytes, and differentiated HL 60 cells. Microchem J 1990; 41: 196-209.
5. Maeda K, Kashiwabara T, Tokuyama M, Mechanism of the chemiluminescence of lucigenin. 2. The charge-transfer structure of lucigenin and reduction of by electron transfer from nucleophiles. Bull Chem Soc Jpn 1977; 50: 473-81.
6. Legg K D, Hercules D M, Quenching of lucigenin fluorescence, J Phys Chem 1970; 74: 2114-8.
7. Papadopoulos K, Triantis T, Tsagaraki K, Dimotikali D, Iftimie N, Meghea A, Studies on the photostorage chemiluminescence of aromatic ketones with reactive oxygen species. Prospects for analytical applications. J Photochem Photobiol A; Chem 2002; 152: 11-6.

PART 10

APPLICATIONS IN MICROBIOLOGY, ECOLOGY, AND ENVIRONMENTAL & FOOD TESTING

USE OF BIOLUMINESCENT *SALMONELLA* TYPHIMURIUM DT104 TO MONITOR UPTAKE AND INTRACELLULAR SURVIVAL WITHIN A HUMAN CELL-LINE

JE ANGELL[1], VC SALISBURY[1], PJ HILL[2], HM ALLOUSH[1]

[1]Faculty of Applied Science, University of the West of England, Bristol, UK
[2]School of Biosciences, University of Nottingham, Nottingham, UK
Email: Johanna.Angell@uwe.ac.uk

INTRODUCTION

Salmonellosis is a significant bacterial enteric disease of both human and animals.[1] Multi-drug resistant *Salmonella enterica* serovar Typhimurium phage type DT104 causes food borne disease[2] and is estimated to account for 30% of the 1.4 million Salmonella cases reported there each year.[1] During host infection, micro organisms are internalised within leucocytes before killing takes place.[3] Some micro organisms have the ability to survive and even multiply within host cells.[4] *S.* Typhimurium is one such pathogen that establishes persistent infection by impairing phagolysosomal function.[5]

Traditionally methods of viable counting and microscopy have been used to study phagocytosis of bacteria and subsequent survival or destruction.[6] However, such indirect methods give an underestimate of bacterial survival within cells.[3,7] It has been shown that bioluminescence acts as a convenient real-time method for monitoring survival of *Bordetella bronchiseptica in vitro*,[8] whilst a dual *gfp-luxABCDE* operon has been used to monitor real-time replication of *Staphylococcus aureus*.[9] The use of clinically important bacteria transformed with the *lux* cassette overcomes the problems with viable but non-culturable bacteria, as the expression of *lux* genes depends on the functional biochemistry of the bacteria.[10, 11]

The aim of this study was to establish an optimum sensitive assay using a bioluminescent reporter to investigate the uptake and intracellular survival of the pathogen *S.* Typhimurium DT104 within a macrophage like cell-line.

MATERIALS AND METHODS

Transformation of *S.*Typhimurium DT104 with lux operon

S. Typhimurium DT104 was obtained from the National Food Laboratory (Dublin) and transformed by electroporation using the broad host range plasmid pBBR1MCS-5[12] with *luxCDABE* cassette from *Photorabdus luminescens*. Transformants were spread on LB agar containing 10 µg/mL gentamicin at 37 °C. A highly sensitive photon-counting camera (Photec) was used to select light emitting colonies (*lux*⁺) from the gentamicin resistant transformants.

Phagocytosis of *S.* Typhimurium DT104 *lux*⁺ by THP-1 cell

A single colony of *S.* Typhimuruim DT104 *lux*⁺ was inoculated into 20 ml RPMI 1640 medium 10 µg/mL gentamicin and grown in a shaking incubator overnight at 37 °C. THP-1 cells were counted, washed once in PBS at 1400 rpm for 5 min and resuspended in fresh pre-warmed RPMI 1640 medium. Cells were then placed at 37 °C for 20 min.

Multiplicities of infection (MOI) of 10^3:1 and 2×10^3:1 (Salmonella: THP-1 cells) were used. Salmonella cells were harvested at 3500 rpm for 8 min, washed 3 times in PBS and resuspended in 200 μL fresh antibiotic free RPMI medium. Bacteria and THP-1 cells were incubated together for 40 min at 37 °C and were then decanted into sterile tubes and washed twice in pre-warmed PBS. [6] Colistin was added at 50 μg/mL to inactivate extracellular bacteria and 200 μL of the cultures were placed into a black, clear-bottomed 96 well-plate. 0.1% saponin was added at time 180 min where appropriate. *Lux*[+] and *lux*[-] cultures of *S.* Typhimurium DT104 were used as controls. Bioluminescence was measured over 24 hours in an automated luminometer (Anthos) at 37 °C.

Assessment of infection time

The above method was repeated this time using a single MOI of 10^3:1 but with incubation times of 40, 60, 90 and 120 min.

RESULTS

Transformation of *S.* Typhimurium DT104 with plasmid pBBR1MCS-5

S. Typhimurium DT104 was successfully transformed with the *luxCDABE* cassette and stably bioluminescent colonies were obtained.

Assay to assess phagocytosis of *S.* Typhimurium DT104 *lux*[+] by THP-1 cells

The results of phagocytosis are shown in Figure 1. *Lux*[+] Salmonella grown in RPMI gave at least a 2 $\log_{10}$ increase in light output than any other of the treatments (p<0.0001). Addition of 50 μg/mL colistin to *lux*[+] Salmonella resulted in a significant drop in light output compared to salmonella grown without colistin (p<0.0001). Addition of 0.1% saponin to *lux*[+] Salmonella within THP-1 cells caused the bioluminescence to drop rapidly to give at least 2 $\log_{10}$ drop in light output than those incubated without 0.1% saponin (p<0.0001). A MOI of 10^3:1 gave significantly more light output than that of 2×10^3:1 (p=0.0004).

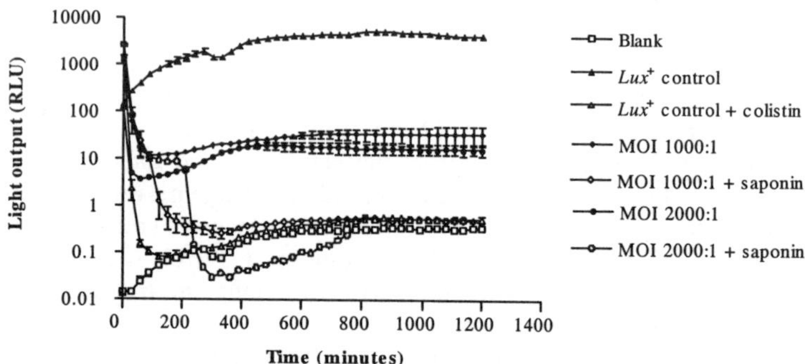

Figure 1. Phagocytosis of *S.* Typhimurium DT104 (pBBR1MCS-5) by THP-1 cells with varying multiplicites of infection

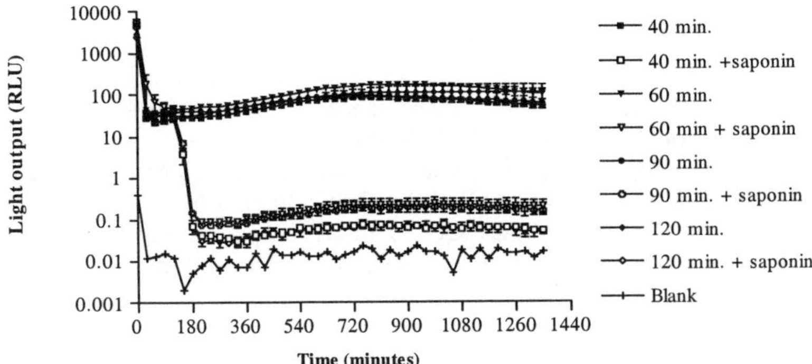

Figure 2. Phagocytosis of *S.* Typhimurium DT104 pBBR1MCS-5 by THP-1 cells with varying incubation times (MOI 10^3:1)

Assessment of Infection times

The results of varying incubations times are shown in figure 2. An incubation time of 60 min gave significantly more light output than that of 40 min (p=0.006) but was not significantly different to that of 90 or 120 min.

DISCUSSION

Detection of bioluminescence from intracellular organisms is affected by several factors. One such problem is optimising the number of phagocytosed bacteria.[8] We found it was necessary to use much higher infection ratios in this work than those in previous studies[13] who had used non-bioluminescence based methods. In contrast to this, evidence suggests that *S.* Typhimurium can cause apoptosis of macrophages by activation of cytokines.[14] The experiment in this study using a MOI at 10^3:1 gave significantly higher levels of continuous light output than the MOI of 2×10^3:1 perhaps suggesting cell death or damage. A MOI of 10^3:1 was therefore chosen as the optimum level for this assay.

A 60 min incubation period was chosen as optimum since it gave the highest level of light output in the shortest period of time.

All cultures that were incubated with saponin, a detergent used to lyse mammalian cells, showed a subsequent fall in bioluminescence to base levels. No such fall was seen in those cultures without saponin thus confirming the intracellular location of the organism.

We have shown that self-bioluminescent *S.* Typhimurium is taken up by THP-1 cells and a stable light output can be monitored over at least 24 hours. These results indicate that this bioluminescence based assay can be used as an effective real-time method to monitor uptake and intracellular survival of *S.* Typhimurium DT104 and, potentially, the effects of antimicrobial agents *in situ* on this clinically important pathogen.

REFERENCES

1. Stephen JM, Toleman MA, Walsh TR, Jones RN. SENTRY Program Participants Group. Salmonella bloodstream infections: report from the SENTRY Antimicrobial Surveillance Program (1997-2001). Int J Antimicrob Agents 2003; 22: 395-405

2. CDR Weekly. National increase in *Salmonella typhimurium* DT104 – update. Commun Dis Rep CDR Rev 2000; 36: 323-6.

3. Hampton MB, Winterbourn CC. Methods for quantifying phagocytosis and bacterial killing by human neutrophils. J Immunological Methods 1999; 232: 15-22.

4. Maurin M, Raoult D. Intracellular organisms. Int J Antimicrob Agents 1997; 9: 61-70.

5. Riber U, Lind P. Interaction between *Salmonella typhimurium* and phagocytic cells in pigs phagocytosis, oxidative burst and killing in polymorphonuclear leukocytes and monocytes. Vet Immunol Immunopathol 1999; 67: 259-70.

6. Greenwood D. Antibiotic effects *in vitro* and the prediction of clinical response. J Antimicrob Chemother 1997; 40: 499-501.

7. Rosen H, Michel BR. Redundant contribution of myeloperoxidase dependant systems to neutrophil-mediated killing of *Escherichia coli*. Infect Immun 1997; 65: 4173-8.

8. Forde CB, Parton R, Coote JG. Bioluminescence as a reporter on intracellular survival of *Bordetella bronchiseptica* in murine phagocytes. Infect Immun 1998; 66: 3198-207.

9. Qazi SNA, Harrison SE, Self T, Williams P, Hill PJ. Real-time monitoring of intracellular *Staphylococcus aureus* replication. J Bacteriol 2004; 186:1065-77.

10. Stewart GSAB, Williams P. Lux genes and the applications of bacterial bioluminescence (review article). J Gen Microbiol 1992; 138: 1289-1300.

11. Salisbury VC, Pfoestl A, Weisinger-Mayr H, Lewis R, Bowker K, MacGowan AP. Use of a clinical *Escherichia coli* isolate expressing lux genes to study the antimicrobial pharmacodynamics of moxifloxacin. J Antimicrob Chemother 1999; 43: 829-32.

12. Kovach ME, Elzer PH, Hill DS, Robertson GT, Farris MA, Roop RM, Peterson KM. Four new derivatives of the broad-host-range cloning vector pBBR1MCS, carrying different antibiotic-resistance cassettes. Gene 1994; 166: 800-2.

13. Chiu CH, Tzou-Yien L, Ou JT. In vitro evaluation of intracellular activity of antibiotics against non-typhoid *Salmonella*. Int J Antimicrob Agents 1999; 12: 47-52.

14. Monack DM, Navarre WW, Falkow S. Salmonella-induced macrophage death: the role of caspase-1 in death and inflammation. Microbes Infect 2001; 3:1201-12.

DEVELOPMENT OF A RANGE OF BIOLUMINESCENT FOOD BORNE PATHOGENS FOR ASSESSING IN-SITU HEAT INACTIVATION AND RECOVERY OF BACTERIA DURING HEAT TREATMENT OF FOODS

A BALDWIN, SM NELSON, RJ LEWIS, A DOWMAN,
HM ALLOUSH, VC SALISBURY

Faculty of Applied Sciences, University of the West of England, Bristol, UK
Email:vyv.salisbury@uwe.ac.uk

INTRODUCTION

Three common food borne pathogenic bacteria were transformed with plasmids carrying the *lux* genes in order to evaluate wet and dry surface pasteurisation of food surfaces. The work was carried out within the 'Bugdeath' programme, an EU Framework V collaborative project to develop predictive models for the surface pasteurisation of raw food materials, based on accurate data obtained from real food samples, heated in standardised, precisely controlled conditions. Previous models, based on indirect viable counts of bacteria, have been shown to be poor predictors of bacterial inactivation and recovery during heat treatment.

It has previously been demonstrated that self-bioluminescent bacteria are highly sensitive reporters of antimicrobial effects.[1,2,3] In order to minimise alteration to the bacterial phenotype, plasmids carrying the *lux* genes were used for bacterial reporter constructs, rather than inserting the *lux* cassette into the chromosome. The plasmids carry the *luxCDABE* genes from *Photorhabdus luminescens*, under the control of a constitutive promoter. The pUC19 derived pLITE27 plasmid[4] was used to transform *Escherichia coli* O157 (tox⁻), whereas the broad host range plasmid pBRRMCS5-LITE[5] was employed for construction of self-bioluminescent *Salmonella enterica* serovar Typhimurium DT104. In order to transform *Listeria monocytogenes* Scott A, a Gram positive vector, pAL2, carrying the modified *lux ABCDE* cassette, was used.[2]

Within the food industry, the use of self-bioluminescent bacterial reporters has been widespread and is well documented.[6,7] In this study the use of self-bioluminescent food borne pathogens enables real time, non-destructive, in-situ monitoring of bacterial inactivation and recovery on food surfaces during and after heat treatment.

MATERIALS AND METHODS
Construction of self-bioluminescent reporter strains
Plasmids were tranformed into bacterial pathogens using electroporation and selected on Luria-Bertani agar containing either 50 mg/L ampicillin (for *E. coli* O157 pLITE27), 10 mg/L gentamicin (for *S.* Typhimurium DT104 pBRRMCS-5LITE) or 10 mg/L erythromycin (for *L. monocytogenes* Scott A pAL2). Resultant

transformants were highly bioluminescent when viewed in a dark room using an ICCD 225 photon counting camera (Photek Ltd., UK).

In-situ monitoring

The *lux* modified bacterial strains were inoculated onto the surface of a range of meat and vegetables, all purchased from local retail shops. Meat samples comprised skinless chicken breast (Ch), turkey breast (Tur), pork (Po), beef (Be), lamb (La), duck (Du), tuna steak (Tu) and chicken breast with skin on (ChS). Vegetables comprised parsnip (Pa), carrot (Ca), courgette/zucchini (Co), celeriac (Ce), leek (Le), pepper/capsicum (Pe) and potato (Po). Food samples, 5cm in diameter and 1cm deep, were cut and fitted into glass sample holders in order to give as flat a test surface as possible. Prepared samples were stored at 4 °C and then inoculated, in the centre without spreading, with 5 µL of an overnight broth culture of the test organism. All heat treatments were carried out on the 'Bugdeath' rig with the ICCD 225 photon camera in place to continuously monitor light output; please see Fig. 1 for details.

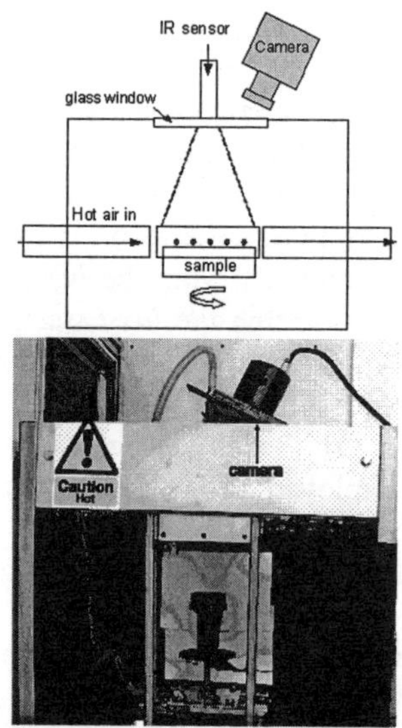

Figure 1. 'Bugdeath' test rig modified to enable real-time quantitation and positional detection of bacterial bioluminescence during heat treatment. The sample holder is shown in the lowered position and light exclusion material is not in place.

RESULTS AND DISCUSSION

The transformed bacteria produced light stably and constitutively at 37 °C, without requiring exogenous substrate. A positive linear correlation between bioluminescence and Optical Density was observed. The results of monitoring light output from bacteria on beef during dry heat treatment are shown in Fig. 2

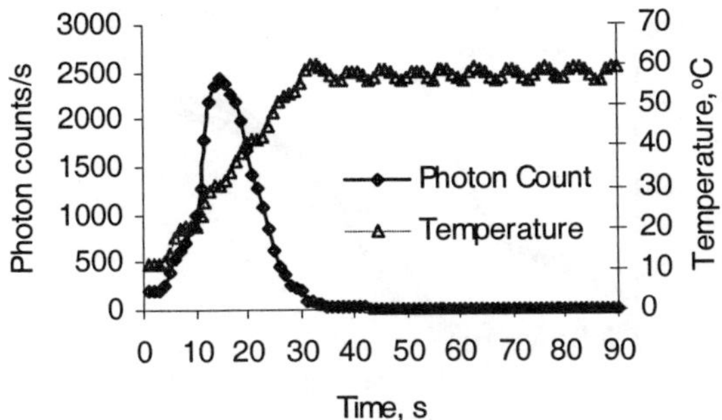

Figure 2. Bioluminescence of *S.* Typhimurium DT104/pBRRMCS-5LITE on the surface of beef and the surface temperature of the meat during a hot air heating cycle to 60 °C and holding for 60 s

Recovery of the bacteria after 21 h at 20 °C is shown in Fig. 3

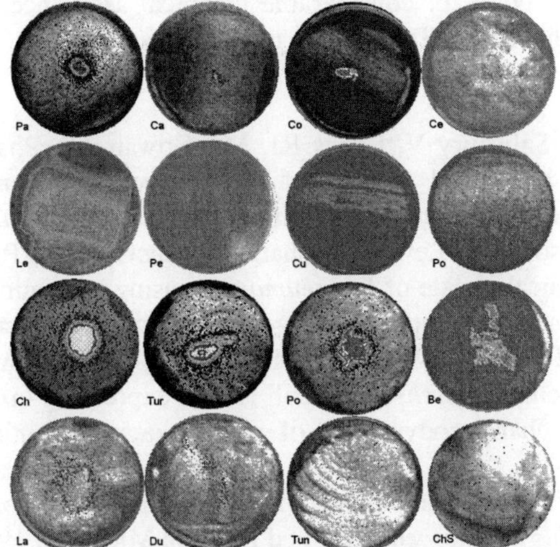

Figure 3. Recovery of *S.* Typhimurium DT104/pBRRMCS-5LITE at 21 h at 20 °C on the surface of foods listed in the methods section, after dry heating to 90 °C and holding for 30 s

The rate of Salmonella recovery on meat surfaces can be seen in Fig. 4. The results show that dry heat treatment at 90 °C had only a limited cidal effect on Salmonella adhering to various food surfaces. In contrast, steam treatment prevented bacterial recovery for at least 24 h on all food surfaces tested.

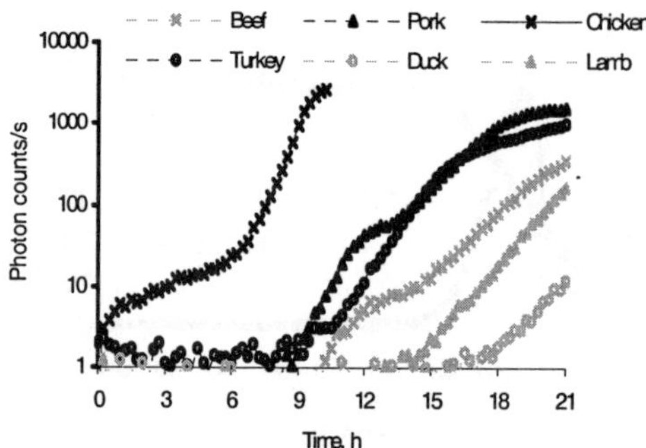

Figure 4. Recovery of Salmonella bioluminescence on the surface of various meats over 21 h at 20 °C after dry heating to 90 °C and holding for 30 s

ACKNOWLEDGEMENTS

We thank Tracey O'Neill for considerable technical assistance during this project and Dr PJ Hill for the *luxABCDE* cassette used in plasmid pAL2.

REFERENCES

1. Alloush HM, Salisbury V, Lewis RJ, MacGowan AP. Pharmacodynamics of linezolid in a clinical isolate of *Streptococcus pneumoniae* genetically modified to express *lux* genes. J Antimicrob Chemother 2003; 52:511-3.
2. Beard S, Salisbury V, Lewis RJ, Sharpe J, MacGowan AP. Expression of *lux* genes in a clinical isolate of *S .pneumoniae*: using bioluminescence to monitor gemifloxacin activity. Antimicrob Agents Chemother 2002;46:538-42.
3. Salisbury V, Pfoestl A, Wiesinger-Mayr H, Lewis RJ, Bowker K, MacGowan AP. Use of a clinical *Escherichia* coli isolate expressing lux genes to study the antimicrobial pharmacodynamics of moxifloxacin. J Antimicrob Chemother 1999; 43:829-32.
4. Marincs F, White D. Immobilisation of *Escherichia coli* expressing *lux* genes of *Xenorhabdus luminescens*. Applied Environ Microbiol 1994; 60:3862-3.
5. Parveen A, Smith G, Salisbury V, Nelson SM. Biofilm culture of *Pseudomonas aeruginosa* expressing lux genes as a model to study susceptibility to antimicrobials. FEMS Microbiol Lett 2001; 199:115-8.
6. Siragusa GR, Nawotka K, Spilman SD, Contag PR, Contag CH. Real-time monitoring of *E. coli* O157:H7 adherence to beef carcass surface tissues with a bioluminescent reporter. Appl Environ Microbiol 1999; 65:1738-45.
7. Maoz A, Mayr R, Scherer S. Temporal stability and biodiversity of two complex antilisterial cheese-ripening microbial consortia. Appl Environ Microbiol 2003; 69:4012-8.

A NOVEL METHOD TO ENHANCE THE SUBCUTANEOUS DETECTION OF BIOLUMINESCENCE IN THE FACULTATIVE ANAEROBE, *STREPTOCOCCUS PYOGENES*, BY DMSO-ASSISTED TRANSDERMAL OXYGEN DELIVERY

DE BUXTON[1], BJ CHILDERS[2], KC OBERG[1,2*]
*Department of Pathology and Human Anatomy[1], Surgery[2],
Loma Linda University, Loma Linda, CA. 92350, USA
Email: koberg@som.llu.edu

INTRODUCTION

Streptococcus pyogenes is a common Gram-positive organism present in the nasopharynx that can cause severe systemic and invasive disease. In necrotizing fasciitis (also known as flesh eating bacteria), this organism (a facultative anaerobe), rapidly advances within ischemic subcutaneous tissues.[1]

To explore the potential pathophysiology of *S. pyogenes*-related necrotizing fasciitis, we are using a bioluminescent strain of *S. pyogenes* harboring the LuxABCDE construct (Xen 20, Xenogen Corporation).[2,3] In aerobic conditions, growth of this strain of *S. pyogenes* is readily monitored by robust bioluminescence. However, under anaerobic conditions, bioluminescence of the oxygen-dependent luciferase system is significantly reduced despite vigorous bacterial growth.

Finley and colleagues[4] utilized dimethylsulfoxide (DMSO) and hydrogen peroxide (H_2O_2) to deliver oxygen to ischemic myocardium via intra-pericardial injection. Although H_2O_2 can be detrimental to bacterial growth, we reasoned that topical delivery of a low concentration of H_2O_2, using DMSO as a transdermal carrier, might provide enough subcutaneous oxygen for luciferase-related bioluminescence. Thus, in this report we describe the use of DMSO and H_2O_2 as a novel method to focally deliver oxygen at targeted subcutaneous sites to enhance the bioluminescent detection of subcutaneous bioluminescent *S. pyogenes*.

METHODS

Subcutaneous injection of bioluminescent *S. pyogenes*

We injected 100 µL of brain heart infusion broth containing 2.6 x 10^6 *S. pyogenes* colony forming units into the subcutaneous tissue of the ventral thigh overlying the femoral vessels in 8 male C57 mice immediately after sacrifice. To enhance bioluminescent detection, all hair from the ventral thigh was removed with a depilatory prior to injection. Following subcutaneous inoculation, carcasses were incubated at 37 °C in sealed plastic bags to maintain humidity and minimize odors.

Bioluminescent detection

The subcutaneous sites were examined for bioluminescence using the Hamamatsu Low Light Imaging System (Hamamatsu Corporation) and the Metamorph imaging software (Universal Imaging Corporation). Development of low levels of bioluminescence at all injection sites was considered time zero for all experiments. Photon counts were captured for 10 min, a standardized boxed area over each injection site was then gated in Metamorph, and the gated bioluminescence integrated by Metamorph. To determine background photon "noise", we captured 10 min photon counts from multiple fields in the absence of bioluminescence using the standardized gated box and averaged the integrated counts. This average was subtracted from the integrated bioluminescence of experimental data points.

Topical application of H_2O_2

50 µL of H_2O_2 solution was applied to a 1.5 X 1.5 mm square of folded Kimwipes tissue (VWR International) resting on the skin overlying the inoculation site. The H_2O_2 saturated Kimwipes tissue was left in position for 10 min. H_2O_2 was diluted in distilled H_2O and mixed with DMSO in various concentrations as described. DMSO and H_2O were used as controls.

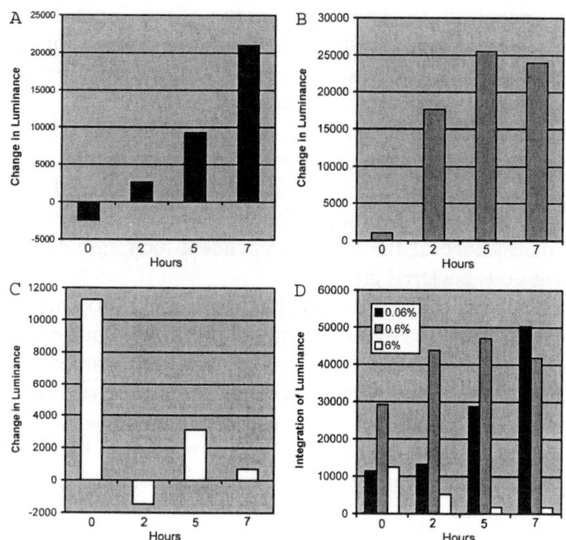

Figure 1. Concentration-Related Effects of H_2O_2.
Change in bioluminescence (post treatment – pre treatment) following
0.06% (A), 0.6% (B) and 6.0% (C) H_2O_2 in 30% DMSO.
(D) Summary of gated integration of data from A, B and C.

RESULTS

Concentration dependent effects of H_2O_2 on bioluminescence

To determine the effect of H_2O_2 concentration on bioluminescence, we topically applied 3 different concentrations of H_2O_2 (0.06%, 0.6% and 6.0%) in 30% DMSO to skin overlying subcutaneous inoculations of bioluminescent *S. pyogenes* (Fig. 1). At both low (0.06%) and moderate (0.6%) H_2O_2 concentrations, treatment promoted increasing bioluminescence with increasing bacterial growth as compared to pre-treatment bioluminescence. High concentrations of H_2O_2 (6.0%) resulted in a dramatic initial increase in bioluminescence, followed by a diminished response, that fell to near zero 7 h later suggesting bacterial toxicity.

DMSO Enhances the delivery of H_2O_2 and the promotion of bioluminescence

To determine whether DMSO enhances the delivery H_2O_2 and promotion of bioluminescence, we applied H_2O_2 (0.6%) in the presence or absence of 30% DMSO to the skin of the ventral thigh following inoculation with S. pyogenes and growth to the point of low level bioluminescent detection (Fig. 2). Baseline bioluminescence was recorded before (Fig. 2.B) and treatment-specific bioluminescence 10 min after (Fig 2.D) application of the topical treatments. Delivery of H_2O_2 via DMSO enhanced bioluminescence nearly 10-fold, compared to a 3-fold increase following H_2O_2 alone. DMSO alone did not induce additional bioluminescence and was similar to application of water. To determine any effect of treatment on bacterial growth, the subcutaneous bacteria was allowed to

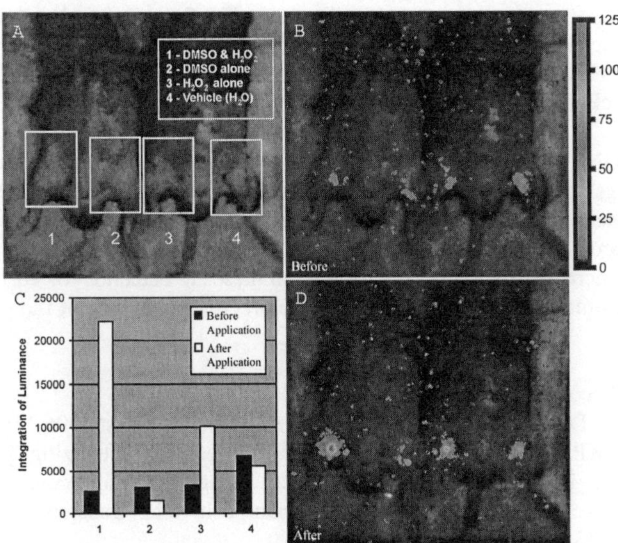

Figure 2. Enhanced Delivery of H_2O_2 by DMSO

(A) Gated box over inoculation sites with legend of topical treatments. (B) Baseline bioluminescence before treatments. (D) Bioluminescence 10 min after topical treatments. (C) Integration of the bioluminescence within the gated boxes of "B", before treatments (black), and "D", 10 minutes after treatment applications (white).

incubate for an additional 7 h and then these same 4 sites previously exposed to the various treatments were all treated with topical H_2O_2 in 30% DMSO. Bioluminescence was again recorded before (Fig. 3.A) and 10 min after (Fig. 3.B) H_2O_2 application. A large increase in bioluminescence was observed at each position except for the site previously treated with H_2O_2 alone suggesting bacterial toxicity and that DMSO may offer some protective effect in the delivery of H_2O_2.

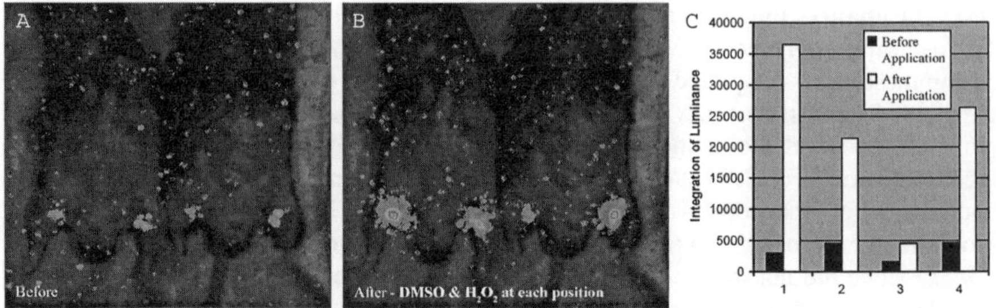

Figure 3. The Effect of H_2O_2 and DMSO on Bacterial Growth.
(A) 7 h after topical treatments indicated in Fig. 2, a low baseline level of bioluminescence is present. (B) Bioluminescence 10 min after H_2O_2/DMSO application. (C) Integration of the gated bioluminescence over the inoculation sites from "A" (Black); and from "B" 10 min after H_2O_2/DMSO application (White).

CONCLUSION

These findings demonstrate a novel approach to enhance the oxygen-dependent bioluminescence of the LuxABCDE luciferase product from bacteria growing in an anaerobic subcutaneous environment. Furthermore, the transdermal delivery of H_2O_2 via DMSO may increase the sensitivity of bioluminescent detection of this strain of *S. pyogenes* in animal models and allow for more precise monitoring of the progression of infection.

REFERENCES

1.	Green RJ, Dafoe DC, Raffin TA. Necrotizing fasciitis.Chest 1996;110:219-29.
2.	Francis KP, Yu J, Bellinger-Kawahara C, et al. Visualizing pneumococcal infections in the lungs of live mice using bioluminescent Streptococcus pneumoniae transformed with a novel Gram-positive lux transposon. Infect Immun 2001; 69:3350-8.
3.	Park HS, Francis KP, Yu J, Cleary PP. Membranous cells in nasal-associated lymphoid tissue: a portal of entry for the respiratory mucosal pathogen group A streptococcus. J Immunol 2003;171:2532-7.
4.	Finney JW, Urschel HC, Balla GA, et al. Protection of the ischemic heart with DMSO alone or DMSO with hydrogen peroxide. Ann N Y Acad Sci 1967;141:231-41.

DEVELOPMENT AND TESTING OF BACTERIOPHAGE-BASED BIOLUMINESCENT BIOREPORTERS FOR THE MONITORING OF MICROBIAL PATHOGENS IN THE SPACECRAFT ENVIRONMENT

KA DAUMER[1], SA RIPP[2], GS SAYLER[2], JL GARLAND[1]

[1]Dynamac Corp., Kennedy Space Center, FL 32899, USA
[2]The Center for Environmental Biotechnology, University of Tennessee, Knoxville, TN 37996, USA

INTRODUCTION

Microbes in enclosed environments pose serious threats; with the recent mandate for the United States to establish a lunar base within the next 10 years and reach Mars by 2025, the need arises for early and rapid detection of bacterial pathogens that may compromise astronauts' immune systems, contaminate food and water supplies, and degrade system materials. The development of bacteriophage-based bioluminescent bioreporters targets NASA's directive for new methods of monitoring and controlling the internal environment of human-occupied spacecraft. We are currently developing a suite of four bacteriophage-based bioreporters for the detection of microbial pathogens within the spacecraft environment. Three of these four pathogens have been recovered from previous missions. *S. aureus* has been recovered from both Apollo[1] and space shuttle[2] flights as well as Skylab[3] and MIR space station habitats.[2,4] This organism causes medical concern due to its association with nasal and skin infections during the Apollo 12 mission.[5] *P. aeruginosa* has been recovered from missions that included Apollo crewmembers and water systems on MIR.[5] It was identified as the cause of a urinary tract infection of an Apollo 13 lunar module pilot[5,1] and has also been found in ground-based studies of wheat-root rhizosperes, an indication that plant-based bioregenerative life support systems may also support pathogenic bacteria once in space.[6]. *E. coli* was repeatedly recovered from crewmembers during the Apollo-Soyuz missions.[6] Though not yet recovered from any space mission, *Salmonella* cause food-related illnesses, and their presence is likely on long-term missions.

Bacteriophage-based bioluminescent bioreporters assimilate the specificity of the phage to infect a certain host pathogen. Many bioluminescent bioreporters are constructed using the lux cassette (*luxCDABE*) derived from the marine bacterium *Vibrio fischeri*. Utilization of all five genes allows for intrinsic, whole-cell bioluminescence that does not require the addition of any exogenous substrate, thus allowing the bioreporter to be totally self-sufficient. The *lux* operon also consists of the *luxI* gene, whose product synthesizes the autoinducer molecule N-acyl-homoserine lactone (AHL). Once a certain concentration of AHL is produced, *luxR*, a transcriptional regulator, is activated and forms a complex with AHL to induce transcription of *luxI* and *luxCDABE*, resulting in the production of bioluminescence.

This mechanism, known as quorum sensing, is a type of cell-to-cell communication employed by *V. fischeri* in the wild. In our systems, the *luxI* gene will be incorporated into the genome of the phage. As phage are metabolically inactive until infection of a suitable host, if the particular host pathogen is present, phage infection will result in *luxI* becoming integrated into the bacterial genome, with the result being the production of AHL molecules. The AHL molecules will diffuse into the neighbouring bioluminescent bioreporter cells resulting in the production of bioluminescence (Fig. 1).

In their final embodiment, each of these phage-based bioreporter systems will be placed on a device known as a bioluminescent bioreporter integrated circuit (BBIC). The BBIC employs integrated circuit optical transducers that directly interface with the bioreporter organisms.[7] BBICs consist of two main components: photodetectors for capturing the on-chip bioluminescent bioreporter signals and signal processors for recording and storing information from the bioluminescence. This will allow for real-time, on-line monitoring of pathogens within the spacecraft environment.

METHODS

All four phage-based bioreporter systems are currently under construction, in varying stages of development. Based on initial testing with a pseudo-biodiagnostic *E. coli* system utilizing temperate phage lambda, it has been determined to use temperate rather than lytic phage for all of the systems. In constructing the pseudo-diagnostic *E. coli* system, the P_L promoter from phage lambda was fused in-frame to the *V. fischeri luxI* gene followed by a T1T2 termination signal. This construct was then inserted into an *E. coli* chromosome. A bioluminescent bioreporter sensitive to the *V. fischeri* specific AHL N-3-(oxohexanol)-L-homoserine lactone (OHHL) was also constructed. The P_L-*luxI*-T1T2 *E. coli* strain along with the bioluminescent bioreporter was inoculated into 24-well microtiter plates. Plates were incubated at room temperature in a Microbeta Victor^2 Multilabel counter with photon counts being measured every hour. Significant bioluminescent responses could be detected from as few as 10 *E. coli* cells within 3 hours of inoculation, 100 cells within 2 hours of inoculation, and >1,000 cells in less than one hour.[8]

CONCLUSIONS

These bioluminescent bioreporter systems utilize the specificity of bacteriophages coupled with the insertion of the quorum sensing *luxI* gene into the phage genome to detect pathogens in the spacecraft environment. The subsequent production of autoinducer molecules from the host pathogen infected triggers the bioluminescent signal from the neighbouring bioreporter cells. This phage-based quorum sensing system may be capable of sensing down to one individual cell. It also allows a measurable amount of bioluminescence to be produced from low-number infection events by using the production of the autoinducer molecules to produce amplification of individual targeted biological agents, thus avoiding the need for target cell growth.

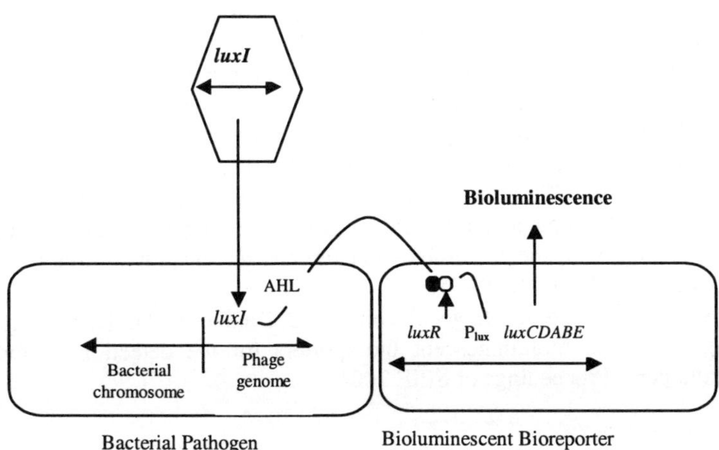

Figure 1. Proposed working model for bacteriophage-based reporter systems. Upon bacterial infection, the bacteriophage *luxI* gene is inserted into the host pathogen chromosome, where it is transcribed together with other phage genes and host cell genes. Subsequent production and diffusion of AHL molecules into neighbouring bioluminescent bioreporter cells triggers *luxCDABE*, thus generating bioluminescence.

ACKNOWLEDGMENTS

Research support was provided by the NASA Advanced Environmental Monitoring and Control Program.

REFERENCES

1. Taylor GR. Recovery of medically important microorganisms from Apollo astronauts. Aerosp Med 1974;45:824-8.
2. Larocca MT, Pierson DL. Deep space exploration: will we be ready? ASM News 1999;12:817-21.
3. Taylor GR, Graves RM, Brock-Ett RM, Ferguson JK, Mieszkuc BJ. Skylab environmental and crew microbiology studies. 1977 NASASP-377.
4. Pierson DL, Chidambaram M, Heath JD, Mallary L, Mishra SK, Sharma B, Weinstock GM. Epidemiology of *Staphyloccus aureus* during space flight. FEMS Immunol Med Microbiol 1996;16:273-81.
5. Ferguson JK, Taylor GR, Mieszkuc, BJ. Microbiological investigations, Biomedical Results of Apollo. NASA SP-368 1975; p 83-103.
6. Taylor GR. Medical microbiology analysis of US crewmembers, The Apollo-Soyuz Test Project Medical Report. NASA SP-411 1977; p 69-85.
7. Bolton EK, Sayler GS, Nivens DE, Rochelle JM, Ripp S, Simpson ML. Integrated CMOS photodetectors and signal processing for very low-level chemical sensing with the bioluminescent bioreporter integrated circuit. Sens Actuators B 2002: 85(1-2):179-85.
8. Ripp SA, Young J, Ozen A, Jeiger P, Johnson C, Daumer K, Garland J, Sayler G. Phage-amplified bioluminescent bioreporters for the detection of food-borne pathogens. Proceedings of SPIE 2004.

A SINGLE-STEP BIOLUMINESCENT ASSAY FOR RAPID DETECTION AND QUANTITATION OF VIABLE MICROBIAL CELLS

F FAN, B BUTLER, KV WOOD

Promega Corporation, 2800 Woods Hollow Road, Madison, WI 53711, USA
Email: frank.fan@promega.com

INTRODUCTION

ATP-based detection of microbial cells represents a key application of luciferase/luciferin bioluminescence assay. Conventional methods require two steps: application of a lysis reagent to release microbial ATP, followed by a detection reagent to elicit bioluminescence. We have developed an assay that combines the lytic reagent with luciferase/luciferin, thus allowing sensitive detection of microbial cells in a single-step. The assay system utilizes a thermostable luciferase to enable extraction of ATP from bacterial cells and to support a stable "glow-type" luminescent signal. Historically, firefly luciferase purified from *Photinus pyralis* has been used in reagents for ATP assays.[1] However, this enzyme has only moderate stability *in vitro* and is sensitive to factors such as pH and detergents, limiting its usefulness in a robust homogeneous ATP assay. We have successfully developed a stable form of luciferase, based on the gene from another firefly, *Photuris pennsylvanica*, using an approach to select for characteristics that improve performance in ATP assays.[2] In addition, we developed a formulation to achieve rapid and efficient extraction of ATP from a variety of microbial cells. The combination of these two essential elements enabled the design of a homogeneous single-reagent system (BacTiter-Glo Reagent) for performing ATP assays on cultured cells.

Discovery and development of new antibiotics are in critical demand to combat infectious diseases caused by microbial pathogens and rapid spreading of antibiotic resistance. The "add-mix-measure" format of the BacTiter-Glo Assay can be easily adapted for high-throughput screening. In this report, we describe the use of BacTiter-Glo Assay to screen and evaluate antimicrobial compounds.

MATERIALS AND METHODS
Bacterial strains and chemical reagents

Bacterial strains *Escherichia coli* ATCC25922, *Staphylococcus aureus* ATCC25923, *Pseudomonas aeruginosa* ATCC27853, and *Bacillus cereus* ATCC10987 were from ATCC (USA). BacTiter-Glo Microbial Cell Viability Assay was from Promega (USA). Antibiotics and Library Of Pharmacologically Active Compound (LOPAC) were from Sigma Chemicals (USA). Mueller Hinton II Broth (MH II) was from Becton, Dickinson and Company (USA).

Bacterial growth and ATP assay
Bacteria were grown in MH II medium at 37°C with shaking at 250 rpm. BacTiter-Glo Assay was performed according to the manufacture's protocol. Briefly, 100 μL of culture sample was mixed with 100 μL of the BacTiter-Glo Reagent and the emitted luminescence was recorded on a Veritas Microplate Luminometer from Turner Biosystems. Bacterial cell numbers were determined by plate counting of colony forming units on Luria-Bertani (LB) agar plates. The signal-to-noise ratio was calculated: S:N = [mean of signal–mean of background]/standard deviation of background].

Antimicrobial compounds screening and evaluation
Overnight culture of *S. aureus* was diluted 100-fold in fresh MH II Broth and used as inoculums for the antimicrobial screen. Working stocks (50X) of LOPAC compounds and standard antibiotics were prepared in DMSO. Each well of the 96-well multiwell plate contained 245 μL of the inoculums and 5 μL of the 50X working stock. The multiwell plate was incubated at 37°C for 5 h. Culture samples were taken from each well and the BacTiter-Glo Assay was performed. The samples and concentrations are: Wells 1–4 and 93–96, negative control of 2% DMSO, wells 5–8 and 89–92, positive controls of 32 μg/mL standard antibiotics tetracycline, ampicillin, gentamicin, chloramphenicol, oxacillin, kanamycin, piperacillin, and erythromycin; wells 9–88, LOPAC compounds at 10 μM.

The dosage effects of oxacillin were examined after 19 h of incubation. The relative percentage of RLU compared to the no-oxacillin control is shown.

RESULTS AND DISCUSSION
We evaluated the BacTiter-Glo Assay on a variety of microbial organisms including the four bacteria shown in Fig. 1. They are Gram-negative bacteria *E. coli* and *P. aeruginosa* and Gram-positive bacteria *S. aureus* and *B. cereus*. There is a linear correlation between luminescent signal and the number of cells over five orders of magnitude for each bacterium. The limit of detection (signal levels greater than three standard deviations above the background signal) for *E. coli, S. aureus, P. aeruginosa*, and *B. cereus* are approximately 40, 150, 70, and 10 cells, respectively. These results indicated that the *B. cereus* has the highest ATP level per cell among the four bacteria, followed by *E. coli, P. aeruginosa* and *S. aureus*. This is consistent with the results from a two-step (extraction, then detection) method (data not shown).

The luminescent signal generated by the BacTiter-Glo Assay has a half-life generally of over 30 min depending on the microbe and medium. High sensitivity and signal stability made the BacTiter-Glo Assay amenable for high-throughput screening. This is reflected by excellent Z'-factor values, which is a measure of assay quality based on the dynamic range and data variability. The BacTiter-Glo Assay has Z'-factor values of 0.90 and 0.87 for 96-well and 384-well formats, respectively.

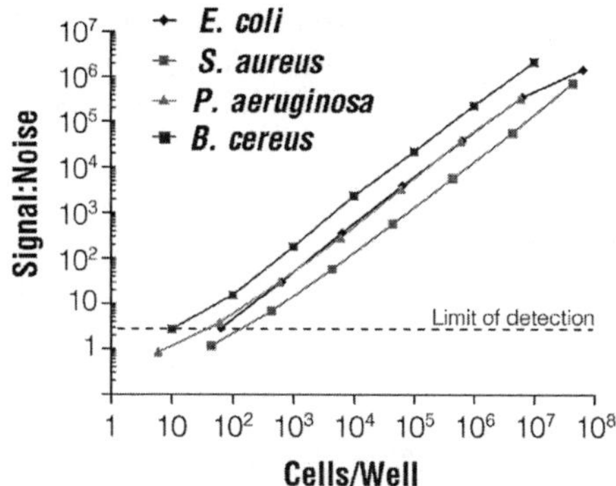

Figure 1. Sensitivity and linearity of the BacTiter-Glo Assay on four bacteria.

To demonstrate this application, we used the BacTiter-Glo Assay to screen some LOPAC compounds (Rack #8, enzyme inhibitors, total of 80 compounds) for antimicrobial activity against *S. aureus*. The results were shown in Fig. 2. All positive controls of standard antibiotics (boxed points) and three LOPAC compounds (circled points) exhibited significant anti-*S. aureus* activity. The three LOPAC hits were D6: emodin; D11: sanguinarine chloride and H7: minocycline. Their anti-*S. aureus* activities were reported in the literature previously.[3-5]

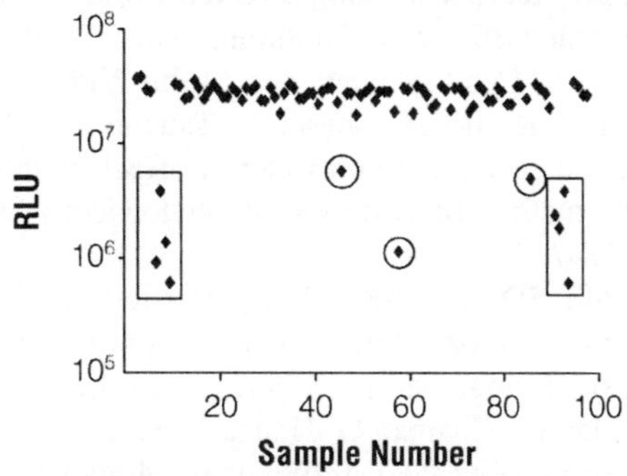

Figure 2. Antimicrobial activity screening against *S. aureus*.

We further examined the dosage effects of oxacillin on *S. aureus* using the BacTiter-Glo Assay. The results are shown in Fig. 3. Oxacillin showed anti-*S. aureus* activity in a dosage dependent fashion. The reported and observed Minimum Inhibitory Concentration (MIC) values for oxacillin on *S. aureus* in MH II broth are 0.125-0.5 μg/mL, corresponding to approximately IC75-IC90 values on the dosage curve determined using the BacTiter-Glo Assay. Compared with traditional MIC determination, which is qualitative (visual examination of bacterial culture) and subjective to variations, the BacTiter-Glo Assay provides a rapid, yet highly quantitative method for evaluating antimicrobial compounds.

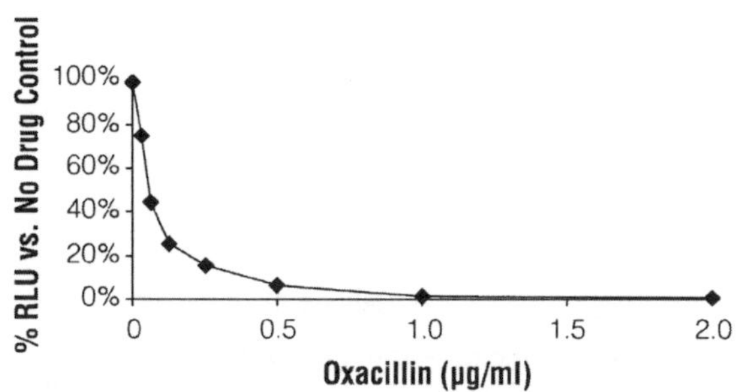

Figure 3. Dosage effects of oxacillin against *S. aureus.*

REFERENCES

1. McElroy WD, Deluca MA. Firefly and bacterial luminescence: Basic science and applications. J Applied Biochem 1983; 5:197-209.
2. Hall MP, Gruber MG, Hannah RR, Jennens-Clough ML, Wood KV. Stabilization of firefly luciferase using directed evolution. In: Roda A, Pazzagli M, Kricka LJ, Stanley PE. eds. Bioluminescence and Chemiluminescence, Perspectives for the 21st Century. New York: John Wiley & Sons, 1998: 392-5.
3. Hatano T, Uebayashi H, Ito H, Shiota S, Tsuchiya T, Yoshida T. Phenolic constituents of Cassia seeds and antibacterial effect of some naphthalenes and anthraquinones on methicillin-resistant Staphylococcus aureus. Chem Pharm Bull 1999; 47: 1121–7.
4. Godowski KC, Wolff ED, Thompson DM, Housley CJ, Polson AM, Dunn RL, Duke SP, Stoller NH, Southard GL. Whole mouth microbiota effects following subgingival delivery of sanguinarium. J Periodontol 1995; 66: 870–7.
5. Radd I, Chatzinikolaou I, Chaiban G, Hanna H, Hachem R, Dvorak T, Cook G, Costerton W. In vitro and ex vivo Activities of Minocycline and EDTA against microorganisms embedded in biofilm on catheter surfaces. Antimicrob Agents Chemother 2003; 47: 3580–5.
6. National Committee for Clinical Laboratory Standards. Methods for dilution antimicrobial susceptibility tests for bacteria that grow aerobically; approved standard-fifth edition M21-A. National Committee for Clinical Laboratory Standards, Wayne, PA. 1999; 19:17.

BIOLUMINESCENT ASSAY
OF TOTAL BACTERIAL CONTAMINATION (TBC) IN FORCE-MEAT USING FILTRAVETTE™

VG FRUNDZHYAN, NN UGAROVA, NA MOROZ

Dept. of Chemistry, Lomonosov Moscow State University, Moscow 119899, Russia
Email: vgf@enz.chem.msu.ru

INTRODUCTION

Bioluminescent assay of TBC is one of the most rapid, simple and economically reasonable method among «rapid microbiology» methods developed for assessment of hygiene quality of food samples. Since the most samples analyzed contain excess of non-bacterial ATP (sum of somatic and free ATP) and/or low number of bacteria, special pretreatment of the sample, laborious and time consuming, is required. We applied special luminometric microcuvette Filtravette™ to simplify the bioluminescent assay of force-meat. Combined application of BCN-reagent for sample pretreatment and highly sensitive ATP-reagent developed in our laboratory and Filtravette™ permitted us to detect 10^4 CFU/g force-meat at the duration of assay ~35 min per sample.

MATERIALS AND METHODS

<u>Instrumentation</u> Luminometer 3550i and Filtravette™ (New Horizons Diagnostics Corp., USA) were used for bioluminescent measurement.

<u>Reagents</u> ATP-reagent[1] (based on soluble *Luciola mingrelica* firefly luciferase) and BCN-reagent (lyophilized mixture of protease, detergent and buffer components) were developed in our laboratory. Dimethyl sulfoxide (DMSO) was from Reakhim (Russia), Neonol-10 was from NPO "Nizhnekamsk" (Russia). Other reagents were analytical grade. Ultrapure deionized water was obtained on Milli-Q (Millipore).

<u>Methods</u> TBC in force-meat was established by the standard Plate Count (30 °C, 48 h). To obtain samples with TBC varied in a wide range, force-meat samples purchased at a local meat market were mixed with a sterile one. To prepare sterile force-meat sample, beef pieces were immersed in ethanol and scorched. The burnt surface was cut off and sterile sample obtained was ground.

To separate bacteria cells, force-meat sample was soaked with saline (5 g in 25 mL) and homogenized in homogenizer (3 min) or incubated in shaker (15 min, 100 min^{-1}, 37 °C). 5 mL of a force-meat suspension obtained were added to the flask with BCN-reagent and incubated (15 min, 100 min^{-1}, 37 °C). After incubation 2-3 mL of the suspension were filtered through double filter paper disk ("blue strip" grade) placed into Swinnex Disk Filter Holder, 25 mm, from Millipore. 0.1-1 mL of the suspension clarified was filtered through Filtravette™. The Filtravette™ was washed with 0.3 mL of saline followed by addition of 0.02 mL of DMSO or 1.5% Neonol-10 to extract bacterial or residual non-bacterial ATP respectively. Finally, 0.18 mL of ATP-reagent was added to the Filtravette™ and bioluminescent signal was recorded. Statistical analysis All samples assayed were analyzed in triplicate both by Plate Count and bioluminescence. The data obtained were first log-transformed. The linear regression (correlation coefficient - R_s) was used to analyze the accuracy of the method.

RESULTS

TBC characterizes hygiene quality of force-meat. Plate Count method generally used in practice to determine the TBC in force-meat takes 48 h. For on-line control we applied bioluminescent assay.

Non-bacterial ATP concentration in force-meat is ~10^{-11} mol/g. This concentration corresponds to the TBC value ~10^7 cell/g. To detect the lower TBC values, destruction or elimination of non-bacterial ATP from the sample analyzed is required.

First, for TBC determination it was necessary to separate bacteria cells from the force-meat particles. We substituted homogenization, generally used for that, with incubation of force-meat in saline under agitation for 15 min. The TBC values in force-meat homogenate and force-meat suspension obtained from the same sample were $(8.46\pm0.60)\times10^6$ and $(1.66\pm0.30)\times10^7$ CFU/g respectively. Thus, incubation under agitation permits to remove more bacteria cells and more convenient in practice use.

To remove non-bacterial ATP, force-meat suspension was filtered through luminometric polystyrene microcuvette (h 13 mm, $\varnothing$ 10 mm) with the bottom made of bacterial membrane filter (pore size 0.45 μm), Filtravette™. Application of Filtravette™ permits to concentrate bacteria cells, extract bacterial ATP and measure bioluminescent signal in the same cuvette. As a result, the assay simplifies, its accuracy enhances and detection limit of bacteria cells increases.

Due to high content of muscle fibers and fat in force-meat suspension, the bacterial membrane filter in Filtravette™ was clogged completely after filtration of 1-5 μl of the sample analyzed. To overcome this obstacle we treated preliminary force-meat suspension with specially designed BCN-reagent. BCN-reagent is a lyophilized, ready to use reagent for deep destruction of muscle fibers in force-meat suspension. In contrast to BPN-reagent[2] described earlier, pretreatment of the sample analyzed with BCN-reagent requires 15 min and 37 °C instead of 60 min and 45 °C. Like BPN-reagent, BCN-reagent does not effect on bacteria cells. After incubation some muscle fibers remained undestroyed in the force-meat suspension and moved away by filtration through filter paper. Less than 10-14 % of bacteria cells retained together with meat fibers on the filter paper.

After pretreatment proposed it was possible to filter up to 1 mL of force-neat suspension through Filtravette™, but the optimal volume for bioluminescence measurement was 0.1 mL. In that case the residual non-bacterial ATP concentration did not exceed 12% of bacterial ATP concentration in Filtravette^{TM}, and bioluminescence signals measured in the samples with high TBC, above 10^8 CFU/g, did not result in overload of luminometer.

We analyzed 20 force-meat samples with TBC varied in a wide range both by Plate Count and bioluminescent assay proposed. Please refer to Table 1 for details.

Table 1. TBC, bacterial and residual non-bacterial ATP concentration in force-meat samples

N	TBC, CFU/g	ATP, mol/g		$\dfrac{ATP_{res} \times 100}{ATP_{bac}}$, %
		Bacterial (ATP_{bac})	Residual non-bacterial (ATP_{res})	
1*	0	0	$(8.9\pm0.4)\times10^{-14}$	-
2*	0	0	$(4.2\pm0.5)\times10^{-14}$	-
3*	0	0	$(4.9\pm0.3)\times10^{-14}$	-
4	$(1.0\pm0.3)\times10^4$	$(1.0\pm0.1)\times10^{-13}$	$(7.4\pm1.4)\times10^{-15}$	7.2
5	$(1.8\pm0.4)\times10^5$	$(5.3\pm0.4)\times10^{-13}$	$(1.7\pm0.2)\times10^{-14}$	3.2
6	$(1.9\pm0.1)\times10^5$	$(5.3\pm0.1)\times10^{-13}$	$(1.2\pm0.0)\times10^{-14}$	2.3
7	$(1.9\pm0.4)\times10^5$	$(3.2\pm0.2)\times10^{-13}$	$(5.6\pm0.9)\times10^{-15}$	1.8
8	$(3.3\pm0.1)\times10^5$	$(8.8\pm1.9)\times10^{-13}$	$(3.6\pm0.3)\times10^{-14}$	4.1
9	$(6.1\pm0.6)\times10^5$	$(5.3\pm0.2)\times10^{-12}$	$(6.6\pm0.2)\times10^{-13}$	12.4
10	$(8.5\pm2.3)\times10^5$	$(4.9\pm0.1)\times10^{-12}$	$(5.6\pm0.2)\times10^{-13}$	11.3
11	$(2.3\pm0.4)\times10^6$	$(4.6\pm0.5)\times10^{-12}$	$(2.6\pm0.3)\times10^{-13}$	5.5
12	$(2.5\pm0.5)\times10^6$	$(5.2\pm0.6)\times10^{-12}$	$(2.6\pm0.0)\times10^{-13}$	5.1
13	$(5.1\pm0.5)\times10^6$	$(2.6\pm0.1)\times10^{-11}$	$(1.8\pm0.8)\times10^{-12}$	6.8
14	$(6.1\pm0.3)\times10^6$	$(4.9\pm0.3)\times10^{-11}$	$(4.6\pm0.2)\times10^{-12}$	9.3
15	$(6.2\pm0.4)\times10^6$	$(6.0\pm0.4)\times10^{-11}$	$(5.5\pm0.3)\times10^{-12}$	9.1
16	$(1.2\pm0.2)\times10^7$	$(1.4\pm0.0)\times10^{-11}$	$(1.7\pm0.3)\times10^{-12}$	11.9
17	$(3.3\pm0.4)\times10^7$	$(3.9\pm0.1)\times10^{-10}$	$(6.9\pm0.6)\times10^{-14}$	0.2
18	$(7.3\pm0.3)\times10^7$	$(3.5\pm0.1)\times10^{-10}$	$(1.4\pm0.2)\times10^{-11}$	3.9
19	$(2.8\pm0.5)\times10^8$	$(4.3\pm0.1)\times10^{-10}$	$(1.8\pm0.2)\times10^{-13}$	0.04
20	$(2.8\pm0.5)\times10^8$	$(8.8\pm1.6)\times10^{-10}$	$(2.3\pm0.0)\times10^{-13}$	0.02

* Sterile force-meat, free of CFU.

According to the data obtained detection limit of bacterial ATP in force-meat, against a background of non-bacterial ATP, was 10^{-13} mol/g. So, the detection limit of bacteria cells was 10^4 CFU/g. The total duration of the assay was ~35 min per sample whereas Plate Count took 48 h. If ATP concentration measured in Filtravette^{TM} was less than 10^{-13} mol/g, the force-meet was free of CFU.

A good correlation between Plate Count and bacterial ATP concentration in force-meat was observed (R=0.96). Please refer to Fig. 1 and equation 1 for details.

$$lg(CFU/g) = (18.22\pm0.24) + (1.06\pm0.07) \times lg ([ATP, mol/g]) \qquad (1)$$

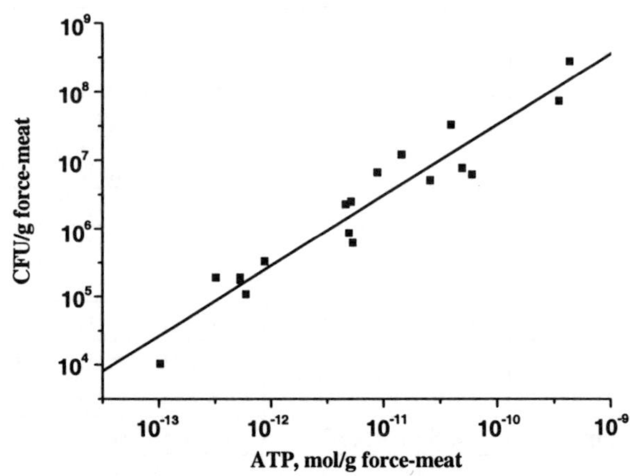

Figure 1. Correlation between bioluminescent assay and
Plate Count in force-meat samples

ACKNOWLEDGEMENTS

We thank for financial support the Civilian Research & Development Foundation (Project RBO-11009-(1)-PNNL) and the Ministry of Education and Science of Russian Federation (State Contract N 43.073.1.1.2505).

REFERENCES

1.   Application for Patent N2164241Rus. Reagent for determination of adenosine5′-triphosphate.
2.   Froundjian V, Brovko L, Ugarova N. Bioluminescent assay of total bacterial contamination (TBC) in food samples and drinking water using Filtravette^{TM}. In: Stanley P, Kricka L. eds. Bioluminescence and Chemiluminescence Progress & Current Applications: Singapore: World Scientific Publishing, 2002: 475-8.

BIOLUMINESCENT ASSAY OF STERILITY OR CLEANLINESS IN A HOSPITAL ENVIRONMENT

VG FRUNDZHYAN[1], NN UGAROVA[1], NI GABRIYELYAN[2], LI AREF'EVA[2], TB PREOBRAZHENSKAY[2]

[1]Dept. of Chemistry, Lomonosov Moscow State University, 119992, Moscow, Russia
[2]Scientific Research Institute of Transplantology and Artificial Organs, 123182, Moscow, Russia
Email: vgf@enz.chem.msu.ru

INTRODUCTION

To control sterility or cleanness in hospital environment microbiology methods, laborious and time consuming (24-120 h), are used at present. To accelerate and simplify the sterility or cleanness control on different surfaces we applied bioluminescent assay of total bacterial contamination (TBC). Since the most surfaces analyzed in hospital contained low number of bacteria, below the detection limit of ATP-reagent used, incubation of the samples in nutritive medium followed by filtration through special luminometric microcuvettes Filtravette[TM][1] was applied.

METHODS

Luminometer 3550i and Filtravette[TM] (New Horizons Diagnostics Corp., USA) were used for bioluminescent measurement. ATP-reagent[2] (based on soluble *Luciola mingrelica* firefly luciferase) was developed in our laboratory. Dimethyl sulfoxide (DMSO) was from Reakhim (Russia). Nutrition broth (NB) from ICN, Tryptic Soy Broth (TSB) from Difco and Thioglycolate broth (TB) from Merck were used. Ultrapure deionized water was obtained on Milli-Q (Millipore). Broth culture of *Escherichia coli LE392* was used in model experiments.

Bacteria cells from the surface analyzed (20x20 cm^2) were gathered using swab wetted in saline followed by incubation in nutritive medium for 6 h (37 °C, 100 min^{-1}). After incubation 0.2 mL of bacteria suspension obtained was filtered through Filtravette[TM] and Filtravette[TM] was washed with 0.2 mL of saline. Finally 0.02 mL of DMSO and 0.18 mL of ATP-reagent were added consecutively to the Filtravette[TM] to extract bacterial ATP and measure bioluminescent signal. All swab samples obtained were analyzed in parallel by the Standard Plate Count[3] (37 °C, 48-120 h).

The ATP and CFU values were determined in triplicate. The CFU were evaluated in CFU/100 cm^2. Data obtained were lg transformed first and liner regression (1) coefficients a, b, R_s were calculated.

$$\lg(\text{CFU}/100 \text{ cm}^2) = a + b \times \lg(\text{ATP, mol/mL}) \quad (1)$$

RESULTS

The samples analyzed for cleanness were contaminated with low number of bacteria or bacteria in depressed energy status resulting in low intracellular ATP concentration. To multiply the number of bacteria and/or increase intracellular ATP concentration we incubated swabs with bacteria in nutritive medium. To select the most appropriate nutritive medium the NB, TSB and TB commonly used in laboratory practice were examined. The highest bacteria growth rate and ATP concentration were determined when the swab samples (floor surface, initial TBC $(1.1\pm0.2)\times10^3$ CFU/100 cm^2) were incubated in NB for 3 h.

To establish the least incubation time in NB required for determination of low number of bacteria we incubated *E. coli* suspensions with cell titer varied in the range ~ 1-100 CFU/mL. Please refer to Table 1 for details.

Table 1 ATP concentration and cell titer in *E. coli* suspensions
incubated in NB up to 6 h

Incuba-tion, h	Parameter	Initial titer of *E. coli* suspension, CFU/mL		
		3.0 ± 0	$(2.0\pm0.1)\times10^1$	$(1.3\pm0.1)\times10^2$
3	CFU/mL	$(1.3\pm0.1)\times10^2$	$(1.2\pm0.0)\times10^2$	$(3.8\pm0.9)\times10^3$
	ATP, mol/mL	no signal	no signal	no signal
4	CFU/ml	$(4.2\pm0.1)\times10^2$	$(4.4\pm0.5)\times10^3$	$(1.7\pm0.2)\times10^4$
	ATP, mol/mL	no signal	$(8.0\pm2.0)\times10^{-15}$	$(8.2\pm0.1)\times10^{-14}$
5	CFU/mL	$(2.4\pm0.2)\times10^3$	$(4.2\pm0.1)\times10^4$	$(2.2\pm0.3)\times10^5$
	ATP, mol/mL	no signal	$(2.2\pm0.4)\times10^{-13}$	$(6.4\pm0.8)\times10^{-13}$
6	CFU/mL	$(1.9\pm0.5)\times10^4$	$(4.6\pm0.4)\times10^5$	$(1.2\pm0.2)\times10^6$
	ATP, mol/mL	$(5.5\pm2.7)\times10^{-13}$	$(9.1\pm0.9)\times10^{-12}$	$(1.8\pm0.1)\times10^{-11}$

According to the data obtained, to detect bioluminescent signal from *E. coli* suspension with initial cell titer 3 CFU/mL incubation in NB for 6 h was required. Therefore for bioluminescent TBC assay the swab samples were incubated in NB for 6 h.

All samples analyzed were categorized into 3 groups: (1) highly contaminated samples (TBC > 100 CFU/100 cm^2), (2) clean samples (TBC < 100 CFU/100 cm^2), (3) sterile samples (free of bacteria). The samples of group (3) were analyzed for monitoring of sterility.

The ATP concentration in the samples from group (1) was determined without incubation. Samples from groups (2) and (3) were incubated before the bioluminescent assay. For details, please, refer to Tables 2-4.

Table 2. ATP and CFU on highly contaminated surfaces (samples of group 1); $a=14.86$, $b=0.93$, $R_s=0.76$

Sample	ATP, mol/mL	CFU/100 cm^2
Jack-lift from autoclave room N1	$(1.6\pm0.1)\times10^{-12}$	$(2.6\pm0.4)\times10^4$
Jack-lift from autoclave room N2	$(1.0\pm0.3)\times10^{-12}$	$(1.8\pm0.3)\times10^4$
Table for not sterile items	$(7.1\pm0.5)\times10^{-15}$	$(2.5\pm0.3)\times10^2$
Oil-cloth from autoclave room	$(1.6\pm0.0)\times10^{-12}$	$(1.8\pm0.5)\times10^4$
Consol for monitor in ward	$(2.8\pm0.1)\times10^{-13}$	$(1.4\pm0.3)\times10^2$
Used bed-sheet	$(1.0\pm0.1)\times10^{-11}$	$(6.4\pm0.8)\times10^4$
Manipulation table	$(3.5\pm0.2)\times10^{-13}$	$(1.5\pm0.2)\times10^2$

Table 3. ATP and CFU on clean surfaces (samples of group 2); $a=14.97$, $b=1.05$, $R_s=0.74$

Sample	ATP, mol/mL	CFU/100 cm^2
Used patient's table	$(1.3\pm0.2)\times10^{-13}$	25 ± 5
Floor in ward	$(1.4\pm0.5)\times10^{-13}$	20 ± 3
Swaddling band	no signal	0
Patient's table in cleaned ward	no signal	0
Doctor's table	no signal	15 ± 4
Used pillow-case	$(3.4\pm0.3)\times10^{-13}$	90 ± 6
Clean pillow-case	no signal	2 ± 1
Table for sterilized items	$(1.2\pm0.1)\times10^{-13}$	50 ± 3

Table 4. ATP and CFU on sterile surfaces (samples of group 3)

Sample	ATP, mol/mL	CFU/100 cm^2
Sterile bandage (lot 1)	no signal	0
Sterile bandage (lot 2)	no signal	0
Napkin (lot 1)	no signal	0
Napkin (lot 2)	no signal	0
Surgical coat (lot 1)	no signal	0
Surgical coat (lot 2)	no signal	0
Operating bed-sheet (lot 1)	no signal	0
Operating bed-sheet (lot 2)	no signal	0
Surgical sponge (lot 1)	no signal	0
Surgical sponge (lot 2)	no signal	0
Operating swaddling band	no signal	0

In spite of correlation coefficient (R_s) between ATP and CFU for the samples of groups (1) and (2) were 0.74 - 0.76 only, the accuracy obtained was enough for cleanness control. As for group (3) both bioluminescent assay and Standard Plate Count showed sterility for the all samples. Thus, the bioluminescent TBC assay with duration 6 h is applicable for cleanness/sterility control in hospital environment while Standard Plate Count takes 24 -120 h.

ACKNOWLEDGMENTS

This work was supported by Ministry of Education and Science of Russian Federation (State Contract N 43.073.1.1.2505).

REFERENCES

1.   Froundjian V, Brovko L, Ugarova N. Bioluminescent assay of total bacterial contamination (TBC) in food samples and drinking water using Filtravette™. In: Stanley P, Kricka L. eds. Bioluminescence and Chemiluminescence Progress & Current Applications: Singapore: World Scientific, 2002: 475-8.
2.   Ugarova N, Maloshenok L. Reagent for determination of adenosine 5′-triphosphate. Application for Patent N 2164241 (Rus.), 2004.
3.   The order N720 of the Ministry of Health of the USSR from 31.07.1978, Appendix N1.

CONTROL OF MICROORGANISMS BY SINGLET OXYGEN

A FUJIMURA[1], Y TOSHITOKU[1], Y MESE [1,] N SUZUKI[1], T NAGAI[2],
I MIZUMOTO[3], H SATO[4], R KANAZAWA[5], A GO[1],
K NAKAGUCHI[1], B YODA[6]

[1]*Graduate School of Biosphere Science and Faculty of Applied Biological Science,
Hiroshima University, Higashi-Hiroshima 739-8528, Japan*
[2]*Dept. of Food Science, Tokyo University of Agriculture, Hokkaido
099-2493, Japan*
[3]*Toyama National College of Maritime Technology, Shin-minato 933-0239, Japan*
[4]*Optec Co., Oshima, Koto-ku, Tokyo 136-0072, Japan*
[5]*Daikin Environmental Institute, Miyukigaoka, Tsukuba 305-0841, Japan*
[6]*Koriyama Women's University, Koriyama 963-8503, Japan*
E-mail: suzukin@hiroshima-u.ac.jp

INTRODUCTION

Recently, control of microorganisms has been increasing in importance, because food poisoning and infectious diseases caused by microorganisms happen often. Additionally, environment-conscious techniques of control of microorganisms, which is less burdensome for the natural environment, are required. Thus, we tried to purify environmental contamination by microorganisms with singlet oxygen ($^1O_2$) that generated from Rose Bengal and phthalocyanine, and visible light irradiation.

In 1999, we reported that white spot syndrome virus (WSSV) attaching on the eggs of Kuruma shrimp (*Penaeus japonicus*) were inactived by $^1O_2$ without injuring the eggs.[1,2]

In this paper, we would like to describe "control of microorganisms by singlet oxygen" that (1) inhibits growth of *Escherichia coli*, (2) decontaminates water-bloom, *Microcystis aeruginosa* and (3) kills *Ichthyophthirius multifiliis*.

METHODS
Chemicals

Rose Bengal (4,5,6,7-tetrachloro-2',4',5',7'-tetraiodofluorescein disodium salt: RB), Polypepton fine granules, magnesium sulfate heptahydrate, agar powder were purchased from Wako Pure Chemical Industries Co., Ltd. (Osaka, Japan) and yeast extract from Kanto Kagaku Co., Ltd. (Tokyo, Japan).

Inhibiting growth of *Escherichia coli* [3-5]

Escherichia coli (HUT 8106 was generously supplied by HUT Culture Collection, Hiroshima University) were used as an example of environmental Gram-negative bacteria. *E.coli* strains were grown aerobically at 37°C in IFO Medium No.802 containing the following constituents (g/L): Polypepton (10); yeast extract (2); $MgSO_4 \cdot 7H_2O$ (1); agar (15), adjusted to pH 7.0. In case where required, RB was added to achieve final concentrations of 40 μmol/L. The suspension cultures were incubated in a water bath for 3 h at 37 °C. After the incubated period, suspension cultures (0.1 mL) were spread over the surface of a dried IFO Medium No. 802 plate using a sterile spreader. These spread plates were exposed to visible light or kept in the dark for 40 min in the presence of RB at room temperature (21 ± 2°C). Irradiation of plates containing *E. coli* using visible light (fluorescent light, FL 15W, TOSHIBA) was carried out at 0, 200, 400, 800, 1500, 3000, and 7500 lux. The intensity of irradiation was measured by a luxmeter (Lux Meter LM-102, Mother Tool Co., Ltd.). After irradiation, these plates were incubated in an aerobic dark incubator at 37 °C for 24 h, and colony forming units (CFU) were counted. CFU values given in Fig. 1 are the means of three independent experiments.

Decontaminating water-bloom

Water-bloom was obtained from Lake Hakuryu (Hiroshima, Japan). Microscope examination revealed that *Microcystis aeruginosa* was the dominating genus. This sample was kept at 20 ± 2°C under a light-dark cycle of 12 h-light period at 2000 lux and 12 h-dark period. In the first irradiation study, the sample was mixed with RB solution to achieve final RB concentrations to be 0, 5, 20, and 50 μmol/L. In the second experiment, the sample was added in test tube, in which a Dye Net was fixed. The Dye Net is mesh containing RB or phthalocyanine (PC) (endowed generously from Optec Co.). The test samples were exposed to visible light (fluorescent light) or kept in the dark for 11 d. Irradiation was carried out at 20 ± 2°C under a light-dark cycle of 12 h-light period at 2000 lux and 12 h-dark period.

Killing *Ichthyophthirius multifiliis*

Goldfish (*Carassius auratus*) obtained from a local fish dealer were kept at 26 ± 1 °C in an aquarium. Caudal fins were obtained from the goldfish stricken with white spot disease. These samples were put on each slide glass, respectively, and a 50 μmol/L RB solution (0.5 mL) were added on the each sample. The samples were exposed to visible light (CH2-100V30W, OLYMPUS) or kept in the dark for 30 through 60 min. Irradiation was carried out at 6000 lux. After irradiation, viability of *I. multifiliis* that causes white spot disease to goldfish was observed under a microscope.

RESULTS

Under this irradiation condition, $^1O_2$, a molecular species of active oxygen with killing activity against microorganisms, was generated from the dye (Type II mechanism). Emission spectra of $^1O_2$ generated from dyes under irradiation with a green laser (532 nm) were measured by an NIR emission spectrometer that was developed and made in our laboratory (Data not shown).

Inhibiting growth of *Escherichia coli*

Exposure of the spread plates to fluorescent light in the absence of RB did not cause a substantial change in CFU. However, in the presence of RB, there was a dramatic decrease in CFU (Fig. 1). RB prevented completely the growth of *E.coli* (reaching 0 % of survival), when illuminated at 7500 lux for 40 min.

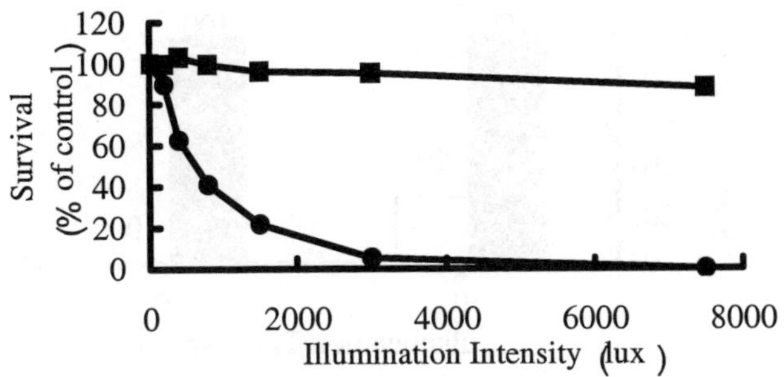

Figure 1. Effect of singlet oxygen on *Escherichia coli*
(■) *E. coli* without RB, (●) *E. coli* with 40 μmol/L RB

Decontaminating water-bloom

As Fig. 2 (A) shows, after irradiation for 11 d, the irradiation effect of RB was detected clearly, over a concentration of 20 μmol/L. The culture of *M. aeruginosa* became clouded after irradiation for 3 d. And then *M. aeruginosa* in the culture was deposited and dissolved slowly. RB showed no toxicity without irradiation.

Fig. 2 (B) shows the effect of fluorescent irradiation of the samples with the Dye Net containing RB or PC. Both RB and PC killed the main part of *M. aeruginosa*.

(A) After 11 days

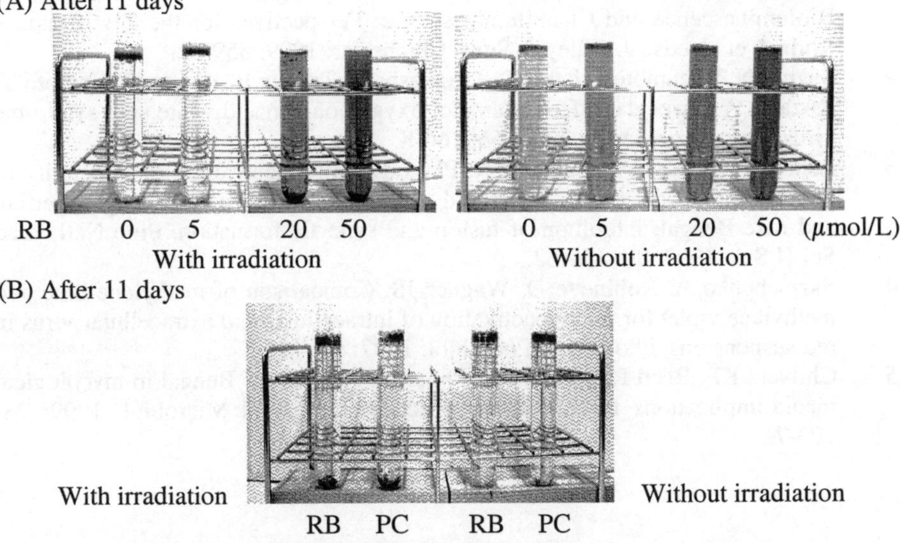

(B) After 11 days

Figure 2. Effect of singlet oxygen on *Microcystis aeruginosa*

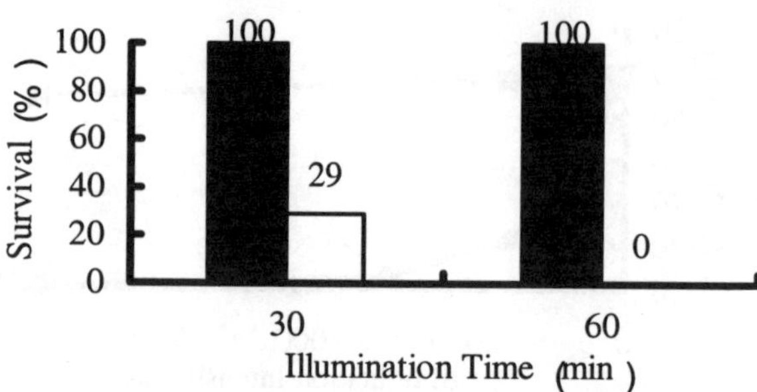

Figure 3. Effect of singlet oxygen on *Ichthyophthirius multifiliis*
(■) without irradiation, (□) with irradiation

Killing *Ichthyophthirius multifiliis*

Fig. 3 shows that *I. multifiliis* was killed by $^1O_2$ generated from RB under irradiation of visible light completely (reaching 0 % of survival), when illuminated for 60 min. These results suggest a possibility that the $^1O_2$ generating system affords a simple, safe and effective technique that is applicable to water purification.

REFERENCES

1. Suzuki N, Mizumoto I, Itami T, Watanabe R, Takahashi Y, Hatate H, Tanaka R, Nomoto T, Kozawa K, Kozawa A. Dye-sensitized inactivation of white spot syndrome virus attached to eggs of crustaceans. In: Bioluminescence and Chemiluminescence: Perspectives for the 21st Century. Roda A et al. eds., J. Wiley & Sons. Chichester: 1999: 559-62.
2. Suzuki N, Mizumoto I, Itami T, Takahashi Y, Tanaka R, Hatate H, Nomoto T, Kozawa A. Virucidal effect of singlet oxygen on penaeid white spot syndrome virus. Fisheries Science. 2000: 66: 166-8.
3. Lenard J, Robson A, Vanderof R. Photodynamic inactivation of infectivity of human immunodeficiency virus and other enveloped viruses using hypericin and rose Bengal: inhibition of fusion and syncytia formation. Proc Natl Acad Sci U.S.A. 1993: 90: 158-62.
4. Skripchenko A, Robinette D, Wagner JS. Comparison of methylene blue and methylene violet for photoinactivation of intracellular and extracellular virus in red suspensions. Photochem Photobiol. 1997: 65: 451-5.
5. Chilvers KF, Reed RH, Perry JD. Phototoxicity of rose Bengal in mycological media-implications for laboratory practice. Lett Appl Microbiol. 1999: 28: 103-7.

QUANTITATIVE ANALYSIS OF CHEMILUMINESCENCE INTENSITY AND TOXICITY *IN SILICO*

TOSHIHIKO HANAI[1], TSUTOMU TACHIKAWA[2]

[1]Health Research Foundation, Institut Pasteur 5F, Sakyo-ku, Kyoto, 106-8225, Japan

[2]Fujitsu Limited, Bio-IT Lab., Nakase, Mihama-ku, chiba, 261-8588, Japan

Email: thanai@attglobal.net

INTRODUCTION

The chemiluminescence detection technique is highly sensitive. Chemiluminescence is produced by a chemical reaction. The efficiency of a chemiluminescence reaction can be expressed as the number of light-emitting molecules related to the number of excited molecules. Peroxyoxalate luminescence is used to assay hydrogen peroxide or the number of fluorophores. Organic reducing compounds, including reducing sugars, ascorbic acid, uric acid, phenacyl alcohol derivatives, and steroids, are detected with the chemiluminescence method using lucigenin and luminol.[1-5] The reaction process is considered the same for similar compounds, but the chemiluminescence sensitivity is thought to be structure-dependent.[6,7] The sensitivity and intensity of phenacyl alcohol derivatives and steroids appears to depend on the reactivity of superoxide. The intensity of chemiluminescence was quantitatively analyzed using computational chemical calculations based on a radical reaction mechanism in which a keto-enol rearrangement produced superoxide, and the superoxide reacted with luminol or lucigenin to produce the chemiluminescence. The partial charge of the carbon atoms of the carbonyl group, calculated using the MOPAC function of the CAChe™ program, changed significantly and strongly correlated with the relative intensity of the chemiluminescence. The square of the correlation coefficient (r^2) was 0.970 (n = 5) and 0.965 (n = 8) for phenacyl alcohol derivatives and steroids, respectively.[8] The r^2 for a variety of compounds including ascorbic acid, saccharides, and aldehyde was 0.922 (n = 12).[9] This computational chemical analytical method can be used to determine the relative sensitivity of the chemiluminescence reaction when using luminol and lucigenin.

In this system, the computational chemical analysis targeted the productivity of superoxide from a keto-enol rearrangement to study chemiluminescence intensity in analytical chemistry. Superoxide is toxic in vivo. The partial charge was therefore related to biologic activities, such as toxicity (rat oral LD50), the efficacy of the steroids as an endermic liniment, and the contraction index of blood vessel by steroids.

EXPERIMENTAL

A variety of molecules were constructed using the molecular editor of the CAChe™ program and their properties were calculated using MOPAC (AM1) after optimizing their structures using the molecular mechanics (MM2) of the CAChe™ program from Fujitsu (Tokyo, Japan). The molecular properties were calculated using Project-Leader™ of the CAChe™ program. The computers used were a Macintosh G3 and a Dell Latitude. The properties were analyzed using the CA Cricket Graph™ program from Computer Associates (San Diego, CA) on a Macintosh G3 computer. A variety of molecules were constructed using the molecular editor of the CAChe™ program and their properties were calculated using MOPAC (AM1) after optimization.

RESULTS AND DISCUSSION

The toxicity (LD50) was calculated using the TOPKAT™ program from Fujitsu. The values for the phenacyl alcohol derivatives with partial charge are summarized in Table 1. The relation between the chemiluminescence intensity (CLI) or the partial charge change (ΔPC) and rat oral LD50 (LD50) for phenacyl alcohol were:

$$\text{LD50} = -0.189\,(\text{CLI}) + 2.231, r = 0.949, n = 5,$$
$$\text{LD50} = -31.730\,(\Delta\text{PC}) + 8.313, r = 0.912, n = 5.$$

The high correlation coefficient indicated that the measurement of chemiluminescence intensity provides a quantitative measurement of the toxicity of an analyte. Furthermore, the calculation of the partial charge change by the computational chemical method can be used to estimate the rat oral LD50.

Table 1. Molecular properties of phenacyl alcohol derivatives

Chemicals	ΔPC [8]	CLI [5]	LD50
Phenacyl alcohol	0.1990	1.00	2.083
2-Acetyl-phenacyl alcohol	0.1986	1.09	1.942
2-Acetyl-4-bromophenacyl alcohol	0.2045	2.07	1.937
2-Acetyl-4-nitrophenacyl alcohol	0.2124	3.61	1.513
2-Acetyl-4-phenyphenacyl alcohol	0.1967	1.11	2.003

These experimental and computational chemical methods will facilitate rapid screening of drug candidates using chemiluminescence assays.

Many steroid-drugs are used for the treatment of skin diseases. Superoxide produced from steroids should also produce chemiluminescence by the same mechanism. Therefore, the above approaches were applied to study the efficacy, i.e., toxicity, of steroid drugs. The properties are summarized in Table 2. These properties were not related to log *P* values, indicating that efficacy does not depend on molecular mass or solubility due to diffusion. The analysis of chemical reactivity was the important factor, and the atomic partial charge contributes to the activity.

Table 2. Properties of steroids

Steroids	$\log P$	EEL	LD_{50}	CIBV
Alclomethasone dipropionate	3.352		3.811	
Amcinonide	3.581	2.71	2.601	360
Beclomethasone 17,21-dipropionate	3.683	3.77	2.153	500
Betamethasone	1.657	4.26	2.643	
Betamethasone butyrate propionate	4.460		2.208	
Betamethasone 17,21-dipropinate	3.559	1.99		1660
Betamethasone 17-valerate	3.572	3.10	1.942	360
Deoxycorticosterone	2.663		2.622	
Dexamethasone	1.657	5.26	2.643	
Dexamethasone acetate	2.145	5.21	2.085	43
Dexamethasone 17,21-dipropionate	3.559		2.149	1700
Dexamethasone 17-valerate	3.572	3.02	1.942	
Diflorasone diacetate	2.825	1.87	3.222	1600
Diflucortolone 21-valerate	4.830	2.93	3.479	500
Difluprednate	4.310	2.19	3.081	1600
Fludroxycortide	1.151	3.45	4.139	
Flumethasone pivarate	3.625	4.79	4.052	361
Fluocinonide	2.966	2.44	3.069	600
Fluocinolone acetonide	2.497	3.73	2.859	100
Hydrocortisone	1.596	5.96	3.493	0.1
Hydrocortisone acetate	2.106	5.79	3.704	
Hydrocortisone 17-butyrate	3.103	4.93	2.858	50
Hydrocortisone 17-butyrate 21-propionate	4.150	3.42	3.215	360
Methylprednisolone acetate	2.231	5.65	3.797	
Predonisolone	1.930	5.49	3.841	0.5
Predonisolone 17-valerate 21-acetate	4.401	3.88	3.716	360
Triamcinolone acetonide	2.293	4.35	2.258	75

The correlation between ΔPC and the efficacy index of steroids as an endermic liniment (EEL)[10] has been found to be 0.80 (n = 23). The computational chemical calculation allowed us to estimate the efficacy of these compounds. The linear relation between ΔPC and the logarithmic contraction index of blood vessel (CIBV)[10] was obtained with a correlation coefficient of 0.73 (n = 19).

The ΔPC did not have a good linear relation with LD50. Dexamethasone valerate, betamethasone valerate, dexamethasone acetate and triamcinolone acetonide were approximately 1-fold more toxic and difluoro-substituted steroids were approximately 1-fold less toxic than that estimated from the partial charge. Specifically, steroids with a fluorine at the 6th position were less toxic. The correlation coefficient between LD50 and Δatomic partial charge was calculated without these positively and negatively affected compounds. The correlation coefficient was 0.90 (n = 14).

The analysis of a molecule with one site of action is a good indicator for the LD50, such as in phenacyl alcohol derivatives. Steroids, however, are complex molecules, and their metabolites contribute to the experimentally measured LD50 values. If the toxicity can be related to superoxide, the measurement of chemiluminescence intensity can be used for a drug candidate screening. Further studies of the substituent effect of steroids are required before LD50 values can be estimated from computational chemical calculations.

REFERENCES

1. Veazey RL, Nieman TA. Chemiluminescence high-performance liquid chromatographic detector applied to ascorbic acid determinations. J Chromatogr, 1980; 200: 153-62.
2. Klopf LL, Nieman TA. Determination of conjugated glucuronic acid by combining enzymatic hydrosys with lucigenin chemiluminescence. Anal Chem, 1985; 57: 46-51.
3. Veazey RL, Nieman TA. Chemiluminescence determination of clinically important organic reductants. Anal Chem, 1979; 51: 2092-6.
4. Maeda M, Tsuji A. Chemiluminescence with lucigenin as post-column reagent in high-performance liquid chromatography of corticosteroids and p-nitrophenacyl esters. J Chromatogr, 1986; 352: 213-29.
5. Toriba A, Kubo H. Chemiluminescence high performance liquid chromatography of corticosteroids and p-nitrophenacylesters based on the luminol reacton. J Liq Chromatog Rel Technol, 1997; 20: 2965-77.
6. Deyl D, Miksik I, Tesarova E. In: Deyl Z. Miksik I. Tagliano F. Tesarova E. eds. Advanced Chromatographic and Electromigration Methods in Biosciences; Amsterdam: Elsevier, 1998: 166-9.
7. Nakashima K, Imai K. LC-chemiluminescence detection. In: Hanai T. Hatano H. eds. Advances in Liquid Chromatography. Singapore: World Scientific, 1996: 99-122.
8. Hanai T. Computational chemical analysis of the sensitivity of phenacylesters and steroids in chemiluminescence detection. Jpn Chem Program Exchange J, 2001; 13: 123-8.
9. Hanai T. Quantitative computational chemical analysis of the sensitivity of chemilulminescence detection. J Liq Chrom Rel Technol. 2002; 25: 2425-31.
10. Nakayama H. Masubuchi K. Sugawara M. eds. SAISHINNNO HIFUGAIYOUZAI (Recent endermic liniment). Tokyo: Namzando, 1991 (in Japanese).

RAPID DETECTION OF MICROORGANISMS IN ASEPTIC PRODUCTS USING AN ATP BIOLUMINESCENT SYSTEM

T IGARASHI

Kikkoman Corporation Japan, 399 Noda Noda-city Chiba pref. 278-0037, Japan
E-mail: t-igarashi@mail.kikkoman.co.jp

INTRODUCTION

There are many pasteurized milk products that can be stored at room temperature. These products are occasionally contaminated by a small number of bacteria, so it is necessary to do sterility testing using agar plates. This sterility test typically takes a long time. For example, pasteurized products are incubated at 30 °C or 37 °C for 2-3 days. Incubation followed by plate counting is time-consuming, yielding results after 3-5 days. For the purpose of shortening the time for this test, many methods have been studied including direct microscopic counting, membrane filtration, and ATP bioluminescence.

Bossuyt and Waes developed a rapid ATP method for milk samples using surfactant reagents and EDTA-apyrase solution.[1,2] They described that the concentrations of bacteria >10^6 CFU/mL could be distinguished with a correlation coefficient of 0.83. Theron et al. studied the selectivity and completeness of removal of non-bacterial ATP by NRS® and Somase® treatment.[3,4] The detection limit of this method was a bacterial concentration of > 10^5 CFU/mL.

The problem of a poor detection limit was caused by high background ATP and by the low sensitivity of the luciferin-luciferase (L-L) reagent. We have already developed an ATP elimination system[5] using two ATP degrading enzymes (adenosine phosphate deaminase and apyrase) and a surfactant tolerant luciferase that was a mutated *Luciola lateralis* firefly luciferase.[6,7] We optimized this elimination system, and investigated its suitability as a detection system.

MATERIALS AND METHODS

Reagents

CheckLite AT100 kit (Kikkoman Corporation, Japan) consisted of the high sensitive L-L reagent, the ATP releasing reagent, the ATP eliminating reagent, the sample buffer and the sample treatment reagent.

Sample

Samples were four kinds of emulsified products, namely two types of "portion milk" (subdivided nondairy creamer for coffee), whipping cream, and cocoa drink. One type of portion milk was emulsified with plant fat (PM-P) and the other was emulsified with butterfat (PM-B). Cream and cocoa drinks were included vegetable-fat and butterfat. Each fat concentration was 25% (PM-P), 27% (PM-A), 35% (Cream), and 3.6% (Cocoa drink). These products were packed under aseptic condition at Moriyamanyugyo corporation.

Cultivation

We used three bacteria cultures of *Pseudomonas fluorescens*, *Bacillus mycoides* and *Klebsiella pneumoniae*, which had been previously isolated from products at Moriyamanyugyo Corporation.

These bacteria were cultivated in each product for 24 h at 37 °C and each cultured broth sample was serially diluted into the corresponding aseptic products. The total colony count was determined with Plate Count Agar (MERCK).

Bioluminescent Assay

0.1 mL of each the milk product sample and the sample treatment reagent were mixed for 10 sec.. 0.1 mL of the ATP eliminating reagent and 0.7 mL of the dilution buffer were added into the mixture. 0.1 mL of the diluted sample was transferred to a new tube and kept at room temperature for 30 min. Then 0.1 mL of ATP releasing reagent was added to the mixture. After 10 sec. waiting, 0.1 mL L-L reagent was pipetted into the test tube and the resulting bioluminescence was measured with Lumitester K-200 (Kikkoman Corporation, Japan).

RESULTS AND DISCUSSION

A standard curve for ATP is shown with the coefficient of variation (Fig. 1). The coefficients of variation (CV value) for measurements were very low. The measurable range of ATP was from 2.0×10^{-13} to 2.0×10^{-8} M and the detection limit was 200 fmol/assay of ATP.

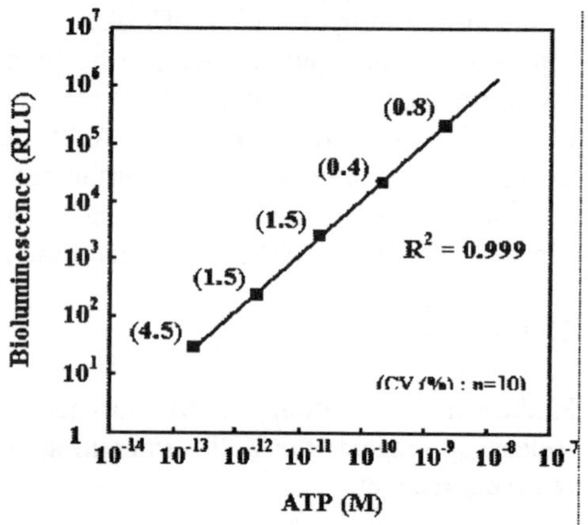

Figure 1. Standard curve for ATP in the releasing reagent
Values in parentheses represent coefficients of variation (%, n=10)

Result for three typical bacteria (*B. mycoides*, *P. fluorescence* and *K. pneumoniae*) are shown in Fig.2. The background luminescence of all four products was below 150 RLU, and became constant. As shown, the detection limit of *B. mycoides*, *P. fluorescence* and *K. pneumoniae* in four products were 1×10^3, 2×10^4, and 5×10^3

CFU/mL, respectively. The differential of detection limit was caused by each kind of bacteria having a different ATP in their cells. *B. mycoides* have much ATP (1.7×10^{17} mol/cell). On the other hand, *P. fluorescence* have less ATP (4.5×10^{-19} mol/cell) than that of *B. mycoides* and *K. pneumoniae* (2.1×10^{-18} mol/cell). In this study, the sample was diluted 10 times with the sample buffer and the diluted mixture was used for detection. If the extracellular ATP was destroyed the effectively and the sample background was low, the sample could be further diluted 5 or 2 times and the detection limit would be correspondingly advanced.

Each of the three bacteria was added into aseptic product and these products were incubated for 24 h at 37 °C. All added bacteria grew in the product and their concentration of the bacteria was over 10^5 cfu/mL (data not shown).

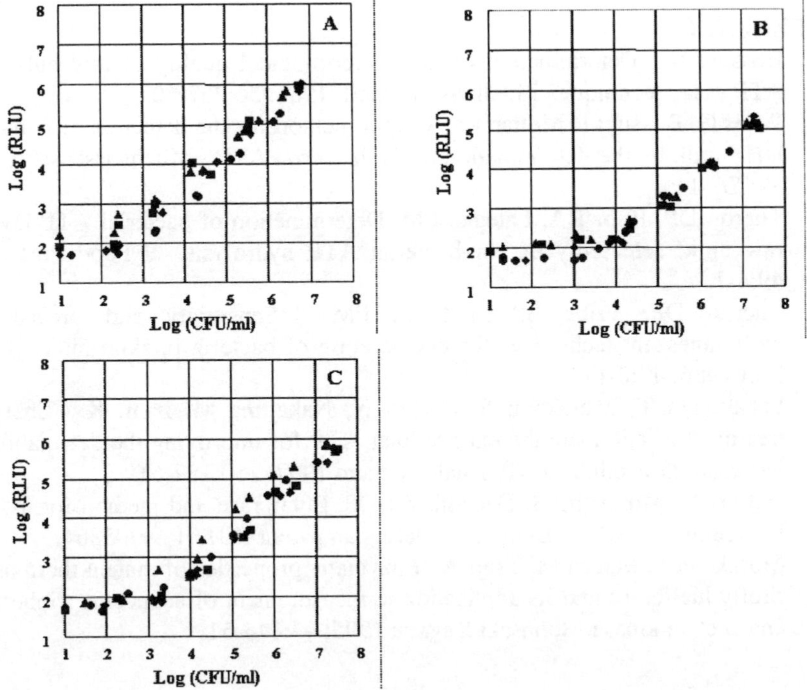

Figure 2. Standard curves for three different bacteria in four samples. A ; *Bacillus mycoides*, B ; *Pseudomonas fluorescens*, C ; *Klebsiella pneumoniae*. PM-P (♦) , PM-B(■), Cream (▲), Cocoa drink (●).

The contamination of pasteurized products is usually caused by specific bacteria species that exist in the raw material and environment. The current sterility test for pasteurized products is that the products are incubated and then applied and cultured to a standard agar plate to check for contamination. But if the contaminated bacteria do not grow on the plate, the contaminated product would be misjudged as sterile. We must keep in mind that all kinds of bacteria have their own conditions for growth (pH, temperature, anaerobic condition, many kind of substrates, etc.). When using only one type of medium plate for culturing, some types of bacteria may not be detected.

The ATP bioluminescence method would detect any contaminated bacteria, if the bacteria were growing over its detection limit in the product. When many samples must be checked, it is possible to select whether each product is aseptic or not in a short time. Dubious samples would be tested by the ordinary plate method. The ATP bioluminescence method is one of the most available sterility tests for initiate screening.

REFERENCES

1. Bossuyt R. Determination of the bacteriological quality of raw milk by an ATP assay technique. Milchwissenschaft. 1981; 36:257-60.
2. Waes G, Bossuyt R Mottar J. A rapid method for the detection of non-sterile UHT milk by the determination of the bacterial ATP. Milchwissenschft 1994; 39:707-11.
3. Theron DP, Prior BA, Lategan PM. Determination of bacterial ATP levels in raw milk: selectivity of non-bacterial ATP hydrolysis. J Food Prot. 1986; 49:4-7.
4. Theron DP, Prior BA, Lategan PM. Sensitivity and precision of bioluminescent techniques for enumeration of bacteria in skim milk. J Food Prot 1986; 49:8-11.
5. Sakakibara T, Murakami S, Hattori N, Nakajima M, Imai. K. Enzymatic treatment to eliminate the extracellular ATP for improving the detectability of bacterial intracellular ATP. Anal Biochem 1997; 250:157-61.
6. Hattori N, Murakami S. December 1998. Luciferase and method for assaying intracellular ATP by using the same. Japan patent PCT/JP 98/05864.
7. Murakami S, Maeda M, Tsuji A. Enzymatic properties of mutant thermostable firefly luciferase and its application to measurement of adenosine triphosphate and acetate kinase. Bunnseki Kagaku 1995; 44:845-51.

CHEMILUMINESCENCE DETECTION OF 3-NITROBENZANTHRONE AND 2-NITROTRIPHENYLENE IN AIRBORNE PARTICLES WITH ON-LINE REDUCTION HPLC SYSTEM

K INAZU[1], T SAITO[1], ND V.U[1], K AIKA[1], Y HISAMATSU[2]

[1]*Department of Environmental Chemistry and Engineering, Interdisciplinary Graduate School of Science and Engineering, Tokyo Institute of Technology, 4259-G1-13 Nagatsuta, Midori-ku, Yokohama 226-8502, Japan*
[2]*Department of Environmental Health, National Institute of Public Health, 4-6-1 Shiroganedai, Minato-ku, Tokyo 108-8638, Japan*
E mail: inazu@chemenv.titech.ac.jp

INTRODUCTION

Nitrated polycyclic aromatic compounds (NPAC) have been extensively investigated as an important environmental direct-acting mutagen especially in the atmosphere.[1] Nevertheless, up to 60% of direct-acting mutagenicity of the soluble organic fraction (SOF) of the collected atmospheric samples has been accounted for by the conventionally studied NPAC such as nitropyrenes in most of the studies,[1] i.e. "excess mutagenicity" has been frequently observed and significant contribution of unknown mutagenic NPAC to the total direct-acting mutagenicity of the atmosphere is suggested.[1] Recently 3-nitrobenzanthrone (3-NBA) and 2-nitrotriphenylene (2-NTP) in airborne particles were reported to be a novel important contributor to the direct-acting mutagenicity of the atmosphere and both diesel emission and atmospheric nitration of parent benzanthrone or triphenylene with nitrogen oxides have been suggested as their source.[2-4] However, sufficient atmospheric observation of these two NPAC to reveal the significance as atmospheric mutagen has not been conducted probably due to their low concentration in the samples to restrict frequent analysis by means of GC-MS techniques.

Hayakawa and co-workers have intensively developed HPLC techniques with on-line reduction of NPAC to aminoPAC (APAC) and chemiluminescence detection of APAC for trace analysis of NPAC, particularly of nitropyrenes.[5] In this study, we examined the HPLC method for the analysis of novel NPAC, 3-NBA and 2-NTP in airborne particles including the interference of coexisting NPAC in the sample in separation and the efficiency of the on-line reduction to selective conversion of 3-nitrobenzanthrone, which has one carbonyl group, to detectable 3-aminobenzanthrone in the HPLC system.

EXPERIMENTAL

The HPLC system employed in this study was essentially the same as previously reported[5] but it was slightly modified to be comprise four feeding pumps, an automated sample injector, a time-programmable six-way switching valve, a reducing column packed with alumina supported Pt-Rh bimetallic catalyst (4.0 mm

"

i.d. × 10 mm, 353 K; RC), a chemiluminescence detector (CLD), two system controllers, and two ODS separation columns (Nacalai Tesque, Cosmosil $5C_{18}$ AR, for NPAC separation (SC1); Cosmosil $5C_{18}$ MS, for resulting APAC separation (SC2); 4.6 mm i.d. × 250 mm for the both) and a concentration column (Cosmosil $5C_{18}$ AR, 4.0 mm i.d. × 10 mm ;CC) in a column oven at 313 K. The mobile phase for SC1, SC2, and CC were 0.02 mol/L acetate ethanol–aqueous buffer solution (75 vol% ethanol; pH = 5.5 at 1.0 mL/min), 0.01 mol/L imidazole–perchloric acid aqueous buffer solution–acetonitrile mixture (1/1, v/v; pH = 7.6 at 1.0 mL/min), and purified water at 4.0 mL/min, respectively. The reagent solution for CLD was 8 mmol/L hydrogen peroxide–0.64 mmol/L bis(2,4,6-trichlorophenyl)oxalate acetonitrile solution at 1 mL/min. 20 μL of the sample solution was injected and the data was stored and analyzed with a PC by Shimadzu Chromatopak Manager.

NBA isomers and NTP isomers were synthesized and purified according to previously reported methods.[2-3] 1-Nitropyrene (1-NP), 1,8-dinitropyrene (1,8-DNP), 2- and 3-nitrofluoranthenes (2- and 3-NF), 6-nitrochrysene (6-NC), and 2-nitrofluoren (2-NFL) were commercially available from several suppliers and used without further purification. 2-NFL was used as an internal standard to calculate the recovery of the target NPAC from the samples and added at a higher level than typical atmospheric concentration (20 fmol/m^3) by factor of 200 to avoid the influence of atmospheric 2-NFL on the calculation.

Sampling of airborne particles was carried out at the rooftop level of a 6-story building of the National Institute of Public Health surrounded with arterial roads in central Tokyo (30 m above the ground) between 16th and 22nd October, 2000. Airborne particles smaller than 10 μm in aerodynamic diameter were collected on quartz fiber filters with a high-volume air sampler with 10-μm cut-off stage for 24 h from 1224 m^3 of the air. Three quarters of the filter samples spiked with 2-NFL internal standard were cut into small pieces and put into dichloromethane to extract SOF from the airborne particle. The extracted SOF solution was isolated by filtration and washed sequentially with 5% sodium hydroxide, 20% sulfuric acid solution, and purified water. After removal of water, the sample solution was concentrated by drying under nitrogen and was dissolved into 1 mL of acetonitrile for subsequent analysis.

RESULTS AND DISCUSSION

Although on-line reduction HPLC with chemiluminescence detection (R-HPLC-CLD) is much more sensitive for nitroarenes such as nitropyrenes up to by a factor of 1000 than conventional GC-MS system[5], the specific sensitivity would decrease when target NPAC have reactive functional groups or heteroatoms in the ring system. These would be easily reductively decomposed, since formation of such reduced products other than corresponding APAC, results a decrease in the number of APAC molecules detected by CLD. Moreover, if non APAC products are formed depending on the amount of NPAC, linear dose-response will be lost. 3-NBA is an NPAC with a reactive carbonyl group. Thus, the feasibility of 3-NBA analysis by R-HPC-CLD was examined for sensitivity and accuracy. Unexpectedly, R-HPLC-CLD employed in this study exhibited quite high sensitivity to 3-NBA (20 fmol for detection limit and 80 fmol for quantification limit) comparable to 1,8-DNP, which is the NPAC most suitable for this method, and excellent linear response between 80–1500 fmol with RSD of 2.3% as shown in Table 1. The conversion of 3-NBA to 3-ABA was estimated to be 88% under the reduction condition employed in this study. On the other hand, the sensitivity to 2-NTP was not as high among the seven NPAC investigated (40 fmol for detection limit and 200 fmol for quantification limit) while linearity of the response and RSD were excellent.

Table 1. Accuracy of on-line reduction HPLC-CLD analysis for NPAC[a]

NPAC	DL[b] (fmol)	QL[c] (fmol)	Range (fmol)	r^2	RSD[d]/%
3-NBA	20	80	80–1500	0.999	2.3
2-NTP	40	200	200–3000	0.998	3.6
2-NF	4	20	20–1000	0.999	2.1
3-NF	10	40	40–1000	0.997	4.2
1-NP	8	30	30–1000	0.998	1.8
6-NC	60	300	300–3000	0.997	3.3
1,8-DNP	20	80	80–2000	0.997	3.8

[a]Injection volume: 20 μL. [b]Detection limit with S/N = 3. [c]Quantification limit with S/N = 10. [d]Relative standard deviation for the analysis of standard solution (50 nmol/L, n=3).

Another important issue for applying R-HPLC-CLD to atmospheric 3-NBA analysis is sufficient separation from coexisting NPAC in the samples especially for NBA isomers. With respect to isomer separation, 2-NBA should be the most and important since it has been found to be much more abundant in airborne particles and atmospheric formation of 2-NBA was also suggested in the same manner as 2-NF.[6] Actually, 2-NBA was eluted closest to 3-NBA (Fig. 1(a)) while other NBA isomers can be separated by operating switching valve before SC2. As a result, coexisting 2-NBA is not a serious problem for 3-NBA analysis even if 2-NBA is more abundant in the sample. This was because good peak separation was achieved and 3-NBA exhibited much higher specific sensitivity than 2-NBA by a factor of about 500 as shown in Fig. 1(c). It can be seen in Fig. 1(b) that trace 3-NBA (below 0.5% in relative concentration) was observed. 2-NTP was also able to be sufficiently separated from other NPAC and analyzed simultaneously with 2-NF and 6-NC.

The performance of R-HPLC-CLD was then verified by applying it to the analysis of 3-NBA and 2-NTP in airborne particles as well as five other conventionally studied NPAC. As shown in Table 2, 3-NBA was found in a comparable concentration level to 1,8-DNP, suggesting that the contribution to the direct-acting mutagenicity of the atmosphere will be also comparable. 2-NTP concentration was also significant and higher than those of 3-NF and 6-NC.

Table 2. Atmospheric concentration of particle-associated NPAC on October, 2000

Concentration (fmol/m^3)						
2-NF	3-NF	1-NP	6-NC	2-NTP	3-NBA	1,8-DNP
789.3	2.0	106.6	3.2	7.3	1.4	1.8

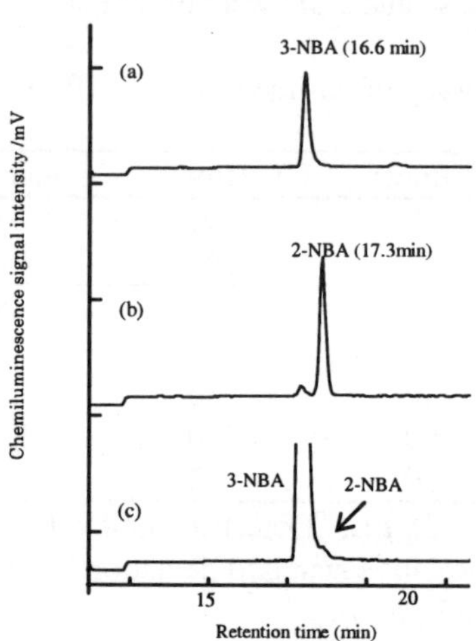

Figure 1. Chromatograms of standard solution (20 μL) of (a) 3-NBA, (b) 2-NBA, and (c) equimolecular mixture of 2- and 3-NBA. Retention time is in parentheses.

REFERENCES

1. Finlayson-Pitts BJ, Pitts, Jr. JN. Chemistry of the Upper and Lower Atmosphere. San Diego, CA: Academic Press, 2000: 440-547.
2. Enya T, Suzuki H, Watanabe T, Hirayama T, Hisamatsu Y. 3-Nitrobenzanthrone, a powerful bacterial mutagen and suspected human carcinogen found in diesel exhaust and airborne particles. Environ Sci Technol 1997; 31: 2772-6.
3. Ishii S, Hisamatsu Y, Inazu K, Kadoi M, Aika K. Ambient measurement of nitrotriphenylenes and possibility of nitrotriphenylenes formation by atmospheric reaction. Environ Sci Technol 2000; 34: 1893-9.
4. Hayakawa K, Murahashi T, Butoh M, Miyazaki M. Determination of 1,3-, 1,6- and 1,8-dinitropyrenes and 1-nitropyrene in urban air by high-performance liquid chromatography using chemiluminescence detection. Environ Sci Technol 1995; 29: 928-32.
5. Phousongphouang PT, Arey J. Sources of the atmospheric contaminants, 2-nitrobenzanthrone and 3-nitrobenzanthrone. Atmos Environ 2003; 37:3189-99.

DETERMINATION OF PARTICLE-ASSOCIATED NITRO-PAH USING HPLC/CHEMILUMINESCENCE DETECTION SYSTEM

T KAMEDA[1], K INAZU[2], Y HISAMATSU[3], N TAKENAKA[1], H BANDOW[1]

[1]*Grdt. Schl. Eng., Osaka Pref. Univ., 1-1 Gakuen-cho, Sakai 599-8531, Japan*
[2]*Interdisciplinary Grdt. Schl. Sci. Eng., Tokyo Inst. Technol., 4259 Nagatsuta, Midori-ku, Yokohama 226-8502, Japan*
[3]*National Institute of Public Health, 4-6-1 Shirokanedai, Minato-ku, Tokyo 108-8638, Japan*
Email: kameda@ams.osakafu-u.ac.jp

INTRODUCTION

Nitrated policyclic aromatic hydrocarbons (nitro-PAH), which have been found in airborne particles, generally have high mutagenic activity, and some of them are known to be carcinogenic.[1] For example, 2-nitrotriphenylene (2-NTP) has strong mutagenicity although the parent triphenylene does not exhibit mutagenic activity, and the concentration of 2-NTP in the atmosphere is relatively high.[2] Nevertheless, sources of atmospheric 2-NTP are still unknown. In order to understand the controlling factors of concentration of atmospheric 2-NTP, it is necessary to observe ambient 2-NTP, other typical nitro-PAH, and major gaseous atmospheric pollutants simultaneously with high time resolution.

In this study, the concentrations of several kinds of nitro-PAH, such as 1-nitropyrene (1-NP), 2-nitropyrene (2-NP), 2-nitrofluoranthene (2-NF), and 2-NTP, in the soluble organic fraction of airborne particles were determined by a column switching HPLC-chemiluminescence detection system in order to clarify the occurrence and behaviour of 2-NTP in the atmosphere.

METHODS

The HPLC system consisted of four pumps, a six-ports switching valve, two separation ODS columns (Wako Pure Chemicals Industries, Wakosil-II 5C18AR and Imtakt, Cadenza CD-C18, each 3.0 mm i.d. x 250 mm), a Pt/Rh column for the reduction of nitro-PAH (Jasco, NPpak-R, 4.6 mm i.d. x 30 mm), a concentration column (Jasco, NPpak-G, 4.6 mm i.d. x 30 mm), and a chemiluminescence detector (Jasco FP2020 with CLKIT C454). An acetonitrile solution containing 0.3 mmol/L of bis(2,4,6-trichlorophenyl)oxalate and 15 mmol/L of H_2O_2 was used as a chemiluminescence reagent. The mobile phase for initial separation and reduction of nitro-PAHs was methanol/water (3/1, v/v) and that for second separation was acetonitrile/imidazole-perchloric acid buffer (1/1, v/v). Sample collection of the airborne particulate was performed every 3 hours to clarify their diurnal variation in a slightly polluted residential area, Sakai, Osaka, Japan using high-volume air samplers (Kimoto Electrics, Model 120) during: (I) September 3-6, 2001 (II) November 26-30, 2001 and (III) May 12-16, 2003. Soluble organic fraction of particles collected on a quartz fiber filter was extracted under sonication for 15-20 min in 200 mL of benzene/ethanol (3/1, v/v), then the solution of extract was filtered with cellulose acetate filter (Advantec MFS, No.2) to remove solid substances. The filtrate was treated with 100 mL of 5% sodium hydroxide solution, 100 mL of 20% (v/v) sulfuric acid solution and then 100 mL of water. By evaporation, the organic layer was reduced to *ca.* 5 mL and it was filtered with a 0.22 μm menbrane filter. 0.5 mL of the sample solution was finally obtained by removing solvent with nitrogen stream. An aliquot of the sample solution was injected into the HPLC system. During the sampling period, concentration of CO was monitored as a typical pollutant gas using NDIR CO analyzer (Thermo Electron, MODEL 48).

RESULTS

Mean concentrations of CO and 1-NP, which were primarily emitted from combustion processes such as diesel-powered vehicles,[3] were 0.5 ppmv and 85.3 fmol m^{-3} in September, 0.9 ppmv and 62.7 fmol m^{-3} in November, and 0.7 ppmv and 63.9 fmol m^{-3} in May, respectively. Mean concentrations of 2-NF and 2-NP, which are produced by atmospheric reactions,[4] were 168.0 and 15.6 fmol m^{-3} in September, 144.7 and 27.6 fmol m^{-3} in November, and 83.2 and 30.6 fmol m^{-3} in May, respectively. Mean concentration of 2-NTP, whose sources are still unknown, in September, November, and May were 17.0, 18.8, and 7.8 fmol m^{-3}, respectively. Clear trends in the seasonal differences of nitro-PAH concentrations were not observed.

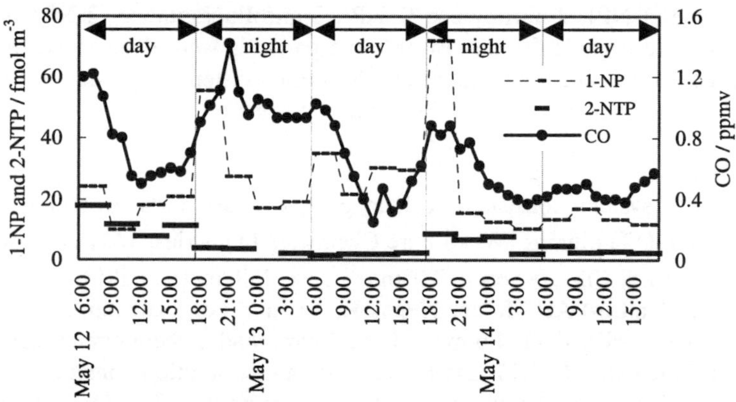

Figure 1. Diurnal changes in concentrations of 1-NP, 2-NTP, and CO during May 12-14, 2003.

Fig. 1 shows the diurnal changes in 3-h averaged concentrations of 1-NP and 2-NTP and in 1-h averaged concentration of CO during May 12-14. The diurnal variation of the concentration of 1-NP was similar to that of CO, while the concentration of 2-NTP showed slightly different pattern in the diurnal variability.

For instance, the concentrations of 1-NP and CO increased early in the evening on May 12 and early in the morning on May 13, while the concentration of 2-NTP was constantly low during these periods of time.   The scatter plot of the 3-h averaged concentration of 2-NTP against that of 1-NP is shown in Fig. 2.   These two factors are not correlated well (correlation coefficient r = 0.29).   These results suggest that atmospheric 2-NTP is not emitted from combustion processes but mainly produced through the secondary formation processes.

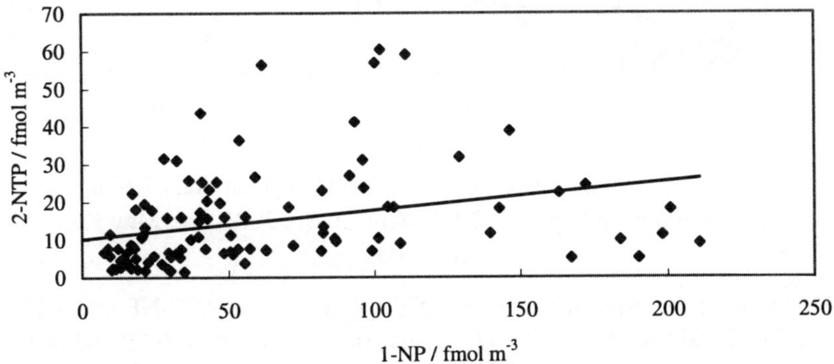

Figure 2. Plot of concentration of 2-NTP against that of 1-NP observed during Sep. 3-6, 2001, Nov. 26-30, 2001, and May 12-16, 2003.

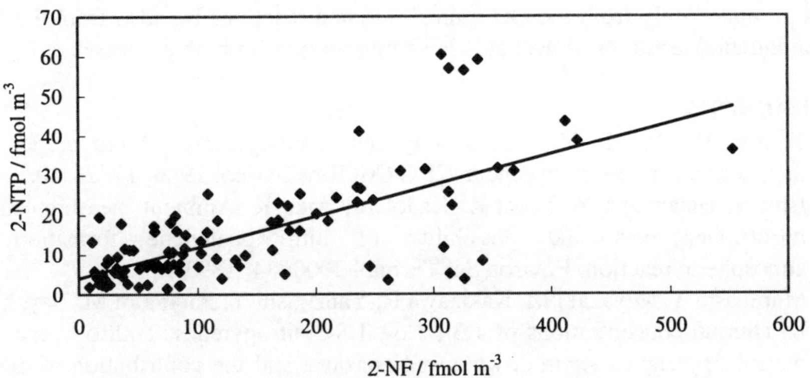

Figure 3. Plot of concentration of 2-NTP against that of 2-NF observed during Sep. 3-6, 2001, Nov. 26-30, 2001, and May 12-16, 2003.

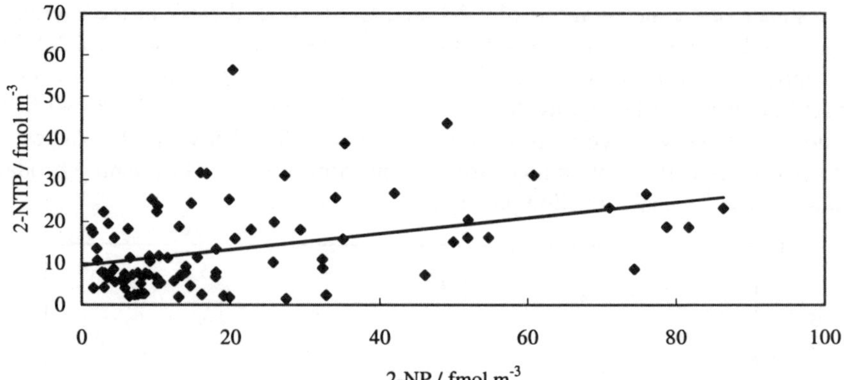

Figure 4. Plot of concentration of 2-NTP against that of 2-NP observed during Sep. 3-6, 2001, Nov. 26-30, 2001, and May 12-16, 2003.

The plots of the concentration of 2-NTP against those of 2-NF and 2-NP are shown in Fig. 3 and Fig. 4, respectively. The concentration of 2-NTP and that of 2-NF were strongly correlated (r = 0.70). On the other hand, an obvious correlation between the concentration of 2-NTP and that of 2-NP was not observed (r = 0.38). 2-NF is known to be formed *via* OH or NO_3 radical initiated reactions in the atmosphere.[4] Although 2-NP is also believed to be formed by atmospheric reaction, its formation is explained only by the reaction of pyrene with OH radicals.[4] Hence, these results obtained from the interrelation analysis suggest that atmospheric 2-NTP is not produced only from the OH radical initiated reactions but also from the NO_3 radical initiated reactions as well as 2-NF being formed *via* both processes.

REFERENCES

1. Tokiwa H, Ohnishi Y. Mutagenicity and carcinogenicity of nitroarenes and their sources in the environment. CRC Crit Rev Toxicol 1986; 17: 23-60.
2. Ishii S, Hisamatsu Y, Inazu K, Kadoi M, Aika K. Ambient measurement of nitrotriphenylenes and possibility of nitrotriphenylene formation by atmospheric reaction. Environ Sci Technol 2000; 34: 1893-9.
3. Murahashi T, Miyazaki M, Kakizawa R, Yamagishi Y, Kitamura M, Hayakawa K. Diurnal concentrations of 1,3-, 1,6-, 1,8-dinitropyrenes, 1-nitropyrene, and benzo[*a*]pyrene in air in downtown Kanazawa and the contribution of diesel-engine vehicles. Jpn J Toxicol Environ Health 1995; 41: 328-33.
4. Atkinson R, Arey J. Atmospheric chemistry of gas-phase polycyclic aromatic hydrocarbons: formation of atmospheric mutagens. Environ Health Perspect 1994; 102: 117-26.

BIOSENSORS BASED ON BACTERIAL BIOLUMINESCENCE FOR ENVIRONMENTAL MONITORING

VA KRATASYUK[1], EN ESIMBEKOVA[2], EV VETROVA[2]

[1] Krasnoyarsk State University, pr.Svobodnii 79, 660041 Krasnoyarsk, Russia
[2] Institute of biophysics SB RAS, Akademgorodok, 660036 Krasnoyarsk, Russia
E-mail: bpl@ibp.ru

INTRODUCTION

To estimate water quality, bioluminescent biosensors have been devised and successfully used. They are characterized by rapidity and simplicity of use, high sensitivity, and accuracy. The Collection of Luminous Bacteria IBSO (http://www.bdt.org.br/bdt/msdn/ibso) is being used to develop bioassays for monitoring the environment, using lyophilized luminous bacteria and the luminescent system isolated from them. Bioluminescent assays have an advantage over other biological assays: luminescence is easy to measure, the method is rapid, and the measurements can be automated.

METHODS

The lyophilized luminous bacteria and lyophilized mixture of luciferase (Lu) from *Photobacterium phosphoreum* and NADH:FMN-oxidoreductase (R) from *P. leiognathi* were produced by the Biotechnology sector of the Institute of Biophysics (Krasnoyarsk). One vial of enzymes contained 0.11 mg of Lu and 0.069 units of activity/mL of R. One unit of R activity was defined as 1 μmol of NADH degraded per min. All the assays were performed in the 0.1 mol/L phosphate buffer solution at pH 6.8 at room temperature.

Before measurements one vial of enzymes was diluted by 0.5 mL of 0.1 mol/L phosphate buffer. For the coupled enzyme system the reaction mixture contained 10 μL Lu+R, 50 μL 0.002 % tetradecanal, 200 μL 0.1 M phosphate buffer (pH 6.8), 200 μL 0.4 mM NADH and 50 μL 0.5 mM FMN. The cuvette was placed into the bioluminometer BLM 8801 (SKTB "Nauka", Krasnoyarsk, Russia) and control light emission (I_c) was recorded. When the light emission reached a steady state 50 μL of a test water was pipetted into cuvette, and the light intensity (I_{ex}) was measured again.

Before measurements one vial of bacteria was diluted by 500 μL 1.5 % NaCl. 20 μL of bacteria solution was added to 1 mL of 3 % NaCl and the control light intensity was recorded using the bioluminometer after 15 min period of incubation. The measurements were repeated with 1 mL 3 % NaCl prepared on water samples and experimental light intensity was measured. The effect of sample water on coupled enzyme system bioluminescence was estimated by the bacterial (BI) or luciferase (LI) index using the following formula: BI $= I_c/I_{ex}$, LI $= I_c/I_{ex}$.[1]

Two triple bioluminescent enzymatic systems were used: with the alcohol dehydrogenase and trypsin. Activity of alcohol dehydrogenase and trypsin was determined by bioluminescent method from the decay constant of bioluminescence.[2-3]

RESULTS
Basis of the bioluminescent ecological monitoring
The preliminary results showed a correlation between physicochemical characteristics of inhibitor (activator) molecules and changes in kinetic parameters of bioluminescent reaction.[4-7] For example the comparison of the effects of the quinones and phenols on three bacterial bioluminescence systems of different complexity indicates that the influence of the compounds on the bioluminescence intensity depends on the structure and redox characteristics. The inhibitory activity of quinones depends on their hydrophobic substituents and the size of the aromatic part.[8-9] Such correlations are closely related to the physical mechanism of bioluminescence; they are the biophysical basis for bioluminescent ecological monitoring.

These data provide a basis for comparing sensitivities and choosing test organisms and enzymatic systems to be included in the sensors for this automated system of bioassays.
Bioluminescent ecological monitoring of salt lake
Bioluminescence bioassays based on luminous bacteria and coupled enzyme system NADH-FMN-oxidoreductase-luciferase were adapted for monitoring the saline-water conditions of Lake Shira (Khakasia, Siberia).[10] The differences in bioluminescence responses have been found to be related to the salt composition and the oxidation-reduction properties of water. Bioluminescent kinetics parameters, which are mostly sensitive to pollution under conditions of saline water, have been observed. Figure 1 shows the typical bioluminescence kinetics of the samples of water due to anthropogenic influence (beach) and control clear water (non-recreational area).

The enzymatic system in the presence of 1,4-benzoquinone is shown to be more sensitive to redox characteristics of the salt water than in the absence of 1,4-benzoquinone. Therefore 1,4-benzoquinone should be applied for the preparation of a model solution for the monitoring of redox properties of the salt water.

Using this technique, the results of bioluminescence analysis are used to construct a heterogeneity map that characterizes the spatial and temporal water quality of lake Shira. A partial map was based on the bioluminescence characteristics of water samples taken along the shoreline, sampling stations in the different places and in different depths of the lake.

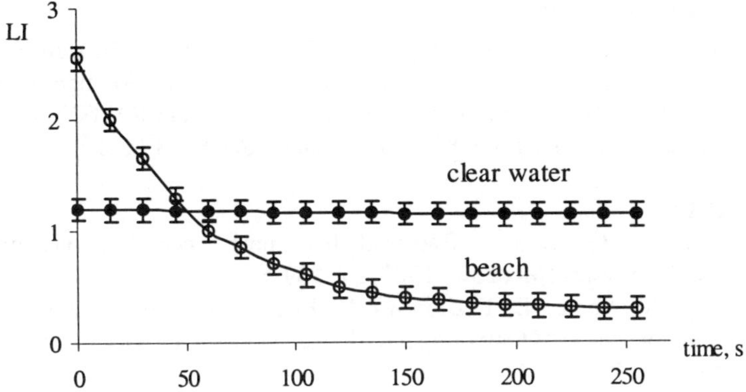

Figure 1. Time dependences of Luciferase Indexes (LI) in the water samples

System of biosensors for ecological monitoring

The approaches to the creation of universal system of biosensors for ecological monitoring using bioluminescent organisms and their enzymes and to devise a laboratory model of a biosensors system are discussed.[11]

The set of bioluminescent tests was developed to monitor water quality in natural and laboratory ecosystems. It consisted of four bioluminescent systems: luminous bacteria, coupled enzyme system NADH:FMN-oxidoreductase-luciferase and triple enzyme systems with alcohol dehydrogenase and trypsin. We investigated their responses to an unpolluted small forest pond, laboratory microecosystems polluted with benzoquinone and a batch culture of blue-green algae.[12] It has been shown that for the unpolluted water body the fluctuations in parameters of the biotests were insignificant and resulted from natural variability of the pond ecosystem. Parameters of the test changed sharply when the water body was contaminated with xenobiotics and in the case of "bloom" of blue-green algae. It is necessary to emphasize that ranges of variability of biotests, which occurred in the unpolluted pond and unpolluted MESs were significantly lower than the degree of response of biotests after the addition of the pollutant (benzoquinone). Therefore we could detect effect of pollutants e.g. quinones, within the variability of responses, caused by natural water.

Hence, the data of a single test cannot provide a basis for a conclusion about the presence or absence of toxic substances in a water body. Only a set of tests, like this ones used in present study can be applied as an alarm system to detect an acute toxicity of aquatic ecosystems.

ACKNOWLEDGMENTS

This work was supported by the Ministry of Education of the Russian Federation (grant PD 02-1.4-316) and the U.S. Civilian Research and Development Foundation for the Independent States of the Former Soviet Union (grant KY-002-X1, Science Education Center "Yenisei", grant Y1-B-02-11 and grant Y1-B-02-12).

REFERENCES

1. Kratasyuk V, Gitelson J. Bacterial bioluminescence and bioluminescent analysis. Biophysics (Moscow) 1982; 27: 937-53.
2. Petushkov V, Shefer L, Rodionova N, Fish A. Bioluminescent method of determination of NAD(P)H-depend dehydrogenase activity. Appl Biochem Biotech 1987; 23: 270-4.
3. Petushkov V, Kratasyuk V, Fish A, Gitelson J. Protease activity determination method. 1983 Patent SU 1027615 A.
4. Kudryasheva N, Kratasyuk V, Belobrov P. Bioluminescent analysis. The action of toxicants: Physical-chemical regularities of the toxicants effects. Anal Lett 1994; 27: 2931-8.
5. Kudryasheva N, Zyuzikova L, Gutnik T, Kuznetsov A. The action of the salts of metals on bacterial bioluminescent systems of various complexity. Biophysics (Moscow) 1996; 41: 264-9.
6. Kudryasheva N, Esimbekova E, Remmel N, Kratasyuk V. Effect of quinones and phenols on a triple enzymic bioluminescent system with protease. Luminescence 2003; 18: 224-8.
7. Kudryasheva N, Kudinova I, Esimbekova E, Kratasyuk V, Stom D. The influence of quinones and phenols on the triple NAD(H)-depend enzyme systems. Chemosphere 1999; 38: 751-8.
8. Kudryasheva N, Vetrova E, Kuznetsov A, Kratasyuk V, Stom D. Bioluminescent assays: effects of quinones and phenols. Ecotox Environ Safe 2002; 53: 221-5.
9. Kudryasheva N, Kratasyuk V, Esimbekova E, Vetrova E, Nemtseva E, Kudinova I. Development of the bioluminescent bioindicators for analyses of environmental pollutions. Field Anal Chem Tech 1998; 2: 277-80.
10. Vetrova E, Kratasyuk V, Kudryasheva N. Bioluminescent characteristics map of the Shira lake water. Aquat Ecol 2002; 36: 309-15.
11. Kratasyuk V. Bioassay for monitoring of ecosystems. In: Materials of 1st International Congress "Biodiversity and dynamics of ecosystems in North Eurasia", Novosibirsk, Russia 2000; Part 5: 13-5.
12. Kratasyuk V, Esimbekova E, Gladyshev M, Khromichek E, Kuznetsov A., Ivanova E. The use of bioluminescent biotests for study of natural and laboratory aquatic ecosystems. Chemosphere 2001; 42: 909-15.

HOSPITAL TESTING OF A RAPID BIOLUMINESCENT ASSAY FOR MRSA

RL LESLIE[1], MJ MURPHY[1], DJ SQUIRRELL[1], SL COTTERILL[2],
SCW MATTHEWS[2], M SKYRME[2]

[1]Detection Dept, Dstl Porton Down, Salisbury SP4 0JQ, UK
[2]Dept of Medical Microbiology, Salisbury District Hospital, Odstock Road,
Salisbury SP2 8BJ, UK
Email: djsquirrell@dstlgov.uk

INTRODUCTION

Methicillin resistant *Staphylococcus aureus* (MRSA) is a major cause of hospital-acquired infections. It is directly responsible for about 1,000 deaths per annum in the UK, is a contributory factor in many more, and imposes a considerable financial burden on health services. Standard microbiological methods take from 2-4 days to determine the presence of MRSA in clinical samples. This limits the value of testing to the monitoring of infection trends rather than in the provision of information to aid in the treatment of patients. A rapid test could be used both to guide the prescription of antibiotics and to identify patients carrying MRSA as a tool in infection control.

Previously,[1] we tested a manual method using antibiotic-mediated lysis of non-target cells followed by immuno-magnetic separation with an adenylate kinase (AK) bioluminescence[2] endpoint determination. This 4 h assay for patient swabs was carried out on a limited number of samples. The work reported here was to introduce automation into the assay and to carry out tests on a larger number of samples in a hospital setting. Initial assay development was done in a non-clinical laboratory using spiked samples. The assay was then adapted to fit in with the standard hospital test. Modifications were introduced as testing in the hospital laboratory proceeded.

There are two points to note: methicillin is no longer available in the UK so testing for MRSA is now carried out using oxacillin; and the "standard" test against which the rapid assay was compared was different to that in the earlier trial[1] where swabs from transport media were streaked directly onto agar containing methicillin, salt, mannitol and a pH indicator.

INSTRUMENTATION, MATERIALS AND METHODS
Instrumentation

A KingFisher ML magnetic particle processor (ThermoLabsystems, Helsinki, Finland), a device that automates immuno-magnetic separation, was modified by replacing the well holder with a purpose built plate to allow samples to be thermostatted at 37 °C whilst being processed. Bioluminescence was measured in a Berthold Detection Systems (Pforzheim, Germany) Sirius luminometer for which a new tube holder was constructed to take KingFisher 5-well strips so that light emission from the final well could be measured.

Materials

Unless specified otherwise, reagents were from Sigma, (Poole, UK). Antibiotic broth for removal of susceptible cells was L-broth (Oxoid, Basingstoke, UK) containing 4 µg mL^{-1} oxacillin and 70 mg mL^{-1} sodium chloride. 0.9 µm sized high carboxyl magnetic beads from Estapor (Pithiviers, France) were coated with either monoclonal antibody C55704M (Biodesign International, Saco ME, US) or fibrinogen. KingFisher 5–well assay strips were prepared with reagents as follows: 10 µL of magnetic beads at 10^8 mL^{-1} in well A; 1 mL of L-broth plus 0.2% Tween 20 in wells B and C; 1 mL of phosphate buffered saline in well D; and 200 µL of either detergent-based extractant plus ADP or lysostaphin (Sigma #L-4402 @ 0.9 units per 100 µL) with 15 mM magnesium acetate in final well E. The tube strips were clean-filled, covered with microtitre plate sealer, refrigerated, and used within 24 h. *Staph. aureus* strain 8588, used in assay development, was obtained from NCIMB (Aberdeen, UK).

Methods

In the "standard" hospital test used here, dry swabs were used. Cells from these were transferred into 5 mL of broth containing 7% sodium chloride and incubated at 30 °C to selectively culture *Staph. aureus*. After overnight growth, a sub-sample was streaked out onto agar for antibiotic sensitivity testing.

For the rapid test, from patient samples that had been prepared from swabs collected up to midday, 1 mL of the broth was transferred to a 5 mL bijou bottle containing 10 µL of 400 µg mL^{-1} oxacillin. Independent identification numbers were assigned at this point to anonymise the samples. Testing of these was carried out at 16:00 each day, allowing up to 4 h incubation for growth of target cells and lysis of antibiotic-susceptible cells. After incubation (see Table 1 for details) the full 1 mL sample was transferred to well A of a pre-prepared KingFisher 5-well strip. The KingFisher system allowed two 15 sample runs to be completed before the lab closed at 17:30. Results were compared with the standard tests after these had all been completed 4 days later.

The KingFisher was programmed to mix sample and beads in well A for 15 min, collect the beads, perform successive bead washes taking a total of 150 s each in wells B to D, mix the beads for 1 min (detergent extraction) or 10 min (lysostaphin-mediated cell lysis) in well E, and then collect the beads and dump them in well D so they did not interfere with light measurements.

AK end-point assays were initiated by the addition of 100 µL magnesium acetate or 100 µL ADP, as appropriate, to well E of each 5-well strip in turn. With 30 s between additions and allowing 7½ min per sample for conversion of ADP to ATP by released AK, a rack of 15 samples could be processed in 15 min. ATP production was determined by the addition of 100 µL of bioluminescence reagent (Celsis, Newmarket, UK) with light output measured over 10 s after a 1 s delay.

Results were determined as positive or negative using a simple threshold (100,000+ RLU = positive) applied to all samples.

RESULTS

In initial optimisation experiments prior to testing in the hospital lab, it was established that *Staph. aureus* cell density after 3 hours incubation could be doubled by using shaking rather than static culture conditions and doubled again at 37 °C compared to 35 °C. It was also established that 95% of *Staph. aureus* cells at $5x10^3$ cfu mL^{-1} could be captured using monoclonal antibody-coated beads at a final concentration of 10^7 mL^{-1}, or 71% could be captured with beads at 10^6 mL^{-1}. The latter was used for cost-effectiveness. The semi-automated KingFisher assay was shown to be capable of detecting about 10^3 cfu mL^{-1} of *Staph. aureus* (Fig. 1).

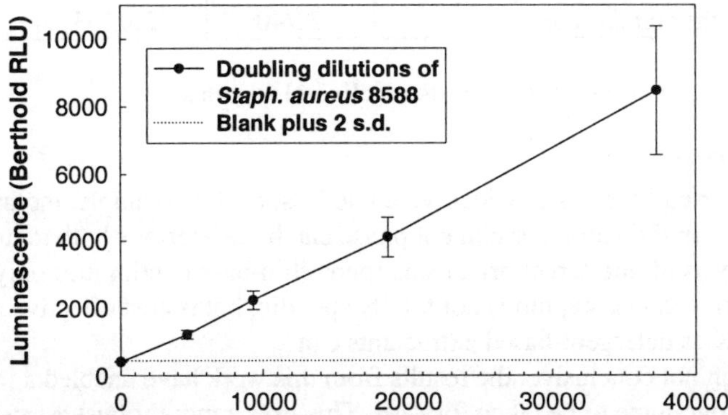

Figure 1. Concentration/response curve for semi-automated assay for *Staph. aureus* using beads at 10^6 mL^{-1}.

When the assay was transferred to testing of clinical samples, only 50% of positives by the standard test were picked up and the number of false positives was unacceptably high. The monoclonal antibody used was found to be affected by the high salt in the antibiotic broth and to cross-react with *Staph. epidermidis*. The capture agent was therefore switched to fibrinogen, and lysostaphin was introduced to add compensating specificity. As the results in Table 1 for test set 2 show, this reduced the number of false positives, but did not help with false negatives.

Up to this point, 30 °C static incubation in antibiotic broth had been adopted to better fit in with the standard assay's conditions. In the light of the pre-trial work, this might have not allowed sufficient growth of *Staph. aureus* from the samples. Optimised growth conditions were therefore adopted with results shown under test set 3 in the Table.

Test set	1	2	3
Incubation conditions for sample in L-broth with salt and 4 μg mL^{-1} oxacillin	30 °C, static	30 °C, static	37 °C, shaken
Capture agent immobilised on beads	antibody	fibrinogen	fibrinogen
Lytic agent used to release AK	detergent	lysostaphin	lysostaphin
Results (%, correct/total)			
Rapid test agreement with positive results from the standard test	**50%** 3/6	**50%** 11/22	**62%** 8/13
Rapid test agreement with negative results from the standard test	**55%** 22/40	**78%** 158/203	**80%** 84/105

Table 1. Summary of results from tests on clinical samples

CONCLUSIONS

The work reported here has provided valuable lessons. For example, inclusion of high salt levels in the initial enrichment phase can be deleterious both to antibody binding and lysis of non-target organisms (penicillin-based antibiotics only lyse growing cells), and lysostaphin is not totally specific, but is cost effective and does not inhibit AK as detergent-based extractants can.

Although not conclusive, the results from this work have enabled a more substantial programme to be taken forward. This uses a more robust assay system including a better monoclonal antibody and a final antibiotic sensitivity test[3].

ACKNOWLEDGMENTS

This work was funded by the UK MoD through the Dstl Technology Transfer Fund.

REFERENCES

1. Leslie RL, Squirrell DJ, White PJ, Green JCD. Rapid detection of MRSA from clinical samples using magnetic separation and AK bioluminescence. In: Stanley PE & Kricka LJ, eds, Bioluminescence & Chemiluminescence: Progress & Current Applications, Singapore:World Scientific, 2002, p.361-4.
2. Squirrell DJ, Price RL, Murphy MJ. Rapid and specific detection of bacteria using bioluminescence. Anal Chim Acta 2002; 457:109-14.
3. O'Hara SP, Murphy MJ, Morant K, Squirrell DJ. Rapid antimicrobial sensitivity testing using adenylate kinase (AK). American Society of Microbiology meeting, New Orleans May 2004, presentation C-144.

MICROCHIP ELECTROPHORESIS WITH CHEMILUMINESCENT DETECTION AND ITS POSSIBLE APPLICATIONS

J-M LIN, R SU

Research Center for Eco-Environmental Sciences, Chinese Academy of Sciences,
Beijing 100085, China
E-mail: jmlin@mail.rcees.ac.cn

INTRODUCTION

Miniaturized total analysis system (μ-TAS),[1,2] known also as "lab-on-a-chip" devices, can dramatically innovate the way chemical and biochemical assays are performed. The miniaturized devices can integrate diversely functional units to accomplish screening of large sample populations or processing of special kinds of samples, and represent the ability to shrink conventional 'bench-top' separation systems with the major advantages of speed, cost, portability and solvent/sample consumption. Therefore they have great potentials in many areas, such as clinical diagnostics, environmental monitoring or forensic investigations.

While microchip technology has grown very rapidly, the development and availability of effective detectors has lagged behind. For the past 10 years, laser-induced fluorescence has dominated the detection of microfluidic devices.[3] Recently, mass spectroscopy has received much attention,[4] in connection with proteomic and protein analysis. However, to realize miniaturization and integration of lab-on-a-chip devices, it is indispensable to employ miniaturized and highly sensitive detectors to match the demands of μ-TAS. Although submicromolar detectability can be readily obtained with these two types of detectors, the high cost and large size of the instruments are quite incompatible with the concept of μ-TAS.

Due to its simple optical devices, wide linear range of response and many well characterized CL systems, chemiluminescence (CL) is uniquely suited to on-line detection for μ-TAS. Some reports have shown that CL is an alternative promising detection method for capillary electrophoresis microchip.[5,6] In the present work, some glass microchips were designed based on the flow injection and electroosmotic flow (EOF)-chemiluminescent devices and applied to determine transition metal ions and organic compounds.

GLASS MICROCHIP FOR TRANSITION METAL ION ANALYSIS

A glass microchip, as shown in Figure 1 (right), was designed according to the principle of flow injection CL devices and fabricated by standard photolithography technology, wet chemical etching and heat bonding technology. Electrophoresis separation with CL detection on the microchip was used, the CL reagents, luminol solution and hydrogen peroxide solution were delivered by a laboratory-made microfluidic pump. In order to study and easily observe EOF and pump flow for the microchip, a Rhodamine B solution was used. Figure 1 (left) shows mixing by pump flow and EOF, both flow streams are steady which indicated that this glass chip is suitable for chip electrophoresis and CL detection. The experimental conditions including buffer, voltage and the flux of the CL reagents were optimized. Under the optimal conditions, Cu^{2+}、 Co^{2+} and Ni^{2+} were separated and detected with the CL method on the chip. The detection limits for Cu^{2+}、 Co^{2+} and Ni^{2+} were 5.0×10^{-11}mol/L , 5.0×10^{-9} mol/L and 1.0×10^{-7} mol/L, respectively.

A MICROCHIP BASED ON THE OXALATE-H_2O_2 CL SYSTEM

The microchip used here, illustrated in Figure 2, was fabricated from soda lime glass using standard photolithography, wet chemical etching, and heat bonding techniques. The Y-shaped layout combined with double-T injection mode was adopted. The channel length is 10.0 mm from the sample reservoir to the injection cross, 10.0 mm from the sample waste reservoir to the injection cross, 10.0 mm from the buffer reservoir to the injection cross, and 90.0 mm from the injection cross to the detection cell. The detection cell is 10.0 mm long. The radius of the three turns including the turn of the CL reagent channel is 2.5 mm. The dimensions of the channels are 40 μm deep and 80 μm wide except that the CL reagent channel and detection cell are 600 μm deep and 800 μm wide.

The double-T geometry allows for high-efficiency sample injection and geometric definition of sample plug size. The branch of Y was used as CL reagent channel, and the CL reagent was delivered by a lab-made micropump. Bis[(2-(3,6,9-trioxadecanyl-oxycarbony)-4-nitrophenyl)]oxalate-H_2O_2 CL system was employed to detect dansyl amino acids. On this microchip, dansyl-phenylalanine and –sarcosine were successfully separated by electrophoresis and detected within 250 s. The detection limits (S/N=3) of dansyl-phenylalanine and –sarcosine were 2.8 μmol/L and 3.2 μmol/L, respectively, due to the vigorous dilution of sample with CL reagent and timely removal of the waste solution from the reaction area.

The double-T injection mode allowed a large volume of sample solution to be injected, and thus the detection limit was significantly decreased. Since the detection cell was a flow-type and the CL reagent was delivered with a micropump, this microchip can be used with almost all the CL systems.

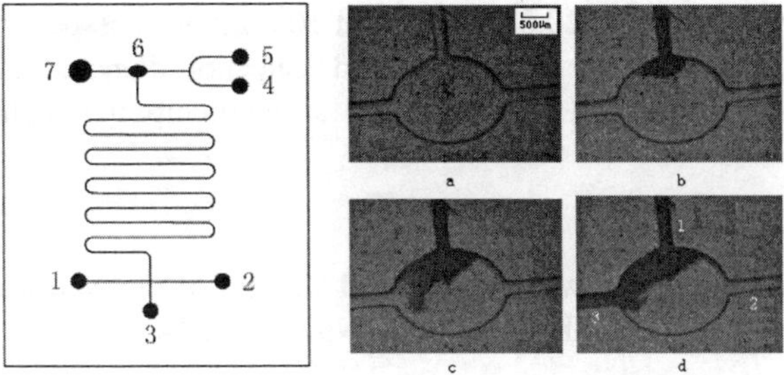

Figure 1. The layout of the channels (right) and mixing by electroosmotic flow and pump flow (left). **Right:** 1. sample cell; 2. sample waste cell; 3. buffer cell; 4. the inlet for luminol solution; 5. the inlet for H_2O_2 solution; 6. reaction cell; 7-waste cell. **Left:** 1. electroosmotic flow; 2. pump flow; 3. waste; a. blank; b. Rhodamine B enters the reaction cell (electroosmotic flow) c. electroosmotic flow and diffusion in reaction cell; d. steady state of electroosmotic flow and pump flow.

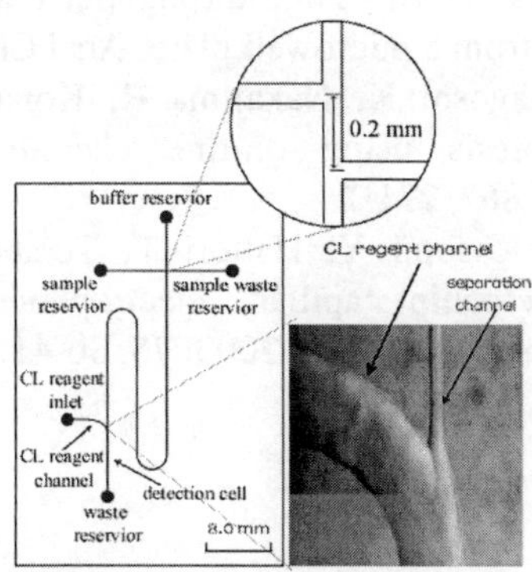

Figure 2. Schematic diagram of the microchip.

SEPARATION AND CL DETECTION OF DOPAMINE AND CATECHOL

The microchip used was similar to the chip shown in Figure 2, which has three main channels, five reservoirs and a detection cell. As model analytes, dopamine and catechol were separated and detected using the permanganate CL system on the microchip. The samples were electrokinetically injected into the double-T cross section and separated in the separation channel, and then oxidized by CL reagent which was delivered by a home-made micropump to produce light in the detection cell. The EOF can be coupled with the micropump flow. The detection limits for dopamine and catechol were 20.0 μmol/L and 10.0 μmol/L, respectively. Successful separation and detection of dopamine and catechol demonstrated the distinct advantages of integrating CL detection on a microchip for rapid and sensitive analysis.

ACKNOWLEDGEMENT

The authors gratefully acknowledge financial support of the National Science Fund for Distinguished Young Scholars of China (No. 20125514).

REFERENCES

1.   Reyes DR, Iossifidis D, Auroux P-A, Manz A. Micro total analysis systems. 1. Introduction, theory, and technology. Anal Chem 2002; 74: 2623-36.
2.   Auroux PA, Iossifidis D, Reyes DR, Manz A. Micro total analysis systems. 2. Analytical standard operations and applications. Anal Chem 2002; 74: 2637-52.
3.   Li H-F, Lin J-M, Su R, Uchiyama K, Hobo T. A compactly integrated laser-induced fluorescence detector for microchip electrophoresis. Electrophoresis 2004; 25: 1907-15.
4.   Zhang B, Foret F, Karger BL. High-throughput microfabricated CE/ESI-MS: automated sampling from a microwell plate. Anal Chem 2001; 73: 2675-81.
5.   Hashimoto M, Tsukagoshi K, Nakajima R, Kondo K, Arai A. Microchip capillary electrophoresis using on-line chemiluminescence detection. J Chromatogr A 2000; 867, 271-7.
6.   Liu B-F, Ozaki M, Utsumi Y, Hattori T, Terabet S. Chemiluminescence detection for a microchip capillary electrophoresis system fabricated in poly(dimethylsiloxane).Anal Chem 2003; 75, 36-41.

OVERVIEW OF NEW ANALYTICAL TOOLS FOR BIOLUMINESCENT BIOMASS ESTIMATION

ARNE LUNDIN, ANNELIE ELVÄNG

BioThema AB, Stationsvägen 17, S-136 40 Haninge, Sweden

Email: arne.lundin@biothema.com

INTRODUCTION

The luciferase assay of ATP has been used for estimation of biomass for several decades. However, it was not until the late 1970s that highly purified and standardized ATP reagents with a stable light emission became commercially available. Over the last 20 years detection limits have improved from around 10^{-14} to 10^{-18} moles.[1,2] Today a detection limit of 10^{-18} moles can be achieved with reasonably priced reagents and luminometers. An overview of ATP monitoring can be found at www.biothema.com. ATP biomass estimations have been used for a variety of purposes: 1) Hygiene control in e.g. food industry. 2) Rapid cell counting of eukaryotic as well as prokaryotic cells. 3) Cell proliferation/cytotoxicity assays.

The most demanding applications are those where bacterial cells (attomol ATP levels per cell) should be determined in the presence of high levels of eukaryotic cells (femtomol ATP levels per cell). Furthermore extracellular ATP must be removed or degraded before bacterial cells can be estimated.

During recent years there has been a number of suggestions on how to improve biomass detection using assays based on firefly luciferase. The purpose of the present paper is to review these suggestions.

IMPROVED ATP REAGENTS

Native luciferase obtained by collecting fireflies was prone to quality variations. Several companies now supply recombinant luciferases in various genetically modified forms. Modification to obtain one desirable property, e.g. thermostability, may change other characteristics, e.g. specific activity, K_m values and pH optimum. The specific activity must be taken into account when comparing prices per mg. A generally agreed unit for measuring luciferase activity is needed, since rlu are dependent on luminometer and reaction conditions.

Ample supply of recombinant luciferase allows us to prepare ATP reagents with a high luciferase activity.[1] Such reagents do not give a completely stable light emission, but will on the other hand degrade their own ATP background below the detection limit. One of our ATP reagents has a decay rate of the light emission around 10% per min. This rate gives a good accuracy even with a manual measurement completed in e.g. 10 seconds. The reagent background is degraded to undetectable levels ($<10^{-18}$ moles/assay) during production. If the user should happen to contaminate the reagent only 1% of the ATP will remain after 46 min.

CALIBRATION OF ATP ASSAYS

Calibration of assays with internal ATP standards is mandatory.[1] The light emission is measured before and after adding a known amount of ATP standard. It is vital that the volume of added ATP standard is a small fraction (1%) of the total reaction volume or the dilution effect will change reaction conditions.[1] Once the linearity of the method has been confirmed by preparing a standard curve, one ATP standard concentration can be used for all samples. This standard should be at least 10x stronger than the highest sample ATP concentration. The use of internal ATP standards compensates for all effects on the signal, e.g. inhibitory compounds in samples, variations in pH, ionic strength, temperature and luminometer sensitivity.[1] Furthermore expressing ATP results in moles rather than rlu enables comparison between different experiments and different laboratories.

Lyophilised ATP standards are commercially available. However, part of the freeze-dried material is often lost with the freeze-drying stopper. We have overcome this by providing ready-made, stabilized liquid ATP standards (Fig. 1).

EXTRACTION OF INTRACELLULAR ATP

Intracellular ATP can be assayed only after its extraction from the cells. As soon as the cells start to lyse, intra- or extracellular ATP degrading enzymes begin to hydrolyse ATP. A variety of extractants can be used to rapidly inactivate all such enzymes. Before deciding on a new extractant it is advisable to compare with 2.5, 5 and 10% trichloroacetic acid (TCA).[1] Many mammalian cells can be extracted with 0.1% Triton X-100 containing EDTA to inhibit ATP degrading enzymes. Most bacteria require more powerful extractants, e.g. quaternary ammonium compounds. Such extractants inactivate all ATP degrading enzymes including luciferase. This problem is overcome by adding neutralizers like cyclodextrins forming strong complexes with e.g. quaternary ammonium compounds.

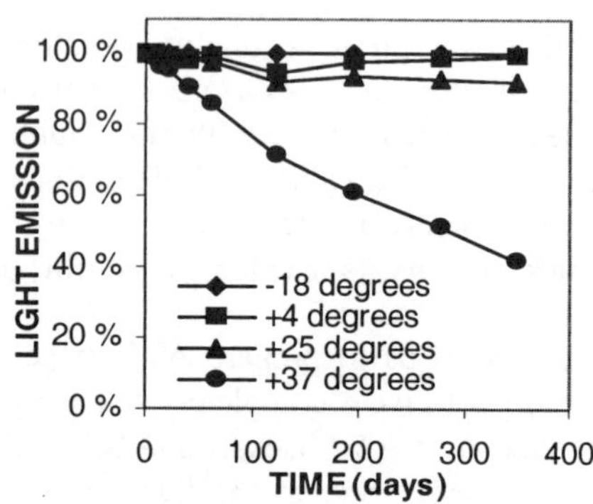

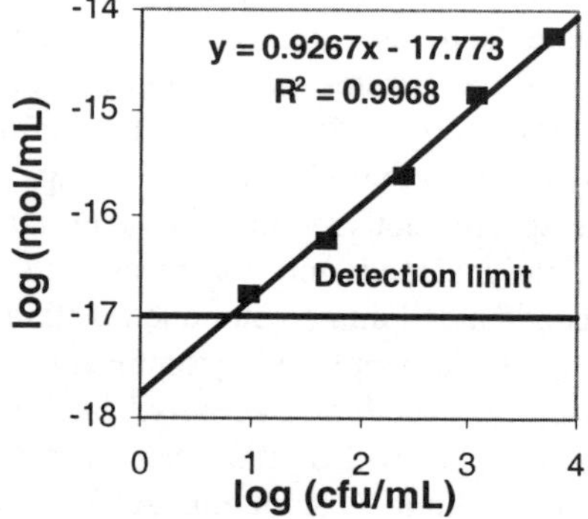

Figure 1. Stability of liquid ATP Standard (10 μmol/L)

Figure 2. Detection of *E.coli* by collecting the cells on a filter

PRETREATMENT OF SAMPLES

Apyrase can be used to degrade extracellular ATP. Different isoenzymes are more or less active against ATP and ADP. With some preparations, the degradation is first order for the first few orders of magnitude of ATP and then changes to a slower degradation rate. The addition of adenosine phosphate deaminase largely overcomes this problem by degrading ATP, ADP and AMP to ITP, IDP and IMP.[3]

Cells can be concentrated by e.g. centrifugation, filtration and immuno-capture. We are presently developing filtration methods for water samples. Fig. 2 shows that we detect 10 cfu/mL using standard reagents (BioThema), disposable filter units and an FB12 luminometer (Berthold Detection Systems). However, filtration will never be an alternative for complex biological samples like urine, blood or orange juice. Immunocapture on magnetic beads seems to be a more promising technique with such samples.[4]

IDENTIFICATION OF BACTERIA

Biotinylated antibodies can via streptavidin be linked to biotinylated enzymes, e.g. firefly luciferase or an ATP producing enzyme like acetate kinase. A genetically fused biotinylated thermostable recombinant luciferase can be detected below attomol levels and is used in a kit for the detection of 10^3 cfu/mL of *Staphylococcus aureus*.[5] In another kit coliforms are detected by their β-galactosidase activity using D-luciferin-*O*-β-galactopyranoside and the firefly reaction.[6]

Strain specific lysis by bacteriophages can be used to release intracellular ATP or adenylate kinase (AK).[7] At the 13th ISBC a paper is presented on identification of bacteria using bacteriophages. There are also several papers on bioluminescent real-time detection of nucleic acid amplification, which may be used for identification of bacteria. In these assays either pyrophosphate or AMP is converted to ATP as a measure of the amplification reaction.

ALTERNATIVES TO TRADITIONAL ATP ASSAYS

A recycling enzymatic system has been designed to measure ATP (and ADP) concentration by the time needed to reach half maximum light emission after addition of sample.[8] A sensitive luminometer is not needed for the assay. However, the assay time is long and the cycling system is sensitive to enzyme inhibitors.

The need for a sensitive luminometer can be obviated using high luciferase activity in a reagent also including pyruvate phosphate dikinase, phosphoenol-pyruvate and pyrophosphate. The cycling system converts AMP to ATP, i.e. the light emission is stable even with a high luciferase activity. The assay measures ATP+AMP rather than ATP. In hygiene monitoring this is an advantage, since ATP in food residues are degraded to ADP and finally AMP. The ATP+AMP assay makes the time for collecting the sample less critical and a higher sensitivity is achieved. The same type of reagent measuring ATP+AMP can be used to detect single bacterial cells on a filter without cultivation using a CCD camera.[2]

The use of adenylate kinase (AK) rather than ATP for biomass estimations has been advocated.[7] AK is a very stable enzyme and the assay does not differentiate between living and dead cells. However, AK may be used in hygiene control. The ATP level in ADP, i.e. the AK substrate, sets the detection limit of the assay. This is a problem since ADP easily disproportionate to ATP and AMP.

CONCLUSIONS

Recombinant luciferases and stable liquid ATP standards have increased the reliability of ATP biomass assays. Detection limits of ATP and ATP+AMP assays have reached the level of a single cell and can, combined with filtration techniques, reach 1 cell/mL. The assay of ATP+AMP is the most reliable method for hygiene monitoring. With immunocapture, bacteriophages or nucleic acid amplification combined with sensitive ATP reagents bacteria can be identified. In the near future we expect to see an abundance of user-friendly systems at acceptable prices.

REFERENCES

1.   Lundin A. Use of firefly luciferase in ATP-related assays of biomass, enzymes, and metabolites. Methods Enzymol 2000; 305:346-70.
2.   Sakakibara T, Murakami S, Imai K. Enumeration of bacterial cell numbers by amplified firefly bioluminescence without cultivation. Anal Biochem 2003; 312:48-56.
3.   Sakakibara T, Murakami S, Hattori N, Nakajima M, Imai K. Enzymatic treatment to eliminate the extracellular ATP for improving the detectability of bacterial intracellular ATP. Anal Biochem 1997; 257-61.
4.   Tu S-I, Patterson D, Uknalis J, Irwin P. Detection of *Escherichia coli* O157:H7 using immunomagnetic capture and luciferin-luciferase ATP measurement. Food Res Int 2000; 33:375-80.
5.   Fukoda S, Tatsumi H, Maeda M. Bioluminescent enzyme immunoassay with biotinylated firefly luciferase. J Clin Ligand Assay 1998; 21:358-362.
6.   Masuda-Nishimura I, Fukuda S, Sano A, Kasai K, Tatsumi H. Development of a rapid positive/absent test for coliforms using sensitive bioluminescence assay. Lett Appl Microbiol 2000; 30:130-5.
7.   Murphy M, Squirrell D, Sanders M, Blasco R. The use of adenylate kinase for the detection and identification of low numbers of microorganisms. In: Hastings J, Kricka LJ, Stanley PE eds. Bioluminescence and Chemiluminescence. Molecular Reporting with Photons. Chichester: John Wiley & Sons, 1997; 319-22.
8.   Chittock R, Hawronsky J-M, Holah J, Wharton C. Kinetic aspects of ATP amplification reactions. Anal Biochem 1998; 255:120-6.

THE USE OF ATP BIOLUMINESCENCE FOR MONITORING BIOCIDE OR DISINFECTANT TREATMENT OF WATER

CM RAMSAY, D WAYMAN, K DAVENPORT, I MICHIE

Biotrace Limited, The Science Park, Bridgend, CF31 3NA,UK

INTRODUCTION

There are many industrial processes where control of microbial contamination is important to maintain the quality and/or safety of water systems; dosing with chemical biocides is generally the method employed. Standard microbiological techniques can be used to monitor the effectiveness of biocide treatment, but the delay in obtaining results limits their usefulness as a control measure. ATP bioluminescence measurement is now a commonly used technique to obtain rapid results in cooling towers,[1] power stations,[2] oil and gas recovery and in paper pulp processing[3]. Simple and easy to use commercial reagent test kits and luminometers are available to allow field testing. However, when considering the application of ATP bioluminescence to rapid biocide assessment it is essential that the mechanism of action of the biocide is taken into consideration[4].

With some biocide treatments there is a simple direct relationship between ATP and viable cell counts. However, where the mechanism of biocide action is related to influences on cell bioenergetics, or the concentration range employed is biostatic and affects cell metabolism, then ATP results may not follow the same pattern as microbial counts. Under these circumstances low ATP results at relatively high microbial levels might be expected but it is also possible that ATP levels increase over a period of time.

In other circumstances performing only a test on 'Total' ATP with microbial extractant can give results that might be misinterpreted. The typical mode of action of many biocides leads to the rupturing of microbes. Cellular ATP is released into the environment and the high ATP reading may be incorrectly interpreted as indicating poor biocide efficiency. In these situations it is beneficial to use two tests, one with microbial extractant (Total ATP) and one without (Free ATP) and thereby obtain a better assessment of the treatment efficacy.

New EU directives may limit the development and application of new biocides and for this reason, and concerns on toxicity or pollution, there is interest in alternative water treatment techniques such as ultrasound, high-pressure vortexing and electro-coagulation. These may also be used where chemical treatment is not appropriate. ATP measurements are likely to prove useful to monitor the effectiveness of these types of treatment but as with chemical biocidal treatment the nature of the ATP results may depend on the mode of action. One type of electro-chemical treatment has been evaluated in our laboratory. The results indicate that as with chemical biocide treatment the use of both Total and Free ATP measurements can be of value in the interpretation of the results, monitoring the process efficacy and in optimisation of this method of water treatment.

MATERIALS & METHODS

Escherichia coli NCIMB 10243 was grown overnight in Tryptone Soya Broth (Oxoid CM129) in an orbital shaker at 35 °C. This was used to inoculate water samples containing biocides (2 mL of culture added to 18 mL sample). Final concentrations of the biocides were 200 ppm. Samples were taken from the biocide/bacteria solutions at various time-points and both ATP assays and plate counts performed.

ATP assays

1 mL of sample was diluted in 9 mL of RO water and assayed using Aqua-*Trace*™ (Biotrace) luciferin/luciferase reagent single dose format devices in a Uni-*Lite*® (Biotrace) luminometer. Results are expressed in Relative Light Units (RLU). Aqua-*Trace* Total and Free sample pick up sticks are coated with microbial extractant and non-ionic surfactant respectively. Aqua-*Trace* assay: dip stick into solution, tap to remove bubbles, 5 s extraction time on stick, activate device, shake for exactly 5 s then read in the Uni-*Lite*. The ATP levels at T_0, that is with no biocide present, were estimated by diluting the neat broth cultures 1 in 100 in RO.

Plate counts

The plate counts at T_0 were estimated by diluting the neat broth culture 1 in 10 with Neutralised Peptone Water (NPW) containing per L 1.0 g Bacteriological Peptone (Oxoid, L37), 8.8 g Sodium Chloride (May & Baker), 3.0 g Amisol 910 (Degussa) and 30.0 g Tween 80 (BDH, 560234H) and then decimally with Phosphate Buffered Saline (Oxoid). For biocide treated samples 1 mL was diluted in 9 mL of NPW, mixed and allowed to stand for 5 min (for neutralisation of the biocide). The solution in NPW was diluted decimally (0.1 mL in 0.9 mL) in PBS. Appropriate dilutions (0.1 mL) were plated out on Tryptone Soya Agar Plates (bioMerieux) and incubated at 37 °C for 24 h.

For the electro coagulation experiment an overnight TSB culture (10 mL) of *E. coli* was added to 1 L of RO water in a conical flask. RO water when tested with the oxidation electrode gave a very high resistance and, on the advice of the supplier of the equipment under test, tap water was added (500 mL *E. coli* in RO + 300 mL tap water in a 1 L beaker). Two beakers of *E. coli* suspension (500 mL *E. coli* in RO + 300 mL tap water) were prepared, one for each type of treatment. Treatment of the water was performed using equipment for bench-top trials from Axonics.

Before and at various time-points after treatment, samples were taken from the bulk liquid and tested with Aqua-*Trace* Free and then with Aqua-*Trace* Total. Plate counts were performed using 0.1 mL of decimal dilutions of samples in RO water plated out on TSA and incubated at 35 °C for 24 h.

RESULTS & DISCUSSION

Table 1 shows the results of a Quaternary ammonium based biocide on *E. coli* where the 24 h plate counts indicate that the biocide is highly effective with a greater than 4 log reduction in 5 min. The Total ATP results initially increase and remain high for a few hours which taken alone might be misleading. But it is clear from the Free

ATP results that the biocide is effective and that after addition the microbes die and release their ATP into solution.

Table 1. Effect of Quat based Biocide

Time following addition of Biocide	Aqua-*Trace* Total (RLU)	Aqua-*Trace* Free (RLU)	Plate Counts (CFU/mL)
T = 0	143709	2270	1.4×10^6
T = 5m	45010	36746	<100
T = 45m	20290	18668	<100
T= 2h	17377	14282	<100
T = 5h	8755	9186	<10
T = 24h	871	921	<10
T = 48h	99	124	<10

The results on Table 2 shows a Methylene Bisthiocyanate based biocide. The mechanism of action of this biocide is to block the transfer of electrons from primary cytochrome dehydrogenase, and thereby cause an uncoupling of oxidative phosphorylation[5]. Here the Total ATP results initially increase on treatment with the biocide; a small increase in Free ATP is also evident along with a reduction in viable counts. With time there is a reduction in Total and Free ATP along with a further reduction in the plate count results.

Table 2. Effect of Methylene Bisthiocyanate based biocide

Time-following addition of Biocide	Aqua-Trace Total (RLU)	Aqua-Trace Free (RLU)	Plate Counts CFU/mL
T = 0	17536	1893	1.1×10^7
T = 5m	128319	3210	$<1 \times 10^5$
T = 45m	110222	3028	7.8×10^5
T = 2h	122525	2905	1.2×10^6
T = 4h	96226	1279	1.3×10^6
T = 24h	4895	66	1.6×10^5
T = 28h	2613	57	9.2×10^4

These patterns in Total and Free ATP are as might be expected from what is known about the mechanism of action of these classes of biocides.

The use of electro-coagulation treatment, thought to cause electroporation, results an initial release of ATP and loss of viability with a residual effect giving further reductions in both ATP and plate count results. This type of treatment can be applied to a range of sample types including waste water, final effluent and sewage.

Table 3. Electro-coagulation Treatment

Treatment	Time of Test	Aqua-*Trace* Total (RLU)	Aqua-*Trace* Free (RLU)	Δ	Plate Count (cfu/mL)
Before Treatment		19442	2058	17384	10^5-10^6
1m	9m	21557	13961	7596	10^5-10^6
	1.22	21932	16681	5257	10^4-10^5
	23	8270	8424	-154	10^1-10^2

The study on the electro-coagulation was limited in scope but illustrates that ATP measurements are a valuable tool in evaluating this type of disinfectant process and may be helpful in its optimisation and in routine monitoring.

ACKNOWLEDGMENT
Electro coagulation treatment equipment was provided and set up by Dr P Morgan, Axonics Limited.

REFERENCES

1.   Czechowski MH. ATP technology a tool for monitoring microbes in cooling systems. Proc Am Power Conf; 1996; 2:893-96. The 1996 58th American Power Conference Part 2 (of 2). Betz Water Management Group 1996

2.   Electric Power Research Institute, Closed Cooling Water Chemistry Guideline, Revision 1: Revision 1 to TR-107396, Closed Cooling Water Chemistry Guideline, April 23, 2004. EPRI, May 2004. Power Research Institute, Palo Alto, 2004.

3.   Barclay RL. ATP (Adenosine Tri-Phosphate) Assay: An Innovative Method for Measuring Biomass Levels in Pulp and Paper Mills, 1994 Papermakers Conference Proceedings. Norcross: Tappi Press, 1994.

4.   Denyer SP. ATP bioluminescence and biocide assessment: effect of bacteriostatic levels of biocide. In: Stanley PE, McCarthy BJ Smither R. eds. ATP Luminescence. Rapid Methods in Microbiology. Oxford: Blackwell Scientific Publications Ltd., 1989: 189-95.

5.   McCoy JW. Microbiology of Cooling Water. Chemical Publishing Co., Inc. New York, 1980: 82-3.

BIOLUMINESCENT BIOREPORTER INTEGRATED CIRCUIT SENSING OF THE CHEMICAL AND BIOLOGICAL SPACECRAFT ENVIRONMENT

SA RIPP[1], JL GARLAND[2], BJ BLALOCK[3], SK ISLAM[3], GS SAYLER[1]

[1]*The Center for Environmental Biotechnology, University of Tennessee, Knoxville, Tennessee 37996, USA*
[2]*Dynamac Corp, Kennedy Space Center, Florida 32899, USA*
[3]*Electrical and Computer Engineering Dept, University of Tennessee, Knoxville, Tennessee 37996, USA*
Email: saripp@utk.edu

INTRODUCTION

Spacecraft environments are particularly vulnerable to upsets due to their closed loop nature. The release and accumulation of toxic chemicals in the breathable atmosphere or the biological contamination of food and water supplies rapidly become detrimental to crew safety. Consequently, instrumentation for environmental monitoring is a prerequisite for mission success. Although seemingly uncomplicated, the additional requirements that these instruments use minimal space, mass, power, and crew time makes their successful development and application challenging. One potential solution may be to use biosensors as environmental monitors. Biosensors consist of a biological component (bioreporter) that recognizes and responds to target analytes to generate an easily discernible phenotypic signal, for example, a colorimetric, fluorescent, or bioluminescent display that is then measured and quantified through an appropriate analytical device interface. Critical to the simplicity of bioreporter assays is the direct and independent transfer of the biochemical signal to the measuring device. The bioreporters described here utilize the bacterial *lux* reporter gene system, which is particularly useful in this regard because all the components necessary for expression of its bioluminescent signal are present within the cell, thus obviating the need for any external manipulations. Thus, no exogenous substrate additions are necessary, allowing the self-generated bioluminescent response to be directly linked to photonic detectors for facile *in vivo*, real time monitoring of select chemical agents within a variety of environmental matrices. We call our device a bioluminescent bioreporter integrated circuit (BBIC), and its low-power, low-mass, and low-maintenance characteristics provides extensive compatibility with current spacecraft monitoring needs.

BIOLUMINESCENT BIOREPORTER INTEGRATED CIRCUITS (BBICS)

The BBIC is an optical application-specific integrated circuit (OASIC) transducer that couples directly to bioreporter matrices to provide a complete, standalone detection system (Fig. 1).[1-3] In a single, small footprint, low power package, the BBIC provides for the detection of the bioreporter optical signal, the distinguishing of this signal from noise, digital processing of the signal, and local communication of the result. In its most basic form, the BBIC measures the amount of light emitted by the bioreporter, digitizes this value and transmits the results to a data receiver. It is fabricated in standard 1.2 μm n-well CMOS and thus features an extremely small footprint (2 mm^2) and associated low power requirements (< 100 milliwatts). Radio frequency telemetry can also be incorporated on-chip for remote, wireless data transmission.

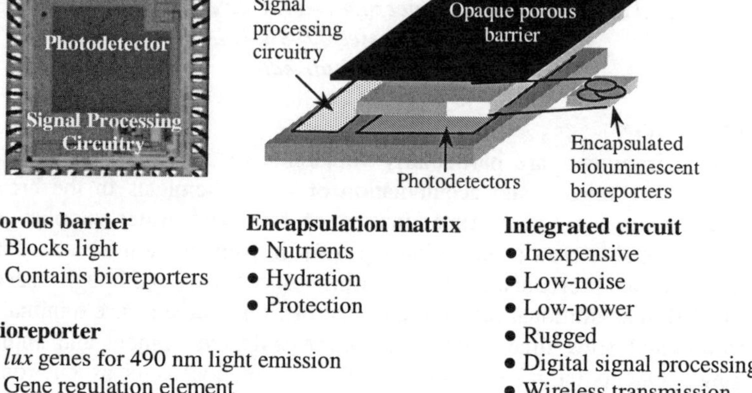

Porous barrier
- Blocks light
- Contains bioreporters

Bioreporter
- *lux* genes for 490 nm light emission
- Gene regulation element

Encapsulation matrix
- Nutrients
- Hydration
- Protection

Integrated circuit
- Inexpensive
- Low-noise
- Low-power
- Rugged
- Digital signal processing
- Wireless transmission

Figure 1. A bioluminescent bioreporter integrated circuit (BBIC)

BBIC SENSING OF THE CHEMICAL ENVIRONMENT

Our fundamental research leading to proof-of-concept for using BBICs for bioluminescence measurement has been reported.[1-3] For initial testing purposes the chip was mounted in a 40-pin ceramic dual inline package. Bioluminescence was determined for cultures containing different concentrations of the bioluminescent bioreporter *Pseudomonas fluorescens* 5RL growing in the presence of the inducer molecule salicylate at 10 ppm. Bioluminescence was determined using the integrated circuit microluminometer and a light-tight enclosure mounted above the chip. Linear regression analysis showed that data fit a linear model indicating that bioluminescence per cell remains constant for cell concentrations ranging from 4×10^5 to 2×10^8 CFU/mL and for detector responses ranging from 0.05 to 20 pA. Using a linear model, it is estimated that 4×10^5 fully induced cells/mL are required for on-chip detection. In other experiments, the lower limit of detection for salicylate exposure was determined to be approximately 50 ppb with a subsequent chip response time of approximately 45 min. Response times decreased to less than 20 min as salicylate concentrations increased to 1 ppm.

We have additionally completed a BBIC testing regimen for the detection of microbial volatile organic compounds (MVOCs).[4] MVOCs are produced as metabolic by-products of bacteria and fungi. Since they are detectable before any visible signs of microbial growth appear, they serve as very early indicators of potential biocontamination problems.[5] Using *p*-cymene as a model MVOC, a bioluminescent bioreporter (*Pseudomonas putida* UT93) was constructed and used in a first generation BBIC platform for remote detection of *Penicillium* growth (Fig. 2). Upon exposure to a single growth plate of *Penicillium* (~ 0.012 – 0.025 ppm *p*-cymene), a significant bioluminescent signal was produced within 3.5 h. This BBIC is housed within an aluminum wand connected to a remote transmitter capable of wireless transmission over line of sight distances of over 150 meters. A new version of the BBIC chip will directly contain an on-chip RF transmitter.

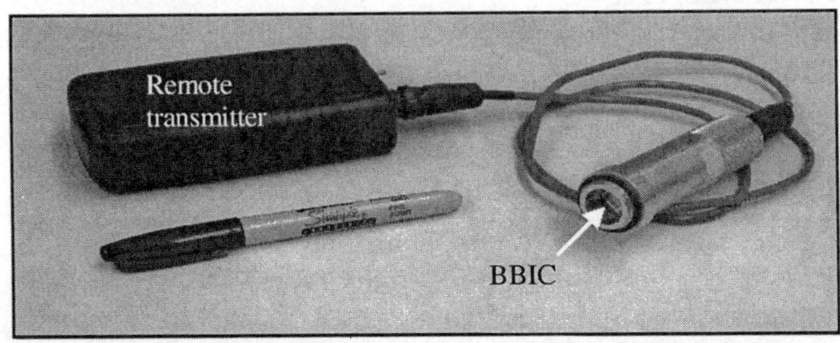

Figure 2. The BBIC sensor probe and remote transmitter

BIOREPORTER SENSING OF BIOLOGICAL AGENTS

In addition to the sensing of chemical analytes, bioreporters are also being developed for direct detection of bacterial pathogens.[6] These bioreporters take advantage of the exquisite specificity of bacteriophages for their bacterial hosts. By genetically engineering the bacteriophage to contain quorum sensing signalling capabilities, the infection event can be indirectly sensed through the generation of bioluminescence for signature identification of the bacterial host (pathogen). And, since the bacteriophage infects and multiplies within its host, there is the added advantage of intrinsic amplification of the infection event, thus, pathogens can be detected at very low levels, possibly avoiding dependence upon target cell growth. Again, as with the bioluminescent bioreporters, detection is accomplished without the addition of any exogenous substrate and, therefore, minimal crew time requirements. Using *Escherichia coli* as a model system, engineered lambda bacteriophages indirectly yielded bioluminescent responses to as few as 10 *E. coli* cells within 8 h of assay initiation.

CONCLUSION

BBICs can offer the specificity and sensitivity required for chemical and biological monitoring of the spacecraft environment within the confines of low-mass, low-power, and minimal crew intervention. We envision the final implementation of BBICs to be in a modular 'plug-and-play' format, wherein single element chips containing their own unique bioreporter would be individually packaged, allowing the end user to custom design a BBIC multiarray sensor based on current monitoring needs. We estimate multiarray BBICs to be approximately 2×2 cm in size with power utilization of less than 20 mW. The influx of this technology represents a novel contribution for assessing exposure risks and maintaining internal environments conducive to crew health and safety.

ACKNOWLEDGEMENTS

This study was supported by the NASA Advanced Environmental Monitoring and Control Program and the U.S. Department of Energy, Office of Science, Laboratory Technology Research Program.

REFERENCES

1.   Simpson ML, Sayler GS, Patterson G, Nivens DE, Bolton E, Rochelle J, Arnott C, Applegate B, Ripp S, Guillorn MA. An integrated CMOS microluminometer for low-level luminescence sensing in the bioluminescent bioreporter integrated circuit. Sens Actuators B 2001;72:135-41.

2.   Bolton EK, Sayler GS, Nivens DE, Rochelle JM, Ripp S, Simpson ML. Integrated CMOS photodetectors and signal processing for very low-level chemical sensing with the bioluminescent bioreporter integrated circuit. Sens Actuators B 2002;85:179-85.

3.   Nivens DE, McKnight TE, Moser SA, Osbourn SJ, Simpson ML, Sayler GS. Bioluminescent bioreporter integrated circuits: potentially small, rugged and inexpensive whole-cell biosensors for remote environmental monitoring. J Appl Microbiol 2004;96:33-46.

4.   Ripp S, Daumer KA, McKnight T, Levine LH, Garland JL, Simpson ML, Sayler GS. Bioluminescent bioreporter integrated circuit sensing of microbial volatile organic compounds. J Ind Microbiol Biotechnol 2003;30:636-42.

5.   Wessen B, Schoeps K-O. Microbial volatile organic compounds-what substances can be found in sick buildings? Analyst 1996;121:1203-5.

6.   Ozen A, Montgomery K, Jegier P, Patterson S, Daumer K, Ripp S, Garland J, Sayler G. Development of bacteriophage-based bioluminescent bioreporters for monitoring of microbial pathogens. Proc SPIE 2004;5270:58-68.

RAPID AND ONSITE BOD SENSING SYSTEM BY LUMINOUS CELLS-IMMOBILISED-CHIP

T. SAKAGUCHI[1], Y. MORIOKA[1], E. TAMIYA[2]

[1]*Department of Biological and Environmental Chemistry, Kinki University, Fukuoka Campus, Fukuoka, Iizuka, 820-8555, Japan*
[2]*School of Materials Science, Japan Advanced Institute Science & Technology (JAIST), Ishikawa, Tatsunokuchi, 923-1292, Japan*

INTRODUCTION

Bioluminescence is widely distributed in various microorganisms, insects, shrimps, squid and fish.[1] The light emission is strongly correlated with energy supplementation owing to carbon source utilization. Bioluminescence can be also used as a reliable reporter for the assessment or monitoring of various aquatic samples containing toxicants such as pesticides, PCBs, polyaromatic hydrocarbons, fuels, and heavy metals.[2-5] The attenuation of photo intensity due to the metabolic inhibition of toxicants, and the *in vivo* gene promoter assay with luciferase gene have been applied in those toxicity assays. Recently, luminescent bacteria have become one of the most important sensing devices for environmental assessments.

Biochemical oxygen demand (BOD) is one of the most widely used environmental indicators in waste water treatment processes. It is proposed and defined as the method for detection of the degree of pollution due to biodegradable substances in aquatic environments.[6] Recently, bioluminescence from a natural luminous marine bacterium, *Photobacterium phosphorous* was applied for the measurement of biodegradable substances.[7] Furthermore, a new rapid BOD sensing system using luminescent recombinants of *Escherichia coli* cells with *lux A-E* genes from *Vibrio fischeri* has been developed. BOD detection of multiple samples in real waste waters can be achieved by using the bioluminescence property of this recombinant *E. coli* and an imaging system containing a charge coupled device (CCD) camera and a photomulti-counter.[8] These reports illustrate the utility of luminous microorganisms in disposable type sensors (reagents), and not necessary to maintain cells-immobilized membrane like BOD_S, for rapid and reproducible BOD measurement.

We describe here the development of a new rapid and onsite BOD monitoring system that arrayed and immobilized luminous cells on the holes (diameter: 1000 μm) of an acrylic chip.

MATERIALS AND METHODS
Bacterial strain and chip manufacturing
A marine lumious bacterium, *Photobacterium phosphoreum* IFO 13896 was grown with the ATCC culture medium no. 1163 at 15 °C for 15 h corresponding to the optimum conditions in order to obtain the maximum bioluminescence. After centrifugation (15000 rpm) of the culture and after twice rinse of harvested cells with the growth medium without the carbon source, cells were immobilised in the holes (diameter: 1000 μm, depth: 100 μm) on an acrylic chip (3cm x 3cm), that were fabricated with a NC micro-fabrication machine. For immobilisation of the cells 3% sodium alginate was used in this procedure.

Detection of bioluminescence
Bioluminescence was measured by a chemi-luminescence detector or our newly developed onsite monitoring system using a digital camera. The data were transferred to a (mobile) PC machine with a smart media card device. Luminescent intensity was numerated by black and white scale using ScionImage soft.

Calibration of BOD values
A glucose-glutamic acid solution (150 ppm glucose and 150 ppm glutamate; GGA solution) equivalent to a BOD_5 value of 220 ppm was prepared for calibration of BOD values. BOD standard solution (5 μL) which was treated with pure oxygen gas was dropped onto the hole of the chip. After 20 min bioluminescence from the well was measured.

Measurement of BOD in wastewater sample
Wastewater samples were collected from the industrial wastewater treatment plant of a confectionary factory in Kyushu. These wastewater samples were applied to the chip. The BOD values were measured and compared with the conventional method (BOD_5).

RESULTS AND DISCUSSION
Detection of bioluminescence in BOD standard solutions
The luminescence of GGA solutions were estimated using the bacterial chip system in order to produce a calibration plot for measurement of pollution containing biodegradable substances. The luminescent intensity was correlated with concentration of BOD standard (GGA) solution. The bioluminescence increased linearly with concentration up to approximately 50 ppm. Measurement of BOD values less than 50 ppm has been achieved. The light intensity reached saturation at concentrations of over 100 ppm. Detection limit was approximately 100 ppm, and the minimum measurable BOD was 1 ppm. Based on the logarithmic curve which was obtained by the measurement with a chemi-imager, two linear calibration curves and equations were estimated and approximated

from the calibration curve for determination of BOD value in actual samples. On the other hand, one linear calibration curve and equation were obtained when the digital camera system was used for the measurement. Both calibrations and equations were adopted for determination of BOD in actual wastewater samples.

Comparison of BOD value with the conventional method in actual wastewater

Measurement of organic pollution due to biodegradable substances in actual samples from the wastewater treatment plant in a confectionery factory was carried out by using the chip system. The images demonstrated that this system allowed us to visualize simultaneously the change of organic pollution in each treatment step. Moreover, BOD values of the multiple samples could be determined by using this chip system in a single small acrylic plate. BOD values that were determined by using our system were compared with those measured by the conventional 5-day method. Although values measured by using the chip system were about 20- 35 % difference from values obtained using BOD_5 (except one case), there was a correlation between the two values. Additionally, in the case of low BOD value in natural environmental samples, the value agreed with that of BOD_5.

CONCLUSION

In this study, we have applied an imaging system containing a commercial digital (CCD) camera, a mobile PC, and luminous photo-bacterial cells-immobilized on an acrylic chip for the measurement of BOD for the first time. Our results showed that it is possible to determine the degree of pollution due to the presence of biodegradable organic substances in multiple samples by using a bacterial chip system within twenty minutes using one drop of just five micro-liter sample. BOD values lower than 50 ppm could be directly detected without dilution of the sample in onsite in principal. These results suggest that the onsite and high-through put determination of BOD can be achieved. Our system may be a powerful tool to determine and monitor specific compounds in wastewater by using a gene promoter assay using recombinant cells that contain bacterial luminescence genes

REFERENCES

1. Lee J. Bioluminescence. In: Smith, K.C. eds. The science of photobiology. Plenum Publishing Corp., New York, 1989: 391-417.
2. Ulitzur S, Lahav T, Ulitzur N. A novel and sensitive test for rapid determination of water toxicity. Environ Toxicol 2002; 17: 291-6.

3. Weitz H J, Campbell C D, Killham K. Development of a novel, bioluminescence-based fungal bioassay for toxicity testing. Environ Microbiol 2002; 4: 422-9.
4. Backhaus T, Grimme L H. The toxicity of antibiotic agents to the luminescent bacterium *Vibrio fischeri*. Chemosphere 1999; 38: 3291-3301.
5. Ruiz M J, Lopez-Jaramillo L, Redondo M J, Font G. Toxicity assessment of pesticides using the microtox test: application to environmental samples. Bull Environ Contam Toxicol 1997; 59: 619-25.
6. Hikuma M, Suzuki H, Yasuda T, Karube I, Suzuki S. Amperometric estimation of BOD by using living immobilized yeasts. Eur J Appl Microbiol Biotechnol 1979; 8: 289-97.
7. Hyun C-K, Tamiya E, Takeuchi T, Karube I, Inoue N. A novel BOD sensor based on bacterial luminescence. Biotechnol Bioeng 1993; 41: 1107-11.
8. Sakaguchi T, Kitagawa K, Ando T, Murakami Y, Morita Y, Yamamura A, Yokoyama K, Tamiya E. A rapid BOD sensing system using luminescent recombinants *of Escherichia coli*. Biosens Bioelectron 2003; 19: 115-21.

SIMULTANEOUS DETERMINATION OF TWENTY-ONE MUTAGENIC NITROPOLYCYCLIC AROMATIC HYDROCARBONS BY HIGH-PERFORMANCE LIQUID CHROMATOGRAPHY WITH CHEMILUMINESCENCE DETECTION

N TANG, R TAGA, T HATTORI, A TORIBA, R KIZU, K HAYAKAWA

*Graduate School of Natural Science and Technology, Kanazawa University,
Kakuma-machi, Kanazawa, Ishikawa 920-1192, Japan
Email: hayakawa@p.kanazawa-u.ac.jp*

INTRODUCTION

Among the carcinogenic and/or mutagenic compounds in the atmosphere, nitropolycyclic aromatic hydrocarbons (NPAHs) such as 1,3-, 1,6- and 1,8-dinitropyrenes (DNPs) show strong direct-acting mutagenicities.[1] Atmospheric NPAHs mainly originate from imperfect combustion of organic matter such as coal and petroleum and are formed in the atmosphere by heterogeneous or homogeneous reactions of their parent polycyclic aromatic hydrocarbons (PAHs) with NOx and OH radicals.[2] Because NPAHs may have effects on eco-systems and human health, it is necessary to develop a method for analyzing them quickly, and to clarify their major contributors and behaviors in the atmosphere. Furthermore, because concentrations of NPAHs in the atmosphere are significantly lower than those of PAHs, the method must be sensitive.

We developed an HPLC method with chemiluminescence detection (HPLC/CLD) for 1,3-, 1,6-, 1,8-DNPs and 1-nitropyrene (1-NP) which were responsible for about one-third of the total direct-acting mutagenicity of diesel-engine exhaust particulate extracts.[3, 4] The sensitivity of this method was two orders of magnitude higher than those by HPLC/FLD or GC/MS. We also developed an automatic HPLC/CLD system for analyzing the above four kinds of NPAHs by adding on-line clean-up, reducer (Pt/Rh-coated alumina) and concentrator columns[5] and improved this system so that it could simultaneously analyze eleven NPAHs in the atmosphere.[6] However, these eleven NPAHs could not explain the direct-acting mutagenicity in the airborne particulates completely, since several other NPAHs such as 3-nitrobenzanthrone showed very strong direct-acting mutagenicities. Therefore, it is necessary to analyze simultaneously more NPAHs in order to clarify the unknown direct-acting mutagens. In this study, an HPLC method for simultaneous determination of twenty-one mutagenic NPAHs in airborne particulates was developed by modifying the operating conditions. As an application of this method, the concentrations of the twenty-one NPAHs were determined in airborne particulates collected at a heavy traffic road site in Kanazawa, Japan.

METHODS
Chemicals

1,3-, 1,6-, 1,8-dinitropyrenes (DNPs), 1-, 4-nitropyrene (NPs), 6-nitrochrysene (NC), 7-nitrobenz[*a*]anthracene (NBaA), 6-nitrobenzo[*a*]pyrene (NBaP), 3-nitroperylene (NPer), 2-nitrofluorene (NF), 2-, 9-nitroanthracenes (NAs), 5-nitroacenaphthene (NAc), 4-, 9-nitrophenanthrenes (NPhs), 2-nitrotriphenylene (NTP), 2-fluoro-7-nitrofluorene (FNF, an internal standard) were purchased from Aldrich Chemical Company (Milwaukee, WI, USA). 1-NPer and 3-nitrofluoranthenes (NFR) were from Chiron AS (Trondheim, Norway), ChemSyn Laboratories (Kansas, USA) and Wako Pure Chemical Industries (Osaka, Japan), respectively. 2-NP and 3-, 10-nitrobenzanthrones (NBAs) were kindly provided by Prof. A. Hirayama of the Laboratory of Public Health, Kyoto Pharmaceutical University and Prof. S. Fujisawa of the Faculty of Science, Toho University, respectively. All other chemicals used were obtained from commercial sources.[3]

HPLC system

The HPLC system consisted of five LC-10A pumps, an SIL-10A auto sample injector, a DGU-14 degasser, a CLD-10A chemiluminescence detector, an SCL-10A system controller, a C-R4A integrator, an HIC-6A and a CTO-10AC column oven (all Shimadzu, Kyoto, Japan). The clean-up column (4.6 i.d. X 150 mm), concentrator column (4.6 i.d. X 30 mm), separator column (4.6 i.d. X 250 + 150 mm), guard columns 1 (4.6 i.d. X 30 mm) and 2 (4.6 i.d. X 50 mm) were packed with Cosmosil 5C18-MS (Nacalai Tesque, Kyoto, Japan), and the reducer column (4.0 i.d. X 10 mm) was a Nitroarene Reactor Column (Shimadzu). All other conditions were the same as those given in our previous report.[6]

Sampling and pretreatment of airborne particulates

Airborne particulates were collected at a heavy traffic site in Kanazawa, Japan by a 123VL high-volume air sampler (Kimoto Electric Company Ltd., Osaka, Japan) with a 2500QAT-UP quartz fiber filter (25 X 20 cm, Pallflex Products, Putnam, CT, USA) for 24 hours at a flow rate of 1.3 m^3/min. The filter was treated according to our previous paper.[3]

RESULTS AND DISCUSSION
Improvement of the HPLC system

We previously observed that, among the 21 NPAHs, 3- and 10-NBAs and 4-NPh were not quantitatively retained on the concentrator column when the loading time was set from 20 to 58 min.[6]. Therefore, the loading time of the new system was set from 16 to 58 min

When the standard solution, a mixture of the 21 NPAHs was injected into the system, 10-NBA and 1,3-DNP, 3-NBA and 5-NAc, 9-NPh and 2-NP and 3-NFR and 1-NP were not completely separated by the separator column (ODS, 4.6 X 250 mm). Although these NPAHs were separated completely by increasing the length of the separator column from 250 to 400 mm, the peak heights of 7-NBaA, 1-, 3-NPers and

6-NBaP, whose retentions were stronger than those of the other NPAHs, became smaller. By increasing the acetonitrile concentration in the mobile phase from 50% to 66% (0.32% per min) after 1-NP was eluted, the retention times of these four NPAHs were reduced from 149 min to 106 min with perfect resolution. Furthermore, the sensitivities of 7-NBaA, 3-NPer and 6-NBaP were more than 3 times higher than those obtained by the previous method.[6]

The 21 NPAHs were determined chemilumigenically with linear calibration graphs from 3 fmol to 20 pmol ($r^2 > 0.899$). The relative standard deviations (n = 3) were less than 5%. The detection limits (S/N = 3) were 1 fmol for the DNPs, 10 fmol for 1-NP, 7-NBaA and 2-NA, 2 fmol for 3-NPer and 6-NBaP, 4 fmol for 9-NA and 1-NPer, 21 fmol for 3-NFR, 30 fmol for 4-NP, 100 fmol for 5-NAc and 4-NPh, 120 fmol for 9-NPh, 150 fmol for 2-NP and 6-NC, 400 fmol for 3-NBA, 450 fmol for 2-NTP, 1 pmol for 2-NF, 5.5 pmol for 10-NBA, when the sample injection volume was 100 μL.

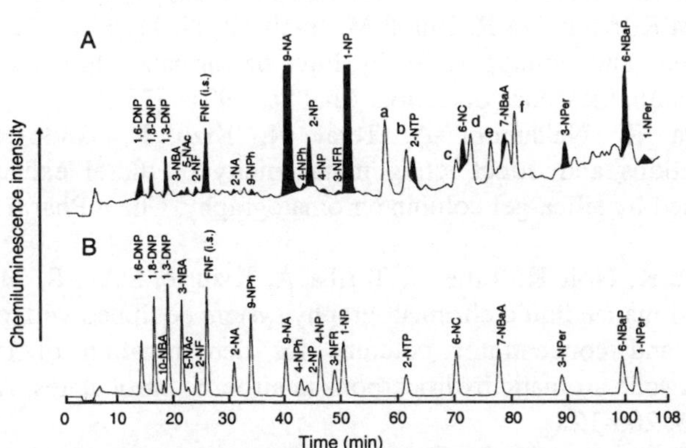

Figure 1. Chromatograms of airborne particulate extracts (A) and NPAHs standard (B)

Application to airborne particulates

NPAHs in airborne particulates collected at a heavy traffic road site in Kanazawa, Japan were analyzed. Fig. 1 shows chromatograms of (A) benzene-ethanol extracts from airborne particulates and (B) 21 standard NPAHs. All NPAHs except for 10-NBA were detected in the extracts. The total atmospheric concentrations of NPAHs were 3.85 pg/m^3 and this value was 1.7 times higher than that obtained by the previous method (2.28 pg/m^3). Nine newly determined NPAHs were measured. The new NPAHs and their atmospheric concentrations (in pg/m^3) were 3-NBA (0.01), 5-NAc (0.28), 4-NPh (0.09), 9-NPh (0.34), 2-NA (0.25), 9-NA (0.20), 3-NFR (0.22), 2-NTP (0.08) and 1-NPer (0.08). These values were lower than that of 1-NP whose concentration was the highest in the atmospheric particulates. Additionally, several compounds of these nine NPAHs, such as 3-NBA[7] and 2-NTP,[8] show very strong mutagenicities. These results suggest that the proposed method is very useful for estimating the contribution of direct-acting mutagenicity of NPAHs to the total mutagenicity.

Several unknown peaks were still observed in chromatogram (A) (Fig. 1). Peroxyoxalate-chemiluminescence detection is highly sensitive and selective for aminopolycyclic aromatic hydrocarbons which are amino derivatives of NPAHs. Therefore, these unknown peaks (a – f), which were detected only after reduction, might have originated from NPAHs. We are presently attempting to identify these unknown peaks.

REFERENCES

1. Rosenkranz H, Mermelstein R. Mutagenicity and genotoxicity of nitroarenes All nitro-containing chemicals were not created equal. Mutat Res 1983; 114: 217-67.
2. Tokiwa H, Ohnishi Y. Mutagenicity and carcinogenicity of nitroarenes and their sources in the environment. Crit Rev Toxicol 1986; 17: 23-60.
3. Hayakawa K, Kitamura R, Butoh M, Imaizumi N, Miyazaki M. Determination of diamino- and aminopyrenes by high performance liquid chromatography with chemiluminescence detection. Anal Sci 1991; 573-7.
4. Hayakawa K, Nakamura A, Terai N, Kizu R, Ando K. Nitroarene concentrations and direct-acting mutagenicity of diesel exhaust particulates fractionated by silica-gel column chromatography. Chem Pharm Bull 1997; 45: 1820-2.
5. Hayakawa K, Noji K, Tang N, Toriba A, Kizu R, Sakai S, Matsumoto Y. A high-performance liquid chromatography system equipped with on-line reducer, clean-up and concentrator columns for determination of trace levels of nitropolycyclic aromatic hydrocarbons in airborne particulates. Anal Chim Acta 2001; 445: 205-12.
6. Tang N, Toriba A, Kizu R, Hayakawa K. Improvement of an automatic HPLC system for nitropolycyclic aromatic hydrocarbons Removal of an interfering peak and increase in the number of analytes. Anal Sci 2003; 19: 249-53.
7. Enya T, Suzuki H, Watanabe T, Hirayama T, Hisamatsu Y. 3-Nitrobenzanthrone, a powerful bacterial mutagen and suspected human carcinogen found in diesel exhaust and airborne particulates. Environ Sci Technol 1997; 31: 2772-6.
8. Ishii S, Hisamatsu Y, Inazu K, Aika K. Environmental occurrence of nitrotriphenylene observed in airborne particulate matter. Chemosphere 2001; 44: 681-90.

BIOENERGETIC CONFIRMATION OF VIABLE PATHOGENS IN FOODS BY ATP-BIOLUMINESCENCE

S TU, A GEHRING, P IRWIN

United States Department of Agriculture, Agricultural Research Service, Eastern Regional Research Center, 600 E. Mermaid Lane, Wyndmoor, PA 19038, USA
Email: stu@errc.ars.usda.gov

INTRODUCTION

Measurement of ATP by the bioluminescence of luciferin/luciferase reaction has been known for years. Since ATP is ubiquitous in all living cells, the measurement does not provide any information on the identity of tested cells. For example, application of this bioluminescence method for bacterial enumeration in milk and sea foods may be complicated by the contamination of non-bacterial cells[1] and metabolic condition-induced variations in ATP level.[2] The specificity of the ATP measurement method has been enhanced by the use of cell-type specific lysis reagents as demonstrated in the work of estimating total microbial contamination on poultry carcasses.[3] The cellular concentrations of ATP are regulated by the bio-energetic status of the cells, i.e. availability of carbon substrates, oxygen, etc. This dependence adds an uncertainty to relate luminescence intensity to the concentration of viable cells. One may easily underestimate microbe concentration by the low luminescence originating from relatively large number of nutrient-deprived cells. In addition, current ATP luminescence procedure lacks the needed specificity for screening of specific pathogenic bacteria in foods.

We have developed a new approach to ascertain the presence of viable bacteria. The bioenergetic status of bacteria was adjusted by the addition of glucose, a carbon energy source, and carbonyl cyanide meta-chlorophenyl hydrazone (CCCP), a membrane protonophore. The addition of glucose restored both the oxygen consumption and the ATP content of the bacteria during cold storage. On the other hand, CCCP enhanced the oxygen consumption and medium acidification but significantly decreased the ATP content. None of the glucose and CCCP effects could be detected with heat-killed bacteria. To develop this pathogen-specific method, *Escherichia coli* O157 specific immunomagnetic beads were applied to capture the bacteria prior to ATP measurement via luciferin-luciferase induced luminescence. Thus, immunomagnetic capture of the *E. coli* followed by testing the bioenergetic responses of captured bacteria could ascertain the presence of viable *E. coli* O157:H7. This CCCP effect allowed the detection of less than one CFU of the *E. coli* per g of ground beef after a 6-h enrichment at 37 °C.

METHODS

Bacterial Samples Cultures of various strains of different types of bacteria were added to suitable growth media and incubated at appropriate temperatures for 18 to

36 h.   At the end of incubation, the cell density (CFU/mL) was determined by standard plate culture and counting techniques.   Stationary phase bacteria were harvested by centrifugation and then suspended in a buffer containing 10 mM Tris, pH 7.5, 2.5mM Mg SO_4 and 150mM NaCl (TBS).   To prepare heat-killed cells, bacterial samples (10^7 CFU/mL) in TBS were heated for 10 min at 100 °C.

Cellular ATP Determination   The intensity of the bioluminescence catalyzed by luciferase was used to estimate the ATP content in bacterial cells.   Bacterial samples were treated with 50% (v/v) of B-PER obtained from Pierce Biotechnology, Rockford, Il.(bacterial protein extraction reagent), a cell lysis reagent specific to bacteria.   The released cellular ATP was determined from the luminescence emitted from luciferase catalyzed oxidation of luciferin.   A Berthold FB12 luminometer was used to record the light output.

Bioenergetic Effects   Freshly harvested bacterial cells of *E. coli* O157:H7 were either used directly or after storage in TBS buffer for 24 h at 4 °C to deplete internal nutrients.   The oxygen consumption rate of ~ 2.0 x 10^8 bacteria, with and without glucose and CCCP in a thermostated Gilson Medical respiration cell filled with 1.9 mL of air-saturated TBS (pH 7.5) was measured by a Clarkson oxygen electrode attached to a YSI 5300 biological oxygen monitor (Yellow Springs, OH).   To test the effects of CCCP on cellular ATP content, ~ 10^6 live or heat-killed bacterial cells in 100 µL of TBS were lysed by B-PER at room temperature for 10 min prior to the incubation with 2 µg of CCCP for 30 min at 37 °C.   At the end of incubation, 100 µL of luciferin-luciferase mixture was added and the luminescence was recorded.   For a second set of bacterial samples, the cells were first incubated with 2 µg of CCCP for 30 min at 37 °C prior to being lysed by B-PER at room temperature for 10 min.   The ATP content was similarly measured by the addition of luciferin-luciferase.   A decreased ATP content in the cells treated with CCCP prior to lysis is an indicator of cell viability.

Immunomagnetic Capture of Bacteria   Cultures of *E. coli* O157:H7 at the stationary phase were centrifuged at 12,000 g for 10 min to remove the growth medium.   The bacterial pellets were then suspended in a solution containing 150 mM NaCl, 2.5 mM $MgSO_4$, 10 mM Tris, pH 7.5 (TBS) with or without 5 mM glucose to 10^9 CFU/mL and then serially diluted to achieve populations from 10^8 to 10^3 CFU/mL.   To 0.9 mL of the bacterial suspensions, a fixed number (~ 6.0 x 10^5) of anti *E. coli* O157 Dynabeads (Dynal A.S, Oslo, Norway) in 10 µL were added and the mixtures were gently shaken for 10 min in plastic Eppendorf centrifuge tubes at room temperature.   A Dynal magnetic concentrator was used to concentrate the beads and to remove the supernatant solutions.   The collected beads with captured bacteria were suspended in 0.9 mL of TBS for ATP analysis.   For experiments with beef hamburger, the same number of Dynabeads were added to 0.9 mL of the filtered cultures obtained at the end of enrichment.

RESULTS

Bioenergetic Effects of Protonophores The cellular level of ATP in viable cells is determined by the rates of ATP synthesis and utilization. It is known that the majority of cellular ATP synthesis is associated with the nutrient-supported and oxygen-dependent membrane electron transfer process. The vectorial membrane proton transport serves as the key intermediate to drive electron transfer-related ATP synthesis[4] and energy-dependent solute transport.[5] Cells utilize the energy released from ATP hydrolysis by membrane H^+-ATPases to actively maintain the internal concentration of ions and small metabolites through the actions of membrane transport systems. Being a membrane protonophore, CCCP has the ability to discharge the proton gradient needed to support ATP synthesis and membrane transport. Consequently, its presence decreases the ATP synthesis by diminishing the proton electrochemical potential ($\Delta\mu_{H+}$) and increases the ATP utilization as the cells are forced to accelerate the consumption of oxygen and the hydrolysis of ATP in an attempt to recover the proton potential. The predicted bioenergetic effects of CCCP are illustrated in Table 1 using *E. coli* O157:H7 as representing bacteria.

Table 1. Bioenergetic Effects of CCCP to *Escherichia. coli* O157:H7

	Oxygen Consumption (nmol min^{-1})			ATP level (RLU)	
	No glucose	-CCCP	+ CCCP	- CCCP	+ CCCP
Normal cells	0	21.44	23.32	8.02×10^4	1.13×10^4
Heat-killed	0	0	0	8.10×10^2	8.89×10^2

The bacteria harvested at the end of 18-h growth were suspended in TBS solution (10^8 CFU per mL) and then stored over ice for 24 h. The oxygen consumption rates for 10^7 cells were measured ($\pm$ 5.0 mM glucose) with and without 2.0 µM of CCCP in TBS. For ATP measurements, 10^5 cells in glucose containing TBS were used.

Verification of CCCP Effects The observed CCCP effects on the cellular ATP level shown in Table 1 may be used as a measure of cell viability. We have expanded the tests on CCCP effects to many other bacteria. As shown in Table 2, the cellular content of ATP was substantially decreased by the actions of CCCP. Although heat treatment decreased the cellular ATP content by ~ 60 to 90%, no CCCP-induced decrease in ATP content of heat-killed bacteria were observed.

Table 2. Effects of CCCP Treatment on Cellular ATP Content

	Live cells (RLU)		Heat-killed (RLU)	
	Total	+ CCCP	Total	+ CCCP
Campylobacter jejuni	54087	13377	9276	10638
Listeria monocytogenes	39806	10309	1574	2256
Salmonella typhimurium	62978	13790	5798	7951
Bacillus cereus	90300	13776	5860	6978
Citrobacter freundii	19827	6786	1477	1917

Detection of Viable *E. coli* O157:H7 in Ground Beef The ultimate goal of the present work was to test whether the developed ATP-bioluminescence method could be applied to complex food systems. Ground beef patties obtained from local supermarket were spiked with different levels of *E. coli* O157:H7 and K-12. After a 6 h incubation in the specific medium at 37 °C, the *E. coli* was captured by the IMB. Captured bacteria were washed and suspended in glucose-containing TBS buffer and were used for ATP-bioluminescence measurement. As shown in Figure 1, the presence of 1 CFU of *E. coli* O157:H7 per gram of beef hamburger was detected after a 5 to 6-h enrichment. In contrast, *E. coli* K-12 yielded insignificant ATP readings above the background. As expected, all the internal ATP associated with the 6-h enriched culture of *E. coli* O157:H7 was discharged by the addition of CCCP indicating that after 6 h at 37 °C, nearly all *E. coli* O157:H7 cells in the enrichment broth were viable.

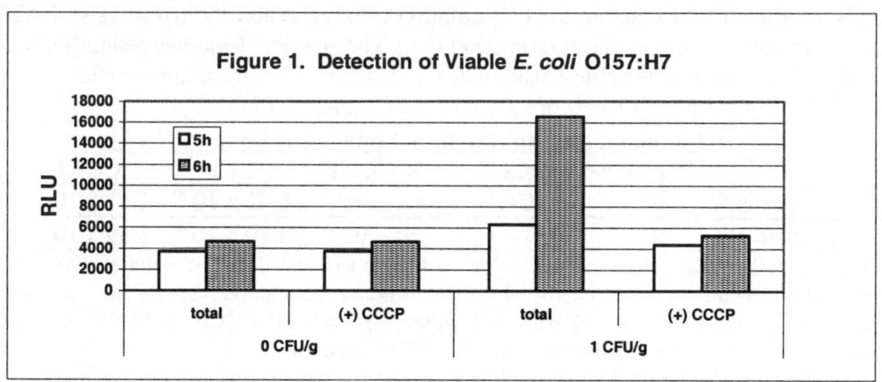

REFERENCES

1. Sharpe AN, Woodrow MN, Jackson AK. 1970. Adenosine triphosphate (ATP) levels in foods contaminated by bacteria. J Appl Bacteriol 1970; 33: 758-67.
2. Knowles CJ. Microbial metabolic regulation by adenine nucleotide pools. *In* Microbial Energetics. In: Haddock B, Hamilton WA, eds. Cambridge, Cambridge University Press, 1977: 241-83.
3. Siragusa GR, Dorse WJ, Cutter CN, Perino LJ, Koohmaraie M. Use of a newly developed rapid microbial ATP bioluminescence assay to detect microbial contamination on poultry carcasses. J Biolumin Chemilumin 1996; 11: 297-301.
4. Mitchell P. Vectorial chemistry and the molecular mechanism of chemiosmotic coupling: Power transmission by proticity. FEBS Lett 1975; 59: 137-9.
5. Epstein W. Bacterial transport ATPases, *In* Bacterial Energetics (series title The Bacteria). TA Krulwich ed. New York Academic Press.1990; 13: 87-110.

APPLICATIONS OF BIOLUMINESCENCE-BASED ASSAY IN MONITORING MICROBIAL BURDEN

KASTHURI VENKATESWARAN[1], ASAHI MATSUYAMA[2], ROGER KERN[1]
*[1]Biotechnology and Planetary Protection Group, Jet Propulsion Laboratory,
California Institute of Technology, Pasadena, CA 91109, USA*
[2]Kikkoman Corp., Noda City, Chiba Prefecture, Japan

INTRODUCTION

NASA has an ongoing research effort to introduce new technologies to evaluate trace biological contaminants on spacecraft outbound from Earth to other planetary bodies. Such contaminants could compromise these pristine environments and interfere with *in situ* life detection experiments. In addition, extraterrestrial sample return procedures stress the importance of avoiding contamination of critical hardware components with terrestrial organisms, their remains, and organic matter in general. Part of assuring that sensitive spacecraft components remain free of terrestrial, biological contamination during assembly is to ensure that the spacecraft clean room environments remain as free of microbial contamination as possible.

The dilemma posed by the cultivability has been known and well appreciated since the time of Winogradsky, meaning that with environmental samples and any given medium only a small fraction of the organisms present will grow. Some microbes are unable to form colonies in normal growth media but are still viable and endowed with metabolic activity, capable of resuming active cell growth under certain conditions. The use of an ATP assay promises to be of value in addressing this problem. An ATP-assay that differentiates free extra-cellular (dead cells, etc.) from intra-cellular (viable microbes) ATP was evaluated to determine the microbial burden of a given sample that contains cultivable, viable and also cells that are in viable but non-cultivable state.

A. NASA APPLICATIONS OF THE ATP-BASED ASSAY

Monitoring microbial burden in clean-room facilities. The use of an ATP-based assay in monitoring all viable cells would facilitate an understanding of the cleanliness of the spacecraft assembling facility (~650 samples). The viable microbial population as evaluated by ATP was one to 3-logs higher than that indicated by aerobic plate counts in these clean rooms[1]. To ascertain the incidence of all viable microbial populations, the samples falling in different categories (see below) were selected and microbial diversity was delineated by modern molecular techniques.

The category 1 samples that showed at least one log higher in ATP content than cultivable counts had more yeast-fungi populations on tryptic soy agar (TSA) plates. Next predominant microbes were Gram-positive groups. About 46% of unclassified clean room samples showed at least one log higher in ATP content than cultivable counts. Microbial diversity of cultivable organisms showed members of *Bacillus, Micrococcus, Staphylococcus*. The DNA extracted from these samples revealed presence of the sequences of the cultivable species such as *Lactobacillus* sp., however, the sequences of commonly occurring *Bacillus* were not seen.

The second category contains samples that had no cultivable counts but at least one log higher ATP content. The 16S rDNA sequences showed similarities to *Peptostreptococcus* sp., *Stenotrophomonas maltophila, Streptococcus thermophilus, Taxeobacter* sp. were amplified from the samples of this second category. It was reported that the optimum growth conditions for *Streptococcus thermophilus* is ~40°C, *Peptostreptococcus* sp. are anerobes, and the *Taxeobacter* species are slow growers and do not grow in nutrient-rich media. In addition to these sequences, sequence similarities to *Ultramicrobacterium* sp. that was reported to contain only 0.3 μm^3 cell volume and only grow efficiently in liquid media rather than in solid media[2] were also sequenced in these samples. Absence of any cultivable microbes but at least one log higher ATP content in about 86% and 60% of samples collected from the clean room that are classified as class 10K and 100, respectively, is interesting. About 57% of the samples collected from working tables of these clean room facilities showed this second category pattern where ethanol was normally used to clean the surface before the start of any activity. However, it is possible that the ATP might have been derived from the microbes that could withstand such harsh treatment.

The third category includes samples that are generating one log lower ATP content than cultivable counts. Such incidence was noticed in 18% of the total samples collected. Actinomycetes were prevalent in these samples. Samples collected from clean room floors (45%) and cabinet tops (67%) fall under this category. The colonies developed in TSA of these samples exhibited higher incidence of microbial populations that produce spore. As the ATP content of purified spores was one-hundredths of its vegetative cell, lower ATP content and higher cultivable count was possible in these samples. Because of this fact, these samples were presumed to contain spores of microbial species (especially *Bacillus*) during sampling and vegetative populations might get germinated from the spores during culture on agar medium. In addition to this, the direct DNA extraction analysis revealed sequences of *Comamonas testosteroni, Streptococcus thermophilus, Taxeobacter* sp. and some unidentified uncultured bacteria.

The fourth category includes samples that showed equal abundance of both cultivable counts and ATP content. Such incidence was noticed in 15% of the total samples collected. No fungal nor yeast colonies were observed in the samples of this category. However, Gram-negative bacteria such as *P. stutzeri, P. graminis* were isolated from these samples.

Measuring microbial burden on the spacecraft surfaces. In order to implement ATP-assay for verifying spacecraft cleanliness, we conducted an extensive QA/QC validation approach by determining the method detection limits, accuracy, reproducibility, and hold times for the assay[3]. The ATP assay detects a minimum of 100 cells and accurately measures ATP to 10^{-14} moles per sample with a confidence level of 99% and is also reproducible with a coefficient of variation <5%. When this ATP-assay was tested in the field ~500 Mars Exploration Rover (MER) surface samples were collected. The results indicated that 27% of the samples were contaminated with >10^{-11} moles of ATP. Since spacecraft achieve very low bioburden levels after rigorous cleaning processes, a simple and rapid method such as the ATP assay is useful to validate "biological cleanliness" and might also help to meet rigid schedules of the spacecraft assembly processes. Although the total ATP assay does not correlate with the spore counts, this 10 minute rapid assay can be used as an indicator of the total microbial burden of the spacecraft. **Estimating spore-burden**. Although the ATP bioluminescence assay has been evaluated to determine the microbial cleanliness of spacecraft and associated- environments, since the level of ATP is very low in spores, the ATP assay is not very sensitive for spore detection. As the AMP level is much higher than the ATP level in spores, AMP may be a better biomarker for spore detection. In a bioluminescence assay developed by Kikkoman Corp., Japan, AMP is converted to ATP using pyruvate orthophosphate dikinase, and ATP is subsequently detected by luciferase. In a recent study, we measured the AMP content of several spores of *Bacillus* species using AMP- bioluminescence assay and optimized conditions suitable for rapid spore detection[5]. Spores of several *Bacillus* type strains as well as spacecraft-associated environmental isolates of the same species were purified and compared for AMP content. Spore samples were heat shocked at 80°C for 15 minutes to kill vegetative cells and extracellular ATP was removed by washing. AMP was released from the spores by heat shock at 100°C for 10 minutes. For the *Bacillus* strains tested, the mean AMP content of a bacteria spore is 10^{-18} moles per colony-forming units (CFU). The limit of this method for spore detection is ~100 CFU. **Enumerating single-spore**. In order to meet rigid schedules of spacecraft assembly, a more rapid, sensitive spore detection assay is being considered as an alternate method for the current three-day NASA standard spore culture assay. We have evaluated the use of the Rapid Micro Detection System (RMDS) as a rapid spore detection tool for NASA applications[4]. This is accomplished by preceding the RMDS incubation protocol with a heat shock step, 15 minutes at 80°C, as a direct selection of spores. Different luciferases and formulations were tested in order to reduce the typical 18-24 hour incubation time required by the RMDS to ~5 hours. Of reagents evaluated, a formulation of high sensitive (HS) bioluminescence reagent was found to be more sensitive than the present commercially available reagents. Assay times of ~5 hours were repeatedly demonstrated along with low image background noise. In order to evaluate the applicability of this method, 7 species of *Bacillus* totaling 9 strains, which have been repeatedly found in clean room environments, were assayed. All strains were detected in ~5 hours.

B. ADVANTAGES AND DISADVANTAGES

Microbial monitoring based on ATP must be used with caution because different microbes possess different amounts of ATP and also metabolically active cells might harbor more ATP. We noticed that highest ATP was synthesized in early logarithmic phase (2 to 6 h) and the ATP pool decreased during early stationary phase (10 to 16-h) for both Gram-positive and Gram-negative strains. Basically, Gram-positive bacteria generate more ATP (1.2×10^{-17} mol ATP/CFU) than by Gram-negative bacteria (2.3×10^{-18} mol ATP/CFU). Likewise, yeast and fungal cells produced about 200 times more ATP than Gram-negative bacteria.

C. CONCLUSION

An issue of evident and continued concern is that the distribution of organisms in any given sampling location is not uniform. Thus there were "hot and cold" outlying spots, and the statistical treatment of data is a real problem. Bearing this in mind and to summarize, the ATP-based assay detected consistently one to 3-logs higher microbial population than the cultured microorganism results. This might be due to the sensitivity of the ATP assay that could detect and differentiate all viable microorganisms.

ACKNOWLEDGEMENTS.

We thank F. Chen, G. Kuhlman, G. Kazarians, S. Chung, N. Hattori, S. Suzuki, M. Pickett, and A. Sage for technical support. Part of the research described was carried out at the Jet Propulsion Laboratory, California Institute of Technology, under a contract with the National Aeronautics and Space Administration.

REFERENCES

1. Venkateswaran K, Hattori N, La Duc MT, Kern R. ATP as a biomarker of viable microorganisms in clean-room facilities. J Microbiol Meth. 2003; 52: 367-77.
2. Iizuka T, Yamanaka S, Nishiyama T, Hiraishi A, Isolation and phylogenetic analysis of aerobic copiotrophic ultramicrobacteria from urban soil. J Gen Appl Microbiol 1998; 44:75–84.
3. Kazarians G, Kuhlman G, Kempf M, Chen F, Venkateswaran K, Kern R. Evaluation of an ATP-based bioluminescence assay for determination of the microbial burden of spacecraft. 104th Annual Meeting of American Society for Microbiology, New Orleans, May 2004. Q-082; page: 518.
4. Chen F, Kuhlman G, Kirshner L, Matsuyama A, Sage A, Pickett M, Venkateswaran K, Kern R. Sensitivity and applicability of a rapid microbial detection system in the enumeration of bacterial spores. 104th Annual Meeting of American Society for Microbiology, New Orleans, May 2004. Q-083; page 519.
5. Chen F, Suzuki S, Kuhlman G, Venkateswaran K, Kern R. AMP based sensitive and rapid spore detection assay. 104th Annual Meeting of American Society for Microbiology, New Orleans, May 2004. Q-081; page 518.

APPLICATION OF IMAGING DETECTION USING XYZ EMISSION SYSTEM FOR FOOD ANALYSIS

Y YOSHIKI, K OKUBO

Laboratory of Biostructural Chemistry, Tohoku
University Graduate School of Life Science,
1-1 Tsutsumidori Amamiyamachi, Aobaku Sendai, Miyagi 981-8555, Japan
Email: yoshiki@bios.tohoku.ac.jp

INTRODUCTION

Low-level chemiluminescence arises in the presence of reactive oxygen species (X), hydrogen donor (Y) and mediator (Z) at room temperature and in neutral conditions (XYZ emission system). The photon intensity (P) in this system shows a high concentration dependence, which can be shown by $[P]=k[X][Y][Z]$ (k=photon constant). This equation suggested the simple detection system for X, Y and Z species by the combination of XYZ reagents (YZ reagent for X detection, XZ reagent for Y and XY reagent for Z). Based on this theory, we developed an image-quantification for ROS, hydrogen donor and mediator using a charge-coupled device (CCD) camera. We selected the H_2O_2/gallic acid/$KHCO_3$/MeCHO system for our standard of imaging detection system. The advantages of our chemiluminescence system are 1) short measuring time (10 min), 2) simultaneous measurement of 10-20 samples, 3) applicability to liquid and solid samples and 4) simplified measurement technique. We also studied a relationship between Y photon emission and ROS scavenging potential of food.

METHODS

Detection of photon emission

Chemiluminescence was detected with a charge-coupled device (CCD) camera (Hamamatsu Photonics, Japan). The CCD camera was connected to an imaging PMP (photocathode microchannel plates) and a position-sensitive detector coupled with an ARGUS-20 image processor for further image enhancement and quantitative analysis. The reaction mixture contained the sample and two reagents selected from 196 mmol/L H_2O_2 (0.5 mL), 5 mmol/L gallic acid (0.5 mL) and satd. $KHCO_3$ in 356 mmol/L MeCHO (0.5 mL). The reaction solution was added in a 12-well multiplate (φ 12 mm). Photon intensity was measured as luminance (cd/m^2).

Measurement of emission spectra

Emission spectra in the visual region were measured with a Simultaneous Multiwavelength Analyzer model CLA-SP2 (Tohoku Electronic, Japan). The wavelength range of the spectroscope was 400-850 nm. Light emission was determined for 180 s.

H_2O_2 scavenging activity

H_2O_2 (0.25 mL of 7.84 mmol/L) was measured by allowing the sample solution (0.1-0.5 mL), 3.6 mol/L $KHCO_3$ in 356 mmol/L MeCHO (0.3 mL) and water (0.95-0.55 mL) to react with 0.25 mL of 20 % H_2SO_4 and 0.15 mL of 1 mol/L $TiSO_4$. The H_2O_2 concentration was determined by absorption at 408 nm (Shimazu UV-1600) and calculated using the H_2O_2 standard curve.

RESULTS

Imaging of ROS, hydrogen donor and mediator

We have reported that the photon emission from the H_2O_2/gallic acid/MeCHO system, which has an emission maximum (E_{max}) at 660 nm, was increased by the addition of $KHCO_3$.[1,2] Increasing emission by addition of $KHCO_3$ (E_{max} 645 nm) enabled us to perform imaging detection of photon emission using a charge-coupled device (CCD) camera with 12-well multiplate (Fig. 1a). We used 196 mmol/L H_2O_2, 5 mmol/L gallic acid and satd. $KHCO_3$/356 mmol/L MeCHO as a standard reagent of ROS, hydrogen donor and mediator, respectively. The photon emission from the H_2O_2/gallic acid/$KHCO_3$/MeCHO system continued over 60 min and showed linear increase. We set the detection time to 10 min for quantification of photon emission. The imaging photon emission has a high concentration dependence for H_2O_2, gallic acid and $KHCO_3$/MeCHO (Fig. 1b). Therefore, it is necessary for image quantification that the concentration of two factors from ROS (X), hydrogen donor (Y) and mediator (Z). X, Y and Z emission from sample was detected with a simple technique that combined two reagents (YZ, XZ and XY) with each sample.

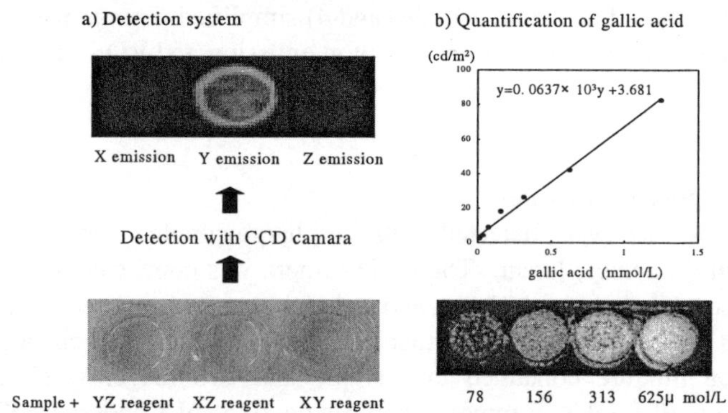

Figure 1.   Detection system (a) and quantification (b) for photon emission

a)X, 196 mmol/L H_2O_2 (0.5 mL); Y, 5 mmol/L gallic acid (0.5 mL); Z, satd.$KHCO_3$ in 356 mmol/L MeCHO (0.5 mL) b)Concentration effects of gallic acid on photon emission intensity from the H_2O_2/gallic acid/$KHCO_3$/MeCHO

Meaning of photon emission

Recently, we reported the unique H_2O_2 scavenging activity of gallic acid in our chemiluminescence system.[3] Gallic acid and $KHCO_3$/MeCHO did not scavenge H_2O_2. In contrast, the gallic acid/$KHCO_3$/MeCHO mixture (chemiluminescence system) showed H_2O_2 scavenging activity, depending on the gallic acid concentration. Tea infusion and soy-food extract showed strong Y emission in the same manner with gallic acid. Therefore, we measured the H_2O_2 scavenging activity by the combination of Y emission food and $KHCO_3$/MeCHO. Tea infusion and soy-food extract showed synergistic H_2O_2 scavenging activity in the presence of $KHCO_3$/MeCHO, while food extracts without $KHCO_3$/MeCHO showed none H_2O_2 scavenging activity. H_2O_2 scavenging activity correlated with photon intensity from the H_2O_2/tea infusion/$KHCO_3$/MeCHO (r^2=0.7516) and H_2O_2/soy-food extract/$KHCO_3$/MeCHO systems (r^2=0.8052), respectively (Fig.2). Photon emission from the food extracts resulted in the H_2O_2 scavenging activity. It has been reported in some studies using CCD camera that the photon emission was associated with auto-oxidation.[4] The results reported here have demonstrated for the first time the use of the CCD camera to measure and monitor photon emission based on H_2O_2 scavenging activity.

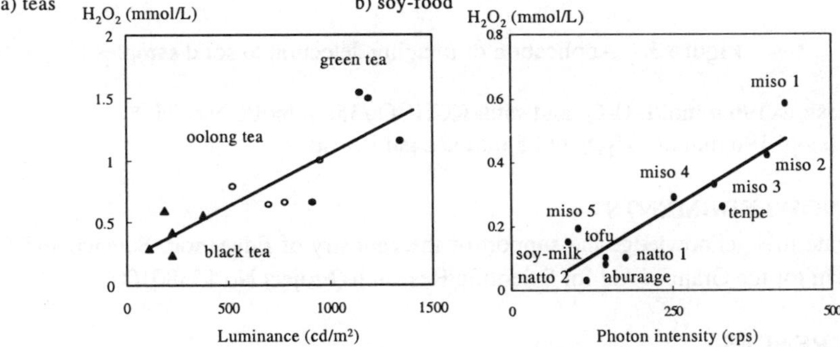

Figure 2. Relationship photon intensity-H_2O_2 scavenging activity of tea infusion and soy-food extracts

Application to food analysis

The imaging of photon emission from the ROS/hydrogen donor/mediator system was applied to solid-type samples. We studied the photon emission of several foods. Hydrogen donor emission (Y emission) was observed from polyphenol rich vegetables and fruits (tea and banana), fermented foods (oyster sauce, soy sauce and miso), alcohol (wine, sake and beer), spices and cereals (wheat and rice). Mediator emission (Z emission) was seen from some vegetables (Japanese radish, Chinese yam and nozawa-na) and fruits (melon), egg white, meat and fish meat. Imaging detection has a potential for visualization of Y and Z component distribution through the Y and Z emission

(Fig. 3). Interestingly, Y and Z emission were effected by heat treatment. Fig. 3b showed the appearance of Z emission from lotus root by heat treatment. Moreover, we observed disappearance of Y emission from some fruits and vegetables. Preparation of food by boiling as important factor effecting efficient intake of ROS scavenging compounds in food. This photon emission system can be applied to the solid sample regardless of water solubility. Detection by CCD camera may be important for food and biomaterials analysis, when total H_2O_2 scavenging activity and distribution of H_2O_2 scavenging compounds is required.

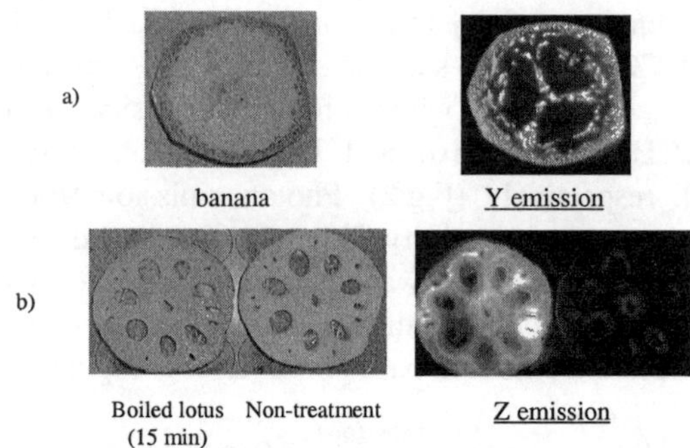

Figure 3.   Application of imaging detection to solid samples

Y emission; 196 mmol/L H_2O_2 and satd. $KCHCO_3$/356 mmol/L MeCHO
Z emission; 196 mmol/L H_2O_2 and 5 mmol/L gallic acid

ACKNOWLEDGMENTS

We gratefully acknowledge the support of the Ministry of Education, Science and Culture of Japan for the Grant in Aid for Scientific Research (Project No.15580100).

REFERENCES

1. Yoshiki Y, Kahara T, Okubo K, Igarashi K, Yotsuhashi K. Mechanism of catechin chemiluminescence in the presence of active oxygen. J Biolumin Chemilumin 1996; 11:131-6.
2. Yoshiki Y, Iida T, Akiyama Y, Okubo K, Matsumoto H, Sato M. Imaging of hydroperoxide and hydrogen peroxide-scavenging substances by photon emission. Luminescence 2001; 16:1-9.
3. Yoshiki Y, Kanazawa T, Okubo K. Approach to a new measurement of reactive oxygen and its scavenging compounds by chemiluminescence system. India: Res signpost, Oxidative Degradation and Antioxidative Activities of Food Constituents 2002:11-7.
4. Slawinska D, Slawinski J. Chemiluminescence of cereal products III. Two dimensional photocount imaging of chemiluminescence. J Biolumin Chemilumin 1998; 13:21-4.

PART 11

LUMINESCENCE IMMUNOASSAYS

TANDEM BIOLUMINESCENT ENZYME IMMUNOASSAY FOR BDNF AND NT-4/5

SEIMEI AKAHANE, KATSUTOSHI ITO[*], HIDETOSHI ARAKAWA,
MASAKO MAEDA

*School of Pharmaceutical Sciences, Showa University, 1-5-8 Hatanodai,
Shinagawa, Tokyo 142-8555, Japan*
Email: itok@pharm.showa-u.ac.jp

INTRODUCTION

We applied the firefly luciferase-luciferin reaction to bioluminescent enzyme immunoassay (BL-EIA) using ATP generating enzymes acetate kinase (AK) and pyruvate phosphate dikinase (PPDK). Then, we developed simultaneous a bioluminescent assay using AK and PPDK. We detected 8.6×10^{-21} mol/assay of AK and 1.4×10^{-20} mol/assay of PPDK respectively, and then we developed BL-EIA for insulin and C-peptide.[1] Recently, Nelson *et al.* have reported that neonatal blood concentration of VIP, CGRP, BDNF, NT-4/5 were higher in autistic than in control children.[2] Therefore, the measurements of these four factors in neonatal blood are useful for diagnose and management of early stage. In this study, we established highly sensitive tandem BL-EIA for BDNF and NT-4/5. Schematic illustration of the proposed tandem BL-EIA for BDNF and NT-4/5 is shown in Fig. 1.

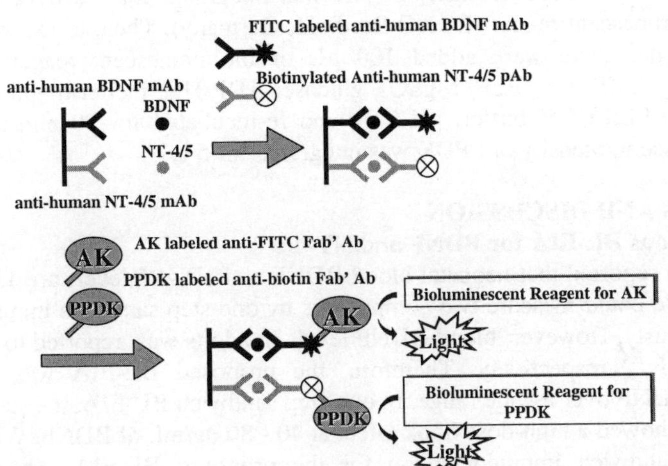

Figure 1. Schematic illustration of the proposed tandem BL-EIA for BDNF and NT-4/5

MATERIALS AND METHODS
Reagent
PPDK (from *Microbispora rosea subsp. Aerata*, EC 2.7.9.1) and thermostable Thr 217 Ile mutant of *Luciola cruciata* firefly luciferase (EC 1. 13. 12. 7) were donated from Kikkoman Co. (Chiba, Japan). AK (from B. *stearothermophilus*, EC 2. 7. 2. 1) was purchased from Seikagakukogyo Co. (Tokyo). ADP-HK (from *Pyrococcus furiosus*, EC 2. 7. 1. 147) was from Asahi Chemical Industly Co. Ltd. (Shizuoka, Japan). D-Luciferin was purchased from Sigma Chemical Co. (St. Louis, MO). Goat anti-FITC IgG and goat anti-biotin IgG were purchased from Vector laboratories, Inc., (Burlingame, CA). Mouse monoclonal anti-BDNF IgG for immobilizing and FITC labeling, biotinylated goat anti-NT-4/5 IgG, mouse monoclonal anti-NT-4/5 IgG, BDNF and NT-4/5 were obtained from R&D Systems Inc. (Minneapolis, MN).

Simultaneous sandwich BL-EIA for BDNF and NT-4/5
The wells of microtiter plate were coated with mouse anti-BDNF and anti-NT-4/5 monoclonal antibody. The wells were then post-coated by adding 1% water soluble gelatin solution containing 0.05 % NaN_3. The plates were stored at 4 °C prior to use. After washing the plate, we added 50 μL of standard or sample solution and 150 μL of assay buffer to each well and incubated for overnight at 4 °C. Then the plate was washed and added 50 μL of FITC labeled anti-BDNF, biotin labeled anti NT-4/5 antibody and 100 μL of assay buffer. After incubation for 3 h at room temperature and then washing the plate, we added 100 μL of AK labeled anti-FITC and PPDK labeled anti-biotin Fab' antibody to the plate and incubated for 1 h at room temperature. The microtiter plate was re-washed. Then, 100 μL of bioluminescent reagent for AK (containing ADP, acetyl-phosphate, $MgSO_4$, luciferine and luciferase in 50 mM HEPES-KOH buffer, pH 7.0) was added and incubated for 20 min at 37 °C. The bioluminescent intensity of AK was integrated for 5 s by a MicroLumat LB96P luminescent reader (EG&G Berthold, Germany). Then, to the same wells of the microtiter plate were added 100 μL of bioluminescent reagent for PPDK (containing AMP, PPi, PEP, $MgSO_4$, glucose, ADP-HK, luciferin and luciferase in 50 mM HEPES-KOH buffer, pH 7.0), and re-incubated for 20 min at 37 °C the bioluminescent intensity of PPDK was integrated for 5 s.

RESULTS AND DISCUSSION
Simultaneous BL-EIA for BDNF and NT-4/5
It has been reported that neonatal blood BDNF and NT-4/5 levels are 13.3 and 28.5 pg/mL.[2] We could measure two components by one-step sandwich immunoreaction, simultaneously. However, blood BDNF levels in adults was reported to be 27.7 and 16.3 ng/mL,[3,4] respectively. Therefore, the proposed BL-EIA was required to measure BDNF over a wide range. In one-step sandwich BL-EIA, the standard curve of BDNF showed a high-dose hook effect at 40 - 80 ng/mL of BDNF. We performed two-step sandwich immunoreaction for the proposed BL-EIA. The measurable ranges of BDNF and NT-4/5 were 4.9 - 40000 and 31.25 - 2000 pg/mL, the

detection limits (blank + 3 SD) of BDNF and NT-4/5 were 1.2 and 11.4 pg/mL, respectively. The intra-assay coefficient of variation of BDNF and NT-4/5 with each standard point were 1.8 - 9.8 % (n=8) and 2.3 - 6.4 % (n=8), respectively.

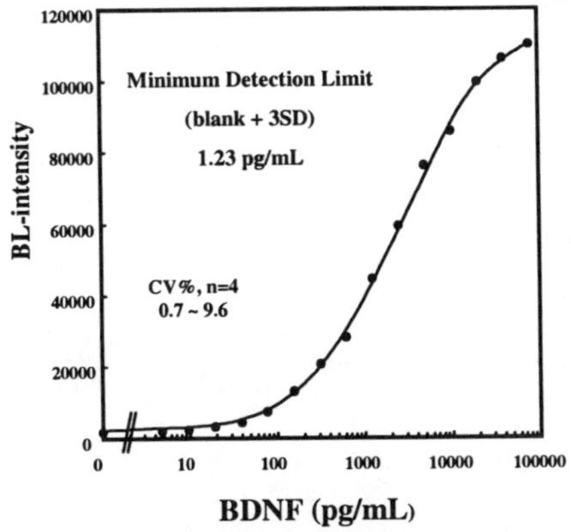

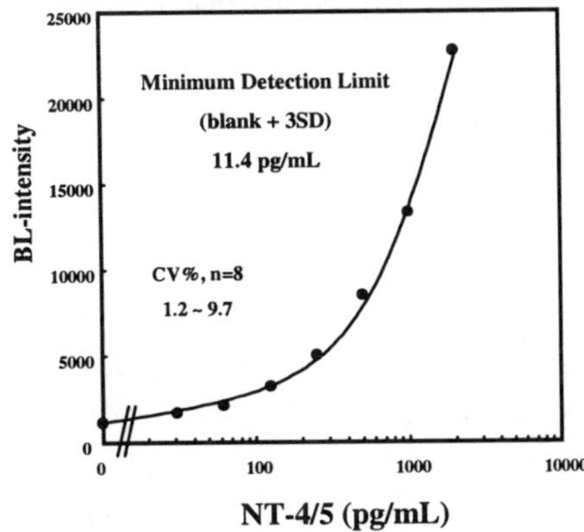

Figure 2. Standard curves for BDNF and NT-4/5 by proposed simultaneous bioluminescent assay

ACKNOWLEDGMENTS

The authors would like to thank Asahi Chemical Industry Co., Ltd., for the gift of ADP-HK. This work is supported by Grants-in-Aid for Scientific Research (C) No. 13672431 from Japan Society for the Promotion of Science, the High-Technology Research Center Project from the Ministry of Education, Culture, Sports, Science and Technology of Japan and Japan Health Sciences Foundation.

REFERENCES

1. Ito K, Nakagawa K, Murakami S, Arakawa H, Maeda M. Highly sensitive simultaneous bioluminescent measurement of acetate kinase and pyruvate phosphate dikinase activities using a firefly luciferase-luciferin reaction and its application to a tandem bioluminescent enzyme immunoassay. Anal Sci 2003; 19: 105-9.
2. Nelson KB, Grether JK, Croen LA, Dambrosia JM, Dickens BF, Jelliffe LL, Hansen RL, Phillips TM. Neuropeptides and neurotrophins in neonatal blood of children with autism or mental retardation. Ann Neurol 2001; 49: 597-606.

3.   Shimizu E, Hshimoto K, Okamura N, Koike K, Komatsu N, Kumakiri C, Nakazato M, Watanabe H, Shinoda N, Okada S, Iyo M. Alterations of serum levels of Brain-Derived Neurotrophic Factor (BDNF) in depressed patients with or without antidepressants. Biol Psychiatry 2003; 54: 70-5.
4.   Lang UE, Hellweg R, Gallinat J. BDNF serum concentrations in healthy volunteers are associated depression-related personality traits. Neuropsychopharmacolojy 2004; 4: 795-8.

CALCIUM-REGULATED PHOTOPROTEIN OBELIN AS A LABEL IN IMMUNOASSAY: AN OUTLOOK FOR APPLICATIONS

LA FRANK, VV BORISOVA, ES VYSOTSKI

Institute of Biophysics, RAS, Krasnoyarsk, 660036 Russia
Email: lfrank@yandex.ru

Obelin – one of the calcium-regulated photoproteins – was isolated from the hydroid *Obelia longissima*. This is a one-chain protein (22.2 kDa) making a stable complex of polypeptide chain, substrate and oxygen. The binding of Ca^{2+} ions initiates conformational changes in a molecule resulting in the oxidation of coelenterazine. The products of the reaction are: coelenteramide, CO_2 and blue light.

The light emission is proportional to the amount of a protein for it participates in the reaction directly. The relationship between obelin concentration and bioluminescence is unlimitedly linear. A high quantum yield of the obelin bioluminescence is 25-30%, and the lack of the background signal due to the high ionic specificity of obelin to Ca^{2+}, make it possible to determine this protein down to 10^{-18} mole by using modern commercially available luminometers.

In addition there is no problem with obelin availability: for the present moment we have the hyper-producing strain *E. coli* of the protein available, and the technology permitting the obtaining of over 50 mg of a highly purified apoobelin per 1 g of raw cell paste.[1,2] The recombinant apoprotein is effectively activated with a synthetic coelenterazine and does not differ from a native product in biochemical and bioluminescent properties. The protein is stable when stored in a soluble or lyophilized state.

So there were obvious implications to apply obelin as a label in immunoassay: the protein accessibility, high sensitivity, an unlimited linear range of bioluminescence, the lack of background, the simplicity with which the reaction is triggered , the absence of hazard and the availability of modern registration devices.

The problem was that the introduction of any enzyme as a label into immune complex implies some modifications – conjugation with the other proteins (antigens or antibodies) or binding of definite anchor groups serving as an effective bridge between the molecules of the immune complex. There are some proteins, being promising for analytical application that lose almost all the specific activity under modifications.

The obelin was conjugated with various molecules, as was dictated by the task set. The modifications were effective and did not lead to considerable loss of obelin bioluminescent activity. Here are some examples:

1. The avidin-biotin system (we used as a bridge between obelin and immune complex) bases on a very high affinity of biotin (vitamin H) to avidin – the protein from hen eggs, or streptavidin – its microbial analog ($k_d=10^{-15}$M). Biotin is introduced into analyzed molecules, while avidin, which has 4 binding sites of biotin serves as a strong and specific bridge between these molecules.

Obelin-biotin complex was successfully obtained using succinimide derivatives of biotin, with less than 30% of its bioluminescent activity lost under the synthesis conditions.[3] The biotinylated obelin is a universal label, suitable for any immunoassay through the avidin bridge. As an example, Fig.1 gives the scheme of a solid-phase microanalysis of alphafetoproteins in standard human sera and displays the results of this analysis.

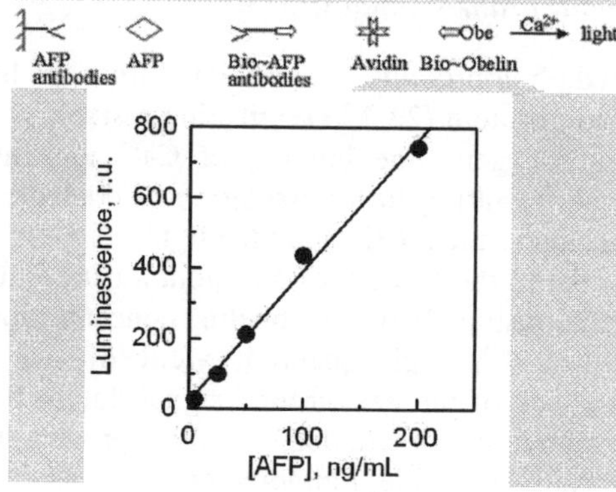

Figure 1. Microtiter-based AFP bioluminescent immunoassay

The main analytical characteristics of obelin and isotope labels at assaying the AFP turned out to be very close.[3] On the one hand, obelin-biotin complex is a universal label, on the other – it extends the analytical chain. The obtaining of obelin-avidin or obelin-antibody conjugates – that's what helps make the analytical chain shorter.

2.   We worked out the effective method for obelin conjugation with the other molecules – avidin, antibodies and so on. To analyze the thyroid gland hormones in sera, we synthesized the obelin conjugates with: antibodies to the thyroid-stimulating hormone (TSH), antibodies to thyroxin (T4), and T4 itself.[4] The bioluminescent labels obtained were applied for different format solid-phase immunoassay. Fig. 2 illustrates the TSH sandwich-type immunoassay, and the calibration curve built as a dependence of bioluminescent response on hormone concentration in standard sera. This curve was used to determine TSH in 34 patients' sera. The results obtained closely corresponded to those provided by the RIA data (r = 0.99).

The T4, total and free, was estimated by using only one antibody via competitive analysis according to schemes (2) and (3), correspondingly. Using the labels obtained, free and total T4 were analyzed in control and patient sera and the results of bioluminescent assay closely correlated with RIA.[4]

3.   Apart from chemical conjugation, the obelin labels were obtained genetically. In 1996 our paper on a chimeric protein "obelin—immunoglobulin-binding fragment of a protein A" was published.[5] The chimera constructed reveals the immunoglobulin-binding ability of the protein A and the obelin bioluminescent

activity. The protein was used to determine the antibodies to the tuberculosis toxin,[5] AFP and luteinizing hormone in human sera.[6]

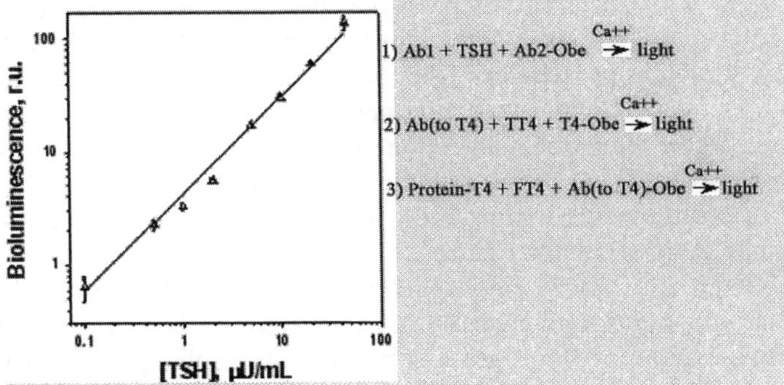

Figure 2. Right: Microtiter-based TSH bioluminescent immunoassay. Left: The schemes of the TSH sandwich-type assay (1), TT4 (2), and FT4 (3) competitive assays.

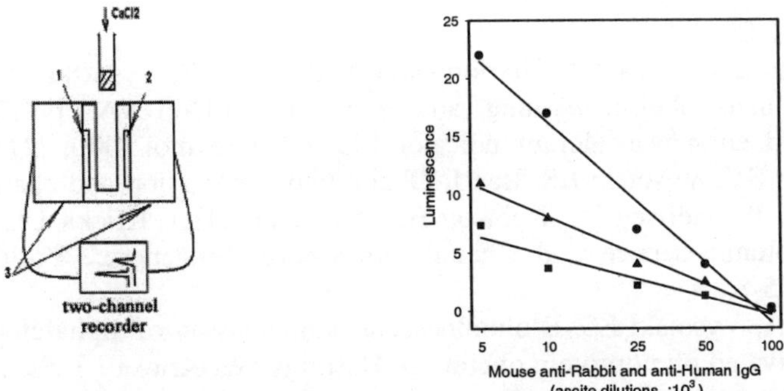

Figure 3. Left: The installation unit scheme: 1,2 – optical filters, 3 – PMTs. Right: The dependence of bioluminescent signals on IgGs titers. (●) – total, (■) – green, and (▲) – violet bioluminescent signals.

Using the genetic approach the biotinylated obelin was obtained.[7] A recombinant apoobelin capable of being biotinylated *in vivo* in *E. coli* cells with BirA was constructed by fusing in-frame a synthetic DNA-fragment encoding the artificial biotin acceptor peptide to the N-terminus of the obelin cDNA gene. The application of the fusion protein as a label in immunoassay was demonstrated.[7]

4.   Of special interest is the application of the so-called "color" mutants of obelin. In our laboratory the mutants of recombinant obelin of unique spectral characteristics of bioluminescence were obtained with a site-directed mutagenesis applied. Among those are "green mutant" (λmax = 493nm) and "violet mutant" (λmax = 390nm), with a small overlapping of signals (unpublished data). It makes possible to use these mutants as bioluminescent labels for simultaneous determination of two antigens. To demonstrate this, the model experiments were

carried out. "Green" and "violet" mutants were conjugated with human and rabbit IgGs as described in.[4] The cells surface was activated with mouse anti-human and anti-rabbit IgGs with different titers. After blocking procedures (1%BSA) the conjugates solutions were placed into the cells and incubated for 1 h at 20 °C following by washing procedures. The bioluminescent signals of obelins were measured simultaneously with two luminometers through optical filters, transmitting violet or green light, after $CaCl_2$ injection. The signals were registered using two-channel pen-recorder. Fig. 3 presents: the installation unit scheme (left) and the dependence of bioluminescent signals on IgGs titers (right). As one may see, the effective signals' separation takes place.

The photoprotein labels, stable and non-hazardous, exhibit high sensitivity comparable to the radioisotope label. As compared to any other enzyme label, basing on chromogenic or fluorogenic substrate, the labels applying photoproteins do not require additional external substrates and incubation periods, thus omitting intermediate steps, and this works especially well under large-scale investigations. The availability of "color" obelin mutants and multiwave registration techniques allows broadening the application scopes.

REFERENCES

1. Illarionov BA, Frank LA, Illarionova VA, Bondar VS, Vysotski ES, Blinks JR. Recombinant obelin: cloning and expression of cDNA, purification and characterization as a calcium indicator. Methods Enzymol 2000; 227: 223-49.
2. Markova SV, Vysotski ES, Lee J. Obelin hyperexpression and characterization. In: Case JF, Herring PJ, Robison BH, Haddock SHD, Kricka LJ, Stanley PE. eds. Bioluminescence and Chemiluminescence. Singapore: World Scientific, 2001; 115-8.
3. Frank LA, Vysotski ES. Bioluminescent immunoassay of alphafetoprotein with Ca^{2+}-activated photoprotein obelin. In: Hastings JW, Kricka LJ, Stanley PE. eds. Bioluminescence and Chemiluminescence: Molecular Reporting with Photons. Chichester: John Wiley, 1997; 435-8.
4. Frank LA, Petunin AI, Vysotski ES. Bioluminescent immunoassay of thyrotropin and thyroxine using obelin as a label. Anal Biochem 2004; 325: 240-6
5. Frank LA, Illarionova VA, Vysotski E.S. Use of proZZ-obelin fusion protein in bioluminescent immunoassay. Biochem Biophys Res Commun 1996; 219: 475-9.
6. Frank LA, Efimenko SA, Petunin AI, Vysotski ES. Chimeric protein proZZ-Obe as a universal label for bioluminescent immunoassay In: Roda A, Pazzagli M, Kricka LJ, Stanley PE. eds. Bioluminescence & Chemiluminescence. Perspectives for the 21-th Century. Chichester: John Wiley, 1998: 111-4.
7. Markova SV, Stepanyuk GA, Frank LA, Vysotski ES. Apoobelin biotinylated *in vivo*: overproduction in *Escherichia coli* cells. In: Stanley PE, Kricka LJ. eds. Bioluminescence & Chemiluminescence: Progress & Current Applications. Singapore: World Scientific, 2002; 107-10.

HIGHLY SENSITIVE CLEIA FOR C-PEPTIDE IN SERUM WITH CHEMILUMINESCENT SUBSTRATE USING A NEW CLEIA SYSTEM

S HAYAMA, K MORIYAMA, S KITAJIMA

[1] Fujirebio Inc, Research and Development Division, 51 Komiya-cho Hachioji, Tokyo, 192-0031, Japan

INTRODUCTION

C-peptide is cosecreted with insulin by the pancreatic β-cells as a by-product of the enzymatic cleavage of proinsulin to insulin. C-peptide and insulin are secreted into the portal circulation in equimolar concentrations.[1,2]

The measurement of C-peptide provides a fully validated means of quantifying endogenous insulin secretion, preventing influence of exogenous insulin or insulin antibody. However, most C-peptide assay kits cannot differentiate C-peptide from proinsulin and proinsulin conversion products. The influence of proinsulin may be significant in cases where serum proinsulin is elevated, as in Type 2 diabetes, familial hyperproinsulinemia and in patients with proinsulin antibody.

To solve the problem, we have developed a highly specific and sensitive assay for C-peptide in serum and plasma using specific monoclonal antibody (MoAb) to the N-terminal of the C-peptide molecule. In this report, we describe the assay performance of C-peptide on the LUMIPULSE system. The system is a fully automated chemiluminescent enzyme immunoassay (CLEIA) system that uses AMPPD as a substrate for alkaline phosphatase and ferrite micro-particles as a solid phase.[3]

MATERIALS AND METHODS

Reagents

MoAbs to C-peptide, 9101 and CPT3-F11, were obtained from OY Medix Biochemica AB and DakoCytomation, respectively. We purchased AMPPD as chemiluminescent substrate for alkali phosphatase (AP) from Applied Biosystems.

Antibody-coated ferrite particles: the ferrite particles (Nippon Paint Co) were coated with MoAb CPT-3F11 and then post-coated by Tris-buffer containing bovine serum albumin.

Enzyme-labeled antibody: The Fab' fragment of antibody was prepared from MoAb 9101 by pepsin digestion, followed by reduction with 2-mercaptoethylamine. The Fab' fragment obtained was conjugated with AP from calf intestine by use of N-(4-Maleimidobutyryloxy)succinimide (GMBS). The Fab' fragment labelled with AP was purified from the coupling mixture by gel filtration chromatography. Appropriate fractions for assay were pooled and used for experiments.

Assay procedure

The assay format for C-peptide was based on the two-step method and the two-site immunometric principle using two MoAbs, which recognize different epitopes.

    All evaluations of this C-peptide assay performance were performed by LUMIPULSE FORTE. The method used on LUMIPULSE FORTE was as follows: 20 µL of serum or standard was mixed with 250 µL of antibody-coated ferrite particle suspension and the mixture was incubated for 37 °C, 10 min. After B/F separation, 250 µL of Enzyme-labeled antibody was added to the ferrite particle suspension and the mixture was incubated for 37 °C, 10 min. After the second B/F separation, 200 µL of AMPPD solution was added. During the washing steps, the ferrite particles were magnetically separated from the bulk solution on the wall of the cartridge. After the enzyme reaction had proceeded for 5 min, the chemiluminescent light emission was measured for 2 s.

RESULTS
Sensitivity
The lower limit of detection calculated from mean +3 SD signal (n=20) of the zero calibrator was 0.002 ng/mL. The functional sensitivity, constructed by assaying 20 replicates of diluted calibrator and determining the dose associated with a 10% intra-assay coefficient of variation (CV), was 0.013 ng/mL (Fig. 1).

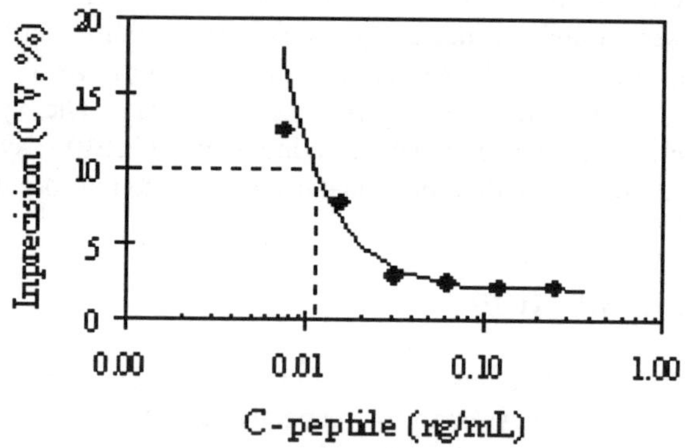

Figure 1. Precision profile for C-peptide assay
Broken lines indicate functional
sensitivity (CV=10%) of the assay

Cross-reactivity
Cross-reactivity with proinsulin and insulin was determined by measuring recombinant human proinsulin (Sigma-aldrich Co.) and WHO approved reference material (International Reference Preparation Code: 66/304, NIBSC).

    We showed reactivity to C-peptide, proinsulin and insulin of the C-peptide assay (Fig. 2). Cross-reactivity to proinsulin and insulin was <1.7% (proinsulin) and <0.03% (insulin) of C-peptide on a molar basis.

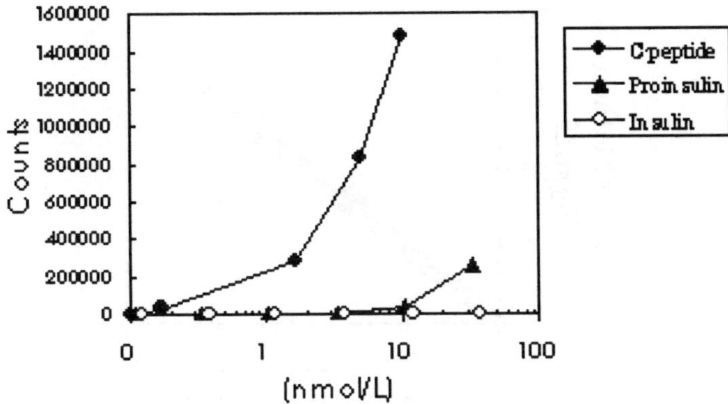

Figure 2. Cross-reactivity to proinsulin and insulin

Precision

The precision data is shown in Table 1. Three serum samples were assayed in replicates of 2, two separate times per day, for 20 testing days. Within-run precision and Total precision were calculated according to the NCCLS EP5-A protocol.

The present C-peptide assay had a within-run CV of 1.1~2.8% and a total CV of 3.6~4.2% at a mean concentration range of 1.66~22.7 ng/mL.

Table 1. Precision study

	mean (ng/mL)	Within-run SD	Within-run CV (%)	Total SD	Total CV (%)
Sample A	1.66	0.025	1.5	0.060	3.6
Sample B	6.60	0.073	1.1	0.25	3.7
Sample C	22.7	0.64	2.8	0.96	4.2

Method comparison

We compared the present method with the EIA method (AIA-PACK C-Peptide, TOSOH Corporation). The correlation of test results by the present method (y) with EIA (x) gave the following regression equation: y = 0.95x + 0.07, r = 0.995 (Fig.3).

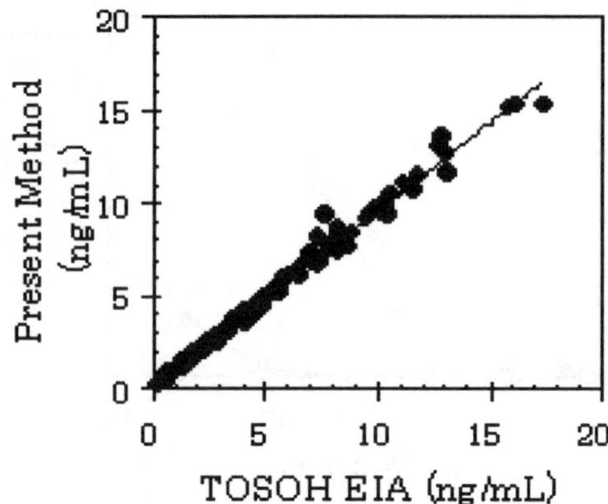

Figure 3. Comparison of present method with EIA method:
n=100, slope=0.95, intercept=0.07, r=0.995

Linearity
Dilution linearity was evaluated with three serum samples. Dilution was conducted by preparing 2 folds serial dilution series. Expected values of the diluted samples were calculated from the concentrations measured in the undiluted sample.

The linear correlation between observed and expected values was: y = 0.992x + 0.008, r = 0.999. For dilutions up to 0.02ng/mL, the deviation from the target value was less than 20%.

CONCLUSION
The use of a monoclonal antibody that recognizes the N-terminal of the C-peptide molecule made it possible to realize a low cross-reactivity to proinsulin. Our results indicate that the C-peptide assay on the LUMIPULSE system shows good specificity, sensitivity, precision and linearity. Accordingly, the C-peptide assay enables for the accurate measurement of C-peptide at low concentrations. Especially, it is useful for the presumption of residual β-cell function of diabetic patients who are nearly depleted of insulin secretion.

REFERENCES

1.   Melani F, Rubenstein AH, Oyer PE, Steiner DF. Identification of proinsulin and C-peptide in human serum by a specific immunoassay. Proc Nat Acad Sci USA 1970; 67: 148-55.
2.   Oyer PE, Cho S, Peterson JD, Steiner DF. Studies on human proinsulin. J Biol Chem 1971; 246: 1375-86.
3.   Nishizono I, Iida S, Suzuki N, Kawada H, Murakami H. eds. Rapid and sensitive chemiluminescent enzyme immunoassay for measuring tumor markers. Clin Chem 1991; 37: 1639-44.

DEVELOPMENT OF TANDEM BIOLUMINESCENT ENZYME IMMUNOASSAY FOR ANGIOTENSIN I AND ENDOTHELIN-1

KATSUTOSHI ITO*, KENTARO OHWAKI, HIDETOSHI ARAKAWA,
MASAKO MAEDA

*School of Pharmaceutical Sciences, Showa University,
1-5-8 Hatanodai, Shinagawa, Tokyo 142-8555, Japan*
**Email: itok@pharm.showa-u.ac.jp*

INTRODUCTION

We developed a simultaneous bioluminescent assay of acetate kinase (AK) and pyruvate phosphate dikinase (PPDK). In the method, the detection limits (blank + 3SD) of AK and PPDK were 1.03×10^{-20} and 2.05×10^{-20} mol/assay, respectively.[1] Previously, we successfully applied a tandem bioluminescent enzyme immunoassay (BL-EIA) for simultaneous measurement of insulin and c-peptide.[1] In this study, we also applied the method to tandem BL-EIA for Angiotensin I and Endothelin-1, which are hypertension related peptides. The tandem BL-EIA used a competitive immuno-reaction for Angiotensin I and sandwich immuno-reaction for Endothelin-1.

MATERIALS AND METHODS

Reagents

PPDK (from *Microbispora rosea subsp. Aerata*, EC 2.7.9.1) and thermostable Thr 217 Ile mutant of *Luciola cruciata* firefly luciferase (EC 1.13.12.7) were donated from Kikkoman Co. (Chiba, Japan). AK (from *B. stearothermophilus*, EC 2.7.2.1) and mouse anti-Endothelin-1 monoclonal antibodies were purchased from Seikagakukogyo Co. (Tokyo, Japan). ADP-HK (from *Pyrococcus furiosus*, EC 2.7.1.147) was from Asahi Chemical Industry Co. Ltd. (Shizuoka, Japan). D-Luciferin was purchased from Sigma Chemical Co. (St. Louis, MO). Goat anti-rabbit IgG antibody was purchased from CHEMICON International, Inc. (Temecula, CA). Endothelin-1 was from PEPTIDE INSTITUTE, INC. (Osaka, Japan). Angiotensin I, biotinylated Angiotensin I and rabbit anti-Angiotensin I antiserum were purchased from Peninsula Laboratories Inc. (San Carlos, CA).

Tandem BL-EIA for Angiotensin I and Endothelin-1

In the proposed BL-EIA, we performed sandwich and competitive immunoreaction for Endthelin-1 and Angiotensin-I in same well, respectively. Schematic illustration of the proposed tandem BL-EIA is shown in Fig. 1.

The wells of a microtiter plate were coated with a purified mouse anti-Endothelin-1 monoclonal antibody and goat anti-rabbit IgG antibody. The wells were then post-coated by adding 1 % water-soluble gelatin solution containing 0.05% NaN$_3$. The plates were stored at 4 °C prior to use. After washing the plate, 50 μL of a standard or sample solution, 50 μL of a FITC labeled anti-Endothelin-1

471

monoclonal antibody, 50 μL of biotinylated Angiotensin I solution and rabbit anti-Angiotensin I antiserum solution were added to each well. The plates were incubated for 72 h at 4 °C. After washing, 100 μL of PPDK labeled rabbit anti-FITC Fab' and 100 μL AK-streptavidin conjugate solution were added and allowed to stand for 1h at room temperature. The microtiter plate was re-washed and assayed by the simultaneous bioluminescent detection method.[1]

Briefly, 100 μL of bioluminescent reagent for AK (containing ADP, acetyl-phosphate, $MgSO_4$, luciferin and luciferase in 50 mM HEPES-KOH buffer, pH7.0) was added and incubated for 15 min at 37 °C. The bioluminescent intensity of AK was integrated for 5 s by a MicroLumat LB96P luminescent reader (EG&G Berthold, Germany). Then, to the same wells of the microtiter plate were added 100 μL of bioluminescent reagent for PPDK (containing AMP, PPi, PEP, $MgSO_4$, glucose, ADP-HK, luciferin and luciferase in 50 mM HEPES-KOH buffer, pH7.0), and re-incubated for 20 min at 37 °C The bioluminescent intensity of PPDK was integrated for 5 s.

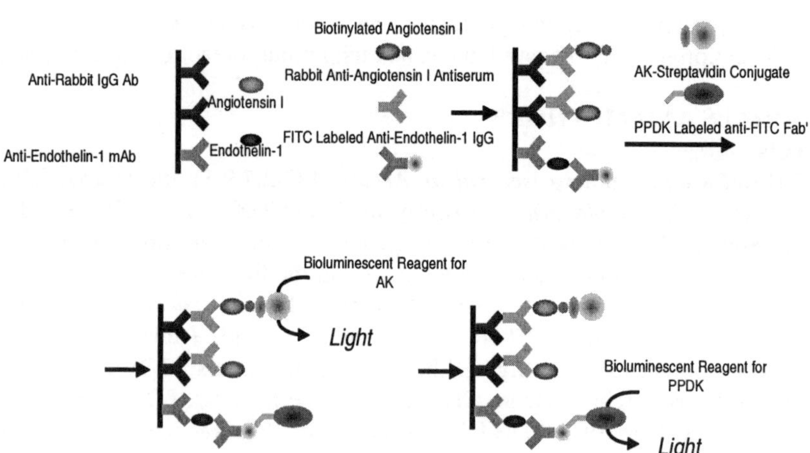

Figure 1. Schematic illustration of the proposed simultaneous BL-EIA for Angiotensin I and Endothelin-1

RESULTS AND DISCUSSION

Using the proposed BL-EIA, the measurable range for Angiotensin I and Endothelin-1 were 7.81 - 1000 pg/mL and 15.63 - 1000 pg/mL, respectively. The intra-assay coefficients of variation of Angiotensin I and Endothelin-1 at each standard point were below 9.7% and 11%, respectively. (Fig. 2)

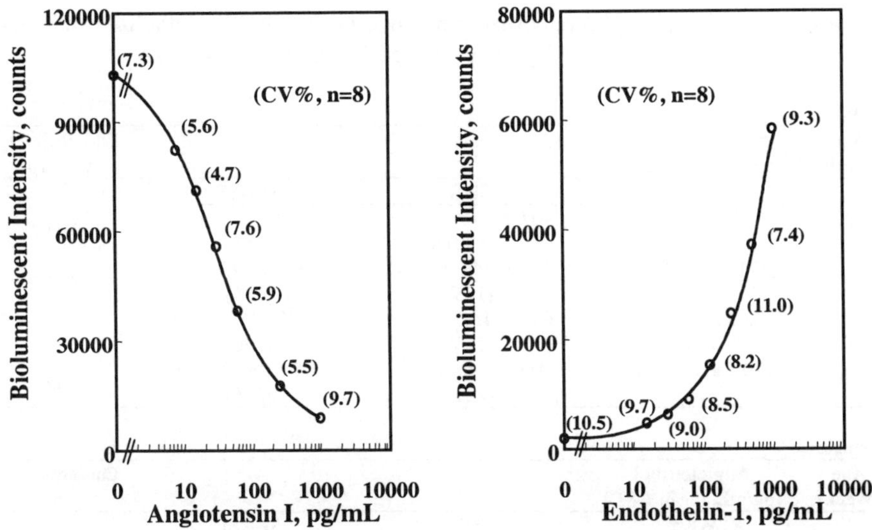

Figure 2. Standard curves for Angiotensin I and Endothelin-1
by the proposed BL-EIA

We carried out the proposed method to determine Angiotensin I and Endothelin-1 in plasma from Wistar Kyoto Rat (WKY), Spontaneously Hypertensive Rat (SHR) and Stroke-prone Spontaneously Hypertensive Rat (SHRSP), which were from Saitama Experimental Animals Supply Co., Ltd. (Sugito, Japan). The plasma samples require pretreatment to removing non-specific interference with the plasma matrix in the BL-EIA. Therefore, we attempted to eliminate the interfering substances using 10 mg of Oasis HLB cartridge (Waters Co., Milford, MA). In the recovery test, the rat pooled plasma were spiked with different levels of Angiotensin I and Endothelin-1 (0 - 500 pg/mL), then pretreated and assayed. The mean recoveries of Angiotensin I and Endothelin-1 were 81.9 ± 15.3 % and 69.0 ± 10.9 % (mean $\pm$ SD, n=18) by the proposed tandem BL-EIA, respectively.

We applied the proposed BL-EIA to the determination of plasma Angiotensin I and Endothelin-1 levels from rats of various ages (Table 1). Angiotensin I levels of SHRSP which have high systolic blood pressure were higher than those of age-matched WKY which is control rat. Further, Angiotensin I levels of SHRSP from 20 weeks old of male and 12 weeks old of female were significantly higher than those of WKY. In SHR subjects, Angiotensin I levels were not found to correlate with systolic blood pressure. Endothelin-1 levels of all rats we examined but could not be detected.

Table 1. Plasma levels of immunoreactive Angiotensin I and Endothelin-1 in WKY, SHR and SHRSP by the proposed simultaneous BL-EIA

Male

Age (Week)	Angiotensin I (pg/mL, mean ± SD)			Endothelin-1 (pg/mL)
	WKY	SHR	SHRSP	
5	151.4 ± 7.5, n=3 (145)	407.2 ± 183.6, n=3 (180)	176.0 ± 34.1, n=3 (185)	ND
8	171.4 ± 72.0, n=3 (140)	264.2 ± 16.0, n=3* (185)	266.9 ± 79.7, n=3 (215)	ND
12	202.3 ± 139.1, n=9 (140)	68.1 ± 16.7, n=9** (210)	226.7 ± 152.8, n=9 (250)	ND
20	170.1 ± 114.1, n=6 (145)	106.4 ± 66.1, n=3 (225)	515.5 ± 338.2, n=3** (275)	ND

Female

Age (Week)	Angiotensin I (pg/mL, mean ± SD)			Endothelin-1 (pg/mL)
	WKY	SHR	SHRSP	
5	175.8 ± 89.3, n=3 (135)	282.5 ± 38.2, n=3 (170)	166.5 ± 18.7, n=3 (180)	ND
8	195.2 ± 54.2, n=3 (145)	165.8 ± 58.0, n=3 (195)	241.8 ± 85.7, n=3 (215)	ND
12	141.4 ± 49.3, n=6 (145)	170.8 ± 38.9, n=3 (205)	240.7 ± 32.5, n=3** (225)	ND

Values in parenthesis are systolic blood pressure levels (mmHg), which are obtained from Saitama Experimental Animals Supply Co., Ltd. ND: not detected (< 15.6 pg/mL), *: $P<0.01$, **: $P<0.05$, compared with age-matched WKY

ACKNOWLEDGMENTS

The authors would like to thank Asahi Chemical Industry Co., Ltd., for the gift of ADP-HK. This work is supported by Grants-in-Aid for Scientific Research (C) No. 13672431 from Japan Society for the Promotion of Science, the High-Technology Research Center Project from the Ministry of Education, Culture, Sports, Science and Technology of Japan and Japan Health Sciences Foundation.

REFERENCE

1.   Ito K, Nakagawa K, Murakami S, Arakawa H, Maeda M. Highly sensitive simultaneous bioluminescent measurement of acetate kinase and pyruvate phosphate dikinase activities using a firefly luciferase-luciferin reaction and its application to a tandem bioluminescent enzyme immunoassay. Anal Sci 2003; 19: 105-9.

NEW METHODS FOR DEVELOPMENT OF FRET-BASED BIOSENSORS WITH EXPANDED DYNAMIC RANGE

T NAGAI[1,2], A MIYAWAKI[1]

[1]*Lab. for Cell Function Dynamics, BSI, RIKEN, 2-1 Wako 351-0198, Japan*
[2]*PRESTO, JST, 4-1-8 Hon-cho, Kawaguchi 332-0012, Japan*
Email: tnagai@brain.riken.jp

INTRODUCTION

Green fluorescent protein (GFP) from jelly fish *Aequorea victoria* and its color variants have revolutionized our ability to uncover the complicated detail of protein dynamics and gene activation. In addition, combination of GFPs with fluorescence resonance energy transfer (FRET) technique allows us to develop genetically-encoded fluorescent indicators that enable visualization localized molecular events within a living cell. To date, increasing number of biosensors that report concentrations of second messenger molecules and activation of signaling components have been developed and successfully used in various cell types.[1] While most indicators have cyan- and yellow-emitting fluorescent proteins (CFP and YFP) as FRET donor and acceptor, respectively, their poor dynamic range often prevents detection of subtle but significant signals. To overcome this drawback, we developed new construction methods. Here, we first show a high-throughput method for making proteolysis indicators by optimising the length of linker regions within the indicator. Second, we show a more rigorous approach that uses circularly permuted GFP variants to optimise the relative orientation of the two chromophores in the indicators. Our methods will provide an important guide for the development and improvement of indicators using GFP-based FRET.

A HIGH-THROUGHPUT METHOD FOR DEVELOPMENT OF FRET-BASED INDICATORS FOR PROTEOLYSIS[2]

SCAT3 is a FRET-based indicator for activity of caspase-3, which is composed of an enhanced cyan fluorescent protein (ECFP), a 18 amino acids linker including DEVD, the sequence for caspase-3 cleavage, and an enhanced yellow fluorescent protein with efficient maturation property (Venus).[3,4] Despite its considerable promise, however, greater responsivity of fluorescence to the proteolysis has been desired for better understanding of spatio-temporal pattern of the activation of caspase-3. To improve the dynamic range of SCAT3, eighty-eight different constructs were prepared by means of a PCR technique. They all contained floppy linkers at both sides of the DEVD sequence. Variation was introduced into the length of both the linkers (Fig. 1A). The eighty-eight different constructs were introduced to *E. coli*, which were grown on an agar plate. The bacterial colonies expressing the constructs were screened for high FRET efficiency using our home-made fluorescence image analyser. The construct with the best FRET-efficiency (64%) was revealed that its structure was ECFPΔC7-DEVD-GT-Venus, and was named SCAT3.1. The FRET signal of SCAT3.1 changed by about ten-fold during apoptotic events in mammalian cells, enabling visualization of caspase-3 activation with better spatial resolution than before (Fig. 1B-E).

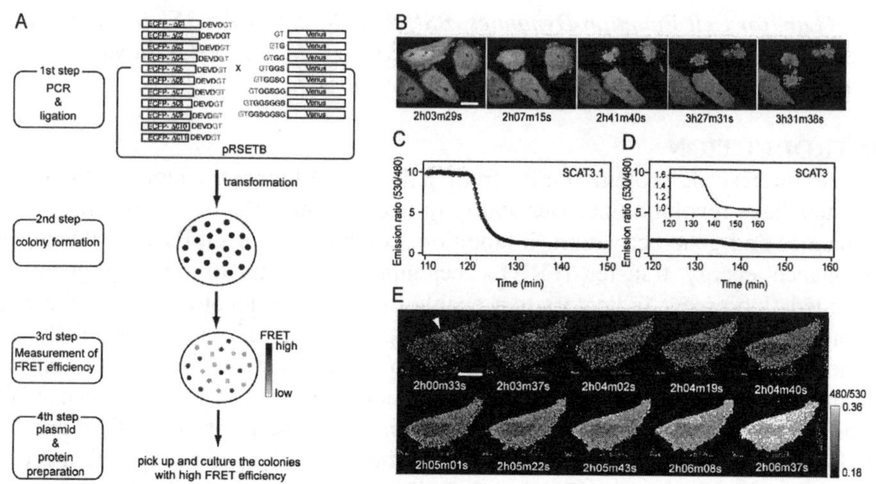

Figure 1. A method for obtaining constructs that show high efficiency of FRET from ECFP to Venus, and the performance of SCAT3.1 for visualization of caspase-3 activation in apoptotic HeLa cells. (A) An outline of the high-throughput screening procedure that is based on PCR and bacterial expression. (B) a series of confocal real color images of HeLa cells that expressed SCAT3.1 showing caspase-3 activation followed by cell-death. Time after the addition of anti-Fas monoclonal antibody is indicated below each image. (C) A time course of 530/480 nm ratio of SCAT3.1 in the experiment of B. (D) A time course of 530/480 nm ratio of SCAT3 in a similar apoptosis experiment using SCAT3-expressing HeLa cells. (E) a series of confocal images of a HeLa cell expressing SCAT3.1 showing an emerging spot of caspase-3 activation in the cytosolic compartment (arrowhead). The 480/530 nm ratio value is shown in pseudo-black to white gradation. Scale bar, 10 μm.

EXPANDED DYNAMIC RANGE OF FRET-BASED Ca²⁺ INDICATORS BY CIRCULARLY PERMUTED YELLOW FLUORESCENT PROTEINS[5]

FRET efficiency depends on the spectral overlap of the donor and acceptor, their distance from each other, and the relative orientation of the chromophore's transition dipoles. The orientation factor is usually assumed to be 2/3, a value that approximates complete random orientation. In contrast to small fluorophores that rotate freely, GFP has slower rotation in comparison with the excited-state lifetime.

Therefore, the orientation factor is not assumed to be 2/3 when using GFPs, indicating the significant impact of the relative orientation between two chromophores when using GFPs.

Yellow cameleons (YCs) are genetically-encoded fluorescent indicators for Ca^{2+} composed of a CFP, calmodulin (CaM), the CaM-binding peptide of myosin light-chain kinase (M13), and a YFP.[6] Ca^{2+} binding to CaM initiates an intramolecular interaction between CaM and M13, which changes the chimeric protein from an extended to a more compact conformation, thereby increasing the efficiency of FRET from CFP to YFP. As in case of other indicators, YCs also suffer from poor dynamic range. The best version available currently, such as YC3.12, exhibits at most a 120% change in the ratio of YFP/CFP upon Ca^{2+} binding *in vitro*.

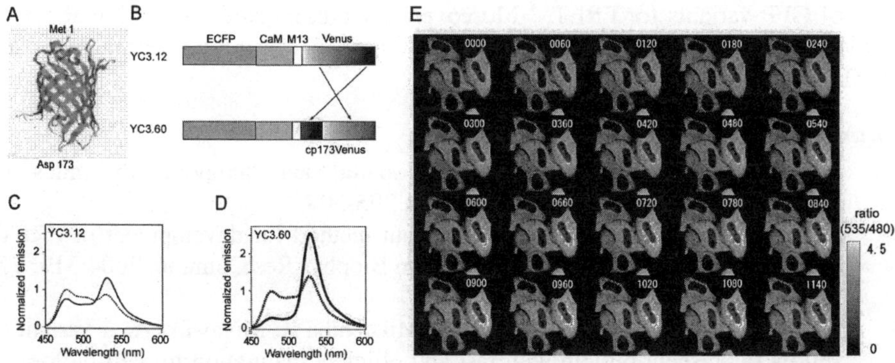

Figure 2. Improvement of yellow cameleon by using a circularly permuted YFP variant. (A) The three-dimensional structure of GFP with the positions of the original (Met1) and new N-termini (Asp173). (B) Domain structures of YC3.12 and YC3.60. CaM, *Xenopus* calmodulin; M13, a CaM binding peptide derived from myosin light chain kinase (C, D) Emission spectra of YC3.12 (C) and YC3.60 (D) (excitation at 435 nm) at zero (doted line) and saturated Ca^{2+} (solid line). (E) A series of confocal pseudo-B/W images showing propagation of $[Ca^{2+}]_c$. These images were taken at video rate (30Hz).

To achieve a Ca^{2+}-dependent large change in FRET signal, we assumed that optimization of relative angle between CFP and YFP is quite important because of the property of GFP-based FRET as mentioned above. Thus, we took a rigorous approach that used a circularly permutated GFP (cpGFP), in which the amino and carboxyl portions were interchanged and reconnected by a short spacer between the original termini.[7] By using cpVenus variants, we attempted to vary the relative orientation of the two chromophores. One of the cpYFPs incorporated in YC absorbes a great amount of excited energy from CFP in its Ca^{2+}-saturated form, thereby increasing the Ca^{2+}-dependent change in the ratio of Venus/CFP by nearly

600% (Fig. 2A-D). Both in cultured cells and in the nervous system of transgenic mice, the new YC (YC3.60) enables visualization of subcellular Ca^{2+} dynamics with better spatial and temporal resolution (100 Hz) (Fig. 2E, data not shown).

The process by which YC3.60 was conceived is a new model for the development of GFP-based indicators. An increasing number of fluorescent indicators based on FRET between CFP and YFP have been developed,[1] in which the relative position between the two chromophores of CFP and YFP is varied. Thus, the cpVenus to be used in combination with CFP should be optimized for each specific application. Also, its combined use with cpCFPs will increase further the variation of the relative position of the two transition dipoles between donor and acceptor. Since cpGFP-based indicators for Ca^{2+} were developed a few years ago, cpGFPs themselves have been expected to become powerful tools comparable to pairs of GFP variants for FRET.[8] Moreover, our present study will bring about an innovation in GFP technology through the marriage of circular permutation and FRET techniques.

REFERENCES

1. Miyawaki A. Visualization of the spatial and temporal dynamics of intracellular signaling. Dev Cell 2003; 4: 295-305.
2. Nagai T, Miyawaki A. A high-throughput method for development of FRET-based indicators for proteolysis. Biochem Biophys Res Commun 2004;319: 72-7.
3. Nagai T, Ibata K, Park ES, Kubota M, Mikoshiba K, Miyawaki A. A variant of yellow fluorescent protein with fast and efficient maturation for cell-biological applications. Nat Biotechnol 2002; 20: 87-90.
4. Takemoto K, Nagai T, Miyawaki A, Miura M. Spatio-temporal activation of caspase revealed by indicator that is insensitive to environmental effects. J Cell Biol 2003; 160: 235-43.
5. Nagai T, Yamada S, Tominaga T, Ichikawa M, Miyawaki A. Expanded dynamic range of fluorescent indicators for Ca^{2+} by circularly permuted yellow fluorescent proteins. Proc Natl Acad Sci USA 2004; 101: 10554-9.
6. Miyawaki A, Llopis J, Heim R, McCaffery JM, Adams JA, Ikura M, Tsien RY. Fluorescent indicators for Ca^{2+} based on green fluorescent proteins and calmodulin. Nature 1997; 388: 882-7.
7. Baird GS, Zacharias DA, Tsien RY. Circular permutation and receptor insertion within green fluorescent proteins. Proc Natl Acad Sci USA 1999; 96: 11241-6.
8. Nagai T, Sawano A, Park ES, Miyawaki A. Circularly permuted green fluorescent proteins engineered to sense Ca^{2+}. Proc Natl Acad Sci USA 2001; 98: 3197-202.

CHAGAS ASSAY USING RECOMBINANT ANTIGENS ON A FULLY AUTOMATED CHEMILUMINESCENCE IMMUNOASSAY ANALYZER

D SHAH, C-D CHANG, K CHENG,
L JIANG, V SALBILLA, A HALLER, G SCHOCHETMAN
Abbott Laboratories, Abbott Park, IL 60064 USA
E-mail: Dinesh.shah@abbott.com

INTRODUCTION

Acridinium (Ac) derivatives have been utilized for highly sensitive immunoassays because of their stability and high chemiluminescence yield. Abbott PRISM® is a high throughput, fully automated serological analyzer to screen plasma or sera for Heptitis B surface antigen, antibodies to Heptitis B core, HCV, HIV, and HTL.V.[1,2] The instrument uses acridinium labels to tag analytes captured on microparticles, buffer to wash nonspecific binding, H_2O_2 to trigger chemiluminescence, and a photon multiplier to collect photon counts. Data are processed by a computer to align with sample barcodes and to sort out the positive samples. The total assay time is under 1 hour and 160 samples/hour can be processed.

The Chagas' disease (American trypanosomiasis), caused by the protozoan *Trypanosoma cruzi*, is endemic to most regions of the Latin Americas. Of the estimated 18 million infected people, approximately 50,000 die from this disease yearly. Transfusion of blood from infected donors has become a major route for contracting the disease. The estimated seroprevalence of the disease in blood donor populations of the United Stated is as high as 0.48% and the trend is increasing with the Hispanic population.[3] Concerns on the safety of blood in the U.S. have been raised because they represent a growing donor population but also a largely "silent" reservoir of *T. cruzi*.

Laboratory diagnosis of Chagas' disease is complex, primarily because of the genetically diverse and polymorphous parasite. PCR is not always able to detect the specific DNA because of intermittent or low levels of parasites in the blood stream during the chronic stage. Radio-immuno-precipitation assay (RIPA), a highly specific test with easily interpreted results, has been a confirmatory test used in the U.S. However, RIPA is not appropriate for ordinary laboratories because it not only requires working with live parasites and radioactive label, but also is labor-intensive and expensive to perform.[4] In contrast, serologic tests detecting antibodies to *T. cruzi* are well-suited for fast and inexpensive diagnosis of the disease. A *T. cruzi* lysate based Chagas assay on the PRISM® analyzer was reported previously.[5] However, due to many advantages with recombinant antigens (rAg) over crude lysate, such as quality control and reproducibility, we switched to the use of rAg of *T. cruzi* for the development of Chagas assay.

MATERIALS AND METHODS
Assay Reagents, Controls, and Samples

<u>Solid-phase</u>: A blend of several different rAg coated microparticles, each kind of microparticle was coated separately with a rAg of *T. cruzi*.

<u>Acridinium Conjugate</u>: Prepared from a Mab anti-human IgG conjugated with 10-(3-sulfopropyl)-N-tosyl-N-(2-carboxyethyl)-9-acridinium carboxamide via N-hydroxysuccinimide ester in a phosphate buffer. After conjugation, the conjugate was sized by HPLC to a pool of fractions with Ac/IgG molar ratios ranging from 3 to 8.

<u>Specimen Diluent Buffer</u> (SDB): SDB contained surfactants, blockers, and buffer salts to enhance specific binding and minimize nonspecific binding.

<u>Controls</u>: Negative Calibrator (NC) is re-calcified negative plasma. Chagas Positive Control (PC) was diluted from pooled plasma of Chagas' patients; each patient was confirmed serologically by at least 2 different tests.

<u>Chagas Positive Specimens</u>: 228 Chagas positive specimens were obtained from American Red Cross, Boston Biomedica, Inc. (BBI, West Bridgewater, MA), BioClinical Partners (Franklin, MA), Teragenix (Ft. Lauderdale, FL) and Goldfinch (Iowa City, Iowa). The human serum or plasma specimens were collected from donors covering many regions, including Central and Southern American countries, as well as United States. These positive specimens were confirmed by RIPA or 2 to 3 different immunotests.

Assay Format

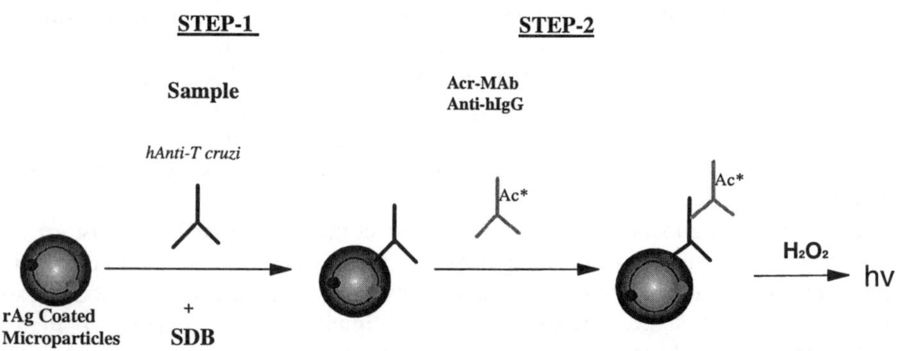

Figure 1. ABBOTT PRISM® Chagas Assay Format

Assay steps and reactions are illustrated in Fig. 1. In Step-1, the assay starts with 100 µL of sample and incubates with 50 µL of SDB and 50 µL of *T. cruzi* rAg coated microparticles in the sample well of a PRISM reaction tray. After the first incubation, the tray is moved to the transfer station where the reaction mixture is flushed into the sandwich reaction well by transfer wash buffer and excessive fluid is absorbed by a blotter underneath. In Step-2, 50 µL of the acridinium anti-human IgG conjugate is dispensed to the reaction well at the conjugate dispensing station. After the 2nd incubation, unbound conjugates are washed into the blotter. 50 µL of an alkaline hydrogen peroxide solution is then injected at the trigger/read station to trigger chemiluminescence from acridinium labels captured in the reaction well. The intensity of chemiluminescent signal is proportional to the amounts of anti-*T. cruzi* in the sample.

RESULTS AND DISCUSSION

The cut-off calculation is based on mean net NC counts plus 0.15 times the mean net PC counts. If a sample response with net counts is equal to or greater than the cut-off, i.e. S/CO value $\geq$ 1.0, it is reactive; if less than the cut-off, it is negative. However, if a sample S/CO value is between 0.9 and 1.0, it is considered "gray zone" and will be re-tested in duplicate.

The Chagas assay was tested on 7,258 unscreened serum and plasma specimens from the southern US, Florida, and California through Gulf Coast Regional Blood Center (Houston, TX) and ProMedDx (Norton, MA). Nine (0.12%) specimens were repeatedly reactive, but non-confirmable; hence, the prototype Chagas assay had specificity of 99.88%. A typical histogram of signal distribution on a negative population is shown in Figure 2.

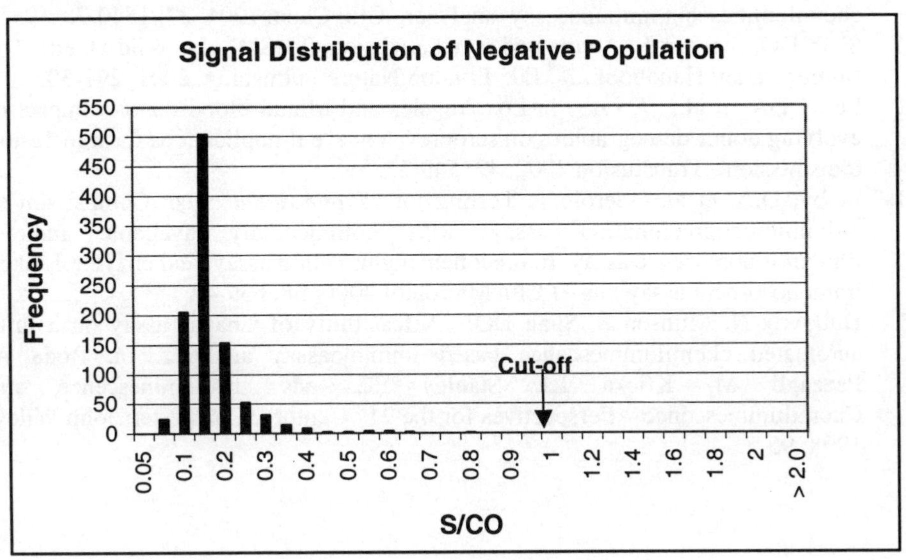

Figure 2. A typical histogram of signal distribution

In further assessment on the specificity, several panels of human sera or plasma specimen with other diseases (e.g. Leishmaniasis, auto-immune, multi-myeloma, toxoplasmosis, and syphilis) or possible interferencing substances (e.g. bilirubin and triglycerides) were tested and found non-reactive in this assay.

Due to the genetically diverse and polymorphous parasite, sensitivity of the prototype assay was assessed with various positive specimens from donors covering many regions of Latin Americas and southern US. 228 out of 228 specimens were detected as reactive or 100% sensitivity. Optimization to further improve the assay performance is in progress. The Abbott PRISM® Chagas Assay is potentially a screening test to improve the safety of the blood supply by reducing the risk of *T. cruzi* transfusion.

ACKNOWLEDGEMENTS

We thank D. Leiby of American Red Cross and L. Kirchhoff of Goldfinch Inc for providing RIPA confirmed Chagas positive human specimens for our evaluation. Furthermore, we appreciate their help in using RIPA to confirm true positive samples with antibodies to T. *cruzi.*

REFERENCES

1. Khalil OS, Zurek TF, et al. Abbott Prism®: A multichannel heterogeneous chemiluminescent immunoassay analyzer. Clin Chem 1991; **37**: 1540-7.
2. Shah DO, Stewart J. Automated panel analyzers PRISM®. In: Wild D. ed. The Immunoassay Handbook, 2nd Ed. London:Nature Publishing, 2001; 297-303.
3. Leiby DA, et al.: *T. cruzi* in Los Angeles and Miami blood donors: impact of evolving donor demographics on seroprevalence and implications for transfusion transmission. Transfusion 2002; 42: 549-55.
4. Leiby, D.A. et al.: Serologic Testing for *Trypanosoma cruzi*: Comparison of radioimmunoprecipitation assay with commercially available indirect immunofluorescence assay, Indirect hemagglutination assay, and enzyme-Linked immunosorbent assay kits. J Clin Microbiol 2000; 38: 639-42.
5. Dubovoy N, Munson S, Shah DO: A feasibility of Chagas assay on a fully automated chemiluminescence based immunoassay analyzer. In: Roda A, Pazzagli M, Kricka LJ, Stanley PE. eds. Bioluminescence and Chemiluminescence – Perspectives for the 21st Century. Chichester: John Wiley, 1998; 95-8.

DEVELOPMENT OF THE ENZYME IMMUNOASSAY USING NEW CHEMILUMINESCENCE SUBSTRATE

M YAMADA[1], M MATSUMOTO[2], N WATANABE[2]

[1] TOSOH Corporation, 2743-1, Hayakawa, Ayase-shi, Kanagawa, 252-1123, Japan
[2] Department of Materials Science, Kanagawa University, Tsuchiya, Hiratsuka-shi, Kanagawa, 259-1293, Japan
Email:ma_yamad@tosoh.co.jp

INTRODUCTION

Various 1,2-dioxetane derivatives have been synthesized and reported. Among them, AMPPD (Adamantyl Methoxy Phenyl Phosphoryl Dioxetane) is a well known dioxetane bearing a spiroadamantyl group at the 3-position and a phenol phosphate at the 4-position, which is now used for chemiluminescence enzyme immunoassay (CLEIA) using alkaline phosphatase with highly sensitive detection. [1-5]

These 1,2-dioxetanes do not, however, necessarily satisfy the demands, such as high thermal stability, easiness for handling, and high light yield to use in the field of clinical applications. [6-9]

METHODS & RESULTS

After an extensive study, we realized a new 1,2-dioxetane derivative having a fused furan ring and a bulky substituent, t-butyl group, and a phenol phosphate, 5-t-butyl-4,4-dimethyl-1-(3'-phosphoryloxy)phenyl-2,6,7-trioxabicyclo[3.2.0]heptane disodium salt (Scheme 1), [10] with superiority to conventional 1,2-dioxetane derivatives.

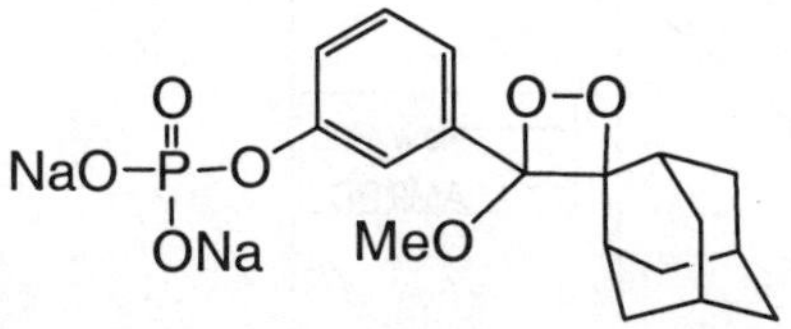

AMPPD

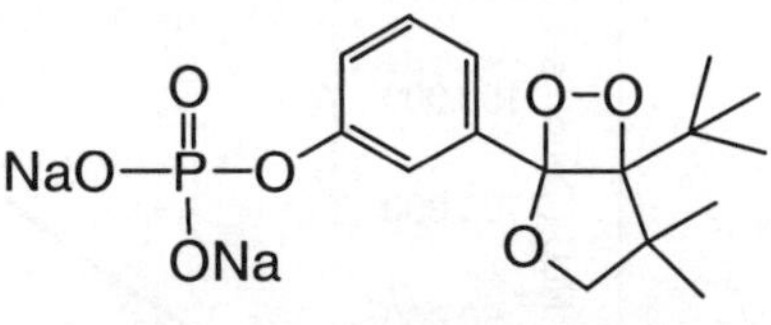

New 1,2-Dioxetane Derivative

Scheme 1. Structure of 1,2-dioxetane derivatives

For a new dioxetane with fused furan ring, steric repulsion between t-butyl and methyls should prevent twisting of the dioxetane ring so that the dioxetane becomes stable.

Enhancer for 1,2-dioxetane derivatives is composed of hydrophobic compound and fluorescent dye. We tested a variety of enhancers and found a better enhancer for this substrate, namely tetra-alkyl phosphonium derivative as hydrophobic compound with fluorescein as fluorescent dye. Using such enhancer, the substrate of the new 1,2-dioxetane derivative was found to have a very high light emitting efficiency and to be quite stable and intact for more than one month at 40 °C and for more than one year at 10 °C as shown in Figure 1.

Furthermore, this substrate was found to have light yield more than six times higher than that of Lumi Phos 530 (AMPPD) after 5 minutes incubation with alkaline phosphatase as shown in Figure 2.

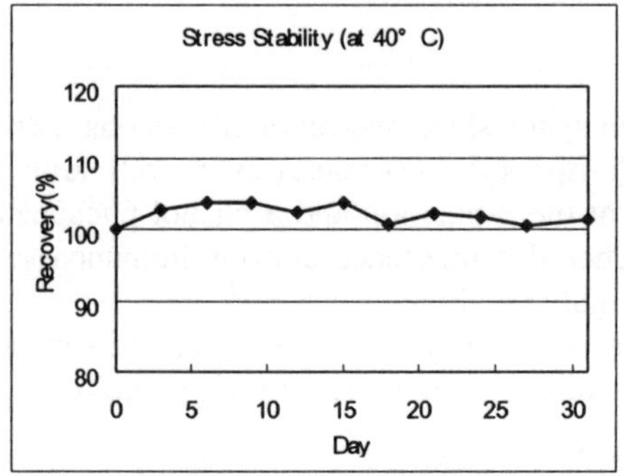
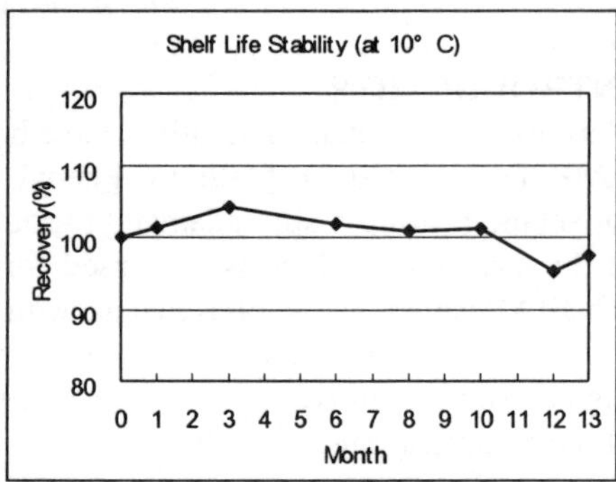

Figure 1. Stress stability (at 40 °C) and shelf life stability (at 10 °C) of the new 1,2-dioxetane derivative

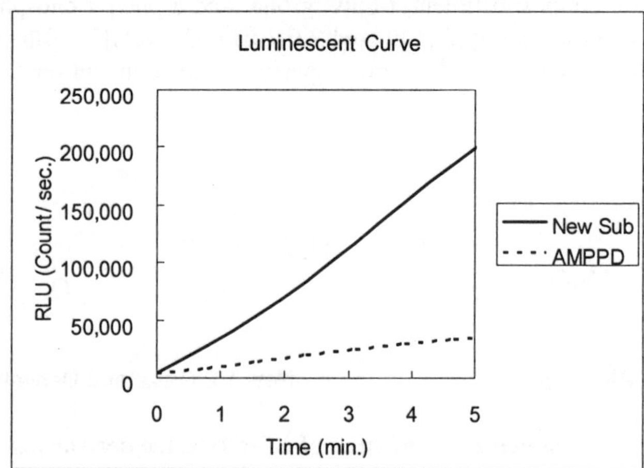

Figure 2. Luminescent curve of new 1,2-dioxetane derivative and AMPPD (During 5 min incubation with alkaline phosphatase assay)

Using the substrate of 1,2-dioxetane derivative in combination with an enhancer, we applied a chemiluminescence enzyme immunoassay (CLEIA) for TSH (Thyroid Stimulating Hormone).

We used an anti-TSH monoclonal antibody bound magnetizable microparticles as solid phase and alkaline phosphatase-labelled anti-TSH monoclonal antibody as a tracer. After 6 min incubation of solid phase, a tracer and sample or calibrator (30 μL), microparticles were washed to remove unbound materials and were then incubated for 5 min with 50 μL of chemiluminescence substrate, new 1,2-dioxetane derivative. The amount of enzyme-labeled monoclonal antibody that binds to the microparticles is directly proportional to the TSH concentration in the test sample.

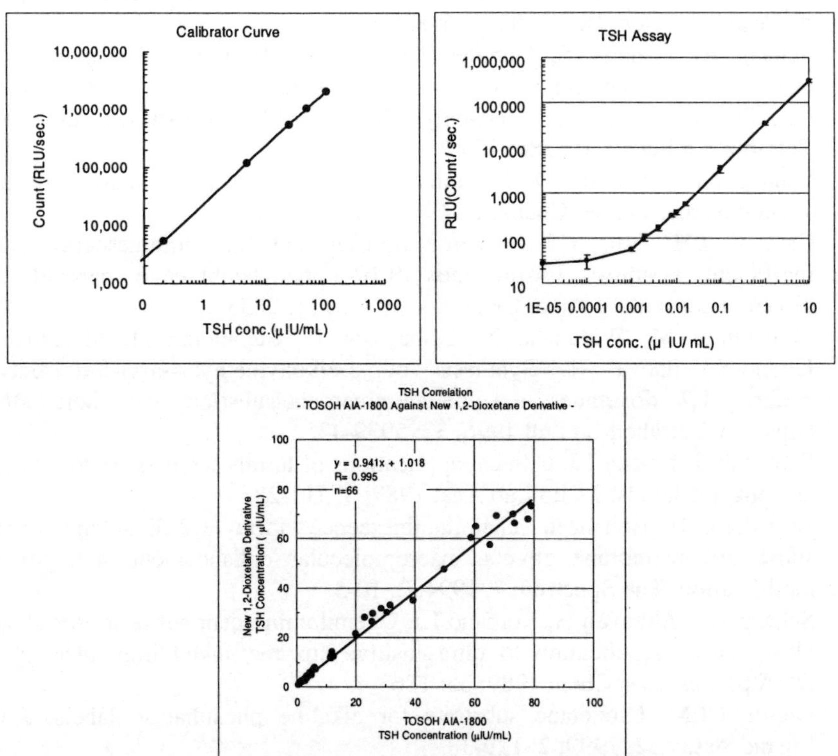

Figure 3. Calibrator curve of TSH, sensitivity of TSH using chemiluminescence enzyme immunoassay and TSH correlation against TOSOH AIA-1800

In this assay, the minimum detectable concentration (MDC) of TSH was estimated to be less than 0.001 μIU/mL TSH. The MDC is defined as that concentration of TSH which corresponds to the relative luminescent unit that is two standard deviations from the mean relative luminescent unit of a zero calibrator.

Furthermore, this chemiluminescence enzyme immunoassay of TSH was found to have a good correlation against TOSOH AIA-1800 as shown in Figure 3.

CONCLUSION

In conclusion, our substrate comprised of a new 1,2-dioxetane derivative and an enhancer was found to have excellent characteristics with a high thermal stability and a very high light yield which can be well applied in clinical usage. CLEIA using new chemiluminescence substrate showed highly sensitive immunoassay.

REFERENCES

1. Adam W, Encamcion L, Zinner K. Thermal stability of spiro[adamantine [1,2] dioxetanes]. Chem Ber, 1983; 116: 839-46.
2. Schaap AP, Chen TS, Handley RS, DeSilva R, Giri BP. Chemical and enzymatic triggering of 1,2-dioxetanes. 2: Fluoride-induced chemiluminescence from tert-butyl dimethylsililoxy-substituted dioxetanes. Tetrahedron Lett 1987; 28: 1155-8.
3. Schaap AP, Gagnon SD. Chemiluminescence from a phenoxide-substituted 1,2-dioxetane. J Amer Chem Soc 1982; 104: 3504-6.
4. Catalani LH, Wilson T. Electron transfer and chemiluminescence. Two inefficient systems: 1,4-dimethoxy-9,10-diphenylanthracene peroxide and diphenyl peroxide. J Amer Chem Soc 1989; 111: 2633-9.
5. Matsumoto M, Watanabe N, Kobayashi H, Suganuma H, Matsubara J, Kitano Y, Ikawa H. Synthesis of 3-alkoxymethyl-4-aryl-3-tert-butyl-4-methoxy-1,2- doxetanes as a chemiluminescent substrate with short half-life emission. Tetrahedron Lett 1996; 37: 5939-42.
6. Bronstein I, Kricka LJ. Clinical applications of luminescent assay for enzymes and enzyme labels. J Clin Lab Anal 1989; 3: 312-22.
7. Bronstein I, Enhanced chemiluminescence form 1,2-dioxetane enzyme substrates: membrane effects, macromolecular enhancement, and structure modification. The Spectrum 7, 1994; 2: 10-5.
8. Schaap AP, Akhavan H, Romano LJ. Chemiluminescent substrate for alkaline phosphatase: Application to ultrasensitive enzyme-linked immunoassay and DNA probes. Clin Chem 1989; 35: 1863-4.
9. Olesen CEM. Dioxetane substrate for alkaline phosphatase labels. J Clin Ligand Assay 22, 1999; 2: 129-38.
10. Matsumoto M, Watanabe N, Kasuga NC, Hamada F, Tadokoro K. Synthesis of -alkyl-1-aryl-4,4-dimethyl-2,6,7-trioxabicyclo[3.2.0]heptanes as a chemiluminescent substrate with remarkable thermal stability. Tetrahedron Lett 1997; 38: 2863-6.

DEVELOPMENT OF A NEW CHEMILUMINESCENCE SUBSTRATE FOR THE ENZYME IMMUNOASSAY

M YAMADA[1], K KITAOKA[1], M MATSUMOTO[2], N WATANABE[2]

[1] TOSOH Corporation, 2743-1, Hayakawa, Ayase-shi, Kanagawa, 252-1123, Japan
[2] Department of Materials Science, Kanagawa University, Tsuchiya,
Hiratsuka-shi,Kanagawa, 259-1293, Japan
Email:ma_yamad@tosoh.co.jp

INTRODUCTION

Various 1,2-dioxetane derivatives have been synthesized and reported. Among them, one bearing a spiroadamantyl group at the 3-position and a phenol phosphate at the 4-position named AMPPD (Adamantyl Methoxy Phenyl Phosphoryl Dioxetane) is well known.[1-5] These 1,2-dioxetanes do not, however, necessarily satisfy the analytical demands, namely high thermal stability, easiness for handling, and high light yield in an aqueous solution, to use in the field of clinicalapplications.[6-9]

METHODS & RESULTS

Upon an extensive study, we found a new chemiluminescence substrate for the enzyme immunoassay using a new 1,2-dioxetane derivative, 5-t-butyl-4,4-dimethyl-1-(3'-phosphoryloxy)phenyl-2,6,7-trioxabicyclo[3.2.0]heptane disodium salt as shown in Scheme 1.[10]

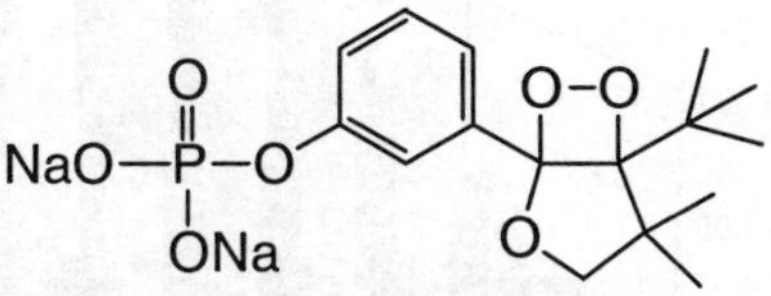

Scheme 1. Structure of new 1,2-dioxetane derivative

Since the light yield of 1,2-dioxetane derivatives are well known to decrease in an aqueous solution, we wanted to optimise the combination of several different hydrophobic compounds and fluorescent dyes for enhancing light yield in an aqueous solution. For hydrophobic compound, we tested inclusion compounds, polymers, proteins, cationic surfactants, anionic surfactants and nonionic surfactants. We selected hexadecyltributyl phosphonium bromide as hydrophobic compound. For fluorescent dye, we tested several fluorescent dyes (fluorescein, uranine, BODIPY, Oregon green 488, Oregon green 514, Rhodol green, Alexa Fluor 488 hydrazide). We selected fluorescein as fluorescent dye. In combination with hexadecyltributyl phosphonium bromide and fluorescein as shown in Scheme 2, the new 1,2-dioxetane derivative gave the best light emitting efficiency among those tested.

488   *Yamada M et al.*

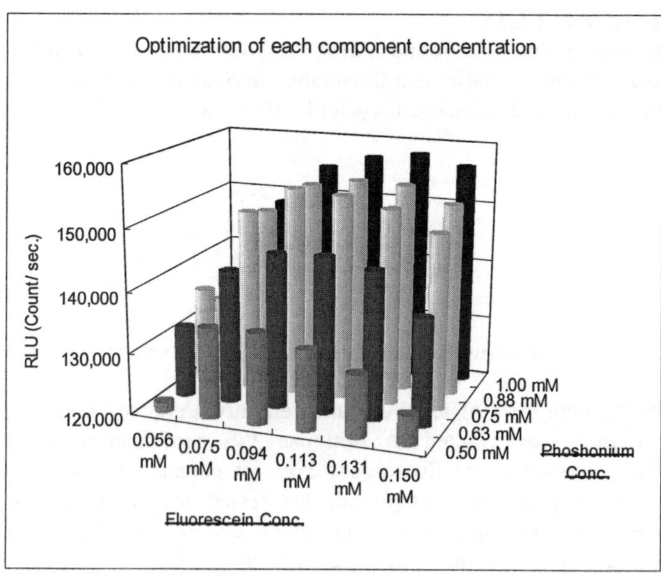

Hexadecyltributyl phosphonium bromide
(Hydrophobic Compound)

Fluorescein
(Fluorescent Dye)

Scheme 2. Enhancer of new chemiluminescence substrate

After selection of the enhancer for the new chemiluminescence substrate, we tested for further optimization. In this study, we found that 1.0 mM hexadecyltributyl phosphonium bromide and 0.13 mM fluorescein are the best concentrations for new chemiluminescence substrate.

Upon this optimization of each component concentration, the new substrate has a high light yield as shown in Figure 1.

Furthermore, we tested pH dependency for the new chemiluminescence substrate and found that pH 10.0 to give the highest light emitting efficiency as shown in Figure 1.

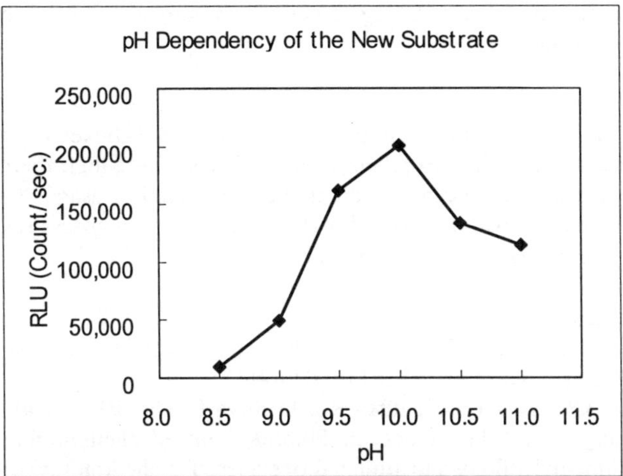

Figure 1. Optimization of each component concentration (previous page) and pH dependency of the new chemiluminescence substrate

For the chemiluminescence enzyme immunoassay (CLEIA) application, high light yield is not an essential requirement but low background is essential.

Since heavy metals cause decomposition of a 1,2-dioxetane ring as non-enzymatic decomposition of substrate, their presence results in a high background, so we used chelating ion exchange resin to remove mainly heavy metals. After the purification, high background is not found as is shown in Figure 2.

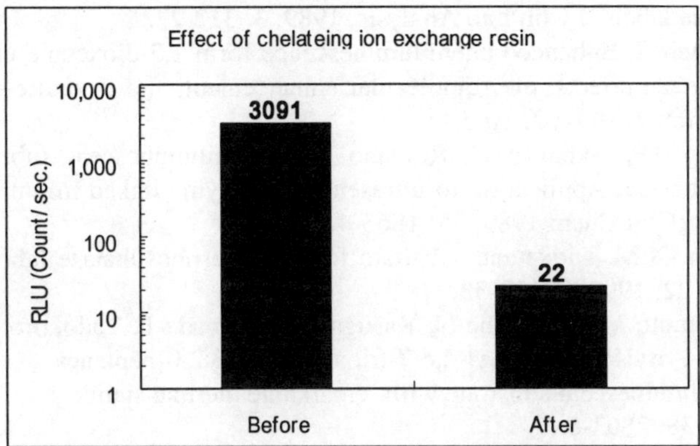

Figure 2. Effect of chelating ion exchange resin
(Background measurement)

Furthermore, this substrate was found to be stable for more than one month at 40 °C and for more than one year at 10 °C as presented in this volume 'Development of the Enzyme Immunoassay Using New Chemiluminescence Substrate', Yamada, Matsumoto & Watanabe.

CONCLUSION
In conclusion, our substrate comprised of a new 1,2-dioxetane derivative and an enhancer was found to have excellent characteristics with a high thermal stability and a very high light yield which can be well applicable in clinical usage. CLEIA using a new chemiluminescence substrate showed highly sensitive immunoassay with TSH (Thyroid Stimulating Hormone) even less than 0.001 μIU/mL.

REFERENCES

1.   Adam W, Encamcion L, Zinner K. Thermal stability of spiro[adamantine [1,2] dioxetanes]. Chem Ber 1983; 116: 839-46.
2.   Schaap AP, Chen TS, Handley RS, DeSilva R, Giri BP. Chemical and enzymatic triggering of 1,2-dioxetanes. 2: Fluoride-induced chemiluminescence from tert-butyl dimethylsililoxy-substituted dioxetanes. Tetrahedron Lett 1987; 28: 1155-8.
3.   Schaap AP, Gagnon SD. Chemiluminescence from a phenoxide-substituted 1,2-dioxetane. J Amer Chem Soc 1982; 104: 3504-6.
4.   Catalani LH, Wilson T. Electron transfer and chemiluminescence. Two inefficient systems: 1,4-dimethoxy-9,10-diphenylanthracene peroxide and diphenyl peroxide. J Amer Chem Soc 1989; 111: 2633-9.
5.   Matsumoto M, Watanabe N, Kobayashi H, Suganuma H, Matsubara J, Kitano Y, Ikawa H. Synthesis of 3-alkoxymethyl-4-aryl-3-tert-butyl-4-methoxy-1,2-doxetanes as a chemiluminescent substrate with short half-life emission. Tetrahedron Lett, 1996; 37: 5939-42.
6.   Bronstein I, Kricka LJ. Clinical applications of luminescent assay for enzymes and enzyme labels. J Clin Lab Analysis, 1989; 3: 312-22.
7.   Bronstein I, Enhanced chemiluminescence form 1,2-dioxetane enzyme substrates: Membrane effects, macromolecular enhancement, and structure modification. The Spectrum 7, 1994; 2: 10-5.
8.   Schaap AP, Akhavan H, Romano LJ. Chemiluminescent substrate for alkaline phosphatase: Application to ultrasensitive enzyme-linked immunoassay and DNA probes. Clin Chem 1989; 35: 1863-4.
9.   Olesen CEM. Dioxetane substrate for alkaline phosphatase labels. J Clin Ligand Assay 22, 1999; 2: 129-38.
10.   Matsumoto M, Watanabe N, Kasuga NC, Hamada F, Tadokoro K. Synthesis of -Alkyl-1-aryl-4,4-dimethyl-2,6,7-trioxabicyclo[3.2.0]heptanes as a chemiluminescent substrate with remarkable thermal stability. Tetrahedron Lett 1997; 38: 2863-6.

CHEMILUMINESCENT IMMUNOMETRIC DETECTION OF SARS-COV IN SERA AS AN EARLY MARKER FOR THE DIAGNOSIS OF SARS

XIAOLIN YANG[1], XUDONG SUN[2]

[1]People's Hospital of Peking University, Beijing, 100044, China
Email: yangzhng@public3.bta.net.cn
[2]Weixiao Biological Technology Development Co. Ltd. Beijing, 100176, China

INTRODUCTION

Severe Acute Respiratory Syndrome (SARS) is a new epidemic with high lethality and infectivity.[1,2,3] It has already had catastrophic consequences last year in China and some other Asian countries,[4] and it still acts as a threat to global public health. Although the SARS-associated Coronavirus (SARS-CoV) has already been identified as its pathogen,[1,2,3,5] the capability of vaccines is still in question. The focus of natural infection is not yet very clear. So the only best way to prevent and control its spread will be to isolate the SARS cases from others before the disease transmission can occur. Unfortunately, none of the current assays for SARS are suitable as they are costly, of long duration and low sensitivity. For example at least 10 days is needed for antibody detection, and the instability for RT/PCR.[6] Hence the development of new laboratory techniques with high sensitivity for early stage diagnosis is essential. We have developed an enhanced chemiluminescent immunoassay with the monoclonal antibodies to Nucleocapsid (N) protein of SARS-CoV, so as to detect the SARS-CoV directly from clinical cases in their early stage.

MATERIALS AND METHODS

The MPC-1 luminometer and software were supplied by Weixiao Biological Technology Development Co. Ltd. (Beijing, China). The monoclonal antibodies to N protein of SARS-CoV were supplied by Central Lab of Pearl River Hospital of No.1 Medical University of PLA (Guangzhou, China). The microwells were purchased from NUNC (Denmark). The deactivated SARS-CoV and other virus were supplied by National Institute of Drug and Biological Products Identification of China. The enhanced chemiluminescence substrates were prepared as the techniques we developed previously.[7] The HRP labeled antibody was prepared by the routine protocol in our laboratory.[7] The EIA kit to detect SARS-CoV antibody was purchased from GBI Biotechnology Co. Ltd. (Beijing, China). The sera of SARS cases of were supplied by Medical Division of Peking University, Center of Disease Control and Prevention (CDC) of Beijing, CDC of Guangzhou, Institute of Microbiology and Epidemiology of the Military Medical Academy of PLA (Beijing, China) respectively. All other reagents were commercial products at AR grade.

The procedure of chemiluminescent immunoassay was as follows: The coating of microwell with 10 μL/mL antibody to N protein was performed as described by us earlier.[7] Then 50 μL of 1% BSA and serum were added into the microwell; after 60 min incubation, all of unbound materials was removed by a washing procedure. Then 100 μL of another HRP labeled antibody to N protein was added, followed by another 60 min incubation. Finally, the 100 μL enhanced chemiluminescent substrates were added after the second washing. The signal was detected for 1 s per well by a luminometer. Normal sera acted as the negative control, the 2.1 in S/N ratio (Sample/negative control) was confirmed as the cut-off to determine the positive results.

All of the operations relevant to the sera of SARS cases were done in Bio-Safe Lab P-3, or P-2 after the sera were deactivated by 30 min incubation at 56 °C.

RESULTS AND DISCUSSION

Table.1 displays the results to detect deactivated SARS-CoV and other viruses, it showed absolutely detectability to all of SARS-CoV species, while no cross reaction to other virus, especially to common human coronal Virus (HCoV-OC43 and HCoV-229E) was found. So the sensitivity and specificity of this immunoassay was fully acceptable. According to the data obtained from US CDC strain, its detectable limit was nearly 3 copies of virus per 50 μL test. This extraordinary high sensitivity seemed to indicate the additional N protein that much more than whole virus was excreted in SARS-CoV culture. Nevertheless it seems that the N protein is a sensitive marker for SARS-CoV proliferation.

Table 1. The results of sensitivity and specificity tests

SARS-CoV species	**Results**	**Other virus**	**Results**
Hong Kong	+	Measles virus	-
Guangzhou 1	+	Mumps virus	-
Guangzhou 2	+	Rubella virus	-
Beijing 1	+	Influenza virus	-
Beijing 2	+	HCoV-OC43	-
WHO: $5*10^4$geq/mL	+	HCoV-229E	-
US CDC: $5*10^2$PFU/mL 2.5$*10^2$PFU/mL 1.3$*10^2$PFU/mL 6.3$*10^1$PFU/mL	+ + + +	Rhinovirus	-
		RSV	-
		rotavirus	-
		IBV	-
		CCV	-

Table 2 shows the results for 19 serial samples of sera from SARS cases that their SARS antibodies were positive. Because the specific antibody detection is always regarded as the golden standard to confirm a case of the infectious disease, so these data indicate that the assay is suitable for this purpose. It is obvious that the immunoassay to SARS-CoV showed very high detectability in early stage of the illness, especially during 6-10 days.

Table 2. The results to measure the serialized sera from the SARS cases confirmed by antibody detection

Days of illness	1-5	6-10	11-15	16-20	21-25	26-30	30—
Positive ratio	75%	100%	71%	50%	0%	0%	0%

Fig. 1 shows the time courses of growth and decline, both for SARS-CoV as well as its antibody to 351 sera of SARS cases that were only identified by clinical symptoms. Comparing with antibody detection, this technique also represented more satisfactory method for early diagnosis of SARS.

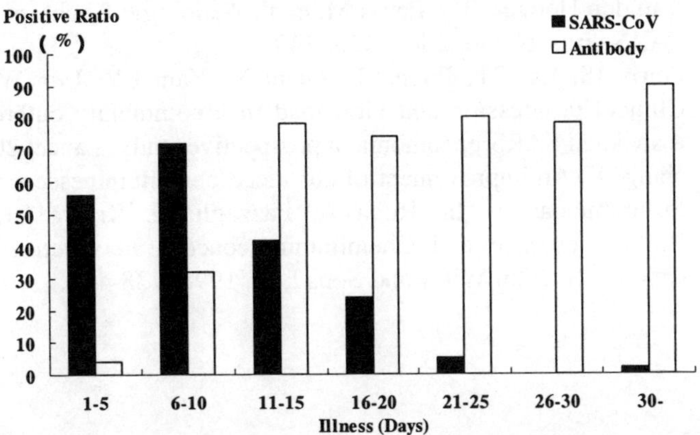

Figure 1. The time courses of SARS-CoV and its antibody in sera of clinical cases

All of the data above strongly suggested that this enhanced chemiluminescent immunoassay perhaps could play a very important role for laboratory diagnosis of SARS in the early stages of infection. In addition this assay can be applied in other fields with great convenience too, such as animal fluids and environmental collections, so probably it also will act as one of essential tools to discover the mechanism of natural infection of SARS.

ACKNOWLEDGEMENTS
Special thanks to Dr. Xiaoyan Che for her kind help, including the supply of antibodies and other important materials.

Recently, this technique was approved for clinical diagnosis by China FDA.

REFERENCES

1. Drosten C, Gunther S, Preiser W, Van Der Werf S, Brodt HR, Becker S, Rabenau H.et al. Identification of a novel coronavirus in patients with severe acute respiratory syndrome. N Engl J Med;2003, 348: 1967-76.
2. Ksiazek TG, Erdman D, Goldsmith CS, Zaki SR, Peret T, Emery S, Tong S. et al. A novel coronavirus associated with severe acute respiratory syndrome. N Engl J Med 2003; 348: 1953-66.
3. Peiris JS, Lai ST, Poon LL, Guan Y, Yam LY, Lim W, Nicholls J. et al. Coronavirus as a possible cause of severe acute respiratory syndrome. Lancet 2003; 361: 1319-25.
4. Jonathan K. Researchers get to grips with cause of pneumonia epidemic. Nature 2003; 422, 547-8.
5. Fouchier RA, Kuiken T, Schutten M, Van Amerongen G, Van Doornum GJ, Van den Hoogen BG, Peiris M. et al. Aetiology: Koch's postulates fulfilled for SARS virus. Nature 2003; 423: 240.
6. Peiris JS, Lai ST, Poon LL, Guan Y, Yam LY, Lim W, Nicholls J. et al. Clinical progression and viral load in a community outbreak of coronavirus-associated SARS pneumonia: a prospective study. Lancet 2003; 361: 1767-72.
7. Yang, X. An improvement of enhanced chemiluminescence and its application to immunoassay. In: Roda A, Pazzagli M, Kricka LJ, Stanley PE. eds. Bioluminescence and Chemiluminescence - Perspectives for 21st Century: Chichester: John Wiley and Sons Ltd., 1998; 138-41.

DEVELOPMENT AND VALIDATION OF AN AVIDIN-BIOTIN CHEMILUMINESCENCE ELISA FOR THE QUANTATIVE DETECTION OF ALBUMIN IN URINE

LX ZHAO[1], J-M LIN[1], YL WEI[1], ZJ LI[2], SJ MA[2]

[1]Research Center for Eco-Environmental Sciences, Chinese Academy of Sciences, Beijing 100085,China,
[2]Kemei Dongya institute of Biotech, Beijing Academy of Science Technology, Beijing 100012, China
E-mail: jmlin@mail.rcees.ac.cn

INTRODUCTION

Immunoassay has been widely applied in clinical diagnostics for many years.[1] Diagnostic tests were mostly done using directly coated tubes, particles or microplates coated with antibodies or antigens via adsorption or chemical bonding.[2] However, the test performance could adversely be influenced by the many parameters. Therefore, the avidin-biotin (AB) systems were introduced into the clinical laboratory to replace directly bound antibodies and antigens as solid phase matrices.

Now, the AB technology has been used in the fields of immunoassay,[3] DNA diagnosis,[4] immunohistichemistry,[5] immunoimaging,[6] DNA sequencing,[7] *in situ* hybridization,[8] and immunohistochemistry. In diagnostic tests for body fluids, the use of AB technology is mainly as an amplified system. However, the hydrophobicity of avidin may give a very strong binding to the solid phase. The interaction of avidin with biotin is one of the tightest binding process known and the very high strength of the bonds is used to improve the binding of the antibody or antigen to the solid phase. This work studied the development of AB system as a solid phase matrix for the determination of human albumin in urine. Compared with CL-ELISA which used directly coated antibody as solid phase, AB has many advantages: low cost, short incubation times and good precision.

METHODS

The purification of antiserum

The rabbit-anti-polyclonal antiserum was purified according to a modified saturated ammonium sulfate (SAS) precipitation method.[9]

Biotinylation of immunoglobulin

The conjugate of biotinylation-anti-albumin was synthesized and purified as described in the paper.[10] Biotinamidohexanoyl-6-aminohexanoic acid N-hydroxysuccinimide ester (BCNHS) with an extended spacer arm was applied to react with primary amines of the protein which can reduce the steric hindrance.

Avidin-biotin chemiluminescence immunoassay procedure

Immunoassay procedures are presented in Figure.1. The wells of the microtiter plates were coated with 150 μL of avidin (3 mg/mL) in a phosphate buffered saline

                    Zhao LX et al.

(pH=7.0). the plates were allowed to stand sealed at 4 °C for 15 h. Then, the solution was removed and the plates were post-coated with 300 μL of 0.5% gelatin in Tris-HCl for 1 h at room temperature. After washing three times with 400 μL of Tris-HCl buffer, 50 μL albumin standard solutions or urine samples and 50 μL diluted enzyme labeled albumin (1:1000) and 50 μL diluted biotinylated anti-albumin antibody (1:8000) in 0.1 M Tris-HCl buffer containing 0.1% (v/v) gelatin, 0.1% NaN_3 and 0.01% Tween-20 were added and incubated for 1 h at 37 °C. After the competitive reaction, five washings were performed with 400 μL of phosphate buffered saline solution (pH 7.4). Then, 50 μL chemiluminescence substrate solution was added and incubated 20 min at 37 °C, the emmited photons were measured. The role of Tween 20 was crucial in both assays due to its property to unfold proteins, enabling antigens and antibodies to interact, thus, increasing enzyme effectiveness and decreasing background signal.

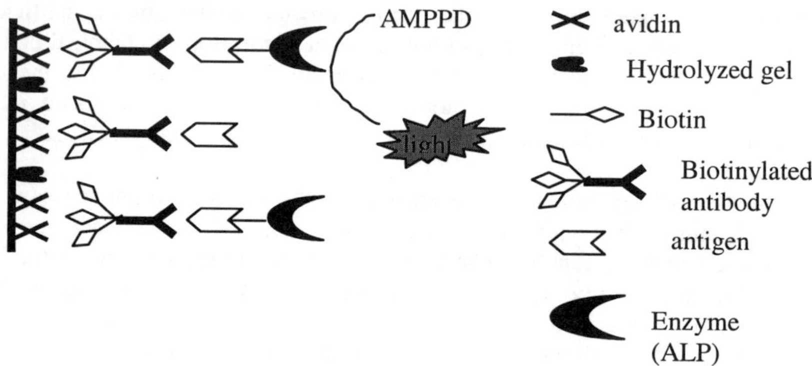

Figure 1.   Schematic illustration of the proposed immunoassay procedures

RESULTS AND DISCUSSIONS

Some experimental parameters (coating and blocking conditions, Tween-20) were studied with two aims: (1) to improve immunoassay sensitivity, (2) to study immunoassay performance under the optimal conditions. These experiments were carried out using the proposed method described above. Criteria used to evaluate the optimization were RLU_{max} and RLU_{max}/I_{50}.

The optimization of solid phase conditions

To develop a highly consistent solid phase, coating and blocking buffer and temperature were studied. It was found that the optimal coating buffer was 10 mM phosphate-buffered saline (PBS), pH 7.0.

The choice of blocking buffer is sometimes critical for sensitive detection. Milk based blocking solutions are not recommended for use avidin-biotin system because milk contains biotin, which may directly cause competition with biotinylated antibody.[11] Bovine Serum Albumin was not selected in this system in order to avoid the cross-reactivity. Therefore, gelatin was chosen as the blocking agent.

The effect of the temperature during coating and blocking steps was also examined. When the plates were coated overnight at room temperature, a large edge effect, i.e. higher standard deviation of signals in the outer wells was observed compared to the plates which were coated at 4 °C. Blocking with gelatin was more efficient at room temperature than 37 °C. Therefore, plates were routinely coated at 4 °C overnight followed by a blocking step at room temperature for 1 h.

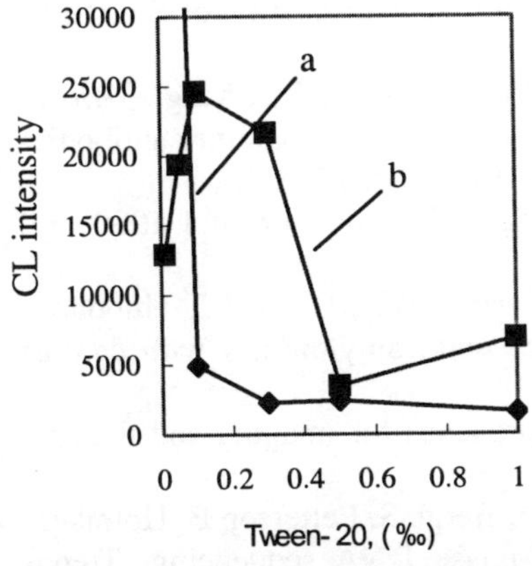

Figure 2. Effect of tween-20 on immunoassay a: RLU_{max} b: RLU_{max}/IC_{50}.

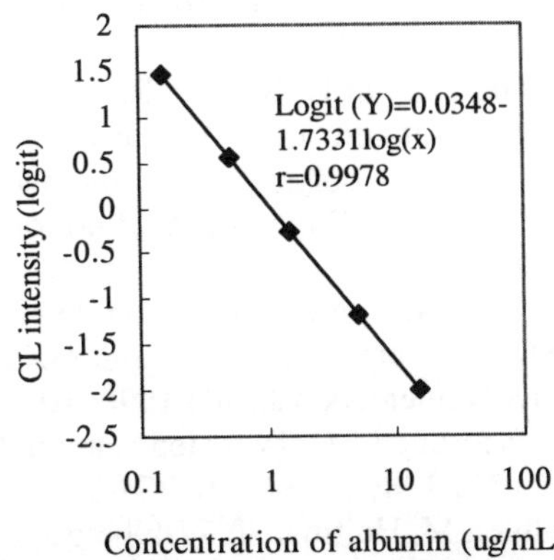

Figure 3. Calibration graph of the avidin-biotin chemiluminescence enzyme immunoassay

The effect of Tween-20

Because surfactants (such as Tween-20) are commonly used in ELISA to reduces nonspecific interactions,[12] their influence on assay performance (IC_{50}, RLU_{max} and RLU_{max}/IC_{50}) was examined in the proposed assay. Figure.2 shows the variation of the RLU_{max}/IC_{50} ratio as a function of the concentrations of the Tween-20. 0.1‰ Tween-20 was selected as the optimal concentration.

Calibration and sensitivity

Dose-response curves obtained with the chemiluminescence detection of enzyme activity under the optimal conditions, are shown in Figure 3. The linear range is 0.15~15 μg /mL. The detection limit, defined as the minimal dose that can be distinguished from zero, the minimum detected concentration (mean - 2SD of zero standard, 10 replicates) of albumin was 0.089 μg /mL.

Comparison with Radioimmunoassay (RIA)

The concentration of albumin in 50 urine samples was determined by the avidin-biotin CL ELISA and RIA. The correlation obtained between the results of the avidin-biotin CL ELISA (Y) and RIA was: Y= 1.9X+ 12 (r= 0.98).

ACKNOWLEDGEMENTS
The authors gratefully acknowledge financial support of the National Science Fund for Distinguished Young Scholars of China (No. 20125514).

REFERENCES

1. Gosling J. A decade of development in immunoassay methodology. Clin Chem 1990; 36: 1408-27.
2. Schetters H. Avidin and streptavidin in clinical diagnostics. Biomol Eng 1999; 16:73-8.
3. Butler JE, Ni L, Nessler R, Joshi KS, Suter M, Rosenberg B, Chang J, Brown WR, Cantarero LA. The physical and functional behavior of capture antibodies adsorbed on polystyrene. J Immunol Methods 1992; 150: 77-90.
4. Wilchek M, Bayer E. Avidin-biotin technology. Methods Enzymol 1990;184: 560-617.
5. Kyle RA, Spittell PC, Gertz MA, Li CY, Edwards WD, Olson LJ, Thibodeau SN. The premortem recognition of systemic senile amyloidosis with cardiac involvement. Am J Med 1996; 101:395-400.
6. Rosebrough SF. Two-step immunological approaches for imaging and therapy. Q J Nucl Med 1996; 40: 234-51.
7. Uhlen M, Hultman T, Wahlberg J, Lundeberg J, Bergh S, Petterson B, Holmberg A, Stahl S, Moks T. Semi-automated solid-phase DNA sequencing. Trends Biotechnol 1992; 10: 52-5.
8. Yu GH, Montone KT, Frias-Hidvegi D, Cajulis RS, Brody BA, Levy RM. Cytomorphology of primary CNS lymphoma review of 23 cases and evidence for the role of EBV. Diagn Cytopathol 1996; 14: 14-20.
9. Zhu LP, Chen XQ. Mianyixie Changyong Shiyan Fangfa. Renmin Junyi Press. p.75
10. Dotsikas Y, Loukas YL, Siafaka I. Determination of umbilical cord and maternal plasma concentrations of fentanyl by using novel spectrophotometric and chemiluminescence enzyme immunoassay Anal Chim Acta 2002; 459: 177-185.
11. Yu H, Raymonda JW, McMahon TM, Campagnari AA. Detection of biological threat agents by immunomagnetic microsphere-based solid phase fluorogenic and electro-chemiluminescence. Biosens Bioelectron 2000; 14: 829-40.
12. Botchkareva AE, Eremin SA, Montoya A, Manclus JJ, Mickova B, Pavel R, Fini F, Girotti S. Development of chemiluminescent ELISAs to DDT and its metabolites in food and environmental samples, J Immun Methods 2003; 283: 45-57.

PART 12

LUMINESCENT ASSAYS FOR ENZYMES, SUBSTRATES, INHIBITORS & CO-FACTORS

USE OF THE PEROXYOXALATE CHEMILUMINESCENT REACTION IN ACETONE IN THE PRESENCE OF NILE RED FOR THE ANALYSIS OF GLUCOSE

PABLO CASTRO-HARTMANN, SILVIA GUERRERO, JOAN-RAMON DABAN
Departament de Bioquímica i Biologia Molecular, Facultat de Ciències,
Universitat Autònoma de Barcelona, 08193-Bellaterra (Barcelona), Spain

INTRODUCTION

In the peroxyoxalate chemiluminescent system, an oxalate ester, usually TCPO [*bis*(2,4,6-trichlorophenyl)oxalate], reacts with H_2O_2 and generates high-energy intermediates capable of producing the chemiexcitation of different fluorophores.[1] In this system the resulting chemiluminescence corresponds to the emission of light produced by the relaxation of the excited fluorophore. The peroxyoxalate chemiluminescent reaction has been used in liquid chromatography and flow-injection analysis for the direct detection of fluorescent molecules and in indirect assays based on the measurement of enzymatically formed H_2O_2.[2-4] This chemiluminescent reaction is remarkably efficient in organic solvents. However, since most bioessays are performed in aqueous solutions, the insolubility and instability of TCPO and other oxalate esters in the presence of water have limited the analytical applications of this system[5]. We have shown previously that the TCPO-H_2O_2 reaction in acetone can be used for the detection of fluorophore-labeled protein and DNA bands on membranes.[6-9] In this work, in order to take advantage of the chemiluminescent properties of this system in organic media, we have developed a procedure based on the peroxyoxalate reaction in acetone for the quantitative analysis of glucose. We have used Nile red as energy acceptor. This hydrophobic fluorophore has been previously employed for the staining of lipid droplets[10] and protein-sodium dodecyl sulfate complexes in electrophoretic gels.[11]

MATERIALS AND METHODS

Since the enzyme glucose oxidase (GOD) has a high degree of specificity for β-D-glucopyranose,[12] the solutions of D-glucose (Merck) in 0.1 M imidazole-HCl (pH 6.4) were prepared two days before they were used to ensure that the mutarotation reaction had reached the equilibrium. For each concentration, 20 μL of the corresponding glucose solution in imidazole-HCl (pH 6.4) was mixed with 20 μL of a solution of GOD (Calbiochem) in the same buffer and incubated at 37 °C for 30 min. The chemiluminescent reaction was initiated by the addition of 20 μL of the resulting oxidized glucose solution to 2 mL of a solution of TCPO (Fluka) and Nile red (Sigma) in acetone at room temperature. The TCPO-Nile red solution was prepared, just before the addition of the glucose solution, in a borosilicate glass tube from concentrated solutions of these reagents (10 mM TCPO, prepared fresh each day; 12.5 mM Nile red) in acetone. The final solution containing the H_2O_2 produced in the enzymatic reaction and the chemiluminescent reagents was vortexed (~1 s) and introduced immediately in the luminometer (Lumat LB-9507, EG&G Berthold). The measurement of the chemiluminescence was initiated 20 s after the initiation of the chemiluminescent reaction. In the final quantitative measurements, each point corresponds to the chemiluminescence accumulated during 1 min. A reagent blank was prepared for each set of samples; the blank values were subtracted from the measured chemiluminescence. Control serum samples containing 82 mg/dL of glucose were prepared from lyophilized bovine serum (BioSystems). The lyophilized serum was dissolved in water or imidazole buffer and treated with GOD and the chemiluminescent reagents as indicated above for pure glucose solutions.

RESULTS AND DISCUSSION

We assayed different concentrations of TCPO and Nile red. As can be seen in Fig. 1A, 2.5 mM TCPO is an adequate concentration to obtain a high and roughly constant chemiluminescence intensity during the first minute of the reaction.

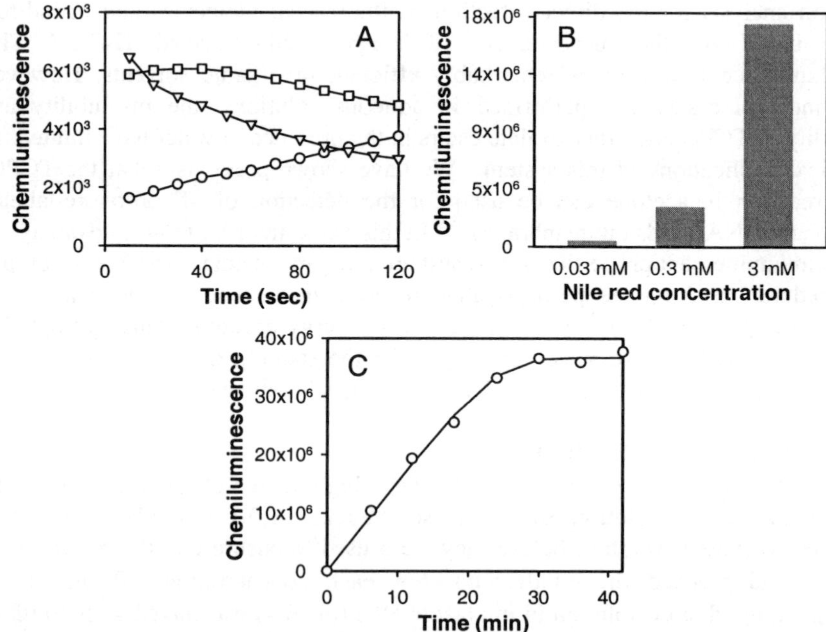

Figure 1. Determination of the optimum conditions for the chemiluminescent detection of glucose. (A) 0.4 (○), 2.5 (□), 4 (▽) mM TCPO. (B) Effect of fluorophore concentration. (C) Time course of the enzymatic oxidation of 600 mg/dL of glucose with 6 mg/mL of GOD at 37 °C; curve obtained using the optimum concentrations of TCPO (2.5 mM) and Nile red (0.3 mM)

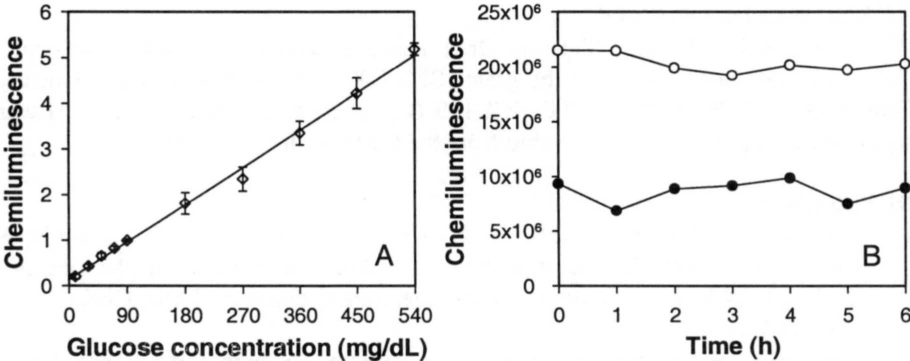

Figure 2. (A) Standard curve for the assay. The chemiluminescence intensity is expressed relative to the value obtained for 90 mg/dL of glucose. The bars indicate the standard deviations calculated from three determinations for each point. (B) Stability of the assay. Glucose concentration: 90 (●), 270 (○) mg/dL

The chemiluminescence intensity is highly dependent on Nile red concentration (Fig. 1B). The most convenient concentration of Nile red is 0.3 mM; when higher concentrations of Nile red were used, the chemiluminescence obtained with samples containing high concentrations of glucose was too intense to be registered with the luminometer. With the indicated concentrations of TCPO and Nile red and 6 mg/mL of GOD, even for the higher concentrations of glucose used in our study (600 mg/dL), the reaction is completed in 30 min at 37 °C (Fig. 1C).

Our results demonstrate that it is possible to determine the concentration of glucose from small volumes (20 µL) of samples having concentrations of clinical interest. In our system, the enzymatic treatment of the sample is carried out in an aqueous solution containing the imidazole buffer. Note, however, that only 1% of water is present in the chemiluminescent detection reaction performed in a large volume of acetone (2 mL).

It can be seen in Fig. 2A that the chemiluminescence intensity is linearly proportional to the concentration of glucose in the range of 10 to 540 mg/dL (correlation coefficient, R^2 = 0.997). Furthermore, the results presented in Fig. 2B show that this chemiluminescent assay is stable. The coefficients of variation corresponding to 13 measurements of samples containing 90 and 270 mg of glucose/dL are, respectively, 11.8 and 7.6%. Unfortunately, the chemiluminescence obtained with control serum samples is significantly higher than that obtained with solutions having the same concentrations of pure glucose. The interference produced by components present in serum precludes the use of this sensitive chemiluminescent method for the direct determination of glucose in blood samples.

ACKNOWLEDGEMENTS
We are grateful to Ferran Oller and Dr. Francesca Canalias for useful comments. This work was supported in part by grants BMC2002-3948 (Ministerio de Ciencia y Tecnología and FEDER) and 2001SGR199 (Generalitat de Catalunya). P C H was supported by a predoctoral fellowship from the Generalitat de Catalunya.

REFERENCES
1. Orlovic M, Schowen RL, Givens RS, Alvarez F, Matuszewski B, Parekh N. A simplified model for the dynamics of chemiluminescence in the oxalate-hydrogen peroxide system: Toward a reaction mechanism. J Org Chem 1989; 54: 3606-10.
2. Imai K. Chemiluminescence detection system for high-performance liquid chromatography. Methods Enzymol 1986; 133: 435-49.
3. Kwakman PJM, Brinkman UATh. Peroxyoxalate chemiluminescence detection in liquid chromatography. Anal Chim Acta 1992; 266:175-92.
4. Emteborg M, Irgum K, Gooijer C, Brinkman UATh. Peroxyoxalate chemiluminescence in aqueous solutions: Coupling of immobilized enzyme reactors and 1,1'-oxalyldiimidazole chemiluminescence reaction to flow-injection analysis and liquid chromatographic systems. Anal Chim Acta 1997; 357: 111-8.
5. Oh SK, Cha SH. Effect of sodium azide on peroxyoxalate chemiluminescence assay method. Anal Biochem 1994; 218: 222-4.
6. Alba FJ, Daban JR. Nonenzymatic chemiluminescent detection and quantitation of total protein on Western and slot blots allowing subsequent immunodetection and sequencing. Electrophoresis 1997; 18: 1960-6.
7. Alba FJ, Daban JR. Detection of Texas red-labelled double-stranded DNA by non-enzymatic peroxyoxalate chemiluminescence. Luminescence 2001; 16: 247-9.
8. Salerno D, Daban JR. Comparative study of different fluorescent dyes for the detection of proteins on membranes using the peroxyoxalate chemiluminescent reaction. J Chromatogr B 2003; 793: 75-81.
9. Castro-Hartmann P, Daban JR. Flow and evaporation cells for the detection of proteins on membranes with the peroxyoxalate chemiluminescent reaction in organic media. Electrophoresis, in press.
10. Greenspan P, Mayer EP, Fowler SD. Nile red: A selective fluorescent stain for intracellular lipid droplets. J Cell Biol 1985; 100: 965-73.
11. Daban JR, Bartolomé S, Bermúdez A, Alba FJ. Rapid and sensitive staining of unfixed proteins in polyacrylamide gels with Nile red. in: Walker JM. ed. The Protein Protocols Handbook. Totowa: Humana Press, 2002: 243-9.
12. Wilson R, Turner APF. Glucose oxidase: An ideal enzyme. Biosensors Bioelectronics 1992; 7: 165-85.

CHEMILUMINESCENCE ASSAY FOR LIPASE ACTIVITY IN HUMAN SERUM BY USING A PROENHANCER SUBSTRATE

T ICHIBANGASE, C HAMABE, Y OHBA, N KISHIKAWA,
K NAKASHIMA, N KURODA

*Graduate school of Biomedical Sciences, Nagasaki University, 1-14 Bunkyo-machi,
Nagasaki 852-8521, Japan*

INTRODUCTION

The determination of pancreatic lipase (EC 3.1.1.3) activity in serum is useful for the diagnosis and monitoring of acute pancreatitis and is generally regarded as providing superior clinical specificity to the determination of amylase.[1,2] Until now, numerous assay methods for lipase (e.g. turbidimetry, titrimetry, colorimetry, immunoassay, etc) have been reported.[2] However, most of them are often laborious and time consuming. Recently, we developed a simple and rapid chemiluminescence (CL) method for lipase activity by using a novel proenhancer type substrate, HDI-laurate, which is a lauric acid ester of 2-(4-hydroxyphenyl)-4,5-diphenylimidazole (HDI) (Fig. 1).[3] In the method, the enzymatic hydrolysis of HDI-laurate by lipase releases HDI, which acts as an enhancer in the luminol-H_2O_2-horseradish peroxidase (HRP) CL system. The method allowed the homogeneous reaction in which the enzymatic hydrolysis of the substrate and the enhanced CL reaction simultaneously occur.

In this study, we tried to apply this method to the determination of pancreatic lipase in human serum sample.

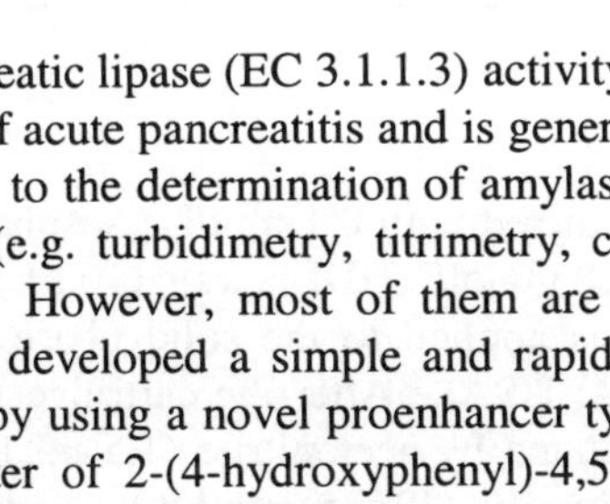

Figure 1. The structure of HDI-laurate

MATERIALS

Reagents and solutions

Lipase from porcine pancreas (activity: 82 U/mg) was purchased from Funakoshi (Tokyo, Japan). HRP (activity: 290 U/mg) was from Sigma (St. Louis, USA). Luminol and H_2O_2 (30%) were obtained from Wako Pure Chemicals (Osaka, Japan). *N,N*-Dimethylformamide (DMF) from Nacalai Tesque (Kyoto, Japan) was of spectrochemical analysis grade. Other chemicals were of analytical reagent grade.

Luminol was dissolved in DMF (0.42 mg/mL) and diluted 20-fold with 0.3 mol/L Tris-HCl solution (pH 6.4) to give a 120 μmol/L solution. Lipase from porcine pancreas was prepared in 0.1 mol/L phosphate buffer (pH 6.5). HDI-laurate and HDI were synthesized according to our previous method.[4] HDI-laurate prepared in DMF is stable for at least one week at 4°C. HRP was dissolved in the Tris solution and then diluted 20-fold with the same solution to give an 80 nmol/L solution. For comparison between the proposed and the colorimetric method,[1] the reagent kit from Roche Diagnostics (Tokyo) was applied to HITACHI Clinical Analyzer 7600 (Tokyo) for the colorimetry.

Specimens

For precision studies, pooled serum from healthy volunteers was used. An inactivated serum used as reference sample was prepared by heating the pooled serum at 100 °C for 30 min after 10-fold dilution with 0.1 mol/L phosphate buffer (pH 6.5).

METHODS

A serum sample (100 μL) mixed with 0.1 mol/L phosphate buffer (pH 6.5, 1.0 mL) and the DMF solution of 2.0 mmol/L HDI-laurate (10 μL) was incubated at 37 °C for 90 min. This solution was applied to the solid phase extraction (SPE) cartridge (Sep-pak C18, Waters, MA, USA). After the cartridge was washed with 10 mL of water, retained HDI was eluted by acetonitrile (1.5 mL). A 100-μL portion of the eluate was placed in a glass tube (12x75 mm i.d.). To this was successively added 0.3 mol/L Tris-HCl solution (pH 6.4) of 120 μmol/L luminol and 80 nmol/L HRP (50 μL each), and of 7.5 mmol/L H_2O_2 aqueous solution (200 μL). The CL emission was measured by a Sirius-Luminometer (Berthold Japan, Tokyo) and integrated from 1.0 to 5.0 min.

RESULTS

Optimization of assay condition

At first, the previous method [3] was applied to the serum sample spiked with porcine pancreas lipase which is morphologically and functionally similar to human lipase. However, significant CL emission was not observed. Due to the fact that the serum sample spiked with HDI did not give any CL signal, serum matrices were considered to be obstacle to the CL reaction. We therefore try to remove serum matrices that interfere in the CL reaction by SPE. The SPE condition was optimized using the serum sample spiked with HDI. After loading sample to the SPE cartridge, washing was performed by various volume of water. The inhibitory components were effectively removed by more than 10 mL of water while the retained HDI was not eluted even with 40 mL of water. Retained HDI was effectively eluted by acetonitrite (1.5 mL); the recovery of HDI from the cartridge was 94%.

The conditions of lipase hydrolysis and enhanced CL reaction were then optimized independently using serum sample spiked with porcine pancreas. Effect of incubation time on CL intensity was investigated; the CL intensity increased with an increase in time up to 150 min. The recommended incubation time was 90 min. The maximum and constant CL intensity was obtained with HDI-laurate concentration above 1.0 mmol/L. The observed Km value estimated from

Lineweaver-Burk plots was 12.5 μmol/L; 18 μmol/L of HDI-laurate (2.0 mmol/L as DMF solution) was selected as a saturating concentration in the reaction mixture. The other optimization results for both hydrolysis and CL reactions are described in "METHODS".

Calibration curves, detection limits, and precision

Before preparing a calibration curve for lipase activity, the detectability of HDI was examined using serum sample spiked with HDI. Linear relationship was obtained (r=0.992) between the concentration of HDI (1.1-17 pmol) and CL intensity. The detection limit of HDI was 0.22 pmol (blank + 3 SD). Calibration curve for lipase activity was prepared by adding porcine pancreas lipase to serum (Table 1). Detection limit for lipase was 1.0 U_{HDI} (blank + 3 SD), where enzyme activity is expressed as U_{HDI}: one U_{HDI} corresponds to the amount which liberates 1 pmol HDI per minute at 37 °C from HDI-laurate as a substrate.

The reproducibilities obtained with within- and between-day assays on 6.8 U_{HDI} were < 3.0% (RSD, n=3) and <4.3% (RSD, n=3), respectively.

Table 1. Calibration curve for lipase activity in serum
spiked with porcine pancreas lipase.

Range (U_{HDI})	y-intercept (U_{HDI})	slope	r
2.6-11	5×10^5	3×10^7	0.999

Application to serum samples

Since the proposed method was optimized for human serum spiked with porcine pancreas lipase, it was applied to real samples to determine its estimated practicability. The seven samples from healthy volunteers were measured by the proposed method and the colorimetric method. The average and median values obtained with the proposed method were 7.76 and 7.67 U_{HDI}, respectively. The colorimetric method is based on the determination of liberated methylresorufin from 1,2-o-dilauryl-rac-glycero-3-glutaric acid-(6'-methylresorufin) ester (DGGR) as substrate.[1] The results obtained with the CL assay demonstrated acceptable correlation with the colorimetric method (r=0.711, Fig. 2).

CONCLUSION

The antioxidants in serum matrices such as ascorbic acid, uric acid and bilirubin have been known to interrupt CL emission.[5] This inhibition of CL emission was completely eliminated by introducing the SPE procedure. HDI-laurate was successfully converted to the active enhancer, HDI, by lipase in serum, and HDI enhanced the luminol-H_2O_2-HRP CL system. The results of the proposed method should good correlation with those of the colorimetric method although the results were obtained with samples from healthy volunteers. Our results indicated the possibility of the CL determination of human serum lipase.

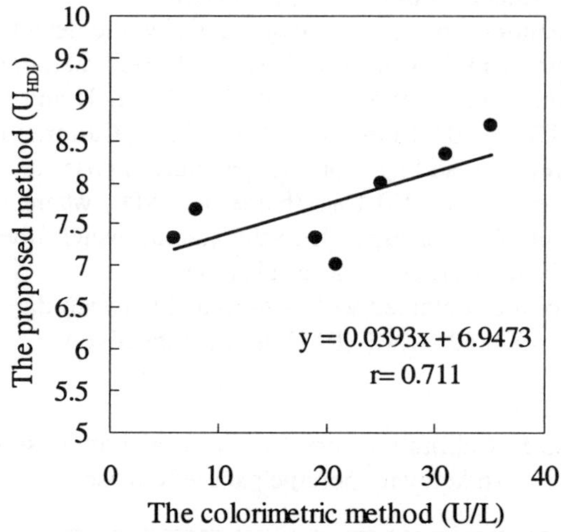

Figure 2. Results of correlation studies

REFERENCES

1. Panteghini M, Bonora R, Pagani F. Measurement of pancreatic lipase activity in serum by a kinetic colorimetric assay using a new chromogenic substrate. Ann Clin Biochem 2001; 38: 365-70.
2. Tietz N W, Shuey D F. Lipase in serum –the elusive enzyme: an overview. Clin Chem 1993; 39: 746-56.
3. Ichibangase T, Ohba Y, Kishikawa N, Nakashima K, Kuroda N. Chemiluminescence assay of lipase activity using a synthetic substrate as proenhancer for luminol chemiluminescence reaction. Luminescence in press.
4. Kuroda N, Takatani M, Nakashima K, Akiyama S, Ohkura Y. Preparation and evaluation of fatty acid esters of 2-(4-hydroxyphenyl)-4,5-diphenylimidazole as fluorescent substrate for measurement of lipase activity. Biol Pharm Bull 1993; 16: 220-2.
5. Whitehead TP, Thorpe GHG, Maxwell SRJ. Enhanced chemiluminescent assay for antioxidant capacity in biological fluids. Anal Chim Acta 1992; 266: 265-77.

RAPID AND SIMULTANEOUS BIOLUMINESCENT ASSAY OF AEQUORIN AND FIREFLY LUCIFERASE

WAKA NISHIMURA[1], KASTUTOSHI ITO[1*], HIDETOSHI ARAKAWA[1], MASAKO MAEDA[1], SATOSHI INOUYE[2], HIROKI TATSUMI[3]

[1]*School of Pharmaceutical Sciences, Showa University, 1-5-8 Hatanodai, Shinagawa, Tokyo 142-8555, Japan*

[2]*Chisso Co., Yokohama Resarch Center, 5-1 Okawa, Kanazawa-ku, Yokohama, 236-8605, Japan*

[3]*Kikkoman Co., 399 Noda, Noda-shi, Chiba, 278-0037, Japan*

Email: itok@pharm.showa-u.ac.jp

INTRODUCTION

We have developed a highly sensitive simultaneous bioluminescent assay of firefly luciferase and aequorin. Firefly luciferin-luciferase reaction is specific and sensitive for the determination of ATP, and this reaction has been widely used, e.g. for hygiene monitoring. Aequorin bounds specifically to Ca^{2+} and then emits blue light, thus aequorin is useful to study intercellular Ca^{2+}.[1,2] We thought that these photoproteins can be measured into same batch, because although these bioluminescent reactions have different mechanisms they are performed at similarly pH conditions. In this paper, the development of simultaneous and high throughput bioluminescent assay for biotinylated luciferase and aequorin is reported.

MATERIALS AND METHODS

Reagents

Biotinylated luciferase[3] was donated from Kikkoman Co. (Chiba, Japan). D-Luciferin was purchased from Sigma Chemical Co. (St. Louis, MO). Cys-type Aequorin (Cys4-AQ) was donated from Chisso Co. Yokohama Research Center (Kanagawa Japan). N-2-hydroxyethylpiperazine-N'-2-ethanesulfonic acid (HEPES) was from Dojindo Laboratories (Kumamoto, Japan).

Simultaneous bioluminescent assay for aequorin and biotinylated luciferase

Each of 10 μL of aequorin and luciferase solution was added to a microtiter plate, and 100 μL of 50 mM Ca^{2+} in 50 mM HEPES-KOH buffer (pH 7.0) was added, and then the bioluminescent intensity was integrated for 1 s by a MicroLumat LB96P luminescent reader (EG&G Berthold, Germany), immediately. Then, 100 μL of the bioluminescence reagent for luciferase (containing 40 mM ATP, 1.4 mM luciferin, 300 mM $MgSO_4$ in 50 mM HEPES-KOH buffer, pH 7.0) was added to the same wells. The bioluminescent intensity from luciferin-luciferase reaction was integrated for 1 s after a delay of 2 s.

                                                          Nishimura W et al.

RESULTS AND DISCUSSION
Time course of light emission of aequorin and biotinylated luciferase
As shown in Fig. 1 (left), light emission of aequorin reached a maximum within 1 s after addition of Ca^{2+} solution and then the light intensity decayed. For measurement of bioluminescent intensity of aequorin, we integrated immediately for 1 s after addition of Ca^{2+} solution. On the other hand, the light emission of biotinylated luciferase reached a maximum about 1 s after addition of the bioluminescent reagent for luciferase and then the light intensity gradually decreased (Fig. 1, right). Therefore, we measured the bioluminescent intensity of biotinylated luciferase for 1 s after a delay of 2 s.

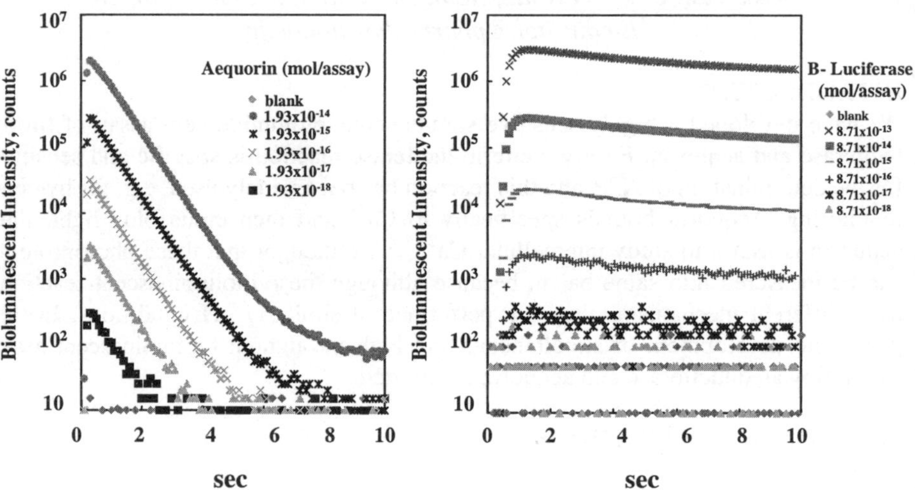

Figure 1. Time course of light emission of aequorin and biotinylated luciferase

Simultaneous bioluminescent assay for aequorin and biotinylated luciferase
In the proposed assay, there was no interference in the measurement of the aequorin activity by biotinylated luciferase. The sensitivity of aequorin was 7.58×10^{-20} mol/assay (blank + 3SD, Fig. 2, left), the calibration linearity of the assay ranged from 1.21×10^{-18} mol/assay to 1.94×10^{-15} mol/assay and the intra-assay coefficients of variation (CV) for 8 replicates with each standard point were from 1.0 to 3.2%. These results are similar to the results from the separate assay.

The sensitivity of biotinylated luciferase was 2.75×10^{-18} mol/assay (blank + 3SD, Fig. 2, right) and the intra-assay CV for 8 replicates with each standard point were from 2.2 to 3.9%. However, the bioluminescent intensity of biotinylated luciferase was about half that obtained in the separate assay. We measured the bioluminescent intensity of biotinylated luciferase with and without 100 μL of 50 mM Ca^{2+} in 50 mM HEPES-KOH buffer (pH 7.0) or 50 mM HEPES-KOH buffer (pH 7.0). The bioluminescent intensity of biotinylated luciferase with 100 μL of 50 mM Ca^{2+} in 50 mM HEPES-KOH buffer (pH 7.0) was similar with 50 mM HEPES-KOH buffer (pH 7.0). The bioluminescent intensities of biotinylated luciferase were reduced by about half and it seems that the bioluminescent intensity of biotinylated luciferase was affected by addition of buffer solution. The sensitivity and CVs are similar to the results of the separate assay. There was no interference with Ca^{2+} and aequorin.

Biotinylated luciferase[3] and aequorin[4] have been used as reported labelling protein for immunoassay. Thus, the proposed simultaneous assay of aequorin and biotinylated luciferase could be applied to immunoassay. We apply the proposed assay to tandem immunoassay using these two proteins as labels.

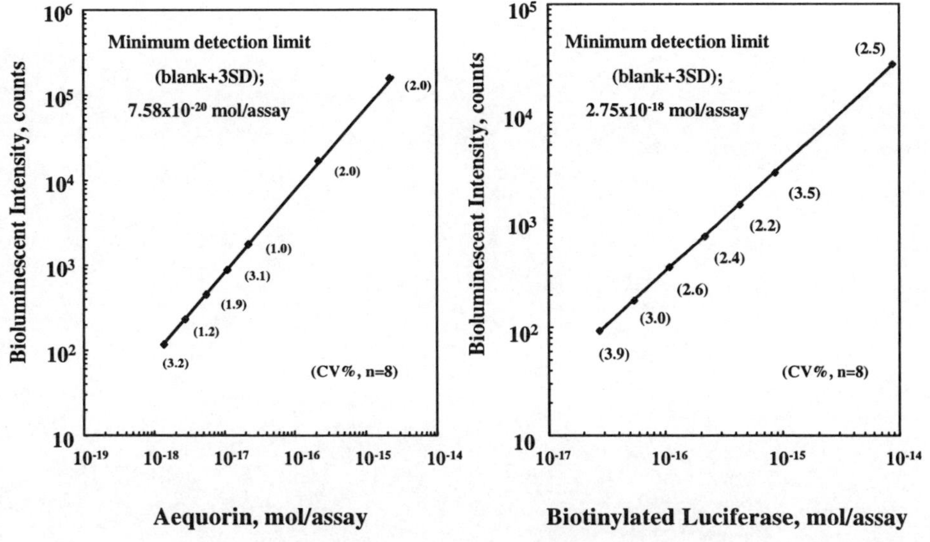

Figure 2. Standard curves of simultaneous assay for aequorin and biotinylated luciferase

REFERENCES
1. James F, Inouye S, Teranishi K, Shimomura O. The crystal structure of the photoprotein aequorin at 2.3Å resolution. Nature 2000; 405: 372-6.
2. Shimomura O, Musicki B, Kishi Y, Inouye S. Light-emitting properties of recombinant semi-synthetic aequorins and recombinant fluorescein-conjugated aequorin for measuring cellular calcium. Cell Calcium 1993; 14: 373-8.
3. Tatsumi H, Fukuda S, Kikuchi M, Koyama Y. Construction of biotinylated firefly luciferase using biotin acceptor peptides. Anal Biochem 1996; 243: 176-80.
4. Desai UA, Wininger JA, Lewis JC, Ramanathan S, Daunert S. Using epitope-aequorin conjugate recognition in immunoassays for complex proteins. Anal Chem 2001; 294: 132-40.

REPETITIVE ASSAY FOR ENHANCED DETECTION OF IMMOBILIZED HORSERADISH PEROXIDASE BY IMIDAZOLE CHEMILUMINESCENCE COUPLED TO THE TECHNIQUE OF ON-LINE REGENERATION OF INACTIVATED PEROXIDASE

O NOZAKI[1], M MUNESUE[2], H KAWAMOTO[3]

[1]*Dept of Clin Lab Med, Kinki Univ School of Med, Osaka 589-8511, Japan*
[2]*Chemco Scientific Co, Osaka 530-0016, Japan*
[3]*Dept of Biol Reg, School of Health Science, Tottori Univ, Tottori 683-8503, Japan*

INTRODUCTION

A method of micro flow injection-horseradish peroxidase (HRP) catalyzed imidazole chemiluminescence for determination of hydrogen peroxide (H_2O_2) was developed by Nozaki et al.[1] The imidazole chemiluminescence (CL) method employed a flow cell reactor packed with HRP immobilized on a gel for starting the CL reaction,[2] and a photo multiplier for detection of light. For good reproducibility of assaying H_2O_2 by HRP catalyzed imidazole CL, it was necessary to keep immobilized HRP active during the assay. We, therefore, developed a method of automatic reactivation of inactivated HRP after reaction with H_2O_2.[3,4]

Miniaturization of systems for assay with a microchip is a recent trend.[5,6] Miniaturization of a system, however, decreases the amount of HRP immobilized in a chip. This caused problems of decreasing detection range of H_2O_2 and poor reproducibility of data. For resolving these problems of narrow detection range of H_2O_2 and poor reproducibility caused by decrease of immobilized HRP, two technologies were required. The first was a technology of on-line reactivation of inactivated HRP with imidazole, and this was already developed.[3,4] The second technology required the accumulation of photons produced by multi-injection of H_2O_2 specimens,[7] and this was the aim of this study.

The principle of accumulation of photons employed in this study is as follows: a full frame transfer (FFT) chip in a CCD camera collects and accumulates photons, then transfers them as electrical signals. This effect of light accumulation by a FFT chip is expected to expand the detection range of very small amount of HRP than use of a photomultiplier for CL detection of H_2O_2, because a photomultiplier cannot accumulate light.

We studied here on accumulation of light collected by flow injection-HRP catalyzed imidazole CL. For the purpose, we have developed the CL monitoring system that consisted of a cooled CCD camera with an FFT-CCD chip as a light detector and the HRP immobilized flow-through chip as a CL reactor. We report the results of image detection and accumulation of light using the novel CL system.

METHODS

The flow-through chip as a reactor

We made the flow-through chip as a flow cell reactor by constructing two acrylic plates (5 cm x 3 cm)) and a spacer plate (5 cm x 3 cm x 1.5 mm) with a window (cell volume: 1.35×10^{-7} m^3). Two ports connecting to Teflon tubes (0.5 mm i.d.) as a inlet and outlet tube was set at the window site on the lower plate, respectively. HRP was immobilized between the two ports at the lower plate after constructing the silica layer (3x5 mm).

The CCD-CL monitor

We used the CCD-CL monitor (Chemco, Osaka, Japan) that consisted of a light tight box, a cooled CCD camera with a full frame transfer typed chip and a lens for a single-lens reflex camera (28 mm f2.8, Tamuron, Tokyo, Japan). The CCD-CL monitor was controlled with a personal computer (e-machines, Tokyo, Japan) with Windows XP (Microsoft, USA). The flow-through chip as a CL reactor on a chip holder was located under the camera. The window of the flow through chip was focused with the lens of the camera. The flow-in and out tubes of the flow reactor connected with a pump for high-performance liquid chromatography (HPLC; PU-980, JASCO, Tokyo, Japan) and a waste bottle outside of the light tight box, respectively.

Assay of H_2O_2

The solution as mobile phase (imidazole 100mmol/L in the Tricine buffer 50 mmol/L, pH 9.4) was flowed at 0.1 mL/min using HPLC pump to the flow cell reactor via inlet connecting tube, and drained via a outlet tube. H_2O_2 specimen (20 μL) was injected with a loop injector (7125, JASCO). The reaction temperature for CL was room temperature. Light from the flow-through chip was detected after the dark frame reduction with the CCD-CL monitor.

Principles for accumulated light detection by HRP catalyzed imidazole CL

HRP catalyzed imidazole CL: The HRP catalyzed imidazole CL consists of three factors (HRP, alkaline imidazole, and H_2O_2). Stable flow of mobile phase is introduced in the flow-through chip where HRP is immobilized. H_2O_2 specimen is injected and reaches the flow cell resulting in light emission after being mixed with the alkaline imidazole and immobilized HRP.

Reactivation of immobilized HRP: The immobilized HRP turns to inactive form by being protonated at the active site after reaction with H_2O_2. The inactivated HRP, however, is reactivated with alkaline imidazole solution by removing the proton from the protonated active site. This allows the immobilized HRP to maintain reactivity with H_2O_2 specimens.

Accumulation of light signal: The signal generated from the light emitted from the reactor chip is accumulated with a CCD chip (full frame transfer type) after dark frame reduction during the assay period. This allows multiple injection of H_2O_2 specimen for enhanced detection of light by increasing the light signal amounts.

RESULTS AND DISCUSSION

Image detection of light by the imidazole CL

Image of light emitted by imidazole CL was not obtained in our previous studies, because a photomultiplier was employed as a detector of the CL. However, imaging of light emitted in the flow through chip by flow injection-HRP catalyzed imidazole CL was detected with the cooled CCD camera in this study. The site where HRP was immobilized in the flow-through chip produced a light emission after reaction with H_2O_2.

Light intensity by single injection of H_2O_2 specimen

Light intensity caused by single injection of the H_2O_2 specimen (9.8, 98, 980 mol/μL; 20 μL) was investigated. As the result, light intensity increased in a dose-dependent manner.

Accumulation of light by multiple injection of H_2O_2 specimen

We tested influence of injection times (one, three and five times) of the two kinds of H_2O_2 specimen (98 μmol/L and 980 μmol/L) with injection volume of 20 μL per sample on light intensity. As the results, intensities of the accumulated light increased corresponding to injection times and concentrations of H_2O_2 in the specimens. The reason of accumulation of light was due to an FFT-CCD chip of the CCD camera. Small amount of HRP can response usually only to small amount of H_2O_2 to emit small amount of light. This method for accumulation of light, however, facilitates reaction with large amounts of H_2O_2 in total, and hence emission of a large amount of light.

In conclusion, a novel flow through chip and a cooled CCD-CL monitor was developed in this study to accumulate lights by repetitive injection of H_2O_2 specimen. The method for light accumulation is expected to detect very small amounts of HRP in a miniaturized assay system.

REFERENCES

1. Nozaki O, Kawamoto H. Determination of hydrogen peroxide by micro-flow injection-horseradish peroxidase catalyzed "imidazole chemiluminescence". In: Stanley PE, Kricka LJ. eds. Bioluminescence & Chemiluminescence- Progress & Current Applications. Singapore:World Scientific, 2002: 335-8.
2. Nozaki O, Kawamoto H. Determination of hydrogen peroxide by micro flow injection - chemiluminescence coupled the flow cell reactor in a chemiluminometer. Luminescence 2000; 15: 137-42.
3. Nozaki O, Kawamoto H. Reactivation of horseradish peroxidase with imidazole for continuous determination of hydrogen peroxide using a micro-flow injection - chemiluminescence detection system. Luminescence 2003; 18: 203-6.

4. Nozaki O, Kawamoto H. Reactivation of inactivated horseradish peroxidase with ethyleneurea and allantoin for determination of hydrogen peroxide by micro-flow injection -horseradish peroxidase catalyzed chemiluminescence. Anal Chim Acta 2003; 495: 233-8.

5. Kricka LJ, Ji X, Nozaki O, Wilding P. Imaging of chemiluminescent reactions in microstructures. J Biolumin Chemilumin 1994; 9: 135-8.

6. Eggers M, Hogan M, Reich RK, et al. A microchip for quantitative detection of molecules utilizing luminescent and radioisotope reporter groups. Biotechniques. 1994; 17: 516-25.

7. Lorimier P, Lamarcq L, Negoescu A, et al. Comparison of 35S and chemiluminescence for HPV in situ hybridization in carcinoma cell lines and on human cervical intraepithelial neoplasia. J Histochem Cytochem. 1996; 44: 665-71.

PART 13

LUMINESCENT DNA PROBE, GENE EXPRESSION & REPORTER GENE ASSAYS

A NEW ASSAY FOR DETERMINING PYROPHOSPHATE USING PYRUVATE PHOSPHATE DIKINASE AND ITS APPLICATION TO DNA ANALYSIS

HIDETOSHI ARAKAWA[1], KOJI KARASAWA[1],
SHIGEYA SUZUKI[2], MASAKO MAEDA[1]
[1]*School of Pharmaceutical Sciences, Showa University, Tokyo 142-8555, Japan*
[2]*Research & Development Division, Kikkoman Corporation, Chiba 278-0037, Japan*
Email: arakawa@pharm.showa-u.ac.jp

INTRODUCTION

We developed a novel bioluminescent assay for pyrophosphate in a PCR assay. The principle of this method is as follows: pyrophosphate released by PCR is converted to ATP by pyruvate phosphate dikinase (PPDK) in presence of pyruvate phosphate as substrate and AMP as coenzyme, and the concentration of ATP is determined using the firefly luciferase reaction. The detection limit for pyrophosphate is 1.5 fmol/assay and time course of light emission was stable for more than 10 minutes. This method is applied to the detection of cariogenic bacteria in dental plaque as prevention diagnosis of dental decay. In this study, the dextranase gene (*dex*) in *Streptococcus mutans* was selected as a marker gene. Allele-specific PCRs were developed for the *dex* genes in *S. mutans* and *S. sobrinus*. The pyrophosphate produced in two allele specific PCRs was measured by the bioluminescent assay. This protocol, which does not require expensive equipment, can be utilized to rapidly monitor cariogenic bacteria in dental plaque.

MATERIALS AND METHODS

Chemicals

PPDK from Microbispora rosea subsp. Aerata (EC2.7.9.1) and thermostable *Luciola cruciata* firefly luciferase (EC 1.13.12.7) were obtained from Kikkoman Co. (Chiba, Japan). Pyrophosphate, Lysozyme and Proteinase K were purchased from Wako Pure Chemical Industries, Ltd (Osaka, Japan). dNTP Mixture, *Taq* DNA Polymerase and 10× PCR buffer were acquired from Takara Shuzo Co., Ltd (Osaka, Japan). Primers were synthesized by Takara Shuzo Co., Ltd (Osaka, Japan). Luciferin was obtained from Sigma Chemical Co (St. Louis, MO). Perfect match (PCR Enhancer) was manufactured by Toyobo. Other reagents were of analytical grade.

PCR

Template DNA (1 µL), which corresponded to 20 ng as bacterial DNA and 1 µL of *Taq* polymerase (2.5U), was introduced to 48 µL of mixed solution containing 5 µL of 10x buffer, dNTP (dATP, dCTP, dGTP and dTTP) and 1 µL of each primer (20 pmol), and H_2O. The reaction was conducted at 94 °C after a 5-minute heating process (94°C for 1 minute), at 55 °C for allele-specific PCR for 1 minute and at 72 °C for 1 minute. The number of PCR cycles was 25. Following the final PCR cycle, the reaction was completed at 72 °C for 7 minutes. Allele-specific PCR product was determined with the bioluminescent assay.

Bioluminescent detection of PCR products and pyrophosphate

Pyrophosphate solution or PCR product (10 μL), which was diluted 10 times with H_2O, was introduced to a test tube; subsequently, 100 μL of PPDK luciferin-luciferase solution (PPDK reagent) (2.34 U/mL PPDK, 0.2 mM luciferin, 5.5 U/mL luciferase, 0.0125 mM AMP, 0.04 mM PEP, 0.005 U/mL apyrase, 0.05 mM DTT, 5% treharose, 1 mM EDTA, 7.5 mM $MgSO_4$, 30 mM BES, pH 8.0) was added. After 150 seconds, emission intensity was measured for 10 seconds using the luminescence reader (Aloka).

RESULTS AND DISCUSSION

Previously, we developed a bioluminescent detection method for the O157 VT gene, which involved the luciferin-luciferase reaction following transformation of pyrophosphate produced during PCR to ATP by adenosine 5'phosphosulfate and ATP sulfurylase.[1] However, the sensitivity of this technique was insufficient due to slight light emission of APS during the luciferin-luciferase reaction, leading to elevation of the blank value. Furthermore, APS and ATP sulfurylase, which are expensive, display poor stability for utility in routine analysis.

It is known that pyruvate phosphate dikinase (PPDK) catalyzes the conversion of phosphoenolpyruvate (PEP) to ATP, phosphate and pyruvate in the presence of AMP and pyrophosphate.[2] And Sakakibara et al used the PPDK-luciferin/luciferase reaction for measurement of AMP and RNA.[3]

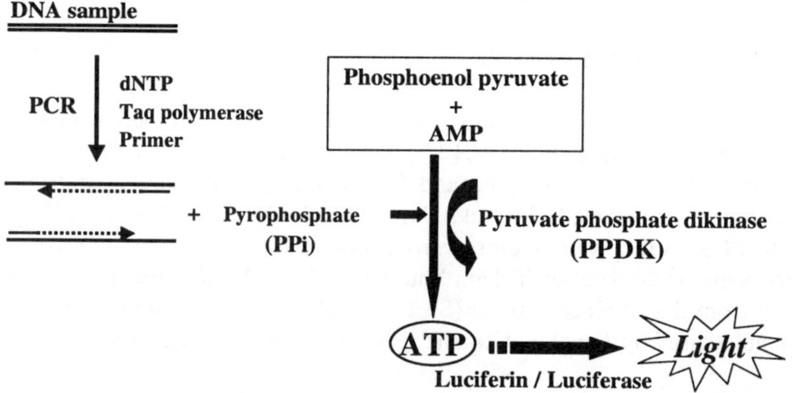

Figure 1. Principle of bioluminescent assay for pyrophosphate by PPDK system

In this study, we developed a novel bioluminescent assay for pyrophosphate using PPDK-luciferin/luciferase reaction. The reaction schema is shown in Fig.1.

Calibration curve of pyrophosphate

In order to develop the PPDK-luciferin/luciferase reaction for measurement of pyrophosphate, the optimal conditions based on the presence of excess AMP were determined as PPDK luciferin-luciferase solution described in Procedure. The calibration curve of pyrophosphate was obtained via this method. Pyrophosphate from 1×10^{-12} to 1×10^{-3} M was examined. The calibration curve of pyrophosphate is illustrated in Fig.2. The detection limit of pyrophosphate was 1.5×10^{-15} mol/assay (as blank + 2SD).

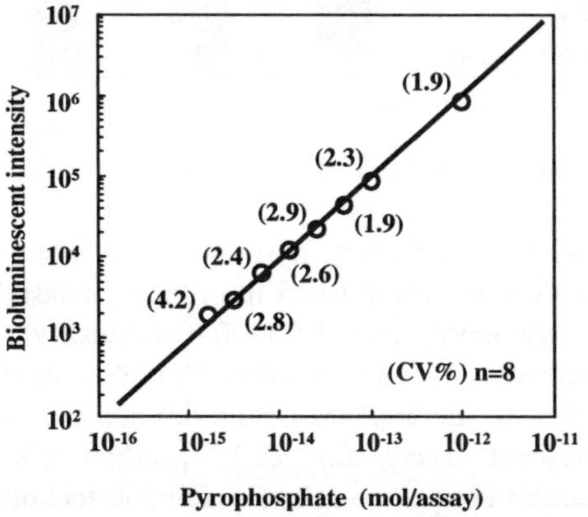

Figure 2. Standard curve for pyrophosphate

Bioluminescent detection of Allele specific PCR products

Streptococcus mutans and *Streptoccocus sobrinus*, which are the primary cariogenic species, play a role in the generation of caries; consequently, these bacteria have been vigorously examined.[4] In this study, the dextranase genes (*dex*) in *Streptococcus mutans* and sobrins were selected as a marker gene for preventive diagnosis of cariogenicity. Allele-specific PCRs were developed for the *dex* genes in *S. mutans* and *S. sobrinus*, pyrophosphate generated during PCR amplification was detected by this novel bioluminescent assay. Allele-specific PCRs (*mutans* and *sobrinus* PCR) employing two specific primer sets for the *dex* gene in *S. mutans* and for the *dex* gene in *S. sobrinus* can be amplified specifically; consequently, *S. sobrinus* and *S. mutans* can be identified readily. Results of bioluminescent detection for the two allele-specific PCR techniques are shown in Fig. 3. Light emission intensity was presented as the S/N ratio (S = signal of PCR product using template DNA and N= noise of PCR product without DNA). When *S. mutans* and *S. sobrinus* DNA which corresponded to 20 ng as bacterial DNA were used, the S/N ratio of *S. mutans*, *S. sobrinus* and the mixture were 13.7, 13.0 and 16.0, respectively. This finding indicates that *S. mutans* and *S. sobrinus* were clearly identified by the respective bioluminescent allele-specific PCR assays.

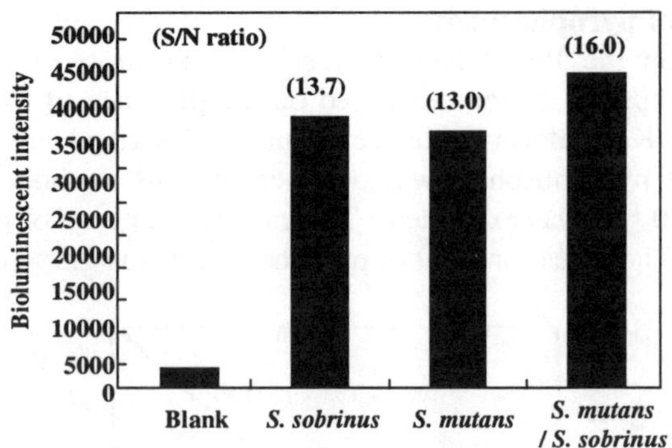

Figure. 3. Bioluminescent intensity obtained by specific PCR products

In conclusion, a novel bioluminescent pyrophosphate assay utilizing the PPDK-luciferin/luciferase reaction was established in order to measure quantitatively PCR products. Detection of pyrophosphate (1.56×10^{-15} mol/assay) was possible with the proposed method. Furthermore, this bioluminescent assay in association with allele-specific PCR was applied to the analysis of the dex gene of *mutans* streptococcus. The novel bioluminescent assay for PCR product based on the PPDK-luciferin/luciferase reaction appears to afford a suitable technique for diagnosis and prevention of bacterial infection and disease.

ACKNOWLEDGEMENTS

We thank Associate Professor T. Igarashi of Showa University, School of Dentistry, for kindly providing with the dex gene of *mutans* streptococcus.

REFERENCES

1.   Imamura O, Arakawa H, Maeda M. Simple and rapid bioluminescent detection of two verotoxin genes using allele specific PCR of *E.coli* O157:H7. Luminescence 2003; 18:107-12.
2.   Wood H.G, O'Brien W.E, Michaels G. Properties of carboxytransphosphorylase; pyruvate,phosphate dikinase; pyrophosphate-phosphofructokinase and pyrophosphate-acetate kinase and their roles in the metabolism of inorganic pyrophosphate. Adv Enzymol 1977; 45:85-155.
3.   Sakakibara T, S.Murakami, Eisaki N, Nakajima M, Imai K. An enzymatic cycling method using pyruvate orthophosphate dikinase and firefly luciferase for the simultaneous determination of ATP and AMP(RNA). Anal Biochem 1999; 268:94-101.
4.   Igarashi T, Yamamoto A, Goto N. Sequence analysis of the *Streptococcusmutans* Ingbritt *dex*A gene encoding extracellular dextranase. Microbiol Imunol 1995; 39:853-60.

BART-NAAT — A NOVEL BIOLUMINESCENT ASSAY FOR REAL-TIME NUCLEIC ACID AMPLIFICATION

OA GANDELMAN, VL CHURCH, JAH MURRAY, LC TISI

Lumora Ltd., Institute of Biotechnology, University of Cambridge,
Tennis Court Road, Cambridge, CB2 1QT, UK
E-mail: o.gandelman@lumora.co.uk

INTRODUCTION

Real-time PCR (RT-PCR) has become a major tool in the life sciences with increasing applications in basic research, medical diagnostics, defence and environmental monitoring. In general, RT-PCR is followed via the detection and quantification of a fluorescent reporter, the signal from which changes in proportion to the amount of amplicon produced in the PCR. As such, sophisticated and expensive hardware is required in order to both thermocycle the samples and follow the fluorescent signals from them. In particular, the ability to follow many thousands of samples simultaneously by RT-PCR, or, to perform RT-PCR in simple portable devices, has proven difficult to do at low cost. We report here the development of a bioluminescent method to follow real-time amplification of DNA, that does not require thermocycling or fluorescence detection of amplicon, and can be adopted for both portable devices and ultra-high throughput with simple hardware at low cost.

In any Nucleic Acid Amplification Technology (NAAT) incorporation of each nucleotide releases one molecule of inorganic pyrophosphate (PP_i) (Scheme 1, reaction 1). Therefore production of amplicon during amplification is characterised by accumulation of PP_i as a by-product. Coupling, in one tube, an isothermal NAAT with quantitative enzymatic conversion of PP_i into ATP and its bioluminometric monitoring using firefly luciferase, allows a nucleic acid amplification to be followed in real-time via bioluminescence (Scheme 1, reactions 1-3). The whole process has exceptionally simple hardware requirements such as a suitable temperature control unit for NAAT and a digital camera for the final step light detection (Fig. 1).

$$\text{DNA/RNA}_{(n)} + \text{dNTP/NTP} \xrightarrow{\text{polymerase}} \text{DNA/RNA}_{(n+1)} + PP_i \quad (1)$$

$$PP_i + \text{APS} \xrightarrow{\text{ATP sulphurylase}} \text{ATP} + SO_4^{2-} \quad (2)$$

$$\text{ATP} + \text{Luciferin} + \text{Oxygen} \xrightarrow{\text{firefly luciferase}} \text{Oxyluciferin} + \text{Light} \quad (3)$$

Scheme 1 $\quad\quad\quad\quad \text{AMP} + PP_i + CO_2$

We demonstrated this principle here by using the isothermal NAAT known as Loop-Mediated Amplification (LAMP) in combination with a modification of the Enzymatic Luminometric Inorganic Pyrophosphate Detection Assay (ELIDA).[1] This novel Bioluminescent Assay for Real-Time NAATs (BART-NAAT) is shown to be quantitative, rapid and cost-effective.

METHODS

The reagents for the latest accelerated-LAMP method[2] were modified along with ELIDA reagent such that the two systems could operate together in the same tube. This 'BART' reagent was used, in conjunction with suitable primers, to detect and quantify a proprietary target sequence of interest.

20 µL samples, containing varying amounts of target template DNA, were placed on a heating block set at 50-65 °C and placed underneath a CCD camera within a Syngene GeneGenius light cabinet (Fig. 1). Over a period of 1 hour, the light emission from the samples was measured 60 times (using Syngene GeneSnap software), each reading being collected with a 60 second integration time. Quantification of the time-dependent light emission of the samples was performed using Syngene GeneTools software.

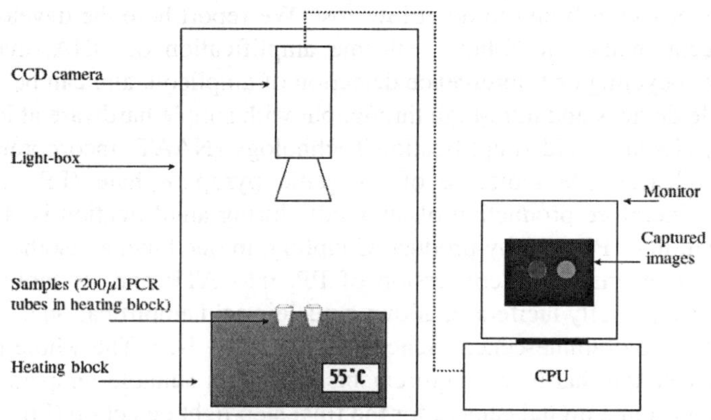

Figure 1. Hardware requirements for BART

RESULTS

A typical time-dependent output of a BART-NAAT reaction is shown in Fig. 2. Initially, the light intensity of samples decreases over time, presumably a result of luciferase reacting with dATP and becoming increasingly inhibited by oxyluciferin. However, as amplification proceeds (as verified by agarose gel electrophoresis) samples producing amplicon start to produce increasing amounts of light. However, soon after this increase in light intensity, a maximum is achieved followed by a rapid decrease in light intensity to a level below that of the control (no amplification). We presume this effect is a result of inhibition of luciferase by high PP_i concentrations.

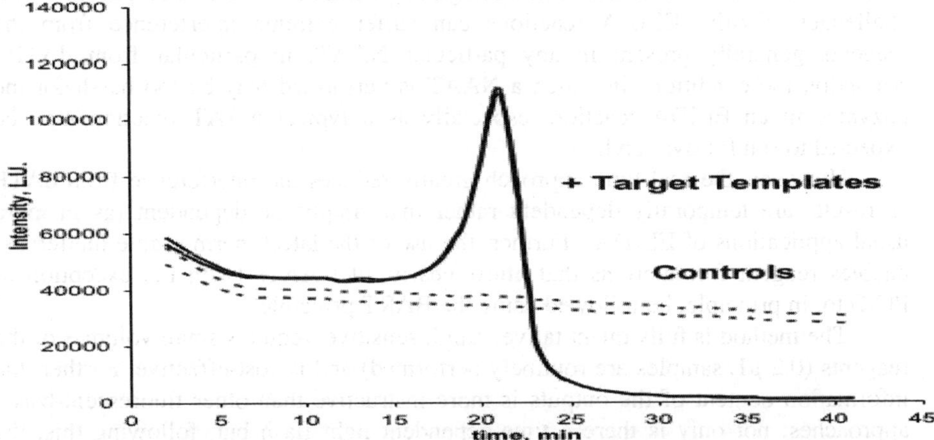

Figure 2. Typical BART outputs, in duplicate: during amplification (+Target Templates) and with no amplification (Controls).

Analysis of the effect of varying the amount of target template demonstrated a key property of BART. As the starting copy number of template decreases, the time taken to reach the maximum of light intensity increases (Fig. 3). Using agarose gel electrophoresis we were able to confirm that the time to light peak is proportional to the amount of DNA amplicon produced. As such, BART is demonstrated to be fully quantitative.

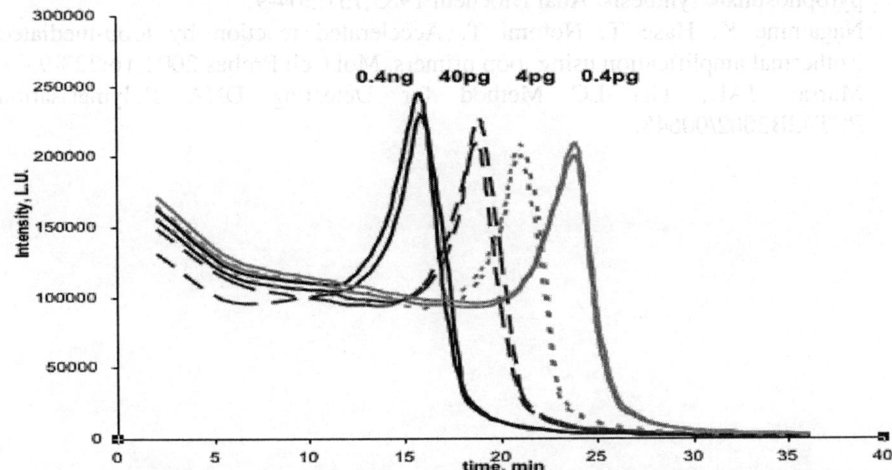

Figure 3. BART reactions with varying amounts of target template

DISCUSSION

In this study we have demonstrated that a bioluminescent reporting system can be used to follow a NAAT in real time. Integrating ELIDA with a NAAT poses several challenges. Firstly, ELIDA reactions can suffer extreme interference from the reagents generally present in any particular NAAT, in particular from dATP.[3] Secondly, the conditions in which a NAAT is performed may be too harsh for the enzymes in an ELIDA reaction, especially as a typical NAAT reaction may be expected to run for over an hour.

However, the real-time approach greatly reduces the interference from dATP as results are temporally dependent rather than amplitude dependent (as in more usual applications of ELIDA). Further, the use of the latest thermostable luciferases enables reagent formulations that allow nearly all NAATs (with the exception of PCR) to, in principle, be followed using the BART principle.

The method is fully quantitative, rapid, sensitive, requires small volumes of the reagents (0.2 μL samples are routinely performed) and is cost-effective. Further, the information content of the outputs is more instructive than other fluorescent-based approaches: not only is there a time-dependent light flash but, following this, the signal then falls to below that of the controls offering a clear end-point indication of whether amplification has occurred or not.

Due to the simplicity of hardware requirements and small volumes, even a gel documentation system could be used for high throughput screening. Further, BART is an ideal format for portable devices using NAATs to detect pathogens and GM.

REFERENCES

1. Nyren P, Lundin A. Enzymatic method for continuous monitoring of inorganic pyrophosphate synthesis. Anal Biochem 1985;151:504-9.
2. Nagamine K, Hase T, Notomi T. Accelerated reaction by loop-mediated isothermal amplification using loop primers. Mol Cell Probes 2002;16:223-9.
3. Murray JAH, Tisi LC Method for Detecting DNA Polymerisation PCT/GB2002/00648.

A SINGLE–STEP BIOLUMINESCENT ENDPOINT ASSAY FOR NUCLEIC ACID AMPLIFICATION TECHNOLOGIES

O GANDELMAN, JAH MURRAY, LC TISI

Lumora ltd., Institute of Biotechnology, Tennis Court Road,
Cambridge, CB2 1QT, UK
l.tisi@lumora.co.uk

INTRODUCTION

Nucleic Acid Amplification Technologies (NAATs) such as PCR, are generally followed (often in real-time) by some means to detect the accumulation of nucleic acid amplicon. However, as pyrophosphate is a by-product of nucleic acid biosynthesis, the generation of pyrophosphate can also be used to follow NAATs.

The Enzymatic Luminometric Inorganic Pyrophosphate Detection Assay (ELIDA) has become well established as a means to follow DNA polymerisation.[1] Further, attempts have been made to use ELIDA as an end-point assay for PCR to determine whether or not amplification has occurred and to what extent.[2-4] In these cases, the level of pyrophosphate detected by ELIDA is shown to reflect the accumulation of amplicon. However, since PCR reagent contains high concentrations of all the dNTPs and since dATP, in particular, is a substrate for firefly luciferase (a key component of ELIDA) careful assay design is required. In particular, if ELIDA reagent is directly mixed with conventional PCR reagent, complicated and time-dependent light outputs may be observed making quantification of pyrophosphate difficult. As a result, some attempts to use ELIDA to quantify PCR have required multi-step processes that helped to minimise the interference from dATP in particular.[2-3]

Here we demonstrate an alternative to multi-step approaches that use ELIDA for endpoint assays of PCR reactions.[4] We show that by optimising PCR reagent formulations to work with d-α–S-ATP instead of dATP, ELIDA reagent can be directly mixed with the resulting PCR reaction giving quantitative results in a one-step assay.

METHODS

Conditions were sought that allowed d-α–S-ATP to successfully replace dATP in PCR reactions where a short 96bp amplicon was amplified from a test plasmid system. The resulting formulation was:

5 μL	Tris-HCl	100 mM	Sigma
4 μL	Magnesium Chloride	50 mM	Gibco
5 μL	d-α-S-ATP	2 mM	Glen research
5 μL	dCTP	2 mM	Pharmacia
5 μL	dGTP	2 mM	Pharmacia
5 μL	dTTP	2 mM	Pharmacia
1 μL	Test plasmid	0.5 ng/μL	
1.25 μL	Primer 1	10 μM	
1.25 μL	Primer 2	10 μM	
0.5 μL	Taq polymerase	5 U/μL	Roche
19 μL	Milli-Q water		

ELIDA reagent was made up as follows:

0.1 M	Tris-acetate (pH7.75)	Sigma
2 mM	EDTA	"
10 mM	Magnesium Acetate	"
0.1 %	Bovine serum albumin	"
5 μM	Adenosine 5' phosphosulphate	"
0.4 mg/mL	Polyvinylpyrrolidone (360,000)	"
0.3 U/mL	ATP Sulphurylase	"
100 mg/mL	D-luciferin	Europa
5.5×10^8 LU	Photinus pyralis luciferase	Promega
1 mM	Dithiothreitol	Melford

PCR reactions were performed using a Perkin-Elmer 'GeneAmp PCR system 2400' and run for multiples of 5 cycles up to 30 cycles. PCR reactions to be assayed were directly mixed 1:1 (20 μL) with ELIDA reagent and light emission immediately followed using a Luminoscan Ascent luminometer over various time periods.

RESULTS

To assess whether a single-step ELIDA-based assay could be used to follow a PCR reaction containing d-α−S-ATP instead of dATP, PCR samples were run for increasing numbers of cycles and assayed as described. The light emission from samples was shown to reflect the accumulation of amplicon as confirmed by agarose gel electrophoresis (Fig. 1).

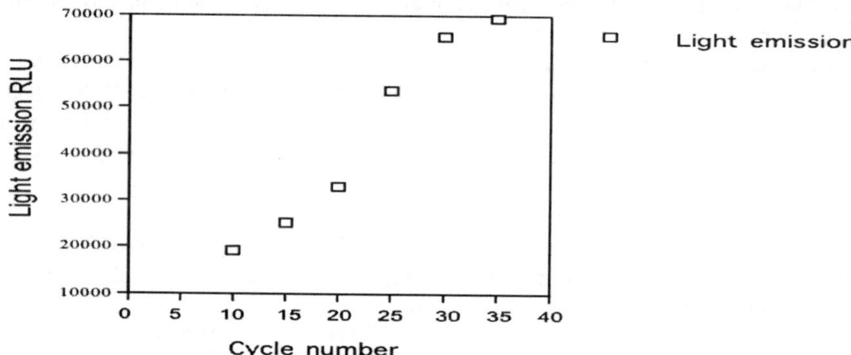

Figure 1. Light emission from single-step 'mix and measure' ELIDA for PCR run with d-α−S-ATP

For comparison, single-step ELIDA assays were used to follow a PCR reaction that had been run using dATP compared to one that had been run with d-α−S-ATP. PCR reactions were run for either 5 or 35 cycles. After mixing the ELIDA reagent with the respective PCR reactions, the light emission from the samples was measured over a period of 50 seconds. As can be seen from Fig. 2, the sample using dATP showed considerably more light emission than the sample using d-α–S-ATP, at both 5 cycles (when very little amplicon would have been generated) and 35 cycles (where amplicon had been produced, as confirmed by agarose gel electrophoresis). Hence the background light emission using dATP was substantially higher than when using d-α–S-ATP. Both the dATP sample and the d-α–S-ATP, at time zero, demonstrated a similar increase in light emission in the sample after 35 cycles, when significant amplicon had been formed. However, it can be seen that the light emission from the dATP sample decays substantially with time, whereas the light emission from the d-α–S-ATP remains relatively constant.

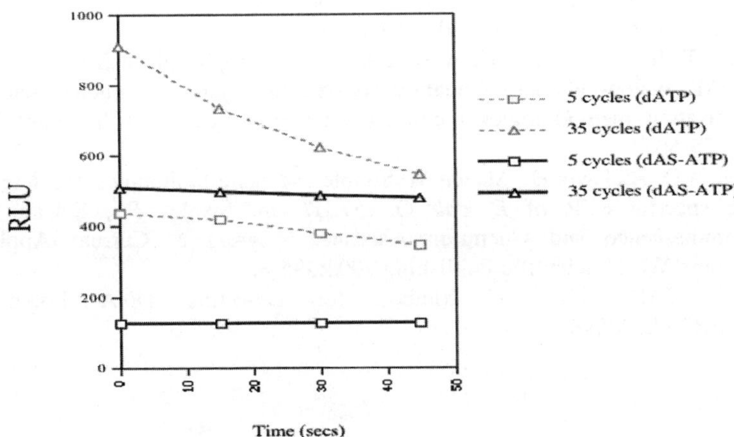

Figure 2. Showing ELIDA assays for PCR reactions run for 5 or 35 cycles using either dATP or d-α–S-ATP

DISCUSSION

The fact that dATP reacts with firefly luciferase has a number of serious consequences for attempts to use an ELIDA reaction to measure PPi in normal PCR reagent. When ELIDA reagent is directly mixed with PCR reagent containing dATP, a high background of light emission from the reaction of dATP with firefly luciferase is observed. When dATP is substituted with d-α–S-ATP in PCR however, the background light emission is greatly reduced.

However, a further consequence of the presence of high levels of dATP when attempting to use an ELIDA to measure PPi in a PCR reaction is the effect on the rate of decay of light emission. Fig. 2 demonstrates that the rate of decrease in light emission is significantly higher in the presence of dATP, compared to d-α–S-ATP where the light output is constant. In fact, in the presence of dATP, the decay in light emission becomes greater as the concentration of PPi increases. This makes single-step measurements extremely time sensitive as depending when the light emission is read, very different results can be obtained.

Previously, the difficulties caused by dATP have, essentially, been addressed by diluting dATP and pyrophosphate before adding ELIDA reagent (or some key part of the ELIDA reagent). This necessarily requires at least two steps to the assay. However, we demonstrate here that an alternative approach is to use d-α–S-ATP instead of dATP in the PCR reaction. As a result, quantitative and stable light outputs can be obtained.

REFERENCES
1.   Nyren P, Lundin A. Enzymatic Method for Continuous Monitoring of Inorganic Pyrophosphate Synthesis. Anal Biochem 1985;151:504-9.
2.   Tabary T, Ju L, Cohen J. Homogeneous phase pyrophosphate (PPi) measurement (H3PIM). A non-radioactive, quantitative detection system for nucleic acid specific hybridisation methodologies including gene amplification. J Immunol Methods 1992; 55-60.
3.   Imamura O, Arakawa H., Maeda M. Simple and rapid bioluminescent detection for allele specific PCR of *E. coli* O 157:H7. In: Stanley PE, Kricka LJ. eds. Bioluminescence and Chemiluminescence: Progress & Current Applications. Singapore:World Scientific Publishing 2002;395-8.
4.   Murray JAH, Tisi LC Method for Detecting DNA Polymerisation PCT/GB2002/00648.

BIOLUMINESCENT DETECTION OF RNA HYDROLYSIS PROBES IN DNA TESTING

O GANDELMAN[1], LC TISI[1], PJ WHITE[2], JAH MURRAY[1], DJ SQUIRRELL[2]

[1]*Lumora Ltd, Institute of Biotechnology, Cambridge, CB2 1QT, UK*
[2] *Dstl Porton Down Salisbury, Wiltshire, SP4 0JQ,UK*

INTRODUCTION

Bioluminescence using firefly luciferase provides great sensitivity and a large dynamic range for the measurement of adenosine triphosphate (ATP). Bioluminescence has recently been used to measure DNA amplification through the inorganic pyrophosphate (PPi) produced from polymerisation.[1] The PPi is reacted with AMP-phosphosulphate (AMP-S) to produce ATP which luciferase can then use to generate light. The process can be summarised as follows:

Enzyme	Reaction
DNA polymerase	template + primers + dNTPs $\rightarrow$ DNA + PPi + dNMPs
Sulphurylase	PPi + AMP-S $\rightarrow$ ATP
Luciferase	ATP + LH$_2$ $\rightarrow$ AMP + PPi + Light

(where dNTPs and dNMPs are deoxynucleotide triphosphates and monophosphates, respectively, and LH$_2$ is D-luciferin). The time profile of the production of the light signal is dependent upon the initial concentration of target DNA and the deoxyadenosine triphosphate (dATP) needed for DNA synthesis does not interfere in real-time measurements.

Specificity and quantification in amplification monitoring could be enhanced through the use of a probe that is complementary to the intended amplification product and which can be coupled to ATP production in proportion to amplicon concentration. A system using an RNA probe is proposed:

Enzyme(s)	Reaction
DNA polymerase / RNAase	template + primers + dNTPs + RNA probe $\rightarrow$ DNA + PPi + dNMPs + NMPs (including AMP)
Phosphotransferase (s)	AMP + phosphate donors $\rightarrow$ ATP
Luciferase	ATP + LH$_2$ $\rightarrow$ AMP + PPi + Light

During DNA amplification, the probe is hydrolysed by 5'-3' exonuclease activity specific to RNA-DNA duplexes and the AMP generated is converted into ATP.

The first step in realising this concept required the development of an assay for the detection of 5'-AMP *via* firefly luciferase that can work solutions containing dNTPs and other components of a nucleic acid amplification reaction. A number of methods to detect NMPs *via* coupling to luciferase are known.[2-4] Taking into account the needs for high sensitivity, a large and linear dynamic range, and good signal stability in relation to nucleic acid amplification reaction conditions, the following AMP-bioluminescent assay was developed:

Enzyme	Reaction
Adenylate kinase (AK)	$AMP + dATP \rightarrow ADP + dADP$
Acetate kinase (AcK)	$ADP + acetate\ phosphate \rightarrow ATP + CH_3COOH$
Firefly luciferase	$ATP + LH_2 \rightarrow AMP + PPi + CO_2 + oxyluciferin + Light$

Although ATP is the main substrate for AK, dATP can substitute for it and react with greater than 60% of the relative velocity.[5] Conveniently, dATP is provided in the reaction mix as one of the four dNTP substrates for DNA synthesis.

To complete the assay, a means with which to degrade DNA-RNA heteroduplexes to give 5'-NMPs is needed. This is not quite as straightforward a task as might be expected. Oligomers and 3'-NMPs rather than 5'-NMPs, are the most common degradation products from the activities of nucleases, but 5'-AMP is the product of choice for coupling ATP-bioluminescence to nucleic acid amplification *via* RNA degradation. At this stage of the work experiments have been carried out with poly A/oligo dT as a model RNA/DNA heteroduplex substrate that provides the maximum amount of AMP.

MATERIALS & METHODS

AK, AcK, 5'-AMP, acetyl phosphate and dNTPs were obtained from Sigma-Aldrich Ltd, Poole, UK. Recombinant firefly luciferase and D-Luciferin were obtained from Promega Corporation, Madison, WI, USA. Bioluminescent assays were performed, at room temperature, in 100 μL volumes in 96 well microtitre plates using a Labsystems Luminoskan Ascent plate luminometer from ThermoLabsystems, Basingstoke, UK. The assay mix contained 100 μmol/L of each dNTP, 10U/mL AK, 2.5mmol/L acetyl phosphate, 1U/mL AcK, 146 ng/mL wild type recombinant luciferase and 500 μmol/L D-luciferin in 50 mmol/L Tris-HCl, 10 mmol/L $MgSO_4$, 1mmol/L ethylenediaminetetraacetic acid and 1mmol/L dithiothreitol, pH 7.9. Poly rA/oligo dT as "poly(rA)·p(dT)$_{12\text{-}18}$", cat. no. 27-7878-01, was obtained from Amersham Biosciences, Chalfont St. Giles, UK.

RESULTS

To test and optimise the coupled reaction, RNA hydrolysis was simulated by the addition of AMP as a preformed product to a complete nucleic acid amplification cocktail. Additions of AMP gave linear bioluminescent responses over a range of concentrations from 0.1 to 10 μmol/L AMP (Fig. 1). The ability to detect RNA-DNA heteroduplex hydrolysis with this assay was tested using a proprietary exonuclease degradation system. 5'-AMP released from the poly rA/oligo dT probe:template heteroduplex was measured using the triple enzyme coupled bioluminescent assay described above. A serial dilution of the poly rA/oligo dT was used to test the linearity of the bioluminescent response. The sensitivity of the assay was sufficient to measure the AMP content of the RNA-DNA heteroduplex over a range of 1-50 μmol/L (Fig. 2).

CONCLUSION

A bioluminescent assay for the detection of amplicon produced during nucleic acid amplification is being developed. The assay relies on the hydrolysis of an RNA probe using a proprietary exonuclease degradation system to release 5'-AMP which can then be converted to ATP *via* AK and AcK. Probe hydrolysis and detection of 5'-AMP have been demonstrated using poly rA/oligo dT in a model system. Currently this method could be used as an end point assay after nucleic amplification however a real-time detection system should be achievable using thermostable enzymes and/or low temperature amplification protocols. Realisation of the full assay system could allow much simpler equipment to be used for quantitative nucleic acid amplification assays and facilitate the parallel processing of multiple assays in, for example, high throughput screening applications.

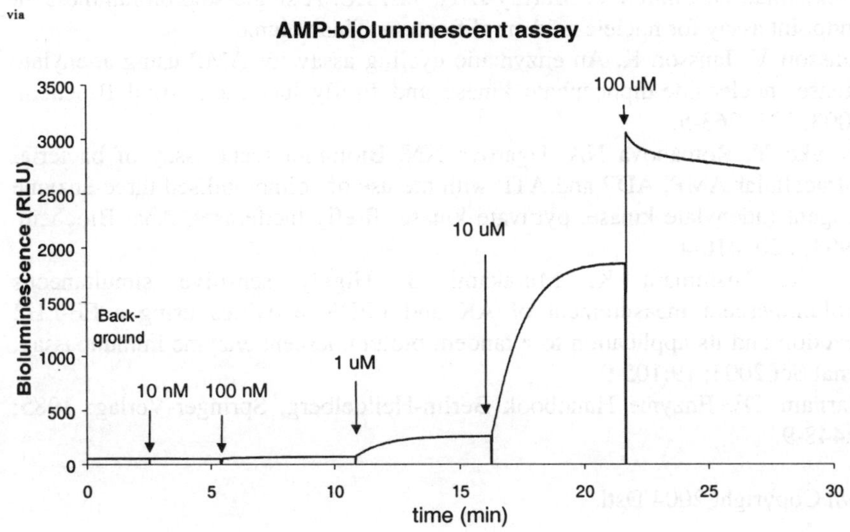

Figure 1. Bioluminescent response of the triple coupled assay to addition of AMP

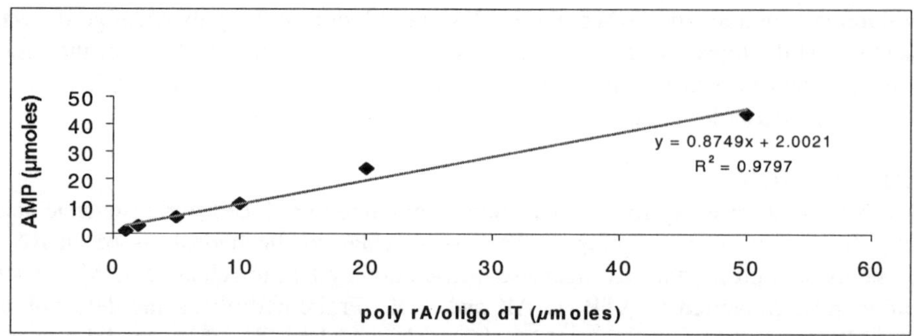

Figure 2. AMP released through exonuclease-catalysed hydrolysis of poly rA/oligo dT and assayed using the AMP bioluminescent assay.

ACKNOWLEDGEMENT

This work was supported financially by the Dstl Technology Transfer Investment Fund.

REFERENCES

1. Gandelman O, Church V, Murray JAH, Tisi LC. A single-step bioluminescent endpoint assay for nucleic acid amplification. This volume
2. Jansson V, Jansson K. An enzymatic cycling assay for AMP using adenylate kinase, nucleoside-diphosphate kinase and firefly luciferase. Anal Biochem. 2003; 321: 263-5.
3. Brovko Y, Romanova NA, Ugarova NN. Bioluminescent assay of bacterial intracellular AMP, ADP and ATP with the use of coimmobilised three-enzyme reagent (adenylate kinase, pyruvate kinase, firefly luciferase). Anal Biochem. 1994; 220: 410-4.
4. Ito K, Nishimara K, Murakami S. Highly sensitive simultaneous bioluminescent measurement of AK and PPDK activities using a FF-LH2 reaction and its application to a tandem bioluminescent enzyme immunoassay. Anal Sci 2003; 19:105-9.
5. Barnam TE. Enzyme Handbook Berlin-Heildelberg, Springer-Verlag: 1985; 1:448-9.

FLASHING A PROTEIN-PROTEIN INTERACTION IN LIVING CELLS VIA SPLIT *RENILLA* LUCIFERASE COMPLEMENTATION

A KAIHARA, Y UMEZAWA

Department of Chemistry, School of Science, The University of Tokyo
Hongo, Bunkyo-ku, Tokyo, 113-0033, Japan
E-mail: akaihara@chem.s.u-tokyo.ac.jp

INTRODUCTION

For spatial and quantitative kinetic analysis of protein-protein interactions (PPIs) in living mammalian cells, we have developed a split *Renilla* luciferase complementation method.[1] It relies on the spontaneous emission of luminescence upon PPI-induced complementation of the split *Renilla* luciferase, with a cell membrane permeable substrate, coelenterazine. Unlike diffusive products involved in other complement enzyme systems, this split *Renilla* luciferase complementation readout is capable of locating the PPIs with emission of bioluminescence only at the sites and time of their occurrence in living cells. *Renilla* luciferase is one of the major reporter proteins for optical imaging studies in living cells and rodents. It catalyzes the oxidation of coelenterazine by O_2 to excited-state oxycolenterazine monoanion that emits light as a broad band (400 nm~630 nm) covering a tissue-transparent near-infrared region.

The split *Renilla* luciferase complementation strategy (Fig. 1) was used for visualizing a known PPI between Y941 peptide and n-terminal SH2 domain (SH2n) upon protein phosphorylation in living Chinese hamster ovary cells overexpressing human insulin receptors (CHO-HIR).[2,3] Cells were expressed with the two separated proteins, the n-terminal half of split *Renilla* luciferase connected to Y941 and the SH2n connected to the c-terminal half of split *Renilla* luciferase. Upon insulin stimulation, Y941 peptide is phosphorylated by insulin receptor and interacts with the SH2n from p85 the subunit ($p85_{330-429}$) of phosphatidylinositol 3-kinase. This interaction simultaneously leads to formation of the complement *Renilla* luciferase, thereby spontaneously emitting bioluminescence by reaction with its cell membrane permeable substrate, coelenterazine *in situ* in living cells.

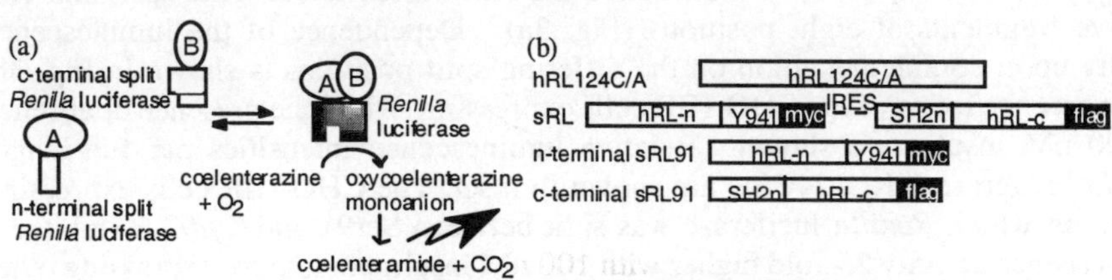

Figure 1(a). A schematic diagram of
the split *Renilla* luciferase complementation strategy
(b). Schematic representation of the plasmid constructs

METHODS

Two separated proteins of sRL were expressed with plasmid construct (Fig. 1b). The

"

amino-acid sequence of Y941 is TEEAYMKMDLGPG. Luminescence intensity was measured with a Minilumat LB9506 luminometer (Berthold, Wildbad, Germany) for 10 s. Supernatant protein concentration was assessed by the Bradford method. Cells were imaged at room temperature on a Carl Zeiss Axioverts100 microscope with a cooled CCD camera MicroMax (Roper Scientific Inc, Tucson, AZ), controlled by Till Vision V3.02 (PHOTONICS, Planegg, Germany: 40 x oil immersion objective).

RESULTS AND DISCUSSION
Luminescent activity of r*Renilla* luciferase in mammalian cells
The synthetic *Renilla* luciferase gene (hRL) was systematically designed by changing codons to those most frequently used in mammals from native *Renilla* luciferase gene (pRL) to increase the expression and reliability of control reporter vectors in mammalian cells. To increase luminescent intensity in accordance with previous reports,[4] 124-cysteine residue in *Renilla* luciferase was replaced with alanine (124C/A). Luminescence of the cells expressed with pRL, pRL124C/A, hRL, hRL124C/A were assessed. The luminescence of the CHO-HIR cells expressed with hRL124C/A (Fig. 2) was 1.4 x 10^3 fold higher than that with pRL. The CHO-HIR cells expressing full-length *Renilla* (hRL124C/A) emitted luminescence uniformly throughout the cells. We used the *Renilla* luciferase mutant hRL124C/A for the split *Renilla* luciferase system to locate protein-protein interaction in mammalian cells.

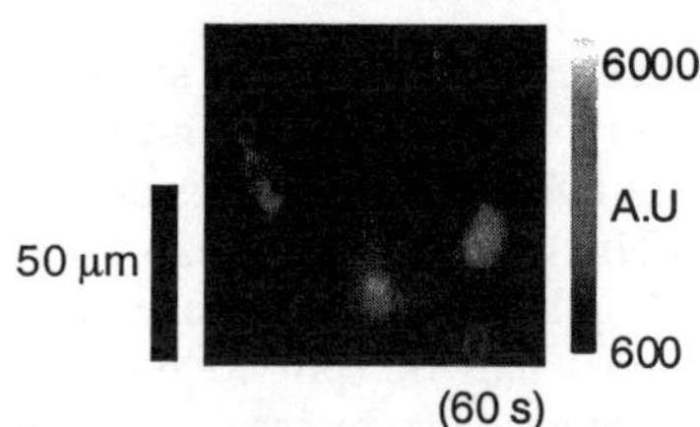

Figure 2. Luminescence image of the cells expressing *Renilla* luciferase (hRL124C/A)

Efficiency of split *Renilla* luciferase complementation
Efficient complementation of a split *Renilla* luciferase fusion protein is needed for it to act as a probe for PPIs. We examined the *Renilla* luciferase gene split into two inactive fragments at eight positions (Fig. 3a). Dependence of the luminescence activity upon complementation on the differing split positions is shown in Fig. 3b. Luminescence ratios of the CHO-HIR cells expressing sRL in the presence or absence of 100 nM insulin are shown. Relative luminescence intensities per full-length *Renilla* luciferase (hRL124C/A) are shown in inset. The CHO-HIR cells expressing sRL91, in which *Renilla* luciferase was split between Ser91 and Tyr92, exhibited a luminescence activity 25-fold higher with 100 nM insulin, but, those expressing other split *Renilla* luciferase fusion proteins exhibited only a 2- to 4-fold increase with 100 nM insulin. It is concluded that PPI-induced complementation of split *Renilla* luciferase was exclusively observed with sRL91, which was demonstrated by an interaction protein pair, Y941 and SH2n, as an example.

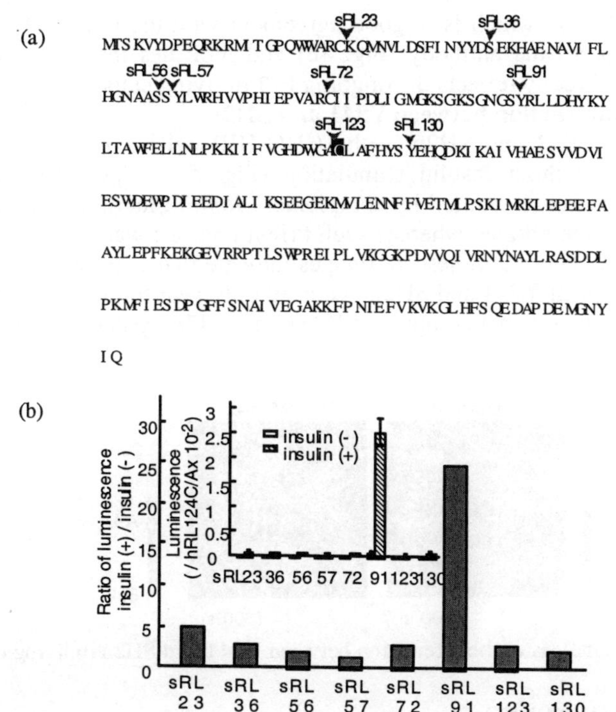

Figure 3(a). Amino-acid sequence of *Renilla* luciferase and split positions

(b). Luminescence analysis of split *Renilla* luciferase fusion proteins
for efficient complementation

Time course of the interaction between Y941 and SH2n

The luminescence intensities of interaction between Y941 and SH2n observed with sRL91 increased within 5 min after insulin stimulation and then gradually decreased (Fig 4a). This time dependence of the interaction is due to tyrosine phosphorylation

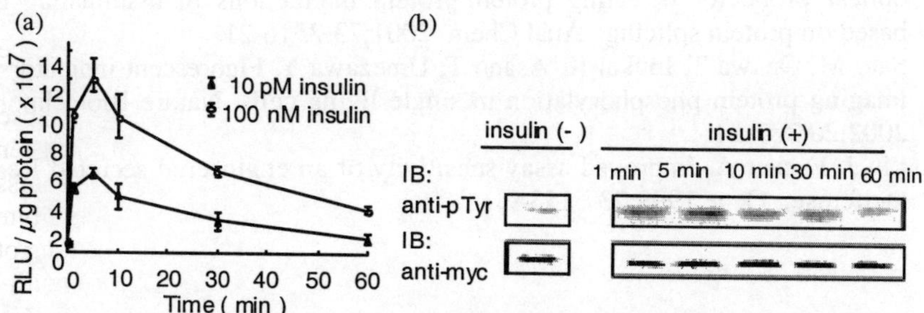

Figure 4(a). Time course of the luminescence upon sRL91 complementation
(b). Immunoblot analysis of tyrosine phosphorylation and dephosphorylation on Y941

and dephosphorylation, which is in good agreement with the immunoblot analysis with anti-phosphotyrosine antibody (Fig. 4b) and indicates that the luminescence activity of sRL91 directly reflects the ongoing PPI in living cells.

Location of the interaction between Y941 and SH2n
The PPI between Y941 and SH2n in the CHO-HIR cells expressing sRL91 was imaged with and without insulin stimulation (Fig. 5). Upon 100 nM insulin stimulation, luminescence emitted by complement *Renilla* luciferase increased only near to the plasma membrane, whereas such bright contrast was not observed in the absence of insulin. These luminescent images indicate that with insulin stimulation, the interaction between Y941 and SH2n occurred only near to the plasma membrane in the cytosol, and the interact complex of Y941 and SH2n existed only there without any diffusion.

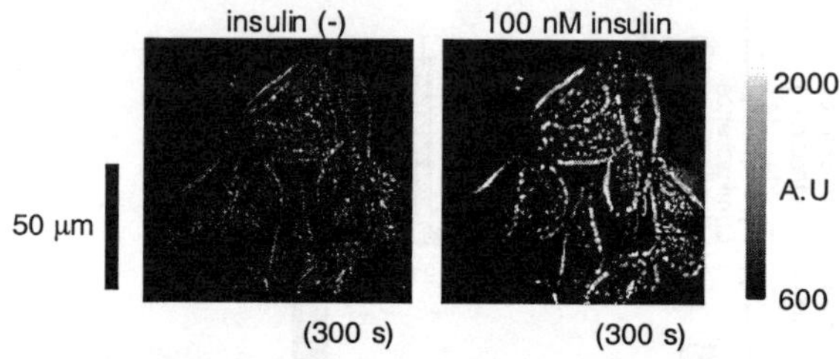

Figure 5. Spatial analysis of the interaction between Y941 and SH2n in living cells.

ACKNOWLEDGEMENTS
This work was supported by Japan Science and Technology Agency (JST) and Japan Society for the Promotion of Science (JSPS). AK thanks JSPS for a fellowship.

REFERENCES

1.   Kaihara A, Kawai Y, Sato M, Ozawa T, Umezawa Y. Locating a protein-protein interaction in living cells via split Renilla luciferase complementation. Anal Biochem 2003;75:4176-81.
2.   Ozawa T, Kaihara A, Sato M, Tachihara K, Umezawa Y. Split luciferase as an optical probe for detecting protein-protein interactions in mammalian cells based on protein splicing. Anal Chem 2001;73:2516-21.
3.   Sato M, Ozawa T, Inukai K, Asano T, Umezawa Y. Fluorescent indicators for imaging protein phosphorylation in single living cells. Nature Biotechnology 2002;20:287-94.
4.   Liu J, Escher A. Improved assay sensitivity of an engineered secreted Renilla luciferase. Gene 1999;237:153-9.

DNA ANALYSIS METHOD BY LUCIFERASE-BASED BIOLUMINESCENCE DETECTION AND A MINIATURIZED LUMINOMETER FOR BIOLUMINESCENCE ASSAY

M KAMAHORI, K HARADA, H KAMBARA

Hitachi, Ltd., Central Research Laboratory,
1-280, Higashi-koigakubo, Kokubunji-shi, Tokyo 185-8601, Japan

INTRODUCTION

Bioluminescence detection is widely used in the biomedical and environmental fields. A luminometer for a bioluminescence assay has several advantages; no excitation light source such as a laser and operation is simple. Also, the background noise of bioluminescence detection is lower than that of fluorescence detection, which is the most common method in DNA analysis. Since the DNA sequencing method based on a real-time PPi assay, "pyrosequencing", was first reported,[1] many researchers have used it for short-sequencing and Single Nucleotide Polymorphisms (SNPs) typing. The commercial luminescence detection system is expensive and bulky because a cooled CCD camera system is utilized. Therefore, miniaturized low-cost luminescence detection systems are needed increasingly in various fields. We have developed a new SNP typing method (BAMPER; bioluminometric assay coupled with modified primer extension reactions).[2] This method is based on specific primer extension reactions combined with bioluminescence assay. A miniaturized luminometer coupled with a photodiode array and an air-driven micro-dispenser has been developed for this method.

METHODS

Reagents:

DNA polymerase I, Klenow Fragment, EXO(-) was obtained from Funakoshi (Tokyo, Japan). Luciferase, Adenosine-5'-triphosphate sulfurylase, adenosine 5'-phosphosulfate sodium salt, D-Luciferin sodium salt, and magnesium acetate Tetra hydrate were obtained from Sigma (MO, USA). Deoxynucleotide and 2'-deoxyadenosine 5'-O-(1-thiotriphosphate) were obtained from Amersham Pharmacia Biotech (UK). Other chemicals were of an analytical-reagent grade.

Instruments:

A miniaturized luminometer consists of four micro-dispensers, four micro-cells, and a photodiode array (Fig.1(a)). The micro-dispensers consisted of capillary tubes placed in each cell. A high photoemission collecting efficiency was about 7% because the photodiode array was closely positioned under the micro-cells. Bioluminescence from the micro-cells was simultaneously detected with the photodiode array (HAMAMATSU S1133-01, Japan) placed on a base plate that had in-house-made amplifiers. A multifunctional DAQ (National Instruments PCI-MIO-16XE-50, TX, USA) and National Instruments LabVIEW 6i were used for

540   *Kamahori M et al.*

acquiring out-put signals from the photodiode array.

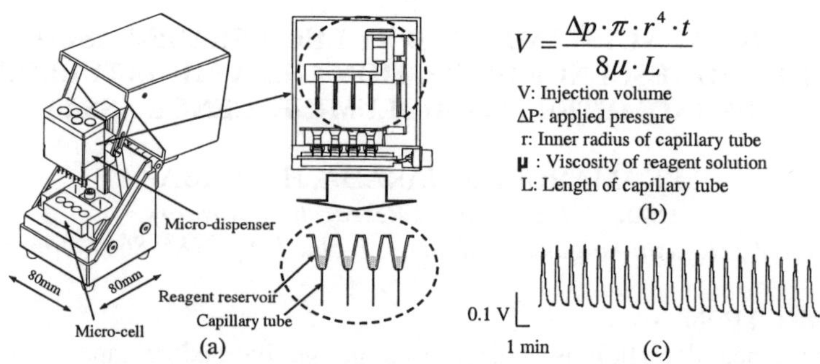

$$V = \frac{\Delta p \cdot \pi \cdot r^4 \cdot t}{8\mu \cdot L}$$

Figure 1.   A schematic view of the luminometer and the principal of dispenser

<u>Bioluminometric assay for SNP typing</u>:
DNA fragments from P53 gene exon 8, as shown in Fig.2, were used as samples. One μL of annealing buffer (100 mmol/L Tris-acetate, pH 7.75, 20 mmol/L magnesium acetate) and 1 μL genome typing primer was added to the template DNA at a total volume of 10 μL. Hybridization was performed by incubating it at 94°C for 20 s at 65 °C for 2 min and then cooling it down to 4 °C. The reaction solution contained 0.1 mol/L Tris-acetate (pH 7.75), 0.5 mmol/L EDTA, 5 mmol/L magnesium acetate, 0.1% bovine serum albumin, 1 mmol/L dithiothreitol, 0.2 U/μL exo(-) Klenow Fragment, 1.0 U/ml ATP sulfurylase, and 2 mg/ml Luciferase. Template DNA/ primer hybridized solution of 1.0 μL and substrate solution (25 μmol/L dNTPs and 1.0 μmol/L APS) of 1.0 μL were added to 4.0 μL of the reaction solution. Bioluminescence from cells was detected by the miniaturized lumonimeter after starting the reaction by adding 20 mmol/L D-Luciferin of 0.1 μL .

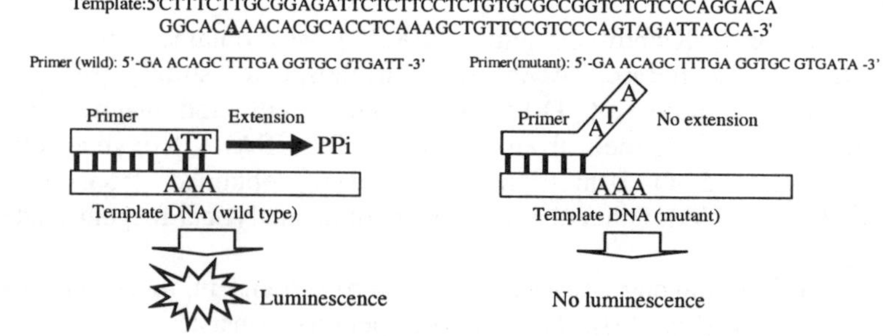

Figure 2.   SNP typing by BAMPER assay

RESULTS

The miniaturized luminescence detection system for SNP typing uses a bioluminometric assay based on the BAMPER method as shown in Fig.2. The key point for the miniaturized luminometer for DNA analysis is how to deliver a small reagent to the reaction device easily and how to detect very weak light emission efficiently.

We designed the delivery system using air-driven force to dispense a sub-μL volume. The principle of the air-driven delivery system for introducing reagents is to use a viscous flow generated by a pressure drop in a narrow capillary. The injection volume can be defined easily by controlling the applied pressure and time from the Hagen-Poiseuille equation as shown in Fig. 1 (b). Each dispenser using a capillary (internal diameter, 25 μm: length, 20 mm) is operated by air-driven force in order of 1×10^4-5×10^4 Pa and can supply 0.01-10 μL of reagent to each cell. The reproducibility (RSD) of the multiple injections (injection volume; 0.05 μL) is 3.4% (n=20) as shown in Figure 1 (c). This reagent delivery system is much simpler and easier than other micro-dispensers such as a piezo-electrically driven pump.

The position of the photodiode array is close to the bottom in each cell for efficient utilization of light and for easy adjustment to the cells. The light from the cell is detected by the photodiode array in the opposite side of the micro-dispenser. The quantitative relationship between signal intensities and ATP, pyrophosphate (PPi) amounts is shown in Fig. 3. The reaction volume was 5 μL under typical analysis condition. The correlation coefficients of the linear regression equations were greater than 0.998 in the range of 1×10^{12}-1×10^{15} mol. The detection limits at a signal-to-noise ratio of 5 were 70 amol for ATP and 100 amol for PPi.

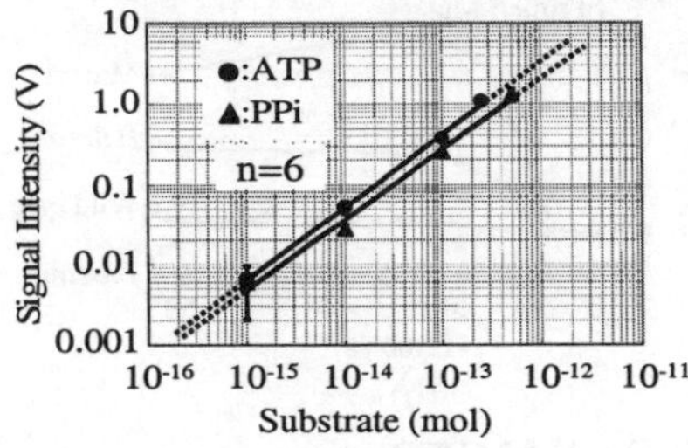

Table 1. Efficiency of each enzyme reaction

Enzyme	Substrate	Product	Yield (%)
DNA Polymerase	dNTP	PPi	48 $\pm$ 10
ATP sulfurylase	PPi	ATP	81 $\pm$ 3

n=6

Figure 3. ATP and PPi substrate curves

The SNP typing using the single base extension method is carried out by observing bioluminescence caused by incorporation of dNTP complementary to the allele species in the target. This produces only one PPi for one-target, and therefore the obtained signals are small. The BAMPER method produces large amount of PPi for one target as shown in Fig. 2. In this method, artificial-mismatch bases at the third position from 3' terminus of the specific primers are introduced to improve the switching characteristics in the primer extension reactions and results in high sensitivity and high selectivity.[2] Though one primer hybridized to a target DNA produces about one hundred PPi in the strand-extension reaction, the same amount of ATP is not produced because the cascade reaction efficiency is lower than 100%. The efficiency of each enzyme reaction is listed in Table 1. The overall efficiency of the enzyme reaction was 40-50%. The detection limit depends on not only the signal but also the background noise. APS acts as the substrates for luciferin. Although the APS reaction efficiency is about 0.16% of ATP (substrate of luciferase), the amount of APS is much larger than the ATP amount produced in the assay. It causes a large background noise. Under the optimized condition, the detection limit is 50 fmol of the target per one base extension. Assuming that the length of a DNA template is 100 bases, the detection limit of the BAMBER method is less than 1 fmol.

The SNPs typing results using a wild type primer and mutant primer are shown in Fig. 4 (a) and (b). Three targets, wild type, mutant, and hetero were measured. The bioluminescence signals of the wild and mutant primers hybridized to the proper targets are clearly distinguished. The signal intensities from the mismatched primer-target pairs are about one-tenth of the intensities obtained with the matched primer-target pairs. When the sample contains a hetero target, the signals are half the intensity of the matched case. It was very easy to determine the types of SNPs using the BAMPER method. As the signal intensities become almost constant after 20 s, the SNPs typing by the BAMPER method should be carried out between 20 and 60 s after a D-Luciferin injection. When the target DNA amount was reduced to 5 fmol, SNP typing was carried out with a good signal to background ratio (data not shown). In conclusion, this miniaturized luminometer holds promise for achieving a low-cost DNA analysis system because of its compact size and easy operation.

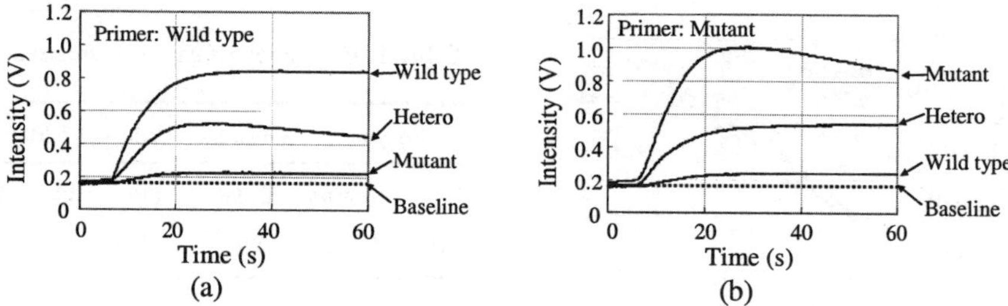

Figure 4.   SNP typing results by BAMPER assay

ACKNOWLEDGEMENTS

This work was performed as a part of the research and development project of Industrial Science and Technology Program supported by New Energy and Industrial Technology Development Organization in Japan.

REFERENCES

1.   Ronaghi M, Uhlen M, Nyren P. A sequencing method based on real-time pyrophosphate. Science 1998; 281: 363, 365.
2.   Zhou G, Kamahori M, Okano K, Chuan G, Harada K, Kambara H. Quantitative detection of single nucleotide polymorphisms for a pooled sample by a bioluminometric assay coupled with modified primer extension reactions (BAMPER). Nucleic Acids Res 2001; 29: e93

OPTIMISATION OF CONDITIONS FOR THE USE OF A NOVEL BIOLUMINESCENT REPORTER SYSTEM IN *MYCOBACTERIUM* SPP.

S WILES, K FERGUSON, B ROBERTSON, D YOUNG

Centre for Molecular Microbiology and Infection,
Imperial College London, SW7 2AZ, UK
Email: siouxsie.wiles@imperial.ac.uk

INTRODUCTION

Bioluminescence serves as an excellent reporter system: as a sensitive marker for microbial detection, as a real-time, non-invasive reporter for measuring gene expression and as a measure of intracellular biochemical function (cell viability).[1] Most widely studied of the bioluminescence systems are those belonging to the luminous bacteria (*Vibrio* sp., *Photobacterium* sp. and *Photorhabdus luminescens*) and the firefly (*Photinus pyralis*). While these systems have proved extremely versatile, there are caveats to their use limiting the array of applications they can be applied to. These caveats mainly surround the nature of the luciferase enzymes, and include temperature and pH stability.

The rapid growth of applications of bioluminescence has stimulated the investigation and exploitation of new bioluminescent systems. The most commonly occurring bioluminescence system in nature is that found in the marine environment, based around the substrate imidazolopyrazine.[2] Coelenterazine is an imidazolopyrazine derivative that when oxidised by the appropriate luciferase produces carbon dioxide, coelenteramide and light. The luciferase from the copepod *Gaussia princeps* has recently been cloned and shown to oxidise coelenterazine to produce light.[3] However, coelenterazine is itself also chemiluminescent, undergoing luciferase-independent oxidation.[4] This limits assay sensitivity by reducing the signal to noise ratio. We are interested in using this as a reporter system in mycobacteria, and this work examines the chemiluminescence of coelenterazine in various bacterial growth media in order to determine the signal to noise ratio and assess its suitability.

MATERIALS AND METHODS

A 10 mmol L^{-1} stock coelenterazine solution was prepared by dissolving coelenterazine (Nanolight™ Technology, Prolume Ltd. Pinetop, AZ, USA) in methanol for use at a final concentration of 10 μmol L^{-1}. All coelenterazine solutions were stored at -20 °C and working solutions were kept on ice in the dark during preparation. Diluent buffers comprised: distilled water (dH$_2$O), Phosphate buffered saline (PBS), Buffer A (10 mmol L^{-1} Tris [pH 7.8], 1 mmol L^{-1} EDTA, 0.6 mol L^{-1} NaCl),[5] 7H9 medium supplemented with Tween-80 with or without 10% OADC (oleic acid, albumin, dextrose, catalase), Luria-Bertani (LB) broth with or without Tween-80, Sauton's (4 g L^{-1} asparagine, 60 g L^{-1} glycerol, 2 g L^{-1} citric acid, 0.5 g L^{-1} magnesium sulphate and 0.05 g L^{-1} ferric ammonium citrate) and Hartmans-de Bont (HdB) minimal medium.[6] Luminescence measurements were obtained using a tube luminometer (Berthold Autolumat LB953) over a 10 s period with an integration time of 1 s. Results are expressed as relative light units (RLU).

RESULTS

The results presented demonstrate the high background luminescence signal exhibited by coelenterazine when diluted in the standard diluents of PBS and dH$_2$O (*ca.* 10^3 RLU mL^{-1}) (Fig. 1). A similar background is exhibited when coelenterazine is diluted in Buffer A. This is in contrast with the use of aldehyde as the substrate for the bacterial luciferase reaction where background levels of light are negligible (ca. 10^1 RLU mL^{-1} [data not shown]). Many mycobacteria are fastidious organisms, however the faster growing strains such as *M. smegmatis* can be grown in LB broth. In general, the detergent Tween-80 is added to stop the bacteria from clumping. Use of LB as a diluent produces a low background signal, however the addition of Tween-80 increases the noise over 10-fold (Fig.1). Using the common media used to grow mycobacterial cultures (namely, 7H9 and HdB) as diluent increases the background luminescence noise 10 to 100-fold to ca. 10^4 to 10^5 RLU mL^{-1}, with 7H9 supplemented with OADC giving the highest noise. The exception to this is dilution of coelenterazine with Sauton's broth. When used, this diluent reduced the background luminescence signal to below that of the standard diluents, PBS and dH$_2$O (ca. 10^2 RLU mL^{-1}) (Fig. 1).

DISCUSSION

Mycobacteria can be found in diverse environments around the world and most appear to exhibit a saprophytic lifestyle. They are commonly found in the aquatic environment, both fresh water and marine, and in soil. However, some have the ability to infect animals, birds and humans, and have evolved mechanisms by which they can invade and grow within host cells.

Mycobacterium tuberculosis, the causative agent of tuberculosis in humans, causes nearly 2 million deaths per year. In their role as pathogens and free-living saprophytes, mycobacteria can encounter a range of acidic environments. For example, acidic conditions often prevail in soil and aquatic habitats. In the host environment, *M. tuberculosis* has been shown to reside in the phagocytic vacuole of host macrophages where the intraphagosomal pH has been shown to be mildly acidic.[7] The luciferase from *Gaussia princeps* is exceptionally resistant to exposure to heat and to strongly acidic and basic conditions that result in denaturation of other commonly used luciferases. We are currently investigating whether this luciferase would be suitable as a reporter in mycobacteria for assessing such systems as those resulting in oxidative damage and the acid tolerance response, which is triggered in response to mild acid and enhances survival at normally lethal pH. We have demonstrated that due to the chemiluminescent nature of coelenterazine, the growth media is an important consideration for use of the *Gaussia* luciferase with *Mycobacterial* spp. Growth in media without the detergent Tween-80 is important in maximising signal to noise ratio. Reducing the background is important for increasing the sensitivity of the luciferase assay.

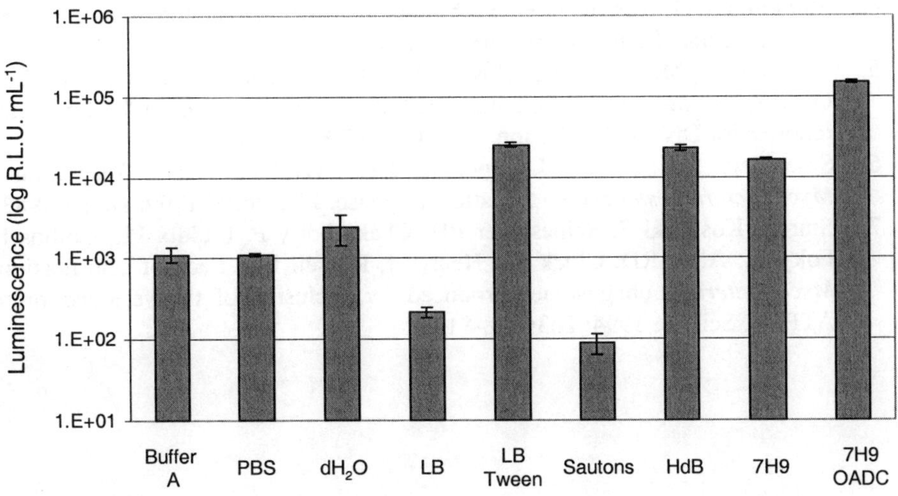

Figure 1. The effects of diluent buffer on coelenterazine chemiluminescence

ACKNOWLEDGEMENTS

This work was supported by the NIH TB Research Unit. The authors wish to thank Bruce Bryan (Nanolight[TM] Technology, Prolume Ltd. Pinetop, AZ, USA) for the kind gift of the colenterazine used in this study.

REFERENCES

1. Greer LF III, Szalay AA. Imaging of light emission from the expression of luciferases in living cells and organisms: a review. Luminescence. 2002; 17: 43-74.
2. Thomson CM, Herring PJ, Campbell AK. The widespread occurrence and tissue distribution of the imidazolopyrazine luciferins. J Biolum Chemilumin 1997; 12: 87-91.
3. Bryan BJ, Szent-Gyorgyi CS. U.S. Patent 6232107, May 2001.
4. Shimomura O, Teranishi K. Light-emitters involved in the luminescence of coelenterazine. Luminescence 2000; 15: 51-8.
5. Verhaegen M, Christopoulos TK. Recombinant *Gaussia* luciferase. Overexpression, purification and analytical application of a bioluminescent reporter for DNA hybridisation. Anal Chem 2002; 74: 4378-85.
6. Smeulders MJ, Keer J, Speight RA, Williams HD. Adaptation of *Mycobacterium smegmatis* to stationary phase. J Bacteriol 1999; 181: 270-83.
7. Sturgill-Koszycki S, Schlesinger PH, Chakraborty P, Haddix PL, Collins HL, Fok AK, Allen RD, Gluck SL, Heuser J, Russell DG. Lack of acidification in *Mycobacterium* phagosomes produced by exclusion of the vesicular proton-ATPase. Science 1994; 263: 678-81.

MONITORING OF COLONIZATION AND CLEARANCE OF LIGHT-EMITTING BACTERIA FROM TUMORS, CUTANEOUS WOUNDS, AND INFLAMMATORY SITES IN LIVE ANIMALS

YA YU[1], S SHABAHANG[2], AA SZALAY[1,3]

[1]*Genelux Corporation, San Diego Science Center, San Diego, CA 92109, USA*
[2]*School of Dentistry, Loma Linda University, Loma Linda, CA 92350, USA*
[3]*Rudolph Virchow Center for Experimental Biomedicine, Institute of Microbiology,
University of Wuerzburg, Am Hubland, Wuerzburg, D97074, Germany
Email: aladar.szalay@virchow.uni-wuerzburg.de*

INTRODUCTION

In previous studies, we demonstrated the monitoring of movement of light-emitting bacteria and viruses in live animals from the time of intravenous (i.v.) injection to their elimination from the body[1]. Our studies showed that bacteria survived exclusively in tumors for weeks without causing bacteremia or infection in normal tissues of the host. In order to understand the mechanisms underlying this finding, here we further examined the factors required for bacteria to colonize tumors in detail. We found that the tumor-specific survival and replication of bacteria is affected by the stage of tumor development, as well as by the number of bacteria injected. Furthermore, tumor development has frequently been compared to wound healing and chronic inflammation. Therefore, we also examined bacterial presence in open wounds of animals and at artificially induced inflammatory sites upon i.v. delivery of bacteria. We found that bacterial colonization in the open wounds is transient, while inflammatory sites alone induced by Sephadex implantation do not allow bacterial survival and replication in live animals. Taken together, we propose that the leaky angiogenic microvasculature allows bacterial entry in tumors; that the apoptotic/necrotic tumor cells provide the bacteria with nutrients; and that the reduced lymphatic activity and impaired immunosurveillance in the tumors prevent the clearance of bacteria from the tumors.

METHODS

Cell line. GI-101A human breast carcinoma cells were cultured in RPMI medium supplemented with 10 mM HEPES, 1 mM sodium pyruvate, 20% fetal bovine serum (FBS), 0.005 μg/mL progesterone (Sigma), and 0.005 μg/mL beta-estradiol (Sigma). C6 rat glioma cells were cultured in DMEM supplemented with 10% FBS.
Analysis of bacterial distribution in tumorous nude mice. GI-101A tumor cells (5×10^6) in 100 μl PBS were implanted subcutaneously (s.c.) in 5-6 week-old male nude mice (Harlan). At different time points after tumor cell implantation, various numbers of attenuated *Vibrio cholerae* transformed with pLITE201 plasmid DNA[1] were injected i.v. into mice. Mice were imaged for luminescence emissions at different time intervals.
Analysis of bacterial distribution in tumorous nude mice with Sephadex-induced inflammation. Sephadex G200-120 (Sigma) was soaked overnight in

PBS. One mL of PBS-treated Sephadex was implanted s.c. on the dorsal flank of nude mice with GI-101A tumors (70 days after tumor cell implantation). Forty-eight hours after Sephadex implantation, a dose of 1×10^8 of *Vibrio cholerae*/pLITE201 was injected i.v. into mice. Three days after bacterial injection, the mice were imaged for luminescence emissions.

Whole-body imaging of luciferase activity in mice. Luminescence emissions from bacteria in mice were imaged under an ARGUS100 low light imager. Mice were placed inside the dark chamber of the Imager. Photon collection was for 1 min. A photographic image of the animal was also recorded, which was then superimposed with the low light image to determine the sites of luminescence emissions.

RESULTS

Colonization of tumors by bacteria was dependent on tumor development. We found that tumor-specific survival and replication of bacteria was achieved when 1×10^8 of bacteria were injected i.v. 43 days after GI-101A tumor cell implantation (Figure 1). At this time the tumor was approximately 1000 mm³ in size. The replication of bacteria in these tumors was followed by imaging for an additional 45 days. In contrast, when the same number of bacteria was injected 30 days after tumor cell implantation (tumor size of 300 mm³), no bacterial survival and replication was observed in tumors (data not shown). Comparable results were also obtained in mice with s.c. rat C6 glioma tumors (5×10^5 tumor cells implanted). In these tumors, bacteria survived and replicated 13 days after tumor development (tumor size 2000 mm³). All rat glioma tumors younger than 8 days (size 500 mm³) were not harboring bacteria. We also showed that bacterial survival and replication in tumors was achieved routinely when at least 1×10^5 bacteria were injected i.v. in mouse recipients. Less than 1×10^4 of bacteria did not result in colonization of tumors in every injected animal.

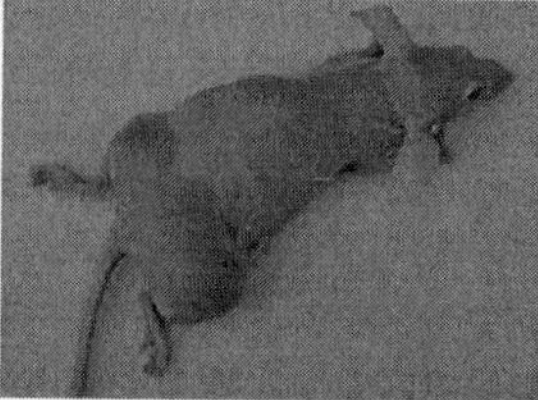

Figure 1. Bacterial survival and replication in s.c. tumors 43 days after tumor cell implantation in nude mice. Left panel, overlay image; Right panel, photographic image.

Bacterial colonization of cutaneous wounds in mice was eliminated upon wound healing. Cutaneous wounds were induced in nude mice by preparing a 5 mm long incision in the femoral region, which was then closed with sutures. A wound was also created by puncturing the mouse ear with ear tags. Bacteria were injected i.v. into mice with cutaneous wounds, and low light imaging of the animals showed bacterial replication in the wounds, whereas the bacteria were cleared from the rest of the body (Figure 2). However, the bacterial colonization of the wounds was found to be transient. After the wounds healed in approximately 7-10 days, light emission also disappeared from the wound region, indicating the clearance of bacteria by the restoration of angiogenic and lymphatic systems.

Figure 2. Bacteria colonize ear tag wounds (indicated by arrow) two days after i.v. injection of bacteria into mice.

No bacterial replication was detected at inflammatory sites caused by injection of Sephadex. We generated artificial inflammation in mice with GI-101A tumors by s.c. implantation of Sephadex beads (see material and methods for experimental details). After i.v. delivery of bacteria, no light emission was seen at the Sephadex implantation site, even though Sephadex-induced inflammation was apparent. However, in the same animal, the tumor was colonized with luminescent bacteria, indicating the importance of a suppressed lymphatic system for bacterial survival and replication.

DISCUSSIONS

This study showed that a minimal number $(1\times10^4 - 10^5)$ of i.v. injected bacteria is required to achieve entry and replication in tumors of a predetermined developmental stage. On the other hand, in cutaneous wounds, the survival and replication of i.v. injected bacteria was transient and the bacteria were cleared in 7-10 days after injection. In contrast, at artificially induced inflammatory sites in a tumor-bearing mouse, no survival and replication of bacteria was observed.

The differences in bacterial survival and replication in implanted tumors may be explained as follows: Blood-borne bacteria enter the tumors through the leaky angiogenic capillaries. It is known that in solid tumors, the center of the tumors is lacking functional lymphatics[2], which may explain why the tumors have very high interstitial fluid pressure[3]. We reported earlier that various bacteria enter and replicate in the "center" of the tumor[1], which is an immunoprivileged site and the clearance of centrally located bacteria through lymphatic drainage is impaired. In tumors at very early developmental stage, the angiogenic capillaries are not well developed and at the same time functional lymphatics may still be present, therefore bacteria may not enter or are immediately cleared from the tumors.

Furthermore, it is known that bacteria do cause local inflammation and dilatation of lymphatics[4]. In the presence of functional lymphatics, the increasing flow of lymph through dilatation of lymphatics may aid the clearance of bacteria from tumors. It is also noted that clonal anergy of lymphocytes occurs in tumors[5]. With such deficiency in immunosurveillance, it is expected that bacteria are able to survive in this "immunoprivileged" environment. In cutaneous wounds, i.v. injected bacteria may enter the surrounding soft tissues through newly formed blood vessels and replicate. However, functional lymphatics are quickly restored during wound healing resulting in active immunosurveillance and elimination of bacteria. The presence of bacteria and the traumatic injury promote the dilatation of lymphatic vessels in the wounded area, which in turn facilitates the removal of bacteria from the infection sites. Both the functional lymphatic drainage and the immunosurveillance also remain intact at the inflammatory sites induced by Sephadex injection. Also little to no blood vessel damage is sustain at these site, and therefore we propose that the bacteria can not escape from circulation before being eliminated.

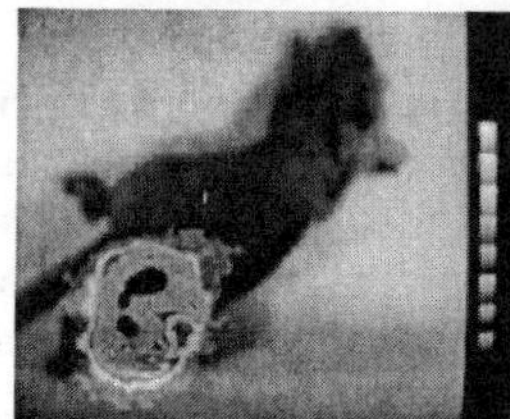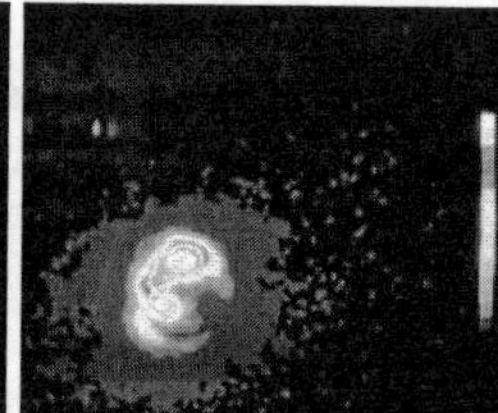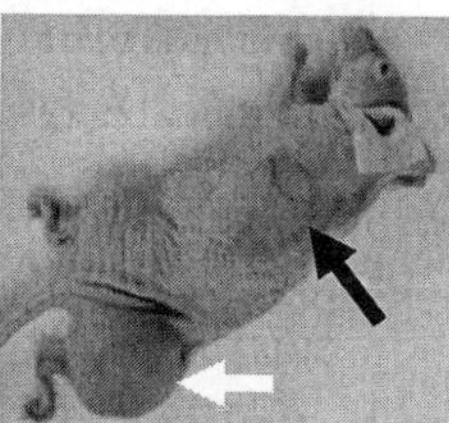

Figure 3. Bacteria colonize tumors (indicated by open arrow) but not inflammatory sites induced by Sephadex (indicated by solid arrow). From left to right, overlay, low light, and photographic images.

REFERENCES

1. Yu YA, Shabahang S, Timiryasova T, Zhang Q, Beltz R, Gentschev I, Goebel W, Szalay AA. Visualization of tumors and metastases in live animals with bacteria and vaccinia virus encoding light-emitting proteins. Nat Biotech 2004; 22:313-20.

2. Leu AJ, Berk DA, Lymboussaki A, Alitalo K, Jain RK. Absence of functional lymphatics within a murine sarcoma: a molecular and functional evaluation. Cancer Res 2000; 60:4324-7.

3. Padera T, Kadambi A, Tomaso E, Carreira C, Brown E, Boucher Y, Choi N, Mathisen D, Wain J, Mark E, Munn L, Jain R. Lymphatic metastasis in the absence of functional intratumor lymphatics. Science 2002; 296:1883-6.

4. Szczesny G, Olszewski WL. The pathomechanism of posttraumatic edema of lower limbs. J Trauma 2001; 52:315-22.

5. Proescholdt MA, Merrill MJ, Ikejiri B, Walbridge S, Akbasak A, Jacobson S, Oldfield EH. Site-specific immune response to implanted gliomas. J Neurosurg 2001; 95:1012-9.